CONCISE ENCYCLOPEDIA OF

WOOD & WOOD-BASED MATERIALS

CONCISE ENCYCLOPEDIA OF WOOD & WOOD-BASED MATERIALS

Editor
ARNO P SCHNIEWIND
University of California
Berkeley, CA, USA

Executive Editor
ROBERT W CAHN
University of Cambridge, UK

Senior Advisory Editor
MICHAEL B BEVER
Massachusetts Institute of Technology
Cambridge, MA, USA

PERGAMON PRESS

OXFORD · NEW YORK · BEIJING · FRANKFURT
SÃO PAULO · SYDNEY · TOKYO · TORONTO

U.K.	Pergamon Press plc, Headington Hill Hall, Oxford OX3 0BW, England
U.S.A.	Pergamon Press, Inc., Maxwell House, Fairview Park, Elmsford, New York 10523, U.S.A.
PEOPLE'S REPUBLIC OF CHINA	Pergamon Press, Room 4037, Qianmen Hotel, Beijing, People's Republic of China
FEDERAL REPUBLIC OF GERMANY	Pergamon Press GmbH, Hammerweg 6, D-6242 Kronberg, Federal Republic of Germany
BRAZIL	Pergamon Editora Ltda, Rua Eca de Queiros, 346, CEP 04011, Paraiso, São Paulo, Brazil
AUSTRALIA	Pergamon Press Australia Pty Ltd, P.O. Box 544, Potts Point, N.S.W. 2011, Australia
JAPAN	Pergamon Press, 5th Floor, Matsuoka Central Building, 1-7-1 Nishishinjuko, Shinjuko-ku, Tokyo 160, Japan
CANADA	Pergamon Press Canada Ltd, Suite No. 241, 253 College Street, Toronto, Ontario, Canada M5T 1R5

First edition 1989

Library of Congress Cataloging in Publication Data

Concise encyclopedia of wood & wood-based materials/
editor, Arno P. Schniewind. —1st ed.
p. cm. —(Advances in materials science and
engineering: 6)
Bibliography: p.
Includes index.
1. Wood—Dictionaries. I. Schniewind, Arno P. II.
Encyclopedia of materials science and engineering.
III. Title: Concise encyclopedia of wood and
wood-based materials. IV. Series
TA419.C64 1989 620.1'1'0321—dc19 88-38474

British Library Cataloguing in Publication Data

Concise encyclopedia of wood and wood-based materials
1. Wood
I. Schniewind, Arno P. II. Cahn, Robert W.
III. Series
620.1'2

ISBN 0-08-034726-6

Distributed in North and South America by
The MIT Press, Cambridge, Massachusetts

Typeset by Macmillan India Ltd, Bangalore

Printed in Great Britain by BPCC Wheatons Ltd, Exeter

CONTENTS

HONORARY EDITORIAL ADVISORY BOARD

FOREWORD

In the short time since its publication, the *Encyclopedia of Materials Science and Engineering* has been accepted throughout the world as the standard reference about all aspects of materials. This is a well-deserved tribute to the scholarship and dedication of the Editor-in-Chief, Professor Michael Bever, the Subject Editors and the numerous contributors.

During its preparation, it soon became clear that change in some areas is so rapid that publication would have to be a continuing activity if the Encyclopedia were to retain its position as an authoritative and up-to-date systematic compilation of our knowledge and understanding of materials in all their diversity and complexity. Thus, the need for some form of supplementary publication was recognized at the outset. The Publisher has met this challenge most handsomely: both a continuing series of Supplementary Volumes to the main work and a number of smaller encyclopedias, each covering a selected area of materials science and engineering, will be published in the next few years.

Professor Robert Cahn, the Executive Editor, was previously the editor of an important subject area of the main work and many other people associated with the Encyclopedia will contribute to its Supplementary Volumes and derived Concise Encyclopedias. Thus, continuity of style and respect for the high standards set by the *Encyclopedia of Materials Science and Engineering* are assured. They have been joined by some new editors and contributors with knowledge and experience of important subject areas of particular interest at the present time. Thus, the Advisory Board is confident that the new publications will significantly add to the understanding of emerging topics wherever they may appear in the vast tapestry of knowledge about materials.

The appearance of Supplementary Volumes and the new series *Advances in Materials Science and Engineering* is an event which will be welcomed by scientists and engineers throughout the world. We are sure that it will add still more luster to a most important enterprise.

Walter S Owen
Chairman
Honorary Editorial Advisory Board

EXECUTIVE EDITOR'S PREFACE

As the publication of the *Encyclopedia of Materials Science and Engineering* approached, Robert Maxwell resolved to build upon the immense volume of work which had gone into its creation by embarking on a follow-up project. This project had two components. The first was the creation of a series of Supplementary Volumes to the Encyclopedia itself. The second component of the new project was the creation of a series of Concise Eneyclopedias on individual subject areas included in the Main Encyclopedia to be called *Advances in Materials Science and Engineering*. These Concise Encyclopedias are intended, as their name implies, to be compact and relatively inexpensive volumes (typically 300-600 pages in length) based on the relevant articles in the Encyclopedia (revised where need be) together with some newly commissioned articles, including appropriate ones from the Supplementary Volumes. Some Concise Encyclopedias will offer combined treatments of two subject fields which were the responsibility of separate Subject Editors during the preparation of the parent Encyclopedia (e.g., dental and medical materials). At the time of writing, nine Concise Encyclopedias have been contracted and others are being planned. These and their editors are listed below.

Concise Encyclopedia of Advanced Ceramic Materials	Prof. Richard J Brook
Concise Encyclopedia of Building & Construction Materials	Prof. Fred Moavenzadeh
Concise Encyclopedia of Composite Materials	Prof. Anthony Kelly CBE, FRS
Concise Encyclopedia of Electronic & Optoelectronic Materials	Dr Lionel C Kimerling
Concise Encyclopedia of Magnetic & Superconducting Materials	Dr Jan E Evetts
Concise Encyclopedia of Materials Economics, Policy & Management	Prof. Michael B Bever
Concise Encyclopedia of Medical & Dental Materials	Prof. David F Williams
Concise Encyclopedia of Mineral Resources	Dr Donald D Carr & Prof. Norman Herz
Concise Encyclopedia of Wood & Wood-Based Materials	Prof. Arno P Schniewind

All new or substantially revised articles in the Concise Encyclopedias will be published in one or other of the Supplementary Volumes, which are designed to be used in conjunction with the Main Encyclopedia. The Concise Encyclopedias, however, are "free-standing" and are designed to be used without necessary reference to the parent Encyclopedia.

The Executive Editor is personally responsible for the selection of topics and authors of articles for the Supplementary Volumes. In this task, he has the benefit of the advice of the Senior Advisory Editor and of other members of the Honorary Editorial Advisory Board, who also exercise general supervision of the entire project. The Executive Editor is responsible for appointing the Editors of the various Concise Encyclopedias and for supervising the progress of these volumes.

Robert W Cahn
Executive Editor

ACKNOWLEDGEMENTS

My involvement with the project that led to the publication of the *Concise Encyclopedia of Wood & Wood-Based Materials* began in 1978 when Professor Michael B Bever asked me to serve as Subject Editor for wood for the *Encyclopedia of Materials Science and Engineering* for which he was Editor-in-Chief. It has been a great privilege to work with Professor Bever and I will always be grateful for the association during which I came to value him as a colleague and as a person. More recently, it has been a great pleasure to have the encouragement and guidance of Professor Robert W Cahn in assembling the present volume, for which I am deeply indebted. Many thanks are due to Dr Colin J Drayton, Mr Michael A Mabe and Ms Deborah A Puleston of Pergamon Press for their assistance and helpful advice. Closer to home, I wish to thank Mrs Mary L Hills and particularly Mr Borjen Yeh for their assistance with a myriad of tasks in editing, assembling the manuscript copy and making the *Concise Encyclopedia of Wood & Wood-Based Materials* a reality. Last but not least my thanks go out to the authors of the articles who have cheerfully and freely given their time and effort.

Arno P Schniewind
Editor

GUIDE TO USE OF THE ENCYCLOPEDIA

This Encyclopedia is a comprehensive reference work covering all aspects of wood and wood-based materials. Information is presented in a series of alphabetically arranged articles which deal concisely with individual topics in a self-contained manner. This guide outlines the main features and organization of the Encyclopedia and is intended to help the reader to locate the maximum amount of information on a given topic.

Accessibility of material is of vital importance in a reference work of this kind and article titles have therefore been selected, not only on the basis of article content, but also with the most probable needs of the reader in mind. In general, the word "wood" has been avoided in titles of articles which describe aspects of wood only, as in the articles *Strength* and *Decay During Use*, for instance. The terms "timber" and "lumber" (as distinct from wood in general) have been permitted in titles (e.g., *Timbers of Africa*, *Lumber: Types and Grades*). Articles on woody materials and wood-based products may be found under the appropriate term (e.g., *Bamboo*, *Laminated Veneer Lumber*). An alphabetical list of all the articles contained in this Encyclopedia is provided.

Articles are linked by an extensive cross-referencing system. Cross-references to other articles in the Encyclopedia are of two types: in-text and end-of-text. Those in the body of the text are designed to refer the reader to articles that present in greater detail material on the specific topic under discussion. They generally take one of the following forms:

> . . . as described in the article *Shrinking and Swelling*.
>
> Above the saturation point (see *Hygroscopicity and Water Sorption*), . . .

The cross-references listed at the end of an article serve to identify broad background reading, and to direct the reader to articles that cover different aspects of the same topic.

The nature of an encyclopedia demands a higher degree of uniformity in terminology and notation than many other scientific works. The widespread use of the International System of Units has determined that such units be used in this Encyclopedia. It has been recognized, however, that in some fields English units are more generally used. Where this is the case, English units are given with their SI equivalent quantity and unit following in parentheses. Where possible the symbol defined in *Quantities, Symbols and Units*, published by the Royal Society of London, have been used.

All articles in the encyclopedia include a bibliography giving sources of further information. Each bibliography consists of general items for further reading and/or references which cover specific aspects of the text. Where appropriate, authors are cited in the text using a name/date system as follows:

> . . . as was recently reported (Smith 1988).
>
> Jones (1984) describes . . .

The contributor's name and the organization to which they are affiliated appear at the end of an article. All contributors can be found in the alphabetical List of Contributors, along with their full postal address and the titles of the articles of which they are authors or co-authors.

A Systematic Outline of the Encyclopedia has been included, which provides a schematic overview of the articles contained in the Encyclopedia and lists all articles under general subject headings.

The most important information source for locating a particular topic in the Encyclopedia is the multilevel Subject Index, which has been made as complete and fully self-consistent as possible.

ALPHABETICAL LIST OF ARTICLES

AN INTRODUCTION TO WOOD AND WOOD-BASED MATERIALS

by Arno P Schniewind

Wood is the hard, fibrous tissue that comprises the major part of stems, branches and roots of trees (and shrubs) belonging to the plant groups known as the gymnosperms and the dicotyledonous angiosperms. Its function in living trees is to transport liquids, provide mechanical support, store food and produce secretions. Virtually all of the wood of economic significance is derived from trees and, as timber, it becomes a versatile material with a multitude of uses. In addition, there are woody materials found in the stems of tree forms in another plant group, the monocotyledonous angiosperms, including the bamboos, rattans and coconut palms. These woody materials are similar to wood in being lignocellulosic in composition, but they differ substantially in their anatomic structure. The principal focus of this volume, and the sole subject of most of the articles in it, is wood as defined above. However, some of the woody materials are dealt with in separate articles because of their present or potential economic importance (see *Bamboo, Rattan and Cane*; *Coconut Wood*).

1. *Wood as a Renewable Natural Resource*

The wood used for purposes other than fuel, that is, wood which is either cut into lumber or used as a raw material for such wood-based materials as plywood, fiberboard or particleboard, is largely derived from the stems of trees.

Woods are classified as either hardwoods or softwoods; the distinction is botanical and is reflected in details of wood structure (see *Macroscopic Anatomy*). Most (but not all) softwoods are needle-bearing evergreen trees, whereas the hardwoods are broad-leaved trees. This classification is not related to the actual hardness of wood, since woods with the least and the greatest known hardness are both found among the hardwoods. Woody materials of the monocotyledons consist of vascular bundles, which are thin strands incorporating elements of both wood and bark, embedded in a mass of soft ground tissue and thus have a structure substantially different from wood. These woody materials are therefore specialty items with properties and uses distinct from wood and lumber.

The annual world production of wood (see *Resources of Timber Worldwide*) is about 2.4×10^9 t, with 10^9 t used as industrial roundwood (e.g., sawlogs, veneer logs, poles and pilings, and pulpwood) and the remainder (about 53%) used as fuelwood and charcoal. Developed and developing countries differ sharply in the ratio of industrial roundwood to fuelwood produced: in developed countries, the percentage of fuelwood and charcoal is, on average, 18% of total wood production, in developing countries it is 80%. In recent years, as energy sources other than petroleum have received increased attention, considerable interest has developed in the use of wood not only as a fuel directly but also as raw material for the production of alcohol and similar sources of energy (see *Chemicals and Liquid Fuels from Wood*; *Energy Generation from Wood*). The forest products industry is in a unique position to use its own wood residue for generating at least part of its energy needs, and is already extensively taking advantage of this possibility. How far the use of wood as an energy source can be developed will depend on economic factors. As the technology for manufacturing wood and wood-based materials continues to develop, the tendency is for a reduction of wood residue in manufacturing operations, which will ultimately affect the availability of such residue in the factories. However, large quantities of presently unused wood that may only be suitable for such uses as fuel are available in the forest. Economic factors will determine if there will ever be conflicting demands for wood resources for fuel on the one hand and as industrial material on the other, but this is not likely in the foreseeable future (see *Future Availability*).

The world production of industrial roundwood is of the same order of magnitude (by weight) as the production of steel and iron. Unlike steel, however, wood represents a renewable resource that can be (and is) constantly replenished as it is being used. Trees can be grown for harvesting in as little as ten years, but may require a hundred years or more depending on climate, species and the purpose the wood is to serve. As long as the rate of removal does not exceed the rate of growth, forests will be a perpetual source of wood and lumber. The composition of the forest with respect to tree species may change, as may the size of trees at the time of harvesting, but under proper management the resource will always be there. Increasing the intensity of management (tree farming) can lead to substantial increases in productivity on a given area of forest land.

2. *Principal Characteristics of Wood*

As a natural product of biological origin, wood is characterized by a high degree of diversity and variability in its properties. No one knows exactly how many species of trees there are in the world, but estimates run from 30000 upwards. A large proportion of these are to be found in the tropical regions, which frequently have an incompletely explored flora. The number of commercial timbers is only a small fraction of the total number of tree species. Since most wood properties are related to wood density, the fact that balsa, the lightest commercial species, has a density of about 160 $kg\,m^{-3}$ as compared with 1280 $kg\,m^{-3}$ for lignum vitae, one of the heaviest, illustrates the diversity that can arise from species differences (see *Density and Porosity*). In addition, considerable variability is found within species due to genetic and environmental factors that influence tree growth. Such variability is found not only from one tree to the next, but also within trees, in part due to natural patterns of growth that make wood properties dependent on radial position and height within a stem. This variability must be taken into consideration if wood is to be used properly. Where wood is used as a load-bearing element, for instance, predictions as to its resistance must be based on near-minimum rather than average values for safety (see *Lumber: Behavior Under Load*). The coefficient of variation (standard deviation expressed as a percentage of the mean) of clear wood strength properties is of the order of 20% (see *Strength*). Further variability is introduced by the presence of strength-reducing characteristics such as knots and cross grain in lumber, major factors which determine the grade of structural lumber (see *Lumber: Types and Grades*). The allowable bending stress of visually graded lumber of mixed Douglas-fir and larch may be as little as 1.9 MPa or as much as 16.9 MPa, depending on grade.

Wood is a highly anisotropic material, that is, its properties depend significantly on the direction in which they are measured. The principal directions in wood are the longitudinal, radial and tangential directions. The longitudinal direction is parallel to the cylindrical axis of the tree stem, and is also referred to as the direction parallel to grain since the majority of the constituent cells of wood are aligned parallel to it. If a tree stem is cut at right angles to its long axis, a series of concentric rings can often be seen on the cross section. These rings are markings produced by annual growth increments. The radial and tangential directions are those which are normal and tangential, respectively, to the growth rings on the cross section. The radial and tangential directions are also referred to collectively as directions perpendicular to grain. The tensile strength and stiffness of wood is highest parallel to grain and very low perpendicular to grain (see *Strength*; *Deformation Under Load*). Conversely, the shrinkage of wood that accompanies loss of moisture is very small parallel to grain and very much higher perpendicular to grain (see *Shrinking and Swelling*). In each of the properties of tensile strength, Young's modulus and shrinkage, the largest and smallest values are in the ratio of about 25:1. The degree of anisotropy is thus extremely high. Wood is often modelled as an orthotropic material by neglecting the curvature of the growth rings. This approximation is most appropriate for pieces of wood cut at some distance from the center of the tree stem.

Wood can be a very durable material. Elaborately inlaid wood furniture from the tombs of the Pharaohs has survived for nearly 4000 years (see *Archaeological Wood*). Being sealed in a tomb in a very dry climate provides almost ideal conditions for preservation of wood. Even in Japan, which has a humid climate, there are wooden temple buildings still in existence after 1300 years of service, and in the harsh climate of the mountains of Norway all-wood stave churches can be found that were built about 800 years ago. The key to durability is proper use and understanding of the factors that destroy wood. Wood is biodegradable, which is essential in nature and can be an advantage when wood in use becomes unserviceable and must be disposed of. The same natural process becomes a severe disadvantage when it occurs as biodeterioration of wood in use. The key to preservation is to create conditions unfavorable to the organisms that cause biodeterioration, principally decay fungi and termites. In the case of fungi this means keeping wood dry. Even wood exposed to the weather will last if the detailing is such that water will run off the surface without being trapped inside joints and interior spaces that cannot dry off readily (see *Weathering*). In the case of termites, contact of wood members with the soil must be avoided. Alternatively, preservative treatments can be effective against either fungi or termites (see *Decay During Use*; *Deterioration by Insects and Other Animals During Use*; *Preservative-Treated Wood*).

Wood is susceptible to damage by fire. In matchstick sizes wood burns readily but it is much more resistant in larger sizes. With suitable construction methods, wood can provide a high degree of fire safety in the early stages of a building fire, giving time for occupants to egress and the fire to be brought under control. This is possible because wood is a poor conductor of heat (see *Thermal Properties*), and the char layer formed beneath a burning wood surface an even poorer one. Wood members can therefore retain a major part of their original strength during the early stages of fire exposure, and thus resist sudden collapse of the building (see *Fire and Wood*).

As a hygroscopic material, wood will take up or give off moisture depending on the temperature and relative humidity of the surrounding atmosphere (see *Hygroscopicity and Water Sorption*). Changes in moisture content below the fiber saturation point, which is the state where the cell walls are saturated with adsorbed water and no free water is present in the cell cavities, have an effect on virtually all properties of wood. Wood also shrinks as it dries, and conversely swells when it is wetted again (see *Shrinking and Swelling*). The total shrinkage in volume of wood from a freshly felled tree dried in an oven will range from about 6 to 20%, depending on species. About two-thirds of the shrinkage will be in the tangential direction and one-third in the radial direction, the longitudinal shrinkage being almost negligible. Wood in use will not ordinarily change moisture content sufficiently to undergo more than a fraction of the total possible shrinkage, but even then the dimensional changes are likely to be significant. The effects of shrinking and swelling can be minimized by drying wood to the approximate moisture content it will reach in service before it is used or installed (see *Drying Processes*). Coatings can be used that will retard moisture changes resulting from short-term fluctuations of temperature and relative humidity, but there are no coatings that would prevent moisture movement entirely. Alternatively, wood can be chemically treated to reduce its hygroscopicity, but such treatments are too expensive for most applications (see *Chemically Modified Wood*). The best method of dealing with shrinking and swelling in wood is to allow for it in proper detailing, by avoiding large dimensions perpendicular to grain, and, where this cannot be done, using floating construction so that dimensional movement perpendicular to grain can occur without constraint.

3. Processing of Wood and Production of Wood-Based Materials

Energy requirements to produce a unit mass of lumber (1000 $kWh\,t^{-1}$) are rather low compared with those for most materials. Steel requires about 4000 and aluminum about 70 000 $kWh\,t^{-1}$. Processing can be done with very simple handtools such as axes, saws and hammers, or on an industrial scale with sophisticated machinery. The principal industrial processes to obtain lumber are extraction (logging), sawing, drying and surfacing (planing or sanding) (see *Machining Processes*). Pressure impregnation is used to treat wood with such chemicals as preservatives or fire retardants. Production of wood-based materials usually involves gluing and pressing, as in the manufacture of plywood, particleboard and glued laminated timber (see *Adhesives and Adhesion*). Members of wood and wood-based materials are given shape mainly by either some type of cutting or by using adhesives to create shapes from smaller pieces. Forming processes by imparting permanent deformation of the material play a minor role, but are possible to a limited extent (see *Forming and Bending*).

Traditionally, wood has not been a subject of intensive study by materials scientists and engineers. Probably the most compelling reason for this is that even if the relationship between various levels of wood structure and properties were perfectly known, the next step, namely that of altering the structure to create properties for certain use requirements, would be very difficult to realize. The quality of wood in trees can be influenced by genetic manipulation (tree breeding) and by certain silvicultural practices that affect conditions of growth, but in a practical sense the opportunities are severely limited. The many factors involved interact in complex ways, use requirements are diverse and sometimes conflicting, and time scales are measured in generations. However, since there are very diverse species of wood available, it is possible to choose the species (and possibly the grade) best suited for a particular application. For example, Douglas-fir and southern pine are the best structural timbers in North America, lignum vitae is uniquely suited for bearing blocks for steamship propeller shafts, teak is excellent for ship's decking, and willow has the right combination of light weight and shock resistance to make it very suitable for artificial limbs.

A further broadening of the range of available wood material is possible by combining wood with various plastics introduced into the wood structure as monomers and polymerized in situ (see *Wood–Polymer Composites*), or by chemical treatments to modify the wood substance (see *Chemically Modified Wood*). By far the most widely practised method, however, is the breaking down of the log into elements that may be as small as a single fiber 2 or 3 mm long or as large as a piece of lumber 40 mm × 300 mm × 6 m, and combining these elements, usually by addition of an adhesive or binder and the application of pressure, into some type of wood-based material. By varying the size and shape of the elements, the way they are oriented with respect to each other, the type of adhesive used and the method by which the product is assembled, a large spectrum of wood-based materials can be created. Single wood fibers or fiber bundles are used as the basic unit for making such sheet or panel products as paper, hardboard, fiberboard or insulation board (see *Hardboard and Insulation Board*; *Paper and Paperboard*). Since the fibers themselves are anisotropic, panels may be isotropic in the plane of the sheet or have directional properties, depending on the degree of orientation of the constituent fibers. Larger elements, collectively referred to as particles, include chips, slivers, strands and flakes. They are generally up to 30 mm long, and sometimes longer. These are assembled into panel products known as particleboards, which again may or may not have a preferred orientation of the elements (see *Particleboard and Dry-Process Fiberboard*). Thin sheets of wood of just a few millimeters thickness can be assembled either into plywood where the grain direction changes by 90° from one layer to the next (see *Plywood*), or into laminated veneer lumber where the grain direction of all layers is parallel (see *Laminated Veneer Lumber*). Finally, dimension lumber, generally about 40 mm in thickness, can be glued up to form much larger structural members, such as beams and arches, than could be produced from solid wood (see *Glued Laminated Timber*).

At present, about 1 m^3 of wood-based panels (fiberboard, particleboard and plywood) is produced for each 4 m^3 of sawn wood on a worldwide basis, which illustrates how important the wood-based materials are. No doubt their relative importance will increase, as inevitable changes in the resource base to smaller logs and advances in technology will lead to further shifts from sawn lumber to wood-based materials.

Finally, it may be pointed out that in addition to the various scientific and technological tangible aspects discussed here, wood has a natural warmth and beauty that are a significant part of its overall character and at times may even be the determining factors in its selection for a particular use.

Bibliography

Forest Products Laboratory 1987 *Wood Handbook: Wood as an Engineering Material*, Agriculture Handbook No. 72. US Department of Agriculture, Washington, DC

Haygreen J G, Bowyer J L 1988 *Forest Products and Wood Science: An Introduction*, 2nd edn. Iowa State University Press, Ames, Iowa

Klass D L (ed.) 1981 *Biomass as a Nonfossil Fuel Source*, ACS Symposium Series No. 144. American Chemical Society, Washington, DC

Koch P 1972 *Utilization of the Southern Pines*: Vol 1, *The Raw Material*; Vol. 2, *Processing*, Agriculture Handbook No. 42. US Department of Agriculture, Washington, DC

Koch P 1985 *Utilization of Hardwoods Growing on Southern Pine Sites*: Vol. 1, *The Raw Material*; Vol. 2, *Processing*; Vol. 3, *Products and Prospective*, Agriculture Handbook No. 605. US Department of Agriculture, Washington, DC

Kollmann F F P, Côté W A Jr 1968 *Principles of Wood Science and Technology*, Vol. 1. Springer, Berlin

Kollmann F F P, Kuenzi E W, Stamm A J 1975 *Principles of Wood Science and Technology*, Vol. 2, *Wood Based Materials*. Springer, Berlin

Marra G G 1981 Overview of Wood as a Material. In: Wangaard F F (ed.) 1981 *Wood: Its Structure and Properties*, Educational Modules for Materials Science and Engineering Project. Pennsylvania State University, University Park, Pennsylvania

Meyer R W, Kellogg R M (eds.) 1982 *Structural Uses of Wood in Adverse Environments*. Van Nostrand Reinhold, New York

Panshin A J, de Zeeuw C 1980 *Textbook of Wood Technology: Structure, Identification, Properties, and Uses of the Commercial Woods of the United States and Canada*, 4th edn. McGraw-Hill, New York

A

Acoustic Emission and Acousto-Ultrasonic Characteristics

Acoustic emission (AE) and acousto-ultrasonics (AU) are useful techniques for determining the integrity of wood and wood-based materials. Acoustic emission arises from material under stress and can be sensed with piezoelectric transducers, usually coupled to the surface of the material. Materials containing flaws or weak areas generate AE at lower stress levels than those of greater integrity. AU combines the sensitivity of AE transducers with ultrasonic transmitters to evaluate the change in energy content of a signal as it passes through the material. In general, a decrease in energy content signifies less material integrity. AE has been used with wood-based materials since the 1960s; AU, which has been in formal existence since the late 1970s, has only recently been used. The application of AE and AU to wood-based materials has been substantially aided by developments in assessing the integrity of fiber-reinforced plastics (FRPs), which have many characteristics similar to those of wood. FRPs, like wood, require special techniques to detect internal flaws, in contrast to metals where x-ray techniques are usually sufficient.

Wood-based materials have AE characteristics similar to those of fiber-reinforced composites. Some indications of weaker (or weakened) material or structure are emissions at low stress levels, and increased numbers and/or rates of events. Another indicator of weakened material is a low Felicity ratio, which is determined from a sequence of load/unload cycles during which AE is monitored. In some materials, no AE occurs during reloading until the previous stress is reached, giving a load ratio (Felicity ratio) of unity (this is referred to as the Kaiser effect). However, many materials, such as composites, have a Felicity ratio of less than unity, not obeying the Kaiser effect.

Acousto-ultrasonics differs from AE in that an active pulser is used to inject stress waves that are received by conventional AE sensors, and the resulting waveform is analyzed to determine the change in signal content from pulser to sensor. AU differs from conventional ultrasonic techniques in that more subtle flaws, such as poor-quality bonding, can be detected. In most applications, the pulsing element is also a conventional AE sensor that is energized with a narrow, high-voltage pulse, but other means of surface stimulation can be used, including mechanical impact, laser bursts and electrical discharges.

1. Acoustic Properties

Typical ultrasonic velocities are 1 $\mathrm{km\,s^{-1}}$ across the grain and 5 $\mathrm{km\,s^{-1}}$ along the grain for lumber and veneer. Reconstituted materials such as composite panels have velocities similar to solid wood across the grain. Wood-based materials are about an order of magnitude greater in attenuation than geological materials and two orders of magnitude greater than metals. Attenuation in wood is much greater across than along the grain. Because of these differences, AE sensor positioning is critical, particularly for solid wood where the grain direction is well-defined.

Since attenuation increases exponentially with frequency, the usable upper frequency level for sensors on wood-based materials is about 100–200 kHz. In this frequency range, attenuation along and across the grain is about 30 and 200 $\mathrm{dB\,m^{-1}}$, respectively. The effects of wood density and moisture content on attenuation have not been clearly determined, but appear to be of lower order as compared with grain angle. Obviously, the more attenuating the material, the greater the limit on sensitivity. However, the more attenuating materials have an advantage in damping out external unwanted signals, permitting the use of higher gains to obtain the needed sensitivity. Attenuation also results from geometric spreading of the signal as it travels through the material; for planar materials, attenuation is proportional to the distance from the source, and for bulk materials, to the square of the distance. For wood-based materials, the geometric spreading may be anisotropic because of the grain direction effect.

2. Coupling of Transducers to Wood-Based Materials

Acoustic impedance (the product of velocity and density) is the determining parameter for acoustic coupling of one material to another. For wood along the grain, acoustic impedance is similar to that of metals; across the grain it is comparable to plastics and water. Coupling presents the greatest source of variability and the major impediment to on-line implementation of AE or AU in processing wood-based materials. For AE laboratory studies, coupling is generally effected using either a couplant (grease type) or by bonding the sensor to the material. Special problems in coupling to wood-based materials include porosity and roughness of the surface. Because of this, grease-type couplants are very difficult to use without them being forced into the material; this changes the pressure at the face of the sensor. Bonded coupling with contact or hot melt adhesives can be used to overcome such problems. Bonded coupling will usually give about a 3 dB increase in efficiency over couplants because of transfer of the shear-wave component. Bonding must be used cautiously since differential expansion of the bonded interfaces can cause failure within the bond or generate stresses that might cause extraneous AE. For many AU applications, the surface must be scanned, which can be done through dry coupling using rigid or elastomeric materials, where coupling pressure must be high enough to "squeeze" out air gaps and maintain consistent local pressure on the material. For both AE and AU, most on-line uses involve materials that are moving, generally requiring a dry coupling system and adding considerable complexity in sensor/coupling design. Dry coupling introduces a loss of about 20–30 dB (at 70 kPa), which can be reduced to about 10 dB with increased pressure (300 kPa). Alternatively, but with some risk with hygroscopic materials, coupling can be achieved by using a water mist at the contact area.

3. Defect Location

Location of the origin of AE emissions can be accomplished by using multiple sensors and measuring time of arrival of the same event. With two sensors, the definition of location is limited to a plane perpendicular to the line between the sensors. The addition of more sensors permits more precise definition of the origin, assuming that the material is reasonably homogeneous and the event rate is not too high to distinguish between events. For solid wood, the relationship of velocity to grain orientation and the high, anisotropic attenuation require special precautions in sensor location and interpretation of time intervals. Some of the more nearly isotropic wood-based composites, such as certain types of fiberboard and particleboard, permit good location resolution. In contrast, plywood presents special problems from both the grain orientation of the plies and boundaries between plies. Oriented strandboard should behave intermediately between particleboard and plywood.

4. Applications

Although there are a limited number of examples of direct application of AE and AU in quality assessment (QA) and quality control (QC), a substantial research effort is underway that will provide the background for development of further industrial systems for nondestructive evaluation and testing. The reported research in AE and AU for wood-based materials can be placed in four major categories: fracture/fracture mechanics, drying (as monitored by formation of checks), decay and machining. Only the first category lacks current QC/QA efforts. Although about 50% of the reported research has been in fracture/fracture mechanics, a major area of emphasis since 1980 has been in wood drying. The most recent research has been in laboratory assessment of wood decay using both AE and AU.

4.1 Fracture and Fracture Mechanics

The first work on wood in which piezoelectric sensors were used was in the mid-1960s for general AE behavior in standard mechanical tests and crack propagation. Local failures from slow extension of intrinsic flaws produced emissions at levels as low as 5–10% of ultimate strength, with irregular "run-and-stop" flaw growth. Subsequently, it was found that longitudinal tension was characterized by early flaw growth at 5–20% of ultimate stress, with a linear increase in events to near failure. In contrast, AE from transverse tension showed a log–log response vs. deformation, but for both types of tensile loading, total events to failure averaged about 300 events per cubic centimeter of specimen. Longitudinal compression produced only about 0.1% of the "AE density" (events/volume) of tensile tests. In flexure, cumulative events increased linearly with deflection to the proportional limit, continuing beyond it in a linear manner but at a much lower slope to near failure. In creep under flexural loading at 80% ultimate stress, AE accumulated stepwise from apparent redistribution of stress concentrations during irregular flaw growth. Mode I cleavage tests produced a uniform distribution of AE over the crack length, with about 8 and 24 events mm^{-1} for soaked and dry specimens, respectively.

In specimens that contain defects, the rate of emissions increases near the proportional limit. Higher emission levels of these specimens occur without corresponding changes in modulus of rupture. The critical load in fracture toughness correlates highly with cumulative AE counts. The initiation of microscopic compression lines has been postulated as the cause of early events at 40–50% ultimate load in static bending. Early AE has been used as a screening test to eliminate flawed specimens from studies.

Solid wood under tensile stress parallel to the grain produces AE in sequential slow-rate and rapid-rate periods. The slow rate is attributed to extension of preexisting microscopic flaws in the cell walls, which has been confirmed microscopically. The cause of the rapid AE rate appears to be brittle failure of tracheids since this rate is proportional to crack velocity. AE output varies with wood species, but has not been related to gross physical or anatomical properties. Also, when the AE rate is high, the observed shearing failure, which reduces crack velocity, is low. AE has been used in static bending to distinguish the effect of compression failures in typhoon-damaged wood that cause a lower modulus of rupture (MOR) but unchanged modulus of elasticity (MOE), from the effect of knots in which both MOE and MOR are lowered. Assuming a linear relationship of total counts to (load-at-failure)^2, the slopes are quite different between clear wood and wood containing either knots or compression failures, providing a much clearer separation of the contributing source than if considering only MOR. Tension fatigue has been observed to cause an exponential increase of AE with cycles; the contribution of AE from loading and unloading, however, was not eliminated.

Clear specimens stressed in flexure generate comparably little AE until near the ultimate stress, in contrast to high activity in material containing defects. The initiation and propagation of failures can be detected by using location techniques with two or more sensors. The initiation of flaws in both laboratory and full-size specimens can be correlated with the location of natural or artificial defects.

4.2 Composites

Wood-based composite materials emit AE at substantially lower stress levels than occurs with solid wood. For example, in static bending tests, composite materials begin emissions at about 10–20% of the ultimate stress, whereas clear solid wood begins at about 40–50%. The smaller the constituent particle in the board, the higher the stress level of initial emissions. In AE monitoring during internal bond (IB) tests on composites, a clear relationship was found between one-half of the total events and ultimate stress. In particleboard with controlled resin levels, total events to failure correlated with resin level. Also, AE data correlated much better with IB values than did specimen density. Some limited data indicate a high correlation of AE with AU measured by transmission. The Felicity ratio of particleboard appears to be reasonably constant (at about 0.9) up to failure.

AE has been measured from internal microfailures during the thickness swelling of particleboard at high relative humidity. The active period of AE coincides with irreversible swelling (springback). Wood-fiber hardboard with controlled pretreatments tested in compression normal to the face produced cumulative AE that correlated well with thickness swelling from boil-swell tests, indicating that cumulative AE may be a good nondestructive predictor of dimensional stability. When the same type of materials were subjected to cyclic water-soak exposure, stress-wave factor (SWF) values decreased with increasing numbers of cycles, corresponding to increasing damage. However, undamaged boards with higher SWF had lower thickness swelling, suggesting that SWF could predict the degree of dimensional stability.

4.3 Machining

AE characteristics have been determined for cutting operations that approximate veneer peeling. The primary source of AE was found to be changes in plastic deformation in the shear zone at the tip of the cutting tool. The AE count rate, while not well correlated with cutting forces, was more sensitive to the cutting process than the rms signal. The AE output was considered potentially more important to monitor tool wear than cutting forces. This work has been extended to preliminary tests on monitoring AE from circular saw cutting, using several types of coupling attachments to the blade.

4.4 Drying and Drying Control

One of the more promising applications of AE is for monitoring and controlling the drying process of lumber (see *Drying Processes*). AE has been used to sense the development of surface checks in both hardwoods and softwoods. The actual fracture of the surface has been directly observed simultaneously with AE emissions. By controlling the environmental conditions to prevent high rates of emissions, it is possible to dry faster with less degradation. Several investigators have demonstrated the feasibility of controlling checking during hardwood drying by using a fixed level of AE to control drying conditions. Hardwoods generally check much more readily than softwoods and consequently have AE rates about an order of magnitude greater. However, among hardwoods, there does not appear to be a good correlation between propensity to check and AE rate. AE has been shown to respond rapidly to changing surface RH conditions, independent of whether end grain or side grain is exposed. Geometry can have a large effect on AE generation, depending on the direction of moisture movement and the degree of stress development. There is some evidence that AE during drying also originates from fracture of water capillaries, analogous to AE from water stress in plants.

4.5 Biological Degradation

Several mechanical tests, including static bending and transverse compression, have shown that AE activity increases substantially in decayed wood (see *Decay During Use*). This effect also occurs within the incipient decay range, which is very difficult to detect with any other analytical technique. A substantial number of variables (including wood and fungus species) and testing configurations must be understood if the technique is to move from laboratory to field testing. Preliminary tests using acousto-ultrasonics show some promise of NDT detection of decayed wood. Marine-borer damage has been simulated by drilled holes in wooden piling and under loading generated AE in proportion to the degree of damage. Coupling through water in which the piling was immersed was found to be more efficient than direct surface contact to the piling.

4.6 Adhesives

The characteristics of AE for adhesively bonded areas have not been well defined. There is some data that indicates lower AE levels for brittle than for flexible adhesives, even for failure at the same ultimate stress. In plywood IB testing, it was found that wood failure could be differentiated from adhesive failure by the shape of the AE vs. strain curves, which were linear for adhesive and curvilinear for wood failure, respectively. Some preliminary plywood work has been done in static bending to understand the effect of voids or poorly bonded areas on AE with the objective of assessing full-size-panel integrity.

Several laboratory studies have been made on the character of AE from fingerjointed structural lumber in static bending. Although the strength of clear wood having fully cured bonds could be predicted with reasonable error at 50% and 80% of ultimate load, the presence of defects in the wood and/or fingerjoints increased the error in prediction. However, location techniques have shown that even in fingerjointed clear wood, the emissions occur predominantly from the area of the joint. In contrast, clear wood without fingerjoints has a fairly uniform distribution of emissions along the length. AE has also been used in an analysis of failure modes of different furniture-joint combinations. The curing of adhesives with wood substrates has been monitored using acousto-ultrasonics. Cure time has been quantified for several types of adhesives by calculating the half-time between initial and final transmission values.

See also: Acoustic Properties

Bibliography

Ansell M P 1982 Acoustic emission from softwoods in tension. *Wood Sci. Technol.* 16: 35–58

Beall F C 1985 Relationship of acoustic emission to internal bond strength of wood-based composite panel materials. *J. Acoust. Emission* 4(1): 19–29

Beall F C 1986 Effect of moisture conditioning on acoustic emission from particleboard. *J. Acoust. Emission* 5(2): 71–76

Beall F C 1987 Acousto-ultrasonic monitoring of glueline curing. *Wood Fiber Sci.* 19(2): 204–14

Beall F C, Wilcox W W 1987 Relationship of acoustic emission during radial compression to mass loss from decay. *For. Prod. J.* 37(4): 38–42

Beattie A G 1983 Acoustic emission, principles and instrumentation. *J. Acoust. Emission* 2(1/2): 95–128

Becker H F 1982 Acoustic emissions during wood drying. *Holz Roh- Werkst.* 40: 345–50

DeBaise G R, Porter A W, Pentoney R E 1966 Morphology and mechanics of wood fracture. *Mater. Res. Stand.* 6(10): 493–99

Dedhia D D, Wood W E 1980 Acoustic emission analysis of Douglas-fir finger joints. *Mater. Eval.* (11): 28–32

dos Reis H L M, McFarland D M 1986 On the acousto-ultrasonic characterization of wood fiber hardboard. *J. Acoust. Emission* 5(2): 67–70

Hamstad M A 1986 A review: acoustic emission, a tool for composite-materials studies. *Exp. Mech.* 26(1): 7–13

Lemaster R L, Klamecki B E, Dornfeld D A 1982 Analysis of acoustic emission in slow speed wood cutting. *Wood Sci.* 15(2): 150–60

Niemz P, Wagner M, Theis K 1983 State and possible applications of acoustic emission analysis in wood research. *Holztechnologie* 24(2): 91–95

Noguchi M, Kitayama S, Satoyoshi K, Umetsu J 1987 Feedback control for drying *Zelkova serrata* using in-process acoustic emission monitoring. *For. Prod. J.* 37(1): 28–34

Sato K, Kamei N, Fushitani M, Noguchi M 1984 Discussion of tensile fracture of wood using acoustic emissions. A statistical analysis of the relationships between the characteristics of AE and fracture stress. *J. Jpn. Wood Res. Soc.* 30(8): 653–59

Vary A, Lark R F 1979 Correlation of fiber composite tensile strength with the ultrasonic stress wave factor. *J. Test. Eval.* 7(4): 185–91

F. C. Beall
[University of California, Berkeley, California, USA]

Acoustic Properties

Anyone who has heard a fine violin, xylophone, piano or almost any other stringed instrument being played has experienced some of the unique acoustic properties of wood. Principles based on these same properties are also used in the nondestructive evaluation of various strength properties of structural wooden members and wood-composite products (see *Nondestructive Evaluation of Wood and Wood Products*) and in tests for decay in existing utility poles and other "in-place" timbers. Investigations are also being carried out into the possibilities of using sound waves to identify various stages of wood drying either by moisture-content determination or by counting the number of acoustic emissions per unit time (see *Acoustic Emission and Acousto-Ultrasonic Characteristics*).

1. Velocity of Sound in Wood and Wood Products

The velocity of sound in wood and wood products can be measured either by using its natural or resonant frequency or by propagating stress waves by physical impact or ultrasonics. The velocity has been found to be influenced by wood species, moisture content, temperature and direction of wave propagation (i.e., longitudinal, radial or tangential). Density does not significantly affect the velocity, but the ratio of the medium's elastic modulus E to its density ρ is important; for the case of rods, the velocity of sound v can be shown to be given by $v=(E/\rho)^{1/2}$.

James (1961) determined the speed of sound in clear Douglas-fir specimens subjected to longitudinal vibration at various moisture contents and temperatures using the specimen's resonant frequency. He calculated velocity from the expression $v=2fl$ where f is the resonant frequency and l is the specimen length. Values from this study are given in Table 1; these show that as either moisture content or temperature is increased the velocity of sound decreases at an increasing rate.

Gerhards (1982) summarized the variables that affect sound velocity as determined by a number of researchers. Velocity was found to decrease substantially as grain angle increased from 0 to 90° with respect to the longitudinal axis. A decrease of 63% was found to occur over this range with the most apparent changes taking place at grain angles less than 20°. Velocity through knots was found to be lower, owing to grain curvature; however, when straight grain was present around the knot, the knot's effect was greatly reduced (Burmester 1965). Another important factor is advanced decay, which has been found to cause a 25% loss in sound velocity (Konarski and Wazny 1977).

The effect of advanced decay on sound velocity forms the basis for a commercial wooden-pole tester marketed under the name Pole-tek. This device uses a spring-loaded impact hammer and a sensing probe, which are placed on opposite sides of the pole to be tested. A comparison is made between sound-velocity values obtained for the test pole and a standard, decay-free pole; if the test pole gives a low value of sound velocity, incremental boring can be used to determine the actual extent of decay (Graham and Helsing 1979).

McDonald (1978), in a study utilizing ultrasonic wave propagation to locate defects in lumber, determined sound velocities in the longitudinal, radial and tangential directions of several wood species. Values for selected species are given in Table 2. These were obtained using a 400 V pulse of 1 μs duration at room temperature. If the radial and tangential values are averaged and compared with the longitudinal values, it can be seen that the longitudinal values are from 2.3 to 3.5 times greater than the transverse values. McDonald also found that a more precise detection system could be obtained if the ultrasonics were transmitted and received in water rather than in air. This results from better surface contact and hence better board adsorption of the induced stress wave.

Dunlop (1980), measuring the velocity of sound in particleboard, found a significant difference between values measured longitudinally (in the plane of the sheet) and those measured normal to this plane or in the transverse direction, as shown in Table 3. He also determined that the higher-density surface layers had different values than the core material as can also be seen in Table 3.

James (1982) investigated the feasibility of monitoring stages in the kiln drying of lumber through changes in sound velocity per unit length as influenced by moisture content. Using ultrasonic pulses and correcting for effects of temperature, the velocity was shown to be well correlated with changes in moisture

Table 1
Velocity of sound parallel to the grain at different temperatures as determined from resonant frequencies for clear Douglas fir (after James 1961)

	Sound velocity ($m\,s^{-1}$)				Percentage decrease in
Moisture content[a]	−18°C	27°C	71°C	93°C	velocity[b]
7.2	5486	5359	5207	5131	6.5
12.8	5512	5334	5182	5080	7.8
16.5	5385	5182	4978	4851	9.9
23.7	5055	4877	4623	4420	12.6
27.2	4978	4775	4496	4293	13.8
Percentage decrease in velocity[c]	9.3	10.9	13.7	16.3	

[a] Oven-dry-weight basis [b] Over temperature range shown [c] Over moisture content range shown

Table 2
Velocity of sound as determined by ultrasonic wave propagation (after McDonald 1978)

	Sound velocity ($m\,s^{-1}$)			
Direction of sound propagation	Douglas fir (dry)	Ponderosa pine (green)	Red oak (green)	Beech (dry)
Longitudinal	4350	4390	4110	6610
Radial	1980	1620	2040	1980
Tangential	1770	1460	1790	1770
Longitudinal:transverse ratio	2.3	2.9	2.9	3.5

Table 3
Velocity of sound in particleboard (after Dunlop 1980)

Specimen	Density ($kg\,m^{-3}$)	Sound velocity ($m\,s^{-1}$) Longitudinal	Transverse
18 mm Board	500	2200–2500	820
6 mm Skin	590	2200	920
8 mm Core	460	1800–1900	620

content. Moisture gradients influenced the readings, however, since the ultrasonic pulse would travel faster through the drier regions, tending to indicate lower moisture content than actually existed. It was found, however, that adjustment for this could be incorporated in the temperature-correction factors. The method shows promise but additional work is needed.

Skaar et al. (1980) made initial examinations in using acoustic emissions to monitor drying of red-oak lumber. Acoustic emissions are thought to originate from check development. The purpose of monitoring emissions is to control the severity of drying conditions to minimize degradation caused by checking (see *Drying Processes*).

As mentioned above, the velocity of sound is related to both its resonant frequency and its elastic modulus. Consequently, either the velocity of sound as measured with longitudinal stress waves or resonant frequency as determined from free- or forced-vibration experiments can be used in the nondestructive evaluation of the elastic modulus of wood, which in turn can be linked to other mechanical properties (see *Nondestructive Evaluation of Wood and Wood Products*).

Two acoustic properties related to sound velocity are the sound wave resistance, which is the product of velocity and density ($v\rho$), and the damping of sound radiation, which is determined essentially by the ratio of velocity to density (v/ρ). Sound wave resistance is important in sound propagation and the reflection of sound at media boundaries, whereas damping of sound radiation determines the energy loss from a vibrating body to surrounding media by radiation.

2. Damping Capacity

The damping capacity of vibrating wood results from energy dissipation through internal friction. A measure of damping capacity is logarithmic decrement δ, which is defined as the natural logarithm of the ratio of two successive decaying wave amplitudes (A_i and A_{i+1}) for a body set in harmonic motion and allowed to vibrate freely:

$$\delta = \ln(A_i/A_{i+1})$$

The values of logarithmic decrement are affected by moisture content, temperature, grain direction and frequency of vibration (Dunlop 1980, Pentoney 1955). The effects of each of these variables are nonlinear. In general, the minimum values for logarithmic decrement increase as moisture content increases and as temperature decreases. Values determined by James (1961) for clear Douglas-fir specimens, based on decay at their resonant frequency, are given in Table 4. These values were determined for free vibration in the longitudinal direction. Other researchers report values of logarithmic decrement parallel to the grain to be two to three times larger than values perpendicular to the grain (Dunlop 1981, Pentoney 1955).

Dunlop (1981) investigated the feasibility of using damping capacity to determine decay in wood poles using a pulse-echo technique. He hypothesized that a more precise method for determining decay in wooden poles would be obtained by sending the signal in the longitudinal direction as opposed to the Pole-tek method which induces the pulse in the transverse direction. The results were promising but the method was not as precise as the Pole-tek technique, especially with respect to decay allocation.

3. Acoustic Properties Affecting Musical Instruments

The acoustic properties of wood make it a preferred material for sounding boards in musical instruments. The velocity of sound in dry wood parallel to the grain is about the same as in steel and most other metals. However, since the density of wood is much lower, it

Table 4
Logarithmic decrement values for clear Douglas fir at various moisture contents and temperatures (after James 1961)

Moisture content[a]	Specific gravity[b]	Logarithmic decrement (10^{-3}) −18 °C	27 °C	71 °C	93 °C
7.2	0.48	36.6	20.2	23.3	25.5
12.8	0.40	27.2	22.2	26.5	34.0
16.5	0.45	27.9	26.4	33.3	43.7
23.7	0.43	29.1	26.3	42.2	84.0
27.2	0.45	32.7	29.2	60.6	108.7

[a] Oven-dry-weight basis [b] Value based on oven-dry-weight and volume at specified moisture content

has low sound wave resistance and high damping of sound radiation. Low sound wave resistance facilitates resonance, and high damping of sound radiation combined with low damping capacity means that less of the sound energy is consumed in internal friction and more is emitted as sound radiation to the surroundings—precisely what is desired of sounding boards in musical instruments. For example, values for sound wave resistance as given by Kollmann and Côté (1968) range from $20–37 \times 10^5\ \mathrm{N\,s\,m^{-3}}$ for wood as compared with 395×10^5 and $258 \times 10^5\ \mathrm{N\,s\,m^{-3}}$ for steel and cast iron, respectively. In Japan, it has been found that the best wood selected by experts in the piano industry for sounding boards had the lowest ratio of damping capacity to elastic modulus (Norimoto 1982).

The environmental conditions in which musical instruments are usually used are optimal for musical acoustic properties. Under these conditions (i.e., room temperature, wood at 8% moisture content), wave dissipation due to sound radiation is increased and damping due to internal friction is at a minimum (see *Stringed Instruments: Wood Selection*).

4. Architectural Acoustics

Architectural acoustics involve two main considerations. The first is the acoustic quality of a room, which is dependent on its sound-absorption characteristics. The second is the isolation of sound in the source room to prevent transmission either to adjoining rooms or to floors below.

Sound absorption is necessary to keep the reflection or bounding of sound waves in a room to a minimum since such reflection or bounding adversely affects the clarity with which the source can be heard. Some reflection is necessary, however, to keep the room from becoming acoustically deadened. The sound absorption of a room is measured by its reverberation time; that is, the time for sound to diminish to one-millionth of its original intensity, equivalent to a reduction of 60 dB. This time is directly proportional to the volume of the room and inversely proportional to the sum of all objects and surfaces absorbing sound. Reverberation time is expressed as

$$t_R = 0.161\ V/\Sigma\ S_i a_i$$

where V is the room volume, S_i is the exposed surface area and a_i is the absorption coefficient of each object. An acoustically well-designed room, therefore, requires the correct amount of sound-absorbing materials. The optimum sound-absorption characteristics for a particular material result from a careful balance of its density, porosity, fineness of fibers, bulk elasticity and thickness. For sound to be absorbed, the surface porosity must be such that a sound wave will enter and be dissipated through internal heat produced by intermolecular friction between the air molecules and the absorbing structure. If the pore size is too small, the wave will be reflected; if it is too large, intermolecular friction will not take place, permitting the wave to pass through.

Owing to the porosity of the surface of normal wood, only 5–10% of sound waves are absorbed and 90–95% are reflected. The percentage absorbed is increased only slightly with a very rough surface. Wood, therefore, is not a good absorber as acoustic materials should absorb at least 50% of the sound waves. Wood can, however, be used for acoustic materials either by product processing as in the manufacture of acoustic tiles or by the design of resonant panels.

Acoustic tile, manufactured from low-density fiberboard, absorbs sound since it contains open pores over at least 15% of its surface. Common fiber insulating board is not as porous and therefore is not a much better absorbent than solid wood. Other solid wood products such as extremely rough and porous particleboard may qualify as sound absorbers, but common particleboard and hardboard do not (Godshall and Davis 1969).

Well-designed resonant panels can absorb and control sound. In this method a panel is constructed to allow the panel skin to vibrate and transmit the sound wave to either a dead air space (at least 2.5 cm) or more efficiently to a sound-absorbing medium in that space such as fiber insulation. As expected, the thicker the air space or absorbing material, the higher the quality of absorbency.

Multifamily dwelling and commercial building codes require sound to be prevented from passing through partitions and floors into adjoining rooms. This is known as sound isolation and is measured as the transmission loss or drop in decibels from one room to the next. A drop of more than 45 dB at midfrequencies is considered necessary for party wall construction (Schultz 1969). In essence, a partition exhibiting a high transmission loss has the same objective as the resonant panels described above; this is to dissipate the vibrations of the wall surface set in motion by the sound waves before these vibrations can reach the wall surface in the adjoining room. If the two wall surfaces are rigidly attached, then both walls vibrate simultaneously and sound is transmitted. If they are allowed to vibrate independently, then the waves can be absorbed within the wall structure.

Jones (1973) has shown that an accurate prediction of the transmission loss for a structural system is dependent on knowledge of the sound frequency, partition mass and partition configuration. The effect of sound frequency has been categorized into three main regions (Schultz 1969). Region I consists of frequencies near the lowest resonance frequency; in this region, the transmission loss is dependent on partition stiffness. Region II consists of midfrequencies (250-1500 Hz); transmission in this region is controlled directly by the mass per unit area of the partition. The midfrequencies are of greatest concern since they are the most common and therefore are used for obtaining standard test values.

As frequencies increase above approximately 1500 Hz, region III is entered where transmission is again controlled by partition stiffness as in region I. This results from the partition "coincidence" frequency being reached, at which point the acoustic wavelength perfectly matches the flexural wavelength of the structure. The transmission loss is greatly reduced at this point (Jones 1973, 1976, 1978, 1979).

There are three main partition systems which use wooden components and are designed to decrease sound transmissions. The most simple involves placement of fiber insulation board between the studs and gypsum board to absorb the vibrations. The second system uses metal-resilient channels placed between the studs and gypsum board. The metal flanges in this system vibrate independently of each other, thereby helping to dissipate the wave. The third and most expensive partition system is a double stud wall with a 2.5 cm separation which allows each wall surface to vibrate independently of the other. All of these systems are enhanced by filling all air spaces with sound-absorbing insulation, using a double layer of gypsum board on the exterior to increase mass, and using adhesives to fasten components instead of nails.

Impacts from the floor above a room can be isolated by increasing the flooring density with foamed concrete or by using a "floating" floor design. The impact isolation for the foamed concrete is greatly improved by using a pad and carpet. The floating design is more complex, consisting of a glass-fiber interface between the flooring and subflooring and a heavy glass-fiber blanket between the joists and a resiliently suspended ceiling.

See also: Building with Wood; Density and Porosity

Bibliography

Bucur V 1984 Relationships between grain angle of wood specimens and ultrasonic velocity. *Catgut Acoust. Soc. J.* 41: 30–35

Burmester A 1965 Relationship between sound velocity and the morphological, physical, and mechanical properties of wood (in German). *Holz Roh- Werkst.* 23(6): 227–36

Dunlop J I 1980 Testing of particle board by acoustic techniques. *Wood Sci. Technol.* 14: 69–78

Dunlop J I 1981 Testing of poles by using acoustic pulse method. *Wood Sci. Technol.* 15: 301–10

Gerhards C C 1982 Longitudinal stress waves for lumber stress grading: Factors affecting applications: State of the art *For. Prod. J.* 32(2): 20–25

Godshall W D, Davis J H 1969 *Acoustical Absorption Properties of Wood-Base Panel Materials*, USDA Forestry Service Research Paper FPL-104. US Department of Agriculture Forest Service, Madison, Wisconsin

Graham R D, Helsing G G 1979 *Wood Pole Maintenance Manual: Inspection and Supplemental Treatment of Douglas-fir and Western Red Cedar Poles*, OSU Forest Research Laboratory Research Bulletin 24. Oregon State University Forest Research Laboratory, Corvallis, Oregon

James W L 1961 Effect of temperature and moisture content on internal friction and speed of sound in Douglas fir. *For. Prod. J.* 9(9): 383–90

James W L 1986 *Effect of Transverse Moisture Content Gradients on the Longitudinal Propagation of Sound in Wood*, USDA Forest Products Laboratory Research Paper FPL-466. US Department of Agriculture Forest Service, Madison, Wisconsin

James W L, Boone R S, Galligan W L 1982 Using speed of sound in wood to monitor drying in a kiln. *For. Prod. J.* 32: 27–34

Jones R E 1973 Improved acoustical privacy in multifamily dwellings. *Sound Vibr.* 7(9): 30–37

Jones R E 1976 How to accurately predict the sound insulation of partitions. *Sound Vibr.* 10(6): 14–25

Jones R E 1978 How to design walls for desired STC ratings. *Sound Vibr.* 12(8): 14–17

Jones R E 1979 Intercomparisons of laboratory determinations of airborne sound transmission loss. *J. Acoust. Soc. Am.* 66(1): 148–64

Koch P 1972 *Utilization of the Southern Pines*, Vol. I: *The Raw Material.* US Government Printing Office, Washington, DC

Kollmann F F P, Côté W A Jr 1968 *Principles of Wood Science and Technology*, Vol. I: *Solid Wood.* Springer, New York

Konarski B, Wazny J 1977 Relationships between ultrasonic velocity and mechanical properties of wood attacked by fungi (in German) *Holz Roh- Werkst.* 35(9): 341–45

McDonald K A 1978 *Lumber Defect Detection by Ultrasonics*, USDA Forestry Service Forest Products Laboratory Research Paper FPL-311. US Department of Agriculture Forest Service, Madison, Wisconsin

Norimoto M 1982 Structure and properties of wood used for musical instruments I. On the selection of wood used for piano soundboards. *J. Jpn. Wood Res. Soc.* 28(7): 407–13

Pentoney R E 1955 Effect of moisture content and grain angle on the internal friction of wood. *Compos. Wood* 2(6): 131–36

Rudder F F Jr 1985 *Airborne Sound Transmission Loss Characteristics of Wood-Frame Construction*, USDA Forest Products Laboratory General Technical Report FPL-43. US Department of Agriculture Forest Service, Madison, Wisconsin

Schultz T J 1969 Acoustical properties of wood: A critique of the literature and a survey of practical applications. *For. Prod. J.* 19(2): 21–29

Skaar C, Simpson W T, Honeycutt R M 1980 Use of acoustic emissions to identify high levels of stress during oak lumber drying. *For. Prod. J.* 30(2): 21–22

Tanaka C, Nakao T, Takahashi A 1987 Acoustic property of wood I. Impact sound analysis of wood. *J. Jpn. Wood Res. Soc.* 33: 811–17

W. R. Smith
[University of Washington, Seattle, Washington, USA]

Adhesives and Adhesion

The first evidence of adhesive-bonded wood products can be traced back over 3500 years, when the joining of wood was undertaken primarily for decorative purposes. The discovery of improved adhesives led to more sophisticated developments in wood composites and greater utilization of these products in structural applications. Until this century most of these adhesives were derived from animal, vegetable or mineral sources.

Projected limitations in the available harvest of wood resources and the overall decrease in wood quality emphasize the importance of adhesives and the science of adhesion in improving wood utilization by allowing smaller trees and present waste wood to be restructured into useful wood products. By proper choice of wood component, wood orientation, adhesive and manufacturing conditions, a variety of engineered composites are possible; these can have greater mechanical strength, decreased variability and improved stress-distributing properties compared to conventional solid wood.

This article outlines the mechanism of wood bonding and pertinent physical and chemical factors involved in achieving good bonds. Both wood-related and adhesive-related characteristics influencing the formation and performance of bonds in products such as laminated lumber, plywood, flakeboard and particleboard are considered.

1. Nature of Adhesion

Bond formation depends upon the development of physical and chemical interactions both within the bulk adhesive polymer and at the interface between adhesive and wood. Interactions within the adhesive accumulate to give cohesive strength while the forces between adhesive and wood provide adhesive strength. Both should exceed the strength of the wood, allowing substantial wood failure during destructive testing of high-quality bonds.

Most wood adhesives are liquids consisting of either 100% polymer material or more frequently an aqueous solution, emulsion or dispersion, wherein one half to one third of the mixture is polymer. Other solvents such as volatile alcohols, esters or aromatics may sometimes be used to replace all or part of the water. For more specialized bonding requirements, adhesives in a solid state (i.e., powders, hot melts) may be utilized with heat applied to transform the polymer into a flowable material.

The adhesive must accomplish three distinct stages during bonding processes: it must wet the wood substrate, flow in a controllable manner during pressing and finally set to a solid phase. Failure of the adhesive to accomplish any of these stages often results in reduced bond quality. The extent of wetting depends on the physical–chemical nature of both adhesive and wood surface. Solvent molecules and low-molecular-weight polymer molecules, being smaller and more mobile, tend to wet and penetrate into the wood substrate most rapidly. Adsorption and diffusion of these liquids into the cell wall often leads to swelling of the lignocellulose wood substance. The porous structure of wood and defect areas created by machining processes facilitate the transport of these materials into the wood. In contrast, higher-molecular-weight polymer material would wet and penetrate the wood substrate at a significantly slower rate, remaining largely as a concentrated polymer on the wood surface. Ultimately, the bonding area should consist of a steadily decreasing concentration of adhesive polymer as one proceeds from the surface to a few micrometers within the wood. Achieving this condition will depend on polymer physical properties, thermodynamic considerations and wood-surface properties. With solid adhesives, an intermediate heating stage which melts or softens the adhesive is necessary to accomplish wetting of the wood surface.

Optimum bond formation requires intimate contact between adhesive and wood substrates to ensure molecular interaction over a large area. This is accomplished using pressure and heat which causes viscous adhesive components to transfer and flow throughout the bonding sites while deforming the wood to achieve better contact between wood surfaces of different texture. Adhesive penetration into the wood structure must not be so great that inadequate amounts of adhesive remain in the glueline (starved glueline) or too little so that fibers and microdefects just below the wood surface are insufficiently bonded together.

Solidification and cure of the wood–adhesive system occurs either by solvent loss, polymerization and cross-linking reactions, or both. This process is time, temperature and pressure dependent with bond quality being a reflection of how well the final solidification stage is accomplished in the particular adhesive and wood substrate present.

2. Mechanisms of Adhesion

A number of theories have been proposed to explain bonding mechanisms in materials science. The unique heterogeneous characteristics of wood limit the relevance of any one particular theory, leading instead to wood bonding being rationalized in terms of a combination of mechanisms involving mechanical, physical and chemical parameters.

The ability of mechanical interlocking to occur readily is a function of the porous nature of most wood surfaces and the ease with which the adhesive can flow into these microcavities under the action of heat and external pressure. The mechanical anchoring of adhesive into wood occurs when the adhesive polymer hardens. This bonding mechanism is of obvious importance in hot melts and does increase the contact area available for bond formation. It cannot, however, account for the documented phenomena of bond-strength reduction with aging of wood surfaces and the limited correlation between adhesive penetration depth and bond-strength development.

Adsorption and diffusion mechanisms allow for intimate molecular contact between adhesive and wood substance over a large area. The accumulative physical interaction arising from attractive Van der Waals, hydrogen bonding and dispersion forces between adhesive polymer and wood components can result in strong bond formation which many consider a primary bonding mechanism. Complete wetting of the adhered surface enhances bonding potential. Both adhesive surface tension and wood surface tension have some bearing on wetting effectiveness. Adhesive spreading methods are additional factors of this mechanism since the mechanical energy input during glue application may overcome some surface force resistance to wetting.

Both adhesive and wood reactive functional groups allow for the high probability of covalent bond formation. In cross-linkable systems these chemical interactions can produce a strongly bonded wood–adhesive matrix capable of maintaining long-term stability under a variety of heat, moisture and stress conditions. While indications of covalent bonding have been found in model systems, conclusive evidence of their presence in wood-bonded products has not yet been made.

3. *Wood-Related Factors Affecting Adhesion*

Wood, being a complex, three-dimensional tissue, has a distinct but variable structural and chemical makeup which must be allowed for in adhesive formulation. The heterogeneous character of wood is reflected in both its bulk and surface properties.

3.1 Bulk Features

Anatomical characteristics, permeability, density and moisture content are among the more important interrelated factors influencing wood-bond formation.

Fiber orientation influences the ease with which the adhesive polymer can wet wood while at the same time retaining sufficient polymer on the bonding surface to form a glueline interface. Rapid adhesive penetration is possible along the grain because of open lumen structures, while penetration across the grain is highly restricted by fiber walls. Anatomical differences between species will affect the permeability of wood to liquids and gases, thus influencing the degree to which adhesive can successfully intermix with wood fibers. In softwoods, fiber lumina are laterally interconnected by small capillaries which allow the ready passage of air and liquid to various depths below the wood surface. In hardwoods, many of these interconnecting routes are blocked or absent, resulting in more restricted flow of the adhesive into bonding sites. Further variations such as knots or grain distortions associated with tree-growth patterns can provide additional impediments to achieving bonding potential. Other structural features of importance are variations imposed by the presence of juvenile and mature wood, and earlywood and latewood effects within one piece of wood. These features are interrelated to wood density.

The density of wood relates directly to its strength and has primarily a physical influence on adhesion. The stronger the wood, the stronger the adhesive must be in order to utilize the full strength of wood. Density variations are associated with the growth ring structure of wood, particularly in the ratio of fiber-wall to fiber-lumen volume. In dense wood, adhesive penetration is restricted while in lower density woods, care must be taken to prevent overpenetration. Where large differences in density can occur (i.e., earlywood–latewood zones) the adhesive must be designed to adapt adequately to both of these conditions. In general wood density in the range 250–450 $kg\,m^{-3}$ (i.e., spruce, pine, fir, aspen) are easier to bond than woods of density greater than 600 $kg\,m^{-3}$ (i.e., birch, oak, maple).

Wood physical properties are highly dependent on the interaction of wood density and moisture content which in turn affect gluing behavior. High-moisture-content conditions often result in increased adhesive flow and penetration while retarding polymer cure by limiting solvent loss into the wood or restricting condensation polymerization reactions. Excessive water at the bonding interface can also act as a barrier to bond formation, especially with hot-melt systems. Typical moisture-content ranges for bonding are 2–16% although successful bond formation has been reported at up to 40% moisture content with some adhesives curing under ambient or slightly elevated temperature conditions. The likelihood of steam formation at higher moisture contents, with its destructive possibilities in the glueline, necessitates wood to be dried to 2–6% moisture content when using high-temperature-curing adhesives. Wood volumetric changes occurring between fiber saturation and oven-dry conditions result in directional shrinkage and swelling forces, directly proportional to density, which can place considerable stress on the glueline. Cured gluelines should partially maintain bond integrity with dimensional-change-induced stresses, even during continual cycling between wet and dry conditions, because they may weaken the wood–adhesive interface considerably in high-density woods. Moisture and density also affect wood deformational properties requiring some adjustments in bonding pressure (i.e., lower density or higher moisture content requires less bonding pressure).

3.2 Surface Features

Both physical and chemical properties of the wood surface can profoundly affect bond-formation potential. Their primary influence impacts on the accessibility of adhesives to wood-fiber bonding sites. From a microscopic perspective, wood surfaces have a rough, physical appearance due to anatomy, in particular grain orientation and fiber size. Surfaces produced from transverse cuts (i.e., cut across the tracheids) expose end-grain areas whose large lumen volume often results in substantial overpenetration of adhesive. Consequently, direct bonding of end-grain surfaces is difficult and not commonly attempted. Cutting along tangential and radial surfaces inevitably excises some fiber walls, exposing different amounts and parts of fiber lumina. Surface topography may vary appreciably between species since lumen diameters may range from as small as 1 μm in fine-grained woods to as large as 200 μm in coarse-grained woods. Presence of resin ducts, vessel elements or knots give added variations to the wood surface. Machine operations during surface formation often result in further irregularities which can be associated with anatomical features. These can include saw-tooth abrasions, indentations from planer or shaper knives, lathe checks and roughness resulting from the peeling of veneer, torn and broken fibers caused during flake cutting and other splits or checks. Although these features may increase available bonding area, they hinder good bond formation by weakening the wood surface. Heat and pressure can be utilized to compress and plasticize the interface to improve contact in lower-density species but high-density woods must rely on smooth machining operations to achieve the intimate contact necessary for bonding. Mechanically weakened wood surfaces can be repaired during gluing by the action of adhesive penetration between and into the wood fiber.

Wood surfaces are chemically heterogeneous. Their composition does not necessarily correspond directly to the chemical makeup of bulk wood but rather is more a function of the conditions and methods of surface formation. Since wood fibers are composed of strands of oriented cellulose molecules intertwined within hemicellulose and lignin–polymer matrices, cleavage of a fiber wall may expose varying amounts of each chemical component at the wood surface layer. The carbohydrate components are highly polar, with fractions existing in both crystalline and amorphous forms and containing numerous functional groups available for reaction. Lignin, an amorphous polymer, offers additional reactive sites on its phenolic components. Both carbohydrate and lignin fractions are present in a range of molecular sizes and polarities which allow for bond formation with polar or nonpolar adhesives.

Extractives, although normally a minor component of the wood, can play a significant role in bond formation. These relatively low-molecular-weight materials are often highly mobile, nonpolar substances of limited reactivity which inhibit bond formation by blocking adhesive accessibility to reactive wood sites. This is accomplished by restricting wetting, altering the wood surface pH or forming complexes with adhesive or wood bonding sites.

Wood-surface history is an additional chemical feature affecting bonding potential. A smooth, freshly-cut wood surface dried at moderate temperatures offers the optimum conditions for bonding. As aging occurs, the surface is chemically modified at a rate depending on temperature and atmospheric conditions. Oxidation or contamination of the wood surface by nonpolar, mobile extractives and airborne pollutants are the primary phenomena occurring during this time. Oxidation reduces the number of active hydroxyl groups available for bonding. Under high-temperature drying conditions the chance of wood-surface oxidative degradation becomes highly probable. Sometimes chemical and thermal treatments can be utilized to change wood-surface chemistry and enhance bonding potential. Strong acids and oxidizing agents are common activators. They create new functional groups through depolymerization and ring cleavage reactions of lignin and carbohydrate molecules, or generate free radicals which can lead to bond formation by oxidative coupling. Resulting gluelines tend, however, to be brittle and have highly variable strength because wood can be degraded by many of the chemicals used.

4. *Wood Adhesives*

Adhesives can be classified as either natural or synthetic. Typical natural adhesives are carbohydrate or protein based, derived from either animal or vegetable sources, or may be an inorganic material. Synthetic adhesives are produced by the controlled polymerization of various monomeric organic molecules. These synthetics can be further divided into thermosetting (most common for wood adhesives) and thermoplastic categories. They can function to produce structural or nonstructural bonds. The present discussion deals only with major adhesives used in the wood industry. The majority of these adhesives are formed by condensation polymerization processes involving the use of formaldehyde as a cross-linking agent. Details of their chemistry have been presented elsewhere (Skeist 1977).

Phenol–formaldehyde (PF) resin is the primary wood adhesive used in the production of strong, durable exterior products. These adhesives are low to moderate molecular weight thermosetting polymers, manufactured in aqueous solutions with acid or alkaline catalysts. By varying reaction time and temperature, catalyst type and amounts and ratio of reactants, a variety of different adhesives can be prepared. Resoles are a class of PF resins produced under alkaline conditions with a molar excess of formaldehyde to phenol. These reactions produce highly branched polymeric structures while remaining soluble in strongly alkaline solutions. Sufficient formaldehyde is present to allow a highly cross-linked network to be formed during cure. Novolacs are PF resins synthesized under mild acid or neutral conditions with a molar excess of phenol to formaldehyde. These polymers have a linear structure, relatively low molecular weight and maintain thermoplastic behavior unless additional formaldehyde is added when curing is undertaken. These are referred to as two-stage resin systems. Plywood, oriented strandboard, particleboard and fiberboard intended for exterior service are manufactured using aqueous resoles tailored to give particular application, flow and cure properties for a specific product. A typical aqueous resole contains a phenol to formaldehyde to sodium hydroxide molar ratio of 1:(1.8–2.2):(0.1–0.8), with a resin-solids range from 35–50% and averaging 5–40 phenolic units linked together. These alkaline resoles require cure temperatures greater than 100 °C to attain adequate cross-linking. Fast cure at ambient temperatures can be achieved by addition of strong acids. Equipment corrosion and long-term acid degradation of the wood interface are, however, deterrents to this procedure.

For waferboard applications, the need for good adhesive distribution on large wafers and the limitations on moisture level in the glueline during high-temperature pressing require the use of powder PF resins. These are produced by spray drying or other solidification techniques.

Significant increases in phenolic resin reactivity are possible by copolymerizing with resorcinol to produce a phenol–resorcinol–formaldehyde (PRF) system. These are often used in specialized applications (i.e., glued laminated timbers, fingerjointed structures) where durable bonds and ambient cure conditions are needed. To ensure adequate pot life, the PRF is synthesized as a novolac system. Additional formaldehyde needed for cure is mixed in just prior to adhesive application. The high cost of resorcinol, relative to phenol, restricts most PRF applications to higher-value and large structural products.

Aqueous urea–formaldehyde (UF) polymers are low-cost, light-color, fast-curing adhesives with poor weather resistance. Their use is limited to products such as decorative plywood panels, particleboard subfloors or fiberboard for furniture or interior applications. Resins are prepared in molar ratios of formaldehyde to urea of (1–2):1 with the higher ratios usually utilized in solid wood or veneer bonding where higher strengths are required. Cure takes place at moderate acid conditions either by addition of acid salts or by use of the inherent acidity of wood species such as oak or cedar. Heat accelerates cure speed; polymer cross-linking, however, can be achieved at ambient temperatures. Although a thermosetting polymer, UF is sensitive to heat and moisture with formaldehyde slowly released as decomposition occurs. Requirements exist restricting formaldehyde emissions in wood products to low concentrations in air. To help control these emissions most particleboard is manufactured with formaldehyde to urea molar ratios less than 1.2:1. Alternatively, fortification of UF to improve its durability and strength is possible by copolymerizing with melamine. This modification adds increased chemical costs while requiring higher cure temperatures.

Polyvinyl acetates (PVAs) are thermoplastic, aqueous emulsions used primarily for furniture assembly and other nonstructural applications. Bond strength develops from water loss into the wood from the adhesive, with a resultant increase in polymer concentration in the glueline. While providing excellent dry adhesion strength and good gap-filling properties, the tendency of PVAs to cold-flow or creep when subjected to sustained loads is a primary disadvantage. Modification of PVA with thermosetting phenolic resin results in cross-linkable PVAs with improved heat and moisture resistance but only moderate alterations in creep behavior.

Other adhesive systems such as epoxy, polyurethanes and polymeric isocyanates have limited applications to wood bonding at the present time. While these adhesives provide improved bond strength with no formaldehyde emissions, they have disadvantages of greater cost and incompatibility with water, which requires the use of organic solvents for applicator cleanup. Application of polymeric isocyanate in particleboard, waferboard and oriented strandboard have recently gained more favor due to their faster cure times and greater tolerance to moisture compared to conventional thermosets. Demand for durable, fireproof wood panels also has resulted in the application of inorganic cement adhesives in particleboard, although cure sensitivity to certain wood species and substantial increase in panel density limits their usage.

5. *Adhesive Factors Influencing Bond Formation*

Adhesive physical and chemical properties play as important a role as wood properties in determining bond performance. Strength, durability, working properties and cost are primary considerations governing adhesive selection. Interrelated polymer parameters which influence the overall performance of an adhesive include reactive functionality, molecular-weight distribution, solubility, rheology and cure rate.

Number and reactivity of adhesive functional groups determine how many active sites are available for further polymerization. These affect directly the reaction speed and degree of cure possible during bond formation. The functionality of the adhesive components must be greater than 2.0 to achieve a cross-linkable system.

Molecular weight and molecular weight distribution is a polymer property affecting viscosity, flow and cure speed of the adhesive. Provided sufficient functional-group reactivity remains, higher-molecular-weight components can more rapidly attain cure but exhibit significantly greater solution viscosities and poorer flow properties than low-molecular-weight components. Optimum adhesive molecular-weight distribution, however, depends upon the wood composite. Solid wood products such as glued laminated lumber and plywood utilize a major portion of higher-molecular-weight components (i.e., $\bar{M}_N$ of 1500–3500) to avoid overpenetration of polymer into wood. Adhesives for waferboard and particleboard have their molecular weight distribution shifted to lower values (i.e., $\bar{M}_N$ of 500–1200). This allows efficient and homogeneous resin application while improving adhesive mobility to facilitate better resin transfer between wood particles during blending operations.

Polymer solubility or dispersibility in solution is an important factor affecting uniform application and cleanup of the adhesive. Improved solubility allows more concentrated adhesive solutions to be used which reduces the amount of solvent to be driven off during cure. Adhesive rheology and cure development, while related to polymer molecular weight and reactive groups present, are highly responsive to heat, pressure and wood moisture content. Various segments of the polymer molecule achieve mobility as plasticization occurs with moisture, or temperature increases to the glass transition (T_g) or softening temperature. Cure results in an increase of T_g. Often compromises are needed to ensure that heat and pressure are applied in a controlled sequence such that the adhesive adequately wets the wood surface and cures before resin overpenetration occurs.

6. *Bond Requirements for Composites*

Performance criteria for wood composites relate directly to product end use. This requires consideration of engineering strength needs, safety and short- and long-term response of the material to the service environment. Structural, exterior-grade products have the most demanding bond-quality requirements, since glueline failure could be catastrophic to these structures. For these situations glueline strength, durability and reliability must be assured, usually by extensive bond-quality testing programs. While conventional bond strength and short-term durability evaluation methodology is well established, there is at present only a limited understanding of the effect of long-term aging on bond integrity. Insight on how product dimensional change and sustained loading affect time-dependent performance properties of wood composites is developed by cyclic testing through simulated environments using variables of heat, moisture and loading stress. The nonlinear deterioration behavior of many adhesive bonds, together with our lack of knowledge of the aging process, are serious obstacles to assessing long-term performance of new wood adhesives.

For conditions where bond performance is less crucial, cost may be the overriding factor—not only in terms of adhesive price but also the impact adhesive costs have on overall production expenses. With more automated wood-composite manufacturing systems, compatibility of the adhesive with the assembly system often is the major consideration.

7. *Future Direction*

Continuing reductions in available harvest volumes and lower wood quality will lead to increased reliance on wood composites and the adhesives used to manufacture them. For commodity products, the importance of manufacturing efficiency will place greater emphasis on adhesives capable of tolerating variable wood properties and moisture contents while having rapid cure speed and reasonable cost. Adhesives will gain prominence in higher-valued wood composites by allowing products of varying shapes, improved engineering strength and greater aesthetic value to be more easily manufactured. Expanded use of continuous pressing techniques and heating methods such as steam pressing or microwave use will create added challenges for adhesive formulation development and wood-handling concepts. Improved knowledge of chemical factors influencing the reactivity of the wood-bonding surface would help enhance functional group reactivity, perhaps leading to reductions in amounts of adhesive required or possibly improved binderless wood composites. Advanced, high-strength composites are likely to increasingly utilize wood fiber or cellulose fiber as a directional, reinforcing matrix surrounded by adhesive polymers which may comprise up to 50% of the composite weight.

Future adhesive improvements will be concentrated in areas of durable, faster curing systems, capable of enduring moderate to long assembly times. Copolymer systems will increase in prominence with emphasis being directed toward multicomponent, cold-setting adhesives possibly applicable to on-site assembly. Isocyanates, because of their extensive reactivity and ability to bond at higher moisture contents are likely to be a component of these newer wood-bonding systems. Reactive hot melts, capable of rapidly curing to thermosetting polymers, will also be a development having significant impact on future bonding systems.

See also: Glued Joints

Bibliography

Collett B M 1972 A review of surface and interfacial adhesion in wood science and related fields. *Wood Sci. Technol.* 6: 1–42

Koch G S, Klareich F, Exstrum B 1987 *Adhesives for the Composite Wood Panel Industry.* Noyes Data Corporation, New Jersey

Kollmann F F 1975 *Principles of Wood Science and Technology*, Vol II: *Wood Based Materials.* Springer, New York

Marian J E 1967 Wood, reconstituted wood and glued laminated structures. In: Houwink R, Salomon G (eds.) 1967 *Adhesion and Adhesives*, Vol II: *Applications.* Elsevier, New York

Oliver J F 1981 *Adhesion in Cellulosic and Wood-based Composites.* Plenum, New York

Pizzi A 1983 *Wood Adhesives Chemistry and Technology.* Dekker, New York

Skeist I 1977 *Handbook of Adhesives*, 2nd edn. Van Nostrand Reinhold, New York

Subramanian R V 1984 Chemistry of adhesion. In: Rowell R (ed.) 1984 *The Chemistry of Solid Wood*, Advances in

Chemistry Series 207. American Chemical Society, Washington, DC
Wellons J D 1983 The adherends and their preparation for bonding. In: Blomquist R F, Christiansen A W, Gillespie R H, Myers G E (eds.) 1981 *Adhesive Bonding of Wood and Other Structural Materials*, Vol 3. Clark C Heritage Series on Wood, Pennsylvania State University, University Park, Pennsylvania

P. R. Steiner
[University of British Columbia, Vancouver, British Columbia, Canada]

Archaeological Wood

Archaeology is often thought to be the study of ancient peoples and their culture, with emphasis on civilizations that existed 2000 or more years ago. Modern archaeology, however, is much broader, and might better be defined as the collection and interpretation of physical evidence of past cultural activities. Thus there need be no time limits, and we can consider that archaeological wood is simply old wood from artifacts that have been out of normal use for a period of time.

Wood can be very long-lasting in normal use under favorable conditions, but it is biodegradable and subject to attack by a host of agents capable of its destruction. Such destruction can take place in a matter of minutes in the case of fire, or a matter of months in the case of decay, or over hundreds or even thousands of years in the case of bacterial or chemical deterioration. Archaeological wood is wood that has been sequestered under special conditions that were unfavorable to agents capable of its more rapid destruction, but it may nevertheless have suffered some degree of degradation.

In order to survive, archaeological wood must be protected from decay by wood-destroying fungi. This can be achieved either by keeping it very dry to deprive the fungi of moisture or completely and continuously wet to deprive them of oxygen (see *Decay During Use*). Archaeological wood that has been preserved under dry conditions is relatively rare and in the main comes from tombs in Egypt. Waterlogged wood, i.e., wood that has been stored under wet conditions such as burial in soil below the permanent water table, at the bottom of rivers or lakes, or in a marine environment, is much more plentiful and is being found in all parts of the world.

1. Dry Archaeological Wood

One of the most spectacular finds of dry archaeological wood was the Solar Boat of Cheops discovered in 1954. Some parts were clearly deteriorated but most of the 4500-year-old wood was in good condition. Even so it was felt necessary to apply a light consolidation treatment with poly(vinyl acetate) (Jenkins 1980). Wood over 4000 years old taken from several Egyptian pyramids still had the outward appearance of new wood except for color changes. Electron microscopy of such wood revealed cracks and fissures not found in recent wood, indicating weakening of the ultrastructure, particularly in the middle lamella region. The samples also showed less birefringence, indicating degradation of the cellulose (Borgin et al. 1975b). Total holocellulose content, however, was higher than for new wood (Van Zyl et al. 1973), apparently because of oxidative degradation of lignin which was present in correspondingly lower percentages (Borgin et al. 1975a).

In spite of these structural and compositional changes, limited tests in bending and in compression parallel to grain indicated that the strength of the wood from the pyramids did not appear to be affected very significantly (Van Zyl et al. 1973).

The aging of more recent, historical wood has been studied by Kohara (1958), using samples taken from various temples in Japan up to 1300 years old. He found that hinoki, a softwood, underwent two simultaneous processes, namely increasing crystallization and degradation of cellulose. The increase in crystallinity is accompanied by increases in certain strength properties but this process terminates after about 350 years, while degradation leading to strength loss is continuous. As a result, some strength properties (bending and compression strengths) show an increase over the first 350 years followed by a continuous decrease in strength thereafter, so that the recent wood and the 1300-year-old wood differed little if at all. Impact bending strength decreased with age but at a decreasing rate. The 1300-year-old wood did show significantly lower hygroscopicity than recent wood. Keyaki, a hardwood, did not have any initial strength increases and lost strength continuously and at a greater rate than the hinoki.

Similar results were found for a roof beam of European spruce of a 300-year-old building. Tensile, compression and bending strengths were the same or slightly higher than those of recent wood of the same density, while impact bending strength was lower (Schulz et al. 1984).

Stabilization of deteriorated, dry archaeological wood requires equilibration to the temperature and relative humidity conditions at the location where it will be stored (see *Hygroscopicity and Water Sorption*). This may entail some shrinkage if the original moisture content was high. The remains of a table discovered in the burial chamber inside a tumulus at Gordion in Turkey were initially damp, and the boxwood underwent subsequent shrinkage of 9.1, 5.5 and 4.3% in the tangential, radial and longitudinal directions, respectively (Payton 1984). Although normal boxwood has relatively large shrinkage, the high longitudinal shrinkage of this material indicates a substantial degree of deterioration.

The wood of the table from Gordion was consolidated by vacuum impregnation with a solution of poly(vinyl butyral) after removing an initial treatment with paraffin wax (Payton 1984). A wide variety of materials has been used for consolidation, ranging from waxes and natural glues and resins to all types of synthetic materials (Unger and Unger 1987). Synthetic resin consolidants can be divided into three groups: thermosetting polymers, polymers that are introduced into wood in the form of monomers and then polymerized in situ (see *Wood–Polymer Composites*), and thermoplastic polymers in solution. An important consideration with all conservation treatments is their reversibility. Although it can be shown that, strictly speaking, no such treatment for wood is truly reversible, consolidation treatments of wood with certain stable, soluble resins can be done in a manner that approaches reversibility (Schniewind 1989a). If only surface layers need be treated to improve handling of artifacts, consolidants can be applied by brushing. More complete penetration can be achieved by vacuum impregnation or by long-term spraying with consolidant solution in a closed system. Care must be taken to select resins with good stability that will not cross-link, embrittle, become yellow or otherwise damage the consolidated object with time.

2. *Buried, Waterlogged Wood*

Buried wood will generally not survive unless it becomes waterlogged to create the anaerobic conditions necessary for protection from decay fungi. This still leaves the wood susceptible to chemical degradation and attack by anaerobic bacteria and, in particular cases, by soft rot (see *Decay During Use*). Wood may have been attacked by decay fungi before burial, particularly if it occurred gradually rather than by a sudden catastrophe. Buried wood, whether it be archaeological wood or wood untouched by man, will undergo processes of aging and fossilization. These may be divided into two types, namely silicification, where the wood substance is gradually replaced by infiltrates of mineral matter; and carbonization, where oxygen and hydrogen are gradually split off approaching an end point of conversion to pure carbon. Both of these are extremely slow processes as measured against the age of man (Fengel and Wegener 1984). Buried wood dating from no earlier than the Pleistocene age (beginning about 2 million years ago) is also referred to as subfossil wood and forms a close parallel to buried archaeological wood. Subfossil oak about 5000 years old has been found in sufficiently good condition to be utilized industrially for furniture manufacture (Scheiber and Wagenführ 1976).

Degradation processes in buried wood usually start with the hemicellulose component, followed by cellulose, while lignin is generally the most resistant (Fengel and Wegener 1984, Sen 1956). Thus there is an apparent increase in lignin content as degradation proceeds, but when composition is expressed on the basis of volume in wood, lignin content is seen to change very little (Hoffmann et al. 1986). Although hemicellulose is generally conceded to be the component to be degraded first, Iiyama et al. (1988) found evidence of greater degradation of cellulose than hemicellulose.

Almost invariably, ash content of buried wood increases, owing to the infiltration of minerals (Fengel and Wegener 1984). This would suggest that initially the processes leading to silicification and to carbonization take place simultaneously and that archaeological wood is affected to some extent by both.

Macroscopically, degradation starts from the outside of wood members, often producing clearly demarcated outer zones of highly degraded wood surrounding a core of virtually intact material (Noack 1969, Hoffmann et al. 1986). Microscopically, evidence of degradation may be circumstantial as in the physical damage incurred in preparation for microscopy (Borgin et al. 1975b). Destruction of the cell wall beginning from the lumen side has been observed (Bednar and Fengel 1974) but a more common pattern appears to be swelling of the secondary wall followed eventually by its complete destruction (Hoffmann and Parameswaran 1982). Barbour and Leney (1986) found that in 2500-year-old waterlogged alder the vessel elements were most resistant to degradation and the fibers the least, which they attributed to a predominance of guaiacyl lignins in the vessels and a high concentration of syringyl lignins in the fibers. Examination with the polarizing microscope has shown that archaeological woods lose birefringence as deterioration progresses (Sen 1956). This loss, which is due to the degradation of cellulose and destruction of its crystal structure, was found to be more pronounced in buried wood than in dry wood from the pyramids (Borgin et al. 1975b).

So far, no one has been able to detect a relationship between the extent of chemical and structural changes and age of the wood. Specimens thousands of years old can be in much better condition than specimens no older than a few hundred years. The degree of deterioration is therefore entirely dependent on species, some species such as European oak being more resistant than others, and the conditions of burial.

The best single indicator of the extent of deterioration is density, preferably expressed as "conventional density," i.e., as density based on oven-dry weight and volume when fully swollen (Hoffmann et al. 1986). For waterlogged wood it is convenient to determine this by the maximum moisture content method, which may be expressed as follows:

$$G_f = (M_{max}/100 + 0.667)^{-1} \quad (1)$$

where G_f is the conventional density in $g\,cm^{-3}$ and M_{max} is the maximum moisture content, in percent, based on oven-dry weight. Equation 1 is based on the assumption that cell wall substance shrinks in an amount equal to the volume of water removed and that the density of dry cell wall substance is $1.5\,g\,cm^{-3}$. For commonly used woods within the density range 0.4–0.8 $g\,cm^{-3}$ the maximum moisture content would vary from 58 to 183%; for highly deteriorated waterlogged wood moisture contents of as much as 800% are not unusual (Noack 1969).

As waterlogged wood deteriorates, its hygroscopicity tends to increase, and equilibrium moisture content values as much as twice those of recent wood have been found in alder (Barbour and Leney 1982). Along with increased hygroscopicity, deteriorated buried wood also exhibits significant increases in shrinkage, reaching values of the order of 70% for volumetric shrinkage. The shrinkage in volume of archaeological oak was found to be related linearly to maximum moisture content (Hoffmann et al. 1986). Most dramatic are the increases in longitudinal shrinkage, presumably because of the destruction of the cellulose microfibrils, leading to a reduction in the degree of shrinkage anisotropy. Shrinkage values of more than 30% in the tangential and more than 10% in the longitudinal directions have been recorded by many investigators (Schniewind 1989b). Such high shrinkage values are the result of extensive collapsing of the cell structure in response to drying stresses. Drying of highly deteriorated waterlogged wood therefore results in severe damage and distortion of artifacts unless special measures are taken. Excessive shrinkage of deteriorated waterlogged wood is its most significant characteristic because it has a major bearing on the conservation of archaeological finds made of wood.

The extensive collapse shrinkage occurs because degradation of carbohydrates not only changes the sorption and shrinkage characteristics, but also has a major effect on strength. Compression strength parallel to grain and static bending strength are the properties most often investigated. Data collected by Schniewind (1989b) show a spectrum of values, ranging from residual strength greater than species averages of recent wood, even for material hundreds of years old, to less than 10% of the presumed original strength. Age alone is clearly not the determining factor for strength losses as was also found for changes in other properties. The data show that in general, strength losses are greater than mass losses, i.e., even very small mass losses can be accompanied by major losses in strength.

3. *Drying and Stabilizing Waterlogged Wood*

Once waterlogged wood is taken from its wet environment and placed into surroundings with less than 100% relative humidity, it will begin to dry. If the object is small and the wood not deteriorated, it may be possible to allow the object to dry with no more than normal wood shrinkage taking place. It may even be possible to dry large objects by using normal precautions employed in industrial wood drying (see *Drying Processes*) provided that the wood is sound. Schweizer et al. (1985) were able to successfully dry some large Roman beams of silver fir dating from the second century AD by carefully controlled air drying. A much more common situation is that buried wood will have undergone deterioration of varying degrees which then requires special methods if drying is to be controlled successfully.

As a first step, the condition of the wood to be treated must be evaluated to determine the best treatment methods. If it is possible to take samples, such evaluation should start with a determination of maximum moisture content or conventional density. To make this most meaningful, the species must also be identified so that mass loss can be estimated by comparison to average density of recent wood of the same species. Additional steps might include microscopic examination and chemical analysis for the major wood constituents including ash content, for which a standard scheme has been proposed (Hoffmann 1982). It is also helpful to take note of wood characteristics such as knots and cross grain which would influence drying recent as well as old wood (Jagels 1982).

Although the above analyses can be performed on relatively small samples, it is often not possible and never desirable to perform destructive tests. Research on nondestructive evaluation for waterlogged wood has focused almost entirely on various types of hardness tests which can leave dents or small holes but in some instances have been reported to be entirely nondestructive (Kazanskaya and Nikitina 1985). An impact needle-hardness test using the Pilodyn instrument has been used successfully in surveying the timbers of the *Mary Rose*, the remains of the flagship of Henry VIII (Jones et al. 1986).

For wood with no more than moderate deterioration which still maintains a firm surface, treatments to control shrinkage may be sufficient, while highly deteriorated woods will require some type of consolidation treatment as well so that the objects may be safely stored and handled. Of the various methods that can be used to modify the shrinkage of wood (see *Shrinking and Swelling*), those that involve the introduction of bulking agents into the cell wall offer at present the only practical approaches. In essence, bulking agents replace some of the water within the cell wall, keeping the wood in a permanently swollen condition. In addition, physical support may be needed to prevent cell collapse from surface tension of the receding water columns in the cell lumina. This support can be provided by loading the cell cavities with materials such as resins and waxes.

When the Swedish warship *Wasa* was resurrected in 1961 after 333 years at the bottom of Stockholm harbor, the decision was made to treat the entire vessel with poly(ethylene glycol) (PEG). The PEG was applied—by spraying in an aqueous solution—to the vessel, which had been placed in a specially constructed building to keep it from drying prematurely. The spraying continued for many years since the PEG must enter the wood structure by the rather slow process of diffusion (Mühlethaler 1973). Various forms of PEG treatment are today the most widely used methods for dealing with degraded, waterlogged wood.

PEG is available in a range of molecular weights. The low molecular weight fractions have the advantage of a higher rate of diffusion into wood and better penetration into the cell wall. At molecular weights of 3000 and more there is little if any penetration into the cell wall, so that cell wall bulking is no longer a factor. As a distinct disadvantage, the hygroscopicity of PEG increases as molecular weight decreases, so that in some cases wood so treated cannot be dried. While low molecular weight PEG is most effective for less degraded wood, high molecular weight fractions tend to be more successful with highly degraded material. In many cases both types of wood are present, even in the same piece. To overcome these problems, Hoffmann (1986) devised a two-step treatment for the wood of the *Bremen Cog*, a medieval merchant vessel resurrected in 1962 after 570 years at the bottom of the river Weser. This involves first treating the entire vessel with PEG 200, followed by a treatment with PEG 3000 at 60 °C. The PEG 200 bulks the cell walls while the PEG 3000, which becomes a hard wax after cooling back to room temperature, serves as a consolidant within the cell lumina. PEG 4000 is also often used, either in aqueous solution or with other solvents such as *t*-butanol or methanol (Grattan 1982). It can take years to obtain sufficient penetration of PEG for large members.

Freeze-drying is also widely used. It has even been proposed to take advantage of the Canadian winter climate for exterior freeze-drying of waterlogged wood (Grattan and McCawley 1978). The effectiveness of freeze-drying can be improved if the wood is pretreated with low molecular weight PEG (Grattan 1982).

Many other methods of bulking the cell wall structure and/or loading the cell cavities have been devised. Among these, various sugars such as sucrose are promising because they afford bulking treatments that are simple and inexpensive. Other methods include treatment with resins by solvent exchange or by in situ polymerization (Mühlethaler 1973, Grattan 1982).

Bibliography

Barbour R J, Leney L 1982 Shrinkage and collapse in waterlogged archaeological wood: Contribution III Hoko river series. In: Grattan D W, McCawley J C (eds.) 1982 *Proc. ICOM Waterlogged Wood Working Group Conf.* International Council of Museums, Ottawa

Barbour R J, Leney L 1986 Microstructural analysis of red alder (*Alnus rubra* Bong) from a 2500 year old archaeological wet site. In: Barry S, Houghton D R (eds.) 1986 *Biodeterioration VI.* CAB International, Farnham Royal, Slough, UK

Bednar H, Fengel D 1974 Physikalische, chemische und strukturelle Eigenschaften von rezentem und subfossilem Eichenholz. *Holz Roh- Werkst.* 32: 99–107

Borgin K, Faix O, Schweers W 1975a The effect of aging on lignins of wood. *Wood Sci. Technol.* 9: 207–11

Borgin K, Parameswaran N, Liese W 1975b The effect of aging on the ultrastructure of wood. *Wood Sci. Technol.* 9: 87–98

Fengel D, Wegener G 1984 *Wood: Chemistry, Ultrastructure, Reactions.* de Gruyter, Berlin

Grattan D W 1982 A practical comparative study of several treatments for waterlogged wood. *Stud. Conserv.* 27: 124–36

Grattan D W, McCawley J C 1978 The potential of the Canadian winter climate for the freeze-drying of degraded waterlogged wood. *Stud. Conserv.* 23: 157–67

Hoffmann P 1982 Chemical wood analysis as a means of characterizing archaeological wood. In: Grattan D W, McCawley J C (eds.) 1982 *Proc. ICOM Waterlogged Wood Working Group Conf.* International Council of Museums, Ottawa

Hoffmann P 1986 On the stabilization of waterlogged oakwood with PEG. II. Designing a two-step treatment for multi-quality timbers. *Stud. Conserv.* 31: 103–13

Hoffmann P, Parameswaran N 1982 Chemische und ultrastrukturelle Untersuchungen an wassergesättigten Eichenhölzern aus archäologischen Funden. *Berl. Beiträge Archäom.* 7: 273–85

Hoffmann P, Peek R- D, Puls J, Schwab E 1986 Das Holz der Archäologen: Untersuchungen an 1600 Jahre altem wassergesättigten Eichenholz der 'Mainzer Römerschiffe.' *Holz Roh- Werkst.* 44: 241–47

Iiyama K, Kasuya N, Tuyet L T B, Nakano J, Sakaguchi H 1988 Chemical characterization of ancient buried wood. *Holzforschung* 42: 5–10

Jagels R 1982 A deterioration evaluation procedure for waterlogged wood. In: Grattan D W, McCawley J C (eds.) 1982 *Proc. ICOM Waterlogged Wood Working Group Conf.* International Council of Museums, Ottawa

Jenkins N 1980 *The Boat Beneath the Pyramid, King Cheops' Royal Ship.* Holt, Rinehart and Winston, New York

Jones A M, Rule M H, Jones E B G 1986 Conservation of the timbers of the Tudor ship Mary Rose. In: Barry S, Houghton D R (eds.) 1986 *Biodeterioration VI.* CAB International, Farnham Royal, Slough, UK

Kazanskaya S Y, Nikitina K F 1985 On the conservation of waterlogged degraded wood by a method worked out in Minsk. In: *Proc. 2nd ICOM Waterlogged Wood Working Group Conf.* International Council of Museums, Grenoble

Kohara J 1958 Study on the Old Timber. *Res. Rep. Faculty Technol. Chiba Univ.* 9(15): 1–55; 9(16): 23–65

Mühlethaler B 1973 *Conservation of Waterlogged Wood and Wet Leather.* Editions Eyrolles, Paris

Noack D 1969 Zur Verfahrenstechnik der Konservierung des Holzes der Bremer Kogge. *Die Bremer Hanse-Kogge*, Monographien der Wittheit zu Bremen, Vol. 8. Röver, Bremen

Payton R 1984 The conservation of an eighth century BC table from Gordion. In: Brommelle N S, Pye E M, Smith P, Thomson G (eds.) 1984 *Adhesives and Consolidants.* International Institute for Conservation of Historic and Artistic Works, London

Scheiber C, Wagenführ R 1976 Subfossiles Eichenrundholz; Aufkommen, Eigenschaften und Verwendung in der DDR. *Holztechnologie* 17: 133–39

Schniewind A P 1989a Consolidation of dry archaeological wood by impregnation with thermoplastic resins. In: Rowell R M, Barbour J (eds.) 1989 *Archaeological Wood:*

Properties, Chemistry, and Preservation, Advances in Chemistry Series. American Chemical Society, Washington, DC (in press)

Schniewind A P 1989b Physical and mechanical properties of archaeological wood. In: Rowell R M, Barbour J (eds.) 1989 *Archaeological Wood: Properties, Chemistry, and Preservation*, Advances in Chemistry Series. American Chemical Society, Washington, DC (in press)

Schulz H, von Aufsess H, Verron T 1984 Eigenschaften eines Fichtenbalkens aus altem Dachstuhl. *Holz Roh- Werkst.* 42: 109

Schweizer F, Houriet C, Mas M 1985 Controlled air drying of large Roman timber from Geneva. In: *Proc. 2nd ICOM Waterlogged Wood Working Group Conf.* Grenoble

Sen J 1956 Fine structure in degraded, ancient and buried wood, and other fossilized plant derivatives. *Bot. Rev.* 22(6): 343–75

Unger A, Unger W 1987 Holzfestigung im musealen und denkmalpflegerischen Bereich. *Holztechnologie* 28: 234–38

Van Zyl J D, Van Wyk W J, Heunis C M 1973 The effect of aging on the mechanical and chemical properties of wood. In: *Proc. IUFRO-5 Meeting*, Vol. 2. International Union of Forestry Research Organizations, Vienna

A. P. Schniewind
[University of California, Berkeley, California, USA]

B

Bamboo

Bamboos are perennial grasses with woody stems or culms which occur mostly in natural vegetation of tropical, subtropical and temperate regions and are abundant in tropical Asia. In these areas bamboos have been used as a main material for house construction, scaffolding, ladders, fencing, containers, furniture and many kinds of handicraft articles, and also for pulp and paper making. Bamboos constitute the *Bambusoideae*, a subfamily of the grass family, the *Gramineae*. There are about 750 species of bamboo in about 45 genera. Many of them are indigenous to the monsoon areas of tropical Asia, and some of them occur only in cultivation (Dransfield 1980).

1. Anatomy

The bamboo culm is characterized by nodes occurring periodically along its length. At the node is a solid cross wall, called the diaphragm, and the internodes are usually hollow. Culms can be up to 36 m long and up to 25 cm in diameter at the base, depending on the species. The outermost part of the culm is composed of a single layer of epidermis and the inside is covered by a layer of sclerenchyma cells. The body of the culm wall consists of vascular bundles embedded in ground tissue made up of parenchyma cells. The vascular bundles are composed of vessels, sieve tubes with companion cells and fibers. Bamboo culms have no cambium and no radial cell elements, such as rays, exist in the internodes. Bamboo is therefore quite different from wood. The culm consists of about 50% parenchyma, 40% fibers and 10% conducting cells (vessels and sieve tubes) on average. Parenchyma and conducting cells are more frequent in the inner-third of the culm wall, whereas in the outer-third the percentage of fibers is higher. In the vertical direction the proportion of fibers increases from the bottom to the top, while that of the parenchyma decreases (Liese 1980).

Parenchyma cells in the ground tissue are mostly elongated vertically (100 × 20 μm) and possess lignified, thick walls. The interspersed shorter cells are characterized by the occurrence of a dense cytoplasm and unlignified, thin walls. The parenchyma cells are connected to each other by small, single pits located on the longitudinal walls (Liese 1980).

Vascular bundles in the culm consists of xylem with two large metaxylem vessels (40–120 μm) and one or two protoxylem elements, and of phloem with thin-walled, unlignified sieve tubes connected to companion cells (Fig. 1). The vessels are large at the inner part of the culm wall and become smaller toward the outer part. Both the vessels and the phloems are surrounded by sclerenchyma sheaths. They differ considerably in size, shape and location according to their position in the culm and to the bamboo species. Most of the species have separate fiber strands on the inner, or the inner and outer, sides of the vascular bundles.

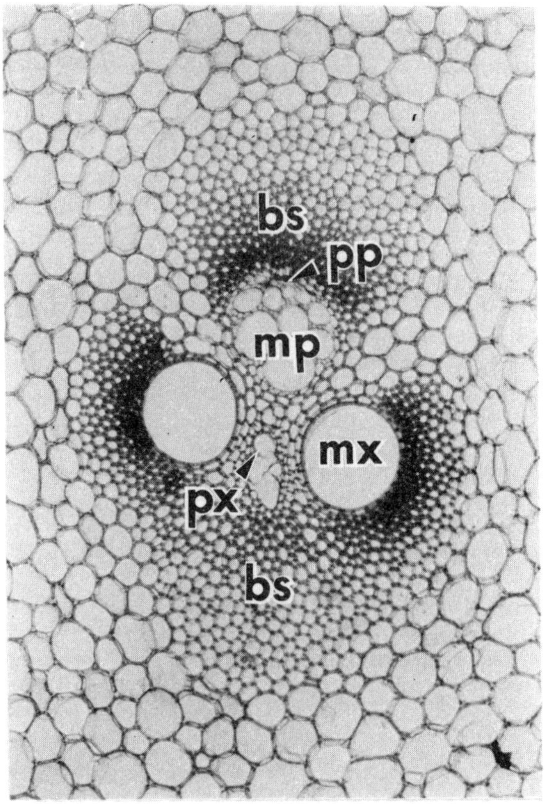

Figure 1
Vascular bundle of bamboo culm (*Phyllostachys pubescens*): mx = metaxylem; px = protoxylem; mp = metaphloem; pp = protophloem; bs = bundle sheath (courtesy of Professor K Shimaji, Wood Research Institute, Kyoto University)

Fibers in the internodes occur as caps of the vascular bundles and in some species also as isolated strands. Their slenderness ratio varies between 150:1 and 250:1. The length shows considerable variations between species and even in the same species. The average fiber length of some bamboo species is given in Table 1. The fiber length often increases from the periphery across the wall, reaches its maximum at

Table 1
Fiber length of bamboos[a]

Bamboo species	Fiber length (mm)
Bambusa arundinacea	1.73
B. blumeana	1.95
B. vulgaris	2.33
Dendrocalamus merrillianus	2.16
D. strictus	2.45[b]
Gigantochloa levis	1.80
Melocanna baccifera	2.78[b]
Oxytenanthera nigrociliata	2.43[b]
Phyllostachys nigra	1.86
Schizostachyum lumampao	2.42
Thyrsostachys oliveri	2.31
Cephalostachyum pergracile	2.20

[a] Tamòlang et al. (1980) [b] Varmah and Bahadur (1980)

about the middle and decreases toward the inner part, or the length may decrease from the outer part to the inside (Liese 1980).

The ultrastructure of most fibers is characterized by thick lamellated secondary walls (Parameswaran and Liese 1981). This lamellation consists of alternating broad and narrow lamella with differing fibrillar orientation. In the broad lamellae, the microfibrils are oriented at a small angle to the fiber axis, but in the narrow ones they are mostly horizontally oriented (Fig. 2). The narrow lamellae contain a higher amount

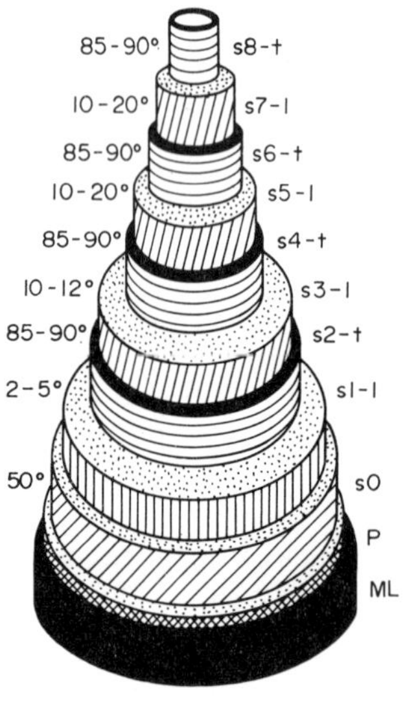

Figure 2
Polylamellated wall structure of bamboo culm fiber: ML = middle lamella; P = primary wall; s = secondary wall; l and t denote longitudinal and transverse orientation of microfibrils (Parameswaran and Liese 1981)

of lignin than the broader ones. A typical tertiary wall is not present, but in some taxa warts cover the innermost layer.

The polylamellated wall structure of the fibers, especially in the outer part of the culm, results in an extremely high tensile strength which is used to advantage in structural applications.

2. *Physical and Mechanical Properties*

Phyllostachys pubescens, one of the typical bamboos of Japan, grows to its full size within about two months, but takes about three years for complete maturation. The fundamental architecture of the cell wall is completed at an early stage but secondary thickening of cell walls and their lignification proceed more gradually. Crystallinity changes of cellulose in the cell walls of internodes from juvenile to mature stages show that the cellulose formed in the early stage is more or less amorphous and becomes oriented subsequently (Nomura 1980a,b)

Bamboo, like wood, is anisotropic and has as its principal directions the longitudinal, radial and tangential as referred to the cylindrical shape of the culm. Shrinkage is least in the longitudinal direction, but in contrast to wood is higher in the radial than in the tangential direction. For *Phyllostachys pubescens* and *P. reticulata* the radial, tangential and longitudinal shrinkage values from green to oven-dry are 6.65, 4.45 and 0.33%, respectively (Ota 1955).

The density of bamboo splint (culm wall material) depends on the species, values between 600 and 1000 $kg\,m^{-3}$ being common. The density of splint is greatest at the surface of the culm and decreases from there inward. In *P. reticulata* the air-dry density of the outer and inner halves of the culm wall has been found to be 918 and 748 $kg\,m^{-3}$, respectively (Ota 1951). Density also depends on vertical position in the culm, increasing from bottom to top. The variations in density are closely correlated with the relative proportions of vascular bundles and ground tissues; especially as the proportion of thick-walled sclerenchyma cells increases, the density increases also (Janssen 1981).

As with wood, density is correlated with mechanical properties of bamboo. Compressive strength σ_c of air-dry splint parallel to the fibers (longitudinally) can be predicted from the equation $\sigma_c = 0.094\,G$, where σ_c is in MPa and the density G in $kg\,m^{-3}$. The constant in the equivalent equation for dry wood is 0.084, which shows that bamboo is somewhat stronger than wood in compression parallel to the fibers, given equal density. This may be due to the higher cellulose content in bamboo (55% as compared with 50% in wood) (Janssen 1981). However, in wood, high compressive strength is often thought to result from high lignin content.

The relationship between compressive strength σ_c and percentage of bundle sheath (sclerenchyma cells)

Table 2
Ultimate compressive strength σ_c and percentage of bundle sheath V[a]

Portion of culm wall	*Phyllostachys reticulata*		*P. edulis*	
	σ_c (MPa)	V(%)	σ_c (MPa)	V(%)
Outer	112	40	105	34
Middle	80	26	76	23
Inner	67	17	68	13
Whole	83	28	82	22

[a] Ota (1951)

V is shown in Table 2, which also illustrates the effect of position within the culm wall. The data can be represented by linear regression equations; for *P. reticulata* this is $\sigma_c = 37.9 + 1.77\,V$, where the compressive strength is in MPa (Ota 1951). Similar relationships can be found for tensile strength, as shown in Fig. 3. The tensile strength parallel to the fibers of bamboo splint is very high, ranging from 2.5 to 3.5 times the compressive strength. The bending strength, or rupture modulus, of bamboo culms and bamboo splint is intermediate between the compressive and tensile strengths.

Shear strength parallel to the fibers of wood is generally about 20–30% of the compressive strength, but in bamboo it is much lower at only about 8% (Janssen 1981). Bending failure of bamboo culms often occurs by horizontal shear. Along with low shear strength, bamboo also has low resistance to splitting, which can be a disadvantage for structural uses. For instance, it is not practical to nail bamboo because it splits too easily. Thus, it is usually used in round form and lashed or tied together. The low splitting resistance can be a great advantage, however, as in the making of baskets and mats, and in many other handicrafts.

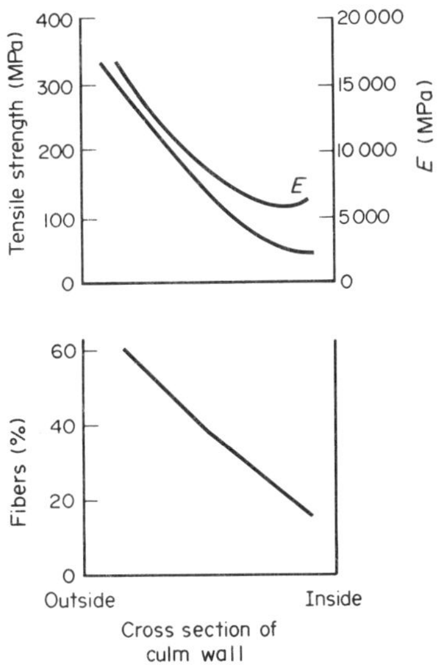

Figure 3
Variation of tensile strength, modulus of elasticity E and percentage of sclerenchyma fibers with position in culm wall (Janssen 1980)

The strength of bamboo splint depends on moisture content, and the same general principles apply as are found for wood. Above the fiber saturation point, moisture content has no effect on strength, and below it the compressive strength increases exponentially as the moisture content decreases. On average, the ratio of compressive strengths in the green and nearly oven-dry conditions is about 2, but the fiber saturation point is much lower than for wood, at 17.2% (Ota 1953). Tensile strength increases more irregularly and reaches a maximum at moisture contents intermediate between the fiber saturation point and the oven-dry condition (Ota 1954).

3. Chemical Properties

The chemical composition of some bamboos is given in Table 3. The proximate chemical compositions of bamboo culms are similar to those of hardwoods, except that alkaline extract, ash and silica contents are higher than in hardwoods. High silica content causes scaling during evaporation of the spent liquor for recovery of the chemicals in pulping.

A xylan that precipitates with Fehling's solution makes up 90% of the bamboo hemicellulose. The xylan is composed of $\beta(1\rightarrow4)$ linked polymer with attachments of single-unit side chains such as residues of L-arabinose and 4-*O*-methyl-D-glucuronic acid. The xylan therefore corresponds to 4-*O*-methyl-D-glucuronoarabinoxylan; the ratio of 4-*O*-methyl-D-glucuronic acid, L-arabinose and D-xylose is 1.0:1.3:24–25. The xylan is different from the arabino-(4-*O*-methyl-D-glucurono)xylan of gymnosperm woods with respect to the degree of branching, molecular properties and its lower content (6–7%) of acetyl groups.

Table 3
Chemical composition of bamboo culms (% oven-dry weight of bamboo)

Bamboo species	Silica	Ash	Cold-water soluble	Hot-water soluble	Alcohol–benzene soluble	Ether soluble	NaOH soluble	Pentosan	Lignin	Holocellulose
Bambusa arundinacea[a]		3.26	4.59	5.25	0.22	0.82	19.35	19.62	30.09	57.56
B. blumeana[b]	3.40	4.80		4.30	3.10		39.50	19.00	20.40	67.40
B. vulgaris[b]	1.50	2.40		5.10	4.10		27.90	21.10	26.90	66.50
Dendrocalamus strictus[a]		2.10	4.20	5.93	0.25	0.56	15.00	19.56	32.20	68.80
Gigantochloa levis[b]	2.80	5.30		4.40	3.20		28.30	18.80	24.20	62.90
Melocanna baccifera[a]		1.87	3.26	6.48	1.43	0.81	18.97	15.13	24.13	62.25
Oxytenanthera nigrociliata[a]		1.25	1.61	3.39	0.24	0.15	17.11	17.41	27.09	66.72
Schizostachyum lumampao[b]	6.40	9.70		4.30	5.00		31.40	20.60	20.40	60.60

[a] Singh and Guha (1981) [b] Tamolang et al. (1980)

In addition, a xylan which gives no copper complexes with Fehling's solution, an arabinogalactan, and an α-glucan are contained in bamboo shoots (Maekawa and Kitao 1973, 1974).

Bamboo lignin is a typical grass lignin composed of mixed dehydrogenation polymer of coniferyl, sinapyl and *p*-coumaryl alcohols (molar ratio 65:25:10 on average) (Higuchi 1958, Higuchi et al. 1967, Higuchi 1980). Bamboo and grass lignins also contain 5–10% of *p*-coumaric acid ester as a unique feature. About 80% of the *p*-coumaric acid is esterified to the γ-hydroxyl groups (terminal hydroxyl) of the lignin side chain; α-linked esters of the acid amount to less than 20%.

Bamboo lignin gives a considerable amount of *p*-hydroxybenzaldehyde in addition to vanillin and syringaldehyde on nitrobenzene oxidation. The molar ratio of the aldehydes obtained is 1:1:2 on average. About two thirds of the *p*-hydroxybenzaldehyde is due to the esterified *p*-coumaric acid of the lignin polymer.

Upon permanganate oxidation of the methylated bamboo lignin, *p*-anisic, veratric and trimethylgallic acids are produced from the uncondensed units, whereas isohemipinic and 4-methoxyisophthalic acids are produced from the condensed units of the lignin.

The amounts of arylglycerol-α- and -β-aryl ether substructures present in bamboo, beech and conifer lignins have been estimated by gas chromatography of acidolysis monomers (as arylacetones) and by spectral analysis (Higuchi et al. 1973). The number of free phenolic hydroxyl groups generated upon cleavage of these two types of ether bonds indicates an occurrence of α-aryl ether bonds of 0.07 per phenylpropane building unit and 0.56 total β-aryl ether linkages (0.51 for beech and 0.35 for thuja). The latter are composed of 0.24 uncondensed and, by difference, 0.32 condensed units. The uncondensed arylglycerol-β-aryl ethers are comprised of 0.02 *p*-hydroxyphenyl, 0.11 guaiacyl and 0.11 syringyl units. Condensed structures (nonhydrolyzable bonds to C_5 and C_6 of the aromatic ring) other than those with *β–O–4* bonds amount to 0.13 per phenylpropane building unit. The relatively low ratio of other condensed units to total condensed units is consistent with the relatively large number of syringyl units in bamboo lignin. It is concluded that bamboo lignin is composed of guaiacyl, syringyl and *p*-hydroxyphenyl units combined via linkages similar to those found in spruce lignin.

4. Pulp and Paper

About 65% of the pulp used in making paper in India comes from bamboo, and there are 35 factories that make paper from bamboo pulp, with an annual production estimated at about 600 000 t in 1980. The Indian paper industry is mainly using *Dendrocalamus strictus* and *Bambusa arundinacea* (Varmah and Bahadur 1980). Pulp and paper from bamboos are also produced in the Philippines and in Bangladesh.

Phyllostachys bambusoides, *Bambusa tulda*, *Dendrocalamus strictus* and *D. gigantus* can be pulped by the kraft process using 14–15% total alkali at 25% sulfidity to produce pulps (40% yield) suitable for making writing paper. The bleaching is carried out using chlorine and calcium hypochlorite with intervening caustic extraction; the total bleach demand is 12–15% of active chlorine and the brightness attained is 75–78%.

In kinetic studies of alkaline (kraft) pulping of *D. strictus*, Singh and Guha (1981) found that (a) at pulp yields of 70% or more, mostly carbohydrate is lost, (b) at yields between 43 and 70%, both carbohydrate and lignin are removed in approximately the same proportions, and (c) chemical pulps of 50% yield (Kappa number 36) could be prepared without any appreciable impairment in strength properties. The pulping conditions for the latter were: alkali 15% as Na_2O, sulfidity 25%, time to 170 °C 1.5 h, time at 170 °C 1.5 h, and chip-to-liquor ratio 1:3.5.

During bleaching of kraft pulp (*D. strictus*), strength properties increased during chlorination, remained almost unaffected in the caustic extraction stage and again improved in the hypochlorite stage. The brightness was improved gradually during chlorination, decreased slowly during caustic extraction and improved greatly during the early stage of hypochlorite treatment. The bleaching conditions were: chlorine 8.0%, consistency 3.0% and temperature 30 °C for chlorination; NaOH 2.0%, consistency 5.0% and temperature 70 °C for extraction; and finally, available chlorine 2.5%, consistency 5.0% and temperature 45 °C for hypochlorite treatment.

The pulping conditions employed by a few paper mills in India are given in Table 4. It is shown that the total cooking chemicals range from 13 to 16% as Na_2O at sulfidities varying between 13 and 22%. The Kappa number ranges from 16 to 21 and pulp yields vary between 38 and 43% with one exception. The total bleach demand varies from 12 to 15% to attain a brightness level of 70–80% (Singh and Guha 1981).

Philippine bamboos are easily digested in the kraft process and produce bleachable pulp. Compared with foreign softwoods and Philippine hardwoods, the bamboo pulps possess higher tear resistance but lower folding endurance, bursting strength and tensile strength. *Gigantochloa levis*, *G. aspera* and *Bambusa blumeana* appear to be suitable raw materials for kraft pulp with respect to pulp strength, pulp yield and acceptable level of silica content (Tamolang et al. 1980).

In alkaline pulping, one of the most difficult problems is in chemical recovery due to the high silica content of bamboo. Desilicification by the addition of calcium oxide to the black liquor and separation by centrifuge have been patented by a pulp company (Oye 1980).

Application of a prehydrolysis sulfate process to *Melocanna baccifera* (Oye 1980) produces quality

Table 4
Pulping and papermaking data for some Indian paper mills

Process details	A	B	C	D	E
Pulping process	kraft	kraft	kraft	sulfite	kraft
Raw materials	bamboo (72–75%) *Bambusa arundinacea* *Bambusa nutans* Dendrocalamus strictus hardwoods (25–28%)	bamboo (70–80%) *Dendrocalamus strictus* hardwoods (20–30%)	bamboo (75%) *Melocanna baccifera* *Dendrocalamus strictus* *bambusa arundinacea* hardwoods (25%)	bamboo (70%) *Dendrocalamus hamiltonii* hardwoods (25%)	bamboo (70%) *Dendraocalamus strictus* hardwoods (30%)
Chemicals (%)	14	13–13.5	14–16	(1) magnesium 7 (2) sulfur 8	bamboo 13 mixed hardwood 16
Sulfidity (%)	17–18	15–20	13–15		for both 22
Cooking schedule	unbleached kraft to 135 °C 1 h at 135 °C $1\frac{1}{2}$ h to 165 °C 1 h at 165 °C $1\frac{1}{2}$ h bleached kraft to 165 °C 3 h at 165 °C 30 min	to 135 °C 1 h at 135 °C 1 h to 165 °C 1 h at 165 °C 1 h	fractional cooking to 142 °C 3 h to 160 °C 3 h	acid filling at 45 °C to 150 °C 3 h at 150 °C 6 h	cooking of bamboo two-stage cooking of mixed hardwoods to 130 °C 45 min at 130 °C 30 min to 167 °C 1 h to 167 °C 1 h
Kappa number	21 ± 1, 16 ± 1	17–18	14–16 ($KMnO_4$ no.)	17 ($KMnO_4$ no.)	18–19 ($KMnO_4$ no. for mixed hardwood)
Bleach demand (%)	15	12–14	10–12	14	14
Pulp yield (%)	39	38	90	42	43
Brightness attained	70–74	78–80	75	78	75–80
Products made	(1) bleached grade: writing, printing, duplex board, poster manila (2) unbleached grade: MG kraft, duplex and gray board mill, wrapper, MF kraft and colored duplex	white printing, cream wove, duplicating kraft, map litho	writing and printing papers	writing and printing papers	writing, printing, duplex board, poster paper

rayon-grade pulp with the following composition: α-cellulose 95.0%, β-cellulose 4.0%, pentosan 3.0%, extractives 0.03%, ash 0.068%, CaO + MgO 0.024% and a brightness of 91.

Dissolving-grade bamboo pulp prepared for viscose rayon production shows significantly different properties from similar wood pulps. While crystallinity of bamboo pulp cellulose is generally not different from that of wood pulps, the bamboo is more difficult to convert into alkali cellulose. A small fraction of the bamboo pulp remains unchanged after mercerization in 16% NaOH solution. This reduced tendency toward swelling is directly related to the cell wall structure as noted (see Fig. 2). The extent of mercerization can be increased by mechanical treatment or heating of the pulp (Fig. 4). Bamboo pulp cellulose also has some desirable features. It is readily pressed to remove excess alkali solution, one of the most important stages in commercial production of viscose. During aging the bamboo alkali cellulose exhibits a higher depolymerization rate and a more uniform degree of polymerization than wood. Also, the spinnability of the viscose and tenacity of the filament are superior.

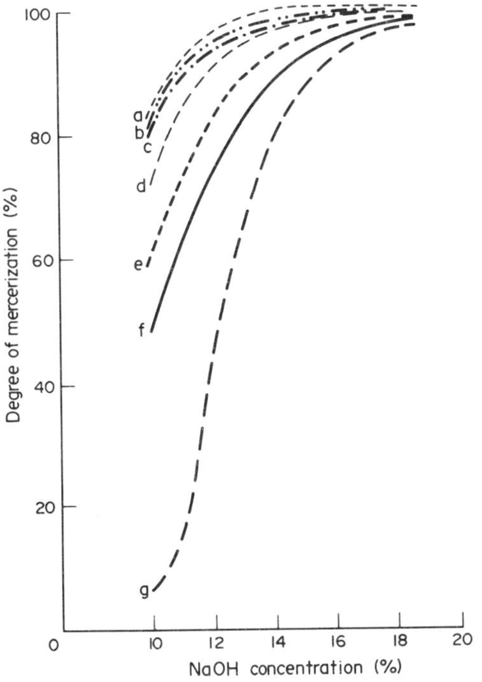

Figure 4
Comparison of mercerization rates of bamboo, linter and wood pulps: a, hardwood DSP; b, conifer DSP; c, conifer BKP; d, hardwood BKP; e, bamboo DKP beaten; f, bamboo DKP unbeaten; g, linter pulp. Wood and inter pulp rates are unaffected by beating. DSP, dissolving sulfite pulp; BKP, bleachable kraft pulp; DKP, dissolving kraft pulp (after Oye 1980)

The papermaking properties of six major species of Burmese bamboo have been investigated by Wai (1984). The strength properties of bamboo pulp sheets lie between those of softwood and hardwood pulps. However, bamboo pulp which contains a substantial amount of thin-walled fibers shows strength properties which are very close to those of softwood pulp sheets. This indicates that the distribution of the Runkel ratio of the fibers is a valuable parameter for the evaluation of bamboo pulp from the various species of bamboo. A remarkable difference between bamboo and wood pulp fibers is found in the morphological change produced by beating; this can be ascribed to the difference in fine structure of the fiber wall, especially orientation and organization of microfibrils. Another unique characteristic of bamboo fibers is the improvement in optical properties of the pulp sheets after beating. Thick-walled bamboo pulp fibers, however, give the sheets a rather bulky and rough surface which leads to undesirable effects when used as printing paper.

5. *Bamboo-Based Materials for Construction*

Bamboos are used for the production of building boards. The process starts with mats woven from bamboo strips, which are then dried, treated with phenol–formaldehyde resin and pressed at 150 °C and 2.7–3.4 MPa pressure. Boards can be prepared as single or multiple mat boards and can also be made in corrugated form. Bamboo boards have potential use as light partitions or ceilings, or for production of molded items like suitcases, table tops and chair seats (Varmah and Bahadur 1980).

Bamboo (*Bambusa arundinacea*) strips, 20 × 9 mm, treated with hot bitumen and subsequently sand blasted can be used as a suitable substitute for steel in reinforced concrete. Bamboo reinforcement in concrete beams increases the load-carrying capacity of members considerably above that to be expected from members of the same dimensions without reinforcement. The load-carrying capacity of beams reinforced with bamboo is increased by the addition of bamboo splints as diagonal tension reinforcement at sections where the vertical shear is high. A safe design stress is 21–28 MPa for concrete members reinforced with bamboo (Varmah and Bahadur 1980, Tamolang et al. 1980). A similar investigation on bamboo reinforcement in concrete is being carried out in Bangladesh (Ali 1981).

Bamboo parquet blocks and laminated bamboo are produced in the Philippines as materials for flooring and building construction. Laminated bamboo is strong and durable and is suitable for structural and decorative parts of houses, boats and furniture (Tamolang et al. 1980).

See also: Coconut Wood; Rattan and Cane

Bibliography

Ali M O 1981 Research and development on the production and utilization of bamboo in Bangladesh. In: Higuchi 1981, pp. 257–70

Atrops J L 1969 Elastizität und Festigkeit von Bambusrohren. *Bauingenieur* 44: 220–25

Dransfield S 1980 Bamboo taxonomy in the Indo-Malaysian region. In: Lessard and Chouinard 1980, pp. 121–130

Higuchi T 1958 Studies on the chemical properties of the lignin of bamboo stalk. *J. Biochem. (Tokyo)* 45: 675–85

Higuchi T 1980 Biochemistry of lignification. *Wood Res.* 66: 1–16

Higuchi T (ed.) 1981 *Bamboo Production and Utilization.* Wood Research Institute. Kyoto

Higuchi T 1987 Chemistry and biochemistry of bamboo. *Bamboo J. (Jpn.)* 4: 132–45

Higuchi T, Ito Y, Shimada M, Kawamura I 1967 Chemical properties of milled wood lignin of grasses. *Phytochemistry* 6: 1551–56

Higuchi T, Shimada M, Nakatsubo F, Tanahashi M 1977 Differences in biosynthesis of guaiacyl and syringyl lignins in woods. *Wood Sci. Technol.* 11: 153–67

Higuchi T, Tanahashi M, Nakatsubo F 1973 Acidolysis of bamboo lignin III: Estimation of arylglycerol-β-aryl ether groups in lignins. *Wood Res.* 54: 9–18

Janssen J J A 1980 The mechanical properties of bamboo used in construction. In: Lessard and Chouinard 1980, pp. 173–88

Janssen J J A 1981 The relationship between the mechanical properties and the biological and chemical composition of bamboo. In: Higuchi 1981, pp. 27–32

Kawase K, Ujiie M 1987 Utilization of sasa-bamboo resources, *Bamboo J. (Jpn.)* 4: 69–80

Lessard G, Chouinard A (eds.) 1980 *Bamboo Research in Asia.* International Development Research Center, Ottawa

Liese W 1980 Anatomy of bamboo. In: Lessard and Chouinard 1980, pp. 161–64

Liese W 1987 Research on bamboo. *Wood Sci. Technol.* 21: 189–209

Maekawa E, Kitao K 1973 Isolation and constitution of a xylan from bamboo. *Agric. Biol. Chem.* 37: 2073–81

Maekawa E, Kitao K 1974 The structure of an arabinogalactan isolated from bamboo shoot. *Argic. Biol. Chem.* 38: 227–29

Nomura T 1980a On the growth of bamboo. *Wood Res. Tech. Notes* 15: 6–33 (in Japanese)

Nomura T 1980b Physical and biophysical properties of bamboo. In: Lessard and Chouinard 1980, pp. 50–51

Ota M 1951 Studies on the properties of bamboo stems, Part 7: The influence of the percentage of structural elements on the specific gravity and compressive strength of bamboo splint. *Bull. Kyushu Univ. For.* 19: 25–47

Ota M 1953 Studies on the properties of bamboo stems, Part 9: On the relation between compressive strength parallel to grain and moisture content of bamboo splints. *Bull. Kyushu. Univ. For.* 22: 87–108

Ota M 1954 Studies on the properties of bamboo stems, Part 10: On the relation between the tensile strength parallel to the grain and the moisture content of bamboo splints. *Bull. Kyushu Univ. For.* 23: 155–63

Ota M 1955 Studies on the properties of bamboo stems, Part 11: On the fiber-saturation point obtained from the effect of the moisture content on the swelling and shrinkage of bamboo splints. *Bull. Kyushu Univ. For.* 24: 61–84

Oye R 1980 Bamboo for pulp and paper. In: Lessard and Chouinard 1980, pp. 53–55

Oye R 1981 The effect of beating on bamboo pulp. In: Higuchi 1981, pp. 39–44

Oye R 1987 Pulping of bamboo. *Bamboo J. (Jpn.)* 4: 61–68

Parameswaran N, Liese W 1981 The fine structure of bamboo. In: Higuchi 1981, pp. 178–83

Singh S V, Guha S R D 1981 Indian experience of papermaking from bamboo. In: Higuchi 1981, pp. 33–38

Suzuki Y 1948 Studies on bamboo, Part 1: Distribution of physical and mechanical properties in the stem of *Phyllostachys pubescens*: specific gravity, fiber content, bending strength, Young's modulus and the shoot strength. *Bull. Tokyo Univ. For.* 36: 135–223

Tamolang F N, Lopez F R, Semana J A, Casin R F, Espiloy Z B 1980 Properties and utilization of Philippine erect bamboos. In: Lessard and Chouinard 1980, pp. 189–200

Varmah J C, Bahadur K N 1980 Country report and status of research on bamboo in India. In: Lessard and Chouinard 1980, pp. 19–46

Wai N N 1984 Papermaking properties of Burmese bamboo pulp. Ph.D. thesis, Kyoto University, Kyoto

T. Higuchi
[Kyoto University, Kyoto, Japan]

Biotechnology in Wood Processing

Biotechnology concerns itself with the use of microorganisms and biological processes that can be manipulated in ways to benefit mankind. Although the term "biotechnology" has only been coined in recent times, in a broad sense it can and has been used to describe industrial processes developed prior to our current "modern" bioindustrial initiate. These industrial processes include grain fermentation for the production of alcohol, the use of yeasts and other microorganisms to enhance the quality of foodstuffs and, more recently, the production of drugs such as penicillin.

The products of biotechnology have been described by Leder and Paul (1983) as falling into three categories: (a) new substances that have never before been available, (b) rare substances that have not been widely available, and (c) existing substances that can be made inexpensively through biotechnology. The current use of the term "biotechnology" in this definition applies to the use of organisms that have been genetically modified, but there is considerable deviation in the way that bioindustrial processes in various fields have been classified by other individuals. In the forestry and forest products field, any new developments related to biochemical or microbiological processes, whether or not they include bioengineering, have been labelled as biotechnological processes, and it is this broad definition that will be used in this article.

In large part, many of the products that have been developed through biotechnological processes have been produced in response to medical needs. Currently, due to high initial development costs, most newly developed biotechnological processes are expensive. To justify the cost of employing such procedures, high-value products are targeted by the biotechnology industries. Few industrial processes have been developed initially for forest products. With new advances and growth of the biotechnology industry, however, more spinoffs from the medical and human health care initiatives will reach the field of agriculture, and from there will move into the fields of forestry and forest products.

1. Tools of the Biotechnologist

1.1 Recombinant DNA

Although only one of several techniques that have application for biotechnological industrial processes, recombinant DNA techniques have received the greatest amount of attention from scientists since the mid-1970s. Recombinant DNA techniques hold the promise for development of new bioprocessing systems through the "modification" of existing organisms for the increased production of useful metabolites or the creation of new substances.

DNA is found within all living cells and contains the genetic blueprint necessary for life. Messenger RNA (mRNA) is transcribed from the DNA and is subsequently translated into proteins. Proteins are composed of 20 common amino acids and it is the order of these amino acids that the nucleic acid sequence dictates. Proteins, either as separate metabolites or combined with sugar molecules (glycoproteins), function in the control and production of cell components. Thus in plants, the production of all cell-wall material and growth-regulating substances is mediated by enzymes. Control over the production of some of these materials is performed through a complex feedback process that "switches" on genes (specific segments of the DNA molecules) that code for proteins allowing the production of cellular metabolites as needed by the organism.

Recombinant DNA techniques enable scientists to select specific genes that code for metabolites of interest. The genes may code for the production of protein subunits, enzymes, vitamins or antibiotics. After a gene has been identified, the goal is then usually to amplify production of the metabolite of interest or to provide for production of this substance outside of the normal environment for the host organism (for example, in the environment of an industrial fermentor or in providing for the production of metabolites not produced by nonengineered organisms). Gene amplification within the same host organism has been accomplished by insertion of the desired gene sequence into the DNA of the host in greater frequency than is found in the nonengineered organism. Supplementary gene sequences must also be inserted into the host's DNA to promote and activate the inserted genes.

To allow production of desired substances under industrial conditions or to allow organisms to produce foreign metabolites, similar genetic engineering techniques are employed. In such cases the genes of interest are identified and the DNA sequence is isolated from the host organism (or potentially is synthesized directly in the laboratory). The DNA sequence potentially can be modified *in vitro* to produce new proteins with modified function. Because only limited information is available on the potential function of novel proteins, this is an area of limited potential at present. New computer modelling methods may provide better information on the function and value of newly synthesized proteins for specific application. When the gene of interest has been isolated, it may then be incorporated through the use of a vector DNA into another organism.

The previous discussion outlines in a very basic manner the techniques involved in genetic engineering. It should be pointed out that although these techniques are now carried out in many laboratories around the world, primarily in animal systems, the skills and knowledge needed to perform this work are not trivial. Although some gene products such as human insulin are now being produced by genetically engineered microorganisms, not all such genetic engineering endeavors have met with success. Not only must the appropriate gene sequences for a metabolite be identified and isolated, but the appropriate promoter DNA sequences and enhancer elements also must be spliced into the new organism's DNA. In addition, mRNA produced by the inserted DNA may require specific modification before it can produce functional proteins.

Another major obstacle is that even after a metabolite is produced, it is still confined by the cell membrane or walls of the host. Many metabolites such as enzymes are often assembled from subunit proteins and frequently have sugar residues attached to help protect their structure or to assist in their function. Specific glycosylation may be required to allow the component of interest to be transported out of the cell. These problems have slowed the development of all but the simpler of bioengineered products. The understanding of the cell's molecular machinery has advanced rapidly, however, and the development of more complex bioengineered products in the 1990s should be anticipated.

One final caveat; the difficulties described in the discussion above have focused primarily on work performed with animal systems. In work with plant-cell systems, additional problems have been encountered which have limited advances in plant and tree bioengineering. Continued progress such as work on plant regeneration from protoplasts suggests that there is promise for the practical application of plant bioengineering systems in the 1990s as well.

1.2 Cell Fusion

An aspect of biotechnology having high potential value to the forest-products field is the production of monoclonal antibodies. Antibodies are produced by higher animals when their immune systems respond to a foreign body (antigen). There is a very high specificity of a particular antibody for its antigen. The foreign body may be a virus or bacterium that infects the animal or, in the context of the biotechnology laboratory, be a specific antigen injected into the animal. Antibodies are produced in response to the presence of the antigen by lymphocytes found in the spleen and other organs, and are released into the blood stream of the animal. Antibodies may be produced in response to many different sites on an antigenic molecule providing for many different antibodies produced to the same antigenic component. Antisera collected from the blood in this manner is known as polyclonal antisera and generally has rather broad specificity.

Greater specificity can be obtained if the antibodies produced by individual lymphocytes can be isolated, since the antibodies produced by a single lymphocyte all have the same specificity. To achieve this, individual lymphocytes are separated and then fused to myeloma cells. Unlike the lymphocytes, myeloma or cancerous cells have the capacity to replicate in culture and thus, when fused to the lymphocyte, produce cell lines which possess both the ability to grow indefinitely in culture and to produce homogeneous antibodies of specificity predetermined by the lymphocyte parent cell. Antibodies produced in this manner are known as monoclonal antibodies. Since not all lymphocytes will produce antibodies specific to the desired antigen, screening of the cell colonies (hybridomas) for production of antibodies of high specific affinity to the selected antigen is necessary. Colonies showing the desired specificity are subsequently propagated (cloned) and large quantities of specific antibody can then be produced.

At present more commercial biotechnology products have been produced through the use of monoclonal antibody technology than have been through recombinant DNA technology. Because of the high specificity of monoclonal antibodies, they have been used to separate and purify proteins, and they have allowed the development of numerous diagnostic tests for detecting viruses, bacteria and fungi. They have also been used to evaluate conditions of pregnancy and to detect early stages of AIDS infection in humans. This latter test employs a technique known as the enzyme-linked immunosorbent assay (ELISA).

2. *Additional Biotechnological Techniques*

Cell-fusion methodology has been employed to produce hybrid plants and thus has the potential for use in the development of improved tree species. However, tissue culture, plant-cell protoplasting and other topics will not be brought out in greater depth in this article as, although within the realm of biotechnology, they are more appropriate for the genetic improvement of trees and hold less direct potential for application to forest products processing.

3. *Forest Products Application of Traditional Biological Processes*

In agriculture there have already been advances in crop and animal production that have resulted from biotechnology. Researchers are working on methods to introduce the genes for nitrogen fixation and disease resistance into crop plants, and growth hormones have been found to improve the quality and production of meat in farm animals. Much of this research and the associated practical advances have "spun off" from work in medical biotechnology research. As stated earlier, the expense of developing and producing a new product considered against the product's value will dictate how biotechnological processes will be employed in the future. In the field of forest products, most currently proposed end-products, in comparison to human hormones, vaccines and antibiotics, are of relatively low value. As the science of biotechnology advances, however, and processes become less expensive, the cost–benefit ratio will improve and new applications of biotechnology will become economically feasible.

In this section we will look at biotechnological processes that have *not* employed recombinant DNA, cell fusion or other advanced technologies, but have been employed as naturally occurring microorganisms in the broad sense of processing wood.

3.1 Protein Production

Protein production, using the spent liquor from the sulfite pulping process as the substrate, was commercially developed in the 1940s. The process employs a traditional type of aerobic fermentation under sterile conditions and a natural strain of *Torula* yeast is provided an environment for growth. The protein (yeast) has been successfully marketed in the USA for both human and animal consumption, with one of the more successful applications being the use of the yeast as a flavor carrier for food products such as imitation bacon bits and flavored potato chips.

In Sweden and in Canada, protein production from hexoses and pentoses, and acetic acid (in spent sulfite liquor) has been experimented with in pilot-scale processes using yeasts, mold fungi and white-rot fungi. However, because of high costs relative to the production of more conventional protein sources, this system has not been developed commercially.

3.2 Edible Mushroom Production

Another wood biomass conversion process of increasing importance is edible mushroom growth on woody substrate. Although the common table mushroom in the USA (*Agaricus bisporus*) is generally produced on agricultural residues, *Lentinus edodes*, the Shiitake mushroom is grown only on wood substrates. *Pleurotus* spp. is another species under commercial production that has been cultivated on wood. Although traditionally, culture of these two fungi has been carried out in sections of logs exposed to the natural environment, more recent developments in production have allowed sawdust or compressed wood-chip logs to be used. Indoors, environmentally controlled cultivation is now also being employed to maximize production of these mushrooms from woody substrates.

4. Modern Bioengineering Applications and Potential

Currently (1988), developmental use of bioengineering applications in the forest products industry is very limited, even when pulp and paper applications are included. Although microbial lignolytic systems have the industrial potential to pulp wood, bleach pulp, modify pulp surface properties, treat wastes and convert wood byproducts to useful chemicals, to date only limited commercial application of a few of these processes has been achieved. These are summarized in this section.

4.1 Biopulping Potential for Microorganisms and Enzymes

There are no commercial processes currently in use that employ isolated cellulases or ligninases from microorganisms to break down woody substrates. To achieve this goal it will be necessary to (a) produce a structurally modified enzyme that can withstand the harsh environment of industrial fermentation processes, and (b) provide for production of this enzyme in large quantities required for industrial processes. An alternative to this is to engineer an organism that produces the enzyme but that also can meet the two criteria. These problems are being researched intensively. Fungal mutants have been produced, such as the cellulose-minus mutants that produce primarily ligninases, and cellulase and ligninase genes have been cloned into recombinant hosts. These recombinant enzymes have the potential for use in certain industrial processes (see Sects. 4.2 and 4.3). Research has also indicated that a promising approach to lignin degradation may be to use only the active site of the enzyme (for lignin-degrading enzymes this is a porphyrin structure). The active site is then chemically modified to stabilize and protect this structure from attack in a harsh environment. A system for scaled-up production of modified enzymes has not yet been developed. Once a system for the production of various industrial enzymes or microorganisms is established, a number of forest products applications will be possible.

4.2 Lignochemical Modification

A ligninase gene from *P. chrysosporium* has been cloned into a recombinant host and an active form of the enzyme has now been successfully produced independently from the natural fungal host. The recombinant enzyme has a lower molecular weight than the *P. chrysosporium*-produced enzyme; however, the activity of this recombinant ligninase has been reported to be equivalent to that of the nonengineered ligninase when assayed on model lignin compounds.

One potential application for the use of recombinant-produced ligninases is the decolorization of lignochemicals formed in the pulping process or to degrade waste materials. A novel use proposed for these enzymes is the modification of kraft lignin when used as a copolymer for a poly(vinyl acetate) adhesive. Stability of the enzyme under industrial conditions and penetration of the enzyme into the substrate, as discussed previously, remain important considerations in the use of any isolated enzyme under process conditions.

4.3 Bleaching of Pulp

Fungal cellulase-minus mutants have been considered for use in bleaching of pulps. These mutants do not produce cellulases and, therefore, cannot degrade the cellulosic cell wall of a pulp fiber. The ligninases produced can theoretically degrade colored phenolic materials in the pulp to improve brightness. Because the wood has already been partially delignified and has been reduced to component fibers, penetration of enzymes into the pulp fibers is less of an obstacle than in larger solid wood materials. However, the fungi still have requirements for growth that must be provided for. The proper balance of nutrients and a suitable environment within a biobleaching system presents a challenge which limits the potential of this process until better organisms or enzymes became available.

4.4 Waste Treatments

Detoxification of pollutants (other than or in addition to lignochemicals) at waste disposal sites through the use of white-rot fungi is another area of focus that is being studied. Currently, nonbioengineered fungi are being used for this process, and fungus infected wood chips are being introduced into waste (pesticide or other pollutant) contaminated soils. Bioengineered organisms capable of breaking down target pollutants in solid-waste-disposal sites or in contaminated soils would allow clean up of these sites with less hazard to laborers and at lower cost than is currently possible. Although this problem would not focus on the utilization of wood, it is likely that in the detoxification process wood chips would be used initially to provide a substrate for growth of the fungi.

4.5 Fermentation Processes for Industrial Chemical Production

Alcohol has been produced through the fermentation of spent sulfite liquor sugars produced in the sulfite pulping process for over 100 years. Ethanol, along with other products such as butanediol, can also be produced in the fermentation of a variety of cellulosic substrates. These fermentation products can be used as portable and easily stored fuels and, in addition, may be used as chemical feedstocks to replace those derived from petroleum. The hexose sugars in the liquor and also in the hydrolyzed cellulosic residues are the primary fermentation substrate. Currently, however, due to the low cost of ethanol production (synthetic production from ethylene), production of industrial alcohols by fermentation is only marginally feasible. To improve the economics of this process, research has focused on the use of new microorganisms and new strains of proven alcohol producers that are capable of both hydrolyzing long-chain carbohydrates and also fermenting the resulting sugars. To date, fungi, yeasts and bacteria have all been used with success for either or both hydrolysis and fermentation.

The goal of developing an economically feasible industrial process is to make the process more efficient by process reduction. Organisms that can more rapidly process cellulosic residues at lower energy costs have been used experimentally. In addition, single organisms that are cellulolytic and also ferment sugars to ethanol such as *Clostridium thermocellum* are being studied in an effort to develop bioconversion processes that can be carried out in a single vessel. The use of a single organism in a single vessel for fermentation provides for greater industrial efficiency that may outweigh the overall rapidity of the actual microbial metabolism. Both aerobic and anaerobic fermentation processes, as well as combinations of the two, are currently being considered for potential use in efficient cellulosic residue fermentation.

Another means of improving the efficiency of chemical production is the fermentation of pentose residues from hemicellulose. Pentose sugars such as xylose are not readily metabolized by most organisms and the high degree of substitution found in most hemicellulose xylans makes fermentation by a single organism difficult. Since the pentose metabolism pathway is not well developed in most organisms, research has focused on screening of selected microorganisms in an effort to isolate naturally occurring strains that may be particularly efficient in this regard. Although the potential for the development of a high-efficiency pentose-metabolizing organism is great, currently no attempts to alter organisms for this purpose through genetic manipulation have been reported. Work has been initiated, however, to clone the genes coding for cellulolytic enzymes into hosts that already possess the capability to ferment the sugars to usable chemicals.

An alternate approach to the direct metabolism of pentose sugars (xylose) is the enzymatic conversion of xylose to xylulose and then to hexose sugars for fermentation. Although the organisms and techniques for developing this process currently exist, inefficiencies in the process and the economics of synthetic ethanol production have not favored the development of this bioconversion pathway.

4.6 Diagnostics

Just as diagnostic tests have been developed to detect the presence of human or animal diseases, it has also been possible, using serological techniques, to develop bioassays for the detection of microorganisms in plant tissues. More recently assays have been developed for the detection of fungi in wood. Polyclonal antisera have been developed to components of both blue stain fungi and decay fungi isolated from wood. These antisera have been used to determine the presence of these fungi prior to the exhibition of signs of deterioration, i.e., staining or significant strength loss.

Monoclonal antibodies have also been produced to a number of extracellular metabolites from brown-rot and white-rot fungi. These antibodies include those produced against cellulases, hemicellulases and lignin-degrading enzymes (ligninase and Mn perioxidases). The monoclonal antibodies may also be effective in decay detection assays.

Both polyclonal and monoclonal antibodies have also been used to study the basic processes of wood degradation caused by these micro-organisms. This is accomplished by tagging specific antibodies with markers and then using the marked antibody to label sites of degradation in the wood. Recent studies with both brown- and white-rot degraded wood indicate that fungal cellulases and lignin-degrading enzymes may not penetrate into the wood cell wall, at least in the early stages of degradation. These observations lend support to the hypothesis that, in brown-rot fungi, a nonenzymatic system small enough to penetrate the microvoids in an intact wood cell wall may exist, and may function in the initial stages of cell wall breakdown. These findings are important because just as they help us to understand the mechanisms of degradation more thoroughly they may also allow us to develop better methods of preventing decay in wood.

See also: Chemically Modified Wood; Decay During Use; Paper and Paperboard

Bibliography

Anon 1985 New horizons for biotechnological utilization of the forest resource. *Marcus Wallenburg Foundation Symposia Proceedings*: 2.S-791 80

Breuil C, Seifert K, Yamoda J, Rossignol L, Saddler J 1988 Quantitative estimation of fungal colonization of wood using an enzyme-linked immunosorbent assay. *Can. J. For. Res.* 18 (in press)

Clark M F, Adams A N 1977 Characteristics of the microplate method of enzyme-linked immunosorbent assay for the detection of plant viruses. *J. Gen. Virol.* 34: 475–83

Goodell B, Jellison J, Hosli J P 1988 Serological detection of wood decay fungi. *For. Prod. J.* 38: 59–62

Goodell B S 1988 Biotechnology for the forest-based industry. *Biomass* (in press)

Joglekar R, Clerman R J, Ouellette R P, Cheremisinoff P N 1983 *Biotechnology in Industry*. Ann Arbor, Science Publishers, Ann Arbor, Michigan

Kirk T K 1987 Biotechnology in wood processing. In: *Biotechnology in Forestry and Wood Products Manufacture*, Planning Session 1A. Forest Products Research Society, Madison, Wisconsin

Kirk T K, Johnsrud S, Eriksson K-E 1986 Lignin degrading activity of *Phanerochaete chrysosporium* burds.: comparison of cellulose-negative and other strains. *Enzyme Microb. Technol.* 8: 75–80

Leder P, Paul W E 1983 The molecular and microbial products of biotechnology. In: Inouie N, Sawa R (eds.) 1983 *Protein Engineering: Applications in Science, Medicine and Industry*. Academic Press, New York, Chap. 2

Mizrohi A, van Wezel A L 1985 *Advances in Biotechnological Processes*, Vol. 5. Liss, New York

Moo-Young M 1987 *Biomass Conversion Technology: Principles and Practice*. Pergamon, Elmsford, New York

Rhoades C, Pierce D, Mettler I J, Mascarenhas D, Detmer J J 1988 Genetically transformed maize plants from protoplasts. *Science* 240: 204–7

Silver S 1985 *Biotechnology: Potentials and Limitations*, Dahlen Workshop Report 35. Springer, Weinhein

Yu E K C, Deschatelets L, Louis-Seize G, Saddler J N 1985 Butandial production from cellulose and hemicellulose by *Klebsiella pneumoniae* growth in sequential coculture with *Trichoderma harzianum*. *Appl. Environ. Microbiol.* 50: 924–29

B. Goodell
[University of Maine, Orono, Maine, USA]

Building with Wood

Wood has a long history of use in construction, dating back for many centuries if not into prehistory. Its applications have virtually covered the range of structural utilization except for buildings beyond a few stories in height. Engineered timber is used widely in such diversified construction as schools, churches, commercial buildings, industrial buildings, residences and farm buildings, highway and railway bridges, towers, theater screens, ships, and military and marine installations. Timber structures were at one time limited in the spans over which they could be used, because of the limited lengths of timber available and the limited cross-sectional dimensions. Modern technology, including glued laminated construction and modern methods of connection, has largely removed many of these limitations. Now, longspan arches and beams, trusses and dome-type structures provide large clear areas for recreational and athletics buildings, churches, auditoriums and bridges. A number of favorable characteristics enhance the suitability of timber for structural use; these will be dealt with in the following sections.

1. Important Characteristics

1.1 Durability

Wood, because it is an organic material, is commonly considered to be shortlived. In truth, its useful life may be measured in centuries if it is not subjected to seriously adverse conditions in service. There are numerous examples of structures several centuries old and, in the USA, examples of houses that date back to the early days of its history, together with covered bridges almost as old. One of the best-known timber structures in the USA is the Tabernacle of the Church of Jesus Christ of Latter Day Saints in Salt Lake City. This structure, whose roof is supported by wood trussed arches, is still in regular use after 100 years with no problems.

The critical question for the engineer is whether or not the structural properties of wood are altered with time. A number of studies, including one on girders from a 100-year-old cathedral in Madison, Wisconsin, have shown that there is no measurable loss in strength or stiffness as a result of age alone. It is known, however, that wood structural members can sustain higher loads for short rather than for long periods, and higher design stresses are permitted for short-term loading. Since the loading history of a structure, particularly one which has been in service over many years, is not likely to be known, caution suggests that for an old structure, design stresses for evaluation of load-carrying capacity should be those applicable to long-term loading (see *Design with Wood*).

Wood is, however, subject to a number of deteriorating influences including decay fungi, insects (especially termites), bacteria and marine organisms (see *Deterioration by Insects and Other Animals During Use*; *Decay During Use*). Fungi, bacteria and many insects require moisture to be active, so their effects can be prevented if moisture is kept from reaching a critical level by careful design of details, by using moisture barriers and by properly maintaining the structure to prevent the intrusion of water from condensation, roof leaks, plumbing or other sources. Where this is not possible, substantial increases in useful life can be obtained by treatment with preservative chemicals.

1.2 Behavior in Fire

Wood is a combustible material. It is also a good insulator, as is the char formed when wood burns. Thus the wood only a short distance from the char–wood interface is at a temperature well below the ignition temperature (see *Fire and Wood*). As a consequence, the rate of charring is low and wood members of large cross section retain significant load-carrying capacity for long periods when exposed to fire. This capability was recognized long ago in buildings with masonry walls and heavy columns and beams—so-called mill-type construction—and led to the acknowledgement that heavy timber construction performed in a manner similar to noncombustible constructions; thus it may be given a one-hour fire rating. Many building codes in the USA permit the use of heavy timber structures in all fire districts with all types of occupancies.

Exposure of wood to elevated temperature results in a loss of strength; long-term exposure causes a permanent loss (see *Thermal Properties*). Because wood just below the char–wood interface has been subject to only a moderate rise in temperature, the largest part of the unburned portion of a wood member retains its original strength. Thus, removal of the char, and perhaps a quarter inch of wood below the char on each fire-exposed face, yields a member whose load-carrying capacity can in some cases be determined by applying original allowable stresses to the section properties determined from its new (reduced) size. In laminated bending members, however, the grade of individual laminations commonly varies within the cross section, the higher grades being used in the areas of highest stress. Laminations at the tensile side, particularly, are commonly required to be of especially high grade. Thus the loss of a large proportion of a tensile lamination may change the overall design stress greatly and this effect, as well as the effect of changes in cross-sectional dimensions, must be considered. In some instances, char removal may be all that is needed to renovate burned structural members.

The effects of fire on wood may be markedly changed by the application of fire retardants. They may be applied in the form of coatings or of chemicals such as mixtures of waterborne salts impregnated into the wood under pressure. Fire retardants are described in some detail in *Fire and Wood*.

1.3 Rate of Load Application; Duration of Load

It has long been known that wood can sustain higher loads for short than for long periods and this characteristic has been accounted for by adjustment of design stresses. Similarly, very rapidly applied loads can be sustained at higher levels than can more slowly applied loads. Design stresses, for example, may be doubled for impact loads. The factors commonly used to account for the effects of load duration or rate of load application have been derived from tests on material that is free from strength-reducing characteristics. Recent data suggest that these effects are different for lumber containing the usual characteristics of knots, cross grain and the like, and that the duration effects may be much less in the lower grades. Work on this problem is continuing, but at the time of writing, design-stress adjustments have not been changed.

1.4 Anisotropy

Wood is not isotropic and has different mechanical and other properties in different directions with respect to the grain of the wood (see *Strength*). This characteristic must be considered in design and special care must be taken to avoid design details that result in large stresses applied at right angles to the grain of the wood.

1.5 Exposure to Extremes of Temperature

Long-term exposure to elevated temperature may result in marked loss in strength, the degree of loss depending on the exposure temperature and the duration of the exposure (see *Thermal Degradation*). Thus the use of wood structures in applications where it is known that the wood will be continuously exposed to markedly elevated temperatures should be approached with caution.

Conversely, wood becomes stronger at very low temperatures and this characteristic has led to its use in vessels designed for transportation of liquefied gas.

1.6 Dimensional Stability

Wood will not change dimension with changes in moisture content unless it is at or below the fiber saturation point (see *Shrinking and Swelling*), commonly considered to be a moisture content of $\sim 30\%$. Below that level, wood shrinks as the moisture content decreases and swells as it increases. These changes may have structural effects. For example, if a connection involves steel plates attached to a wood member by widely separated bolts, a change in moisture content may cause splitting because the bolts are fixed in position by the steel plates and, as the wood attempts to shrink, it is restrained between the bolts by their inability to move. Thus tensile stress across the grain develops; if this stress becomes higher than the perpendicular-to-grain strength of the wood, the wood will split.

An unseasoned member of large cross section may develop deep checks because the outer layers of wood reach moisture contents below the fiber saturation point before the center of the member does. Since shrinkage does not occur until the fiber saturation point is reached, the interior will remain at its original dimension while the surface layer attempts to shrink. Thus checking will occur if the shrinkage tendency is great enough to create stresses beyond the across-the-grain strength of the wood.

Laminated members (see *Glued Laminated Timber*) are made from laminations which are generally not > 5 cm (nominal) in thickness, so that they can readily be dried to a suitable moisture content without serious drying defects. Thus the high initial moisture contents that are normal in a large solid sawn member are not present in a laminated member of large cross section; problems such as splitting and checking are therefore not to be expected in the same degree.

Change in dimension along the grain with a change in moisture content is, for normal wood, only a fraction of that across the grain. However, certain types of wood such as compression wood or juvenile wood (wood from near the pith of the tree) may show substantial dimensional changes along the grain with changes in moisture content. Thus, although it is generally true that problems from changes in length of a member with changes in moisture content are not to be expected, there may be instances when it can cause significant problems.

Coefficients of thermal expansion for wood are in the region of only one-tenth to one-third those of other structural materials. This difference must be taken into account when other materials are used in conjunction with wood.

1.7 Resistance to Chemicals

While wood can be degraded by exposure to certain chemicals, it may generally be considered highly resistant to many others. Wood is superior to cast iron and ordinary steel in resistance to mild acids and solutions of acidic salts. Thus, in the chemical processing industry, it is the preferred material for tanks and other containers and for structures adjacent to or housing chemical equipment. Wood is used widely in cooling towers where the hot water to be cooled contains boiler-conditioning chemicals as well as dissolved chlorine for suppression of algae. It is also used widely in the construction of buildings for bulk chemical storage, where it may be in direct contact with chemicals.

Highly acidic salts tend to hydrolyze wood and cause embrittlement when present at high concentrations. Alkaline solutions are more destructive than acidic solutions. Iron salts develop at points of contact with tie plates, bolts and the like. They may cause softening and discoloration of the wood around corroded iron fastenings. This is particularly likely in acidic woods such as oak and in woods such as redwood that contain tannin and related compounds.

The natural resistance of wood to degradation may be improved by treatment with a variety of substances. Pressure impregnation with a viscous coke-oven coal tar may be used to retard liquid penetration. Resistance to acids may be increased by impregnation with phenolic resin solutions followed by appropriate curing. Resistance to alkaline solutions may be increased by treatment with furfuryl alcohol. Impregnation by a monomeric resin followed by polymerization by radiation or other means is a relatively new development.

1.8 Strength and Structural Design

Strength, resistance to deformation, design and related factors are discussed in other articles (see *Strength*; *Deformation Under Load*; *Lumber: Behavior Under Load*; *Design with Wood*), but it seems desirable to point out here that wood generally has a favorable ratio of strength to weight and that this is one of the characteristics that has led to its use in a wide variety of structural applications.

Each species (or species group) has its own characteristic mechanical properties and not all species are classified according to the same set of grade descriptions. Design allowables, therefore, are presented by species or species groups and by grades under each species. In the USA and Canada, the dimension lumber used in light frame construction is commonly graded under a single rule called the National Grading Rule for Dimension Lumber. That is, the characteristics such as knots or cross grain permitted or limited in the National Grading Rule are the same regardless of species, although design stresses differ.

Design stresses for lumber, plywood and glued laminated timber are commonly developed by either of two methods. In one, fundamental mechanical properties based on data from tests of clear, straight-grained specimens are modified by a series of factors. For example, a property at a near-minimum level is used as a starting point. This is then modified by factors to account for moisture level, the effect of duration of load and the effect of such things as knots and cross grain permitted by the grade; finally, a safety factor is applied. The second method starts with data from full-sized pieces representative of the various grades. The base data are again modified by a series of factors except that relating to grade. In North America, such methods are contained in standards of the American Society for Testing and Materials.

Design stresses developed as above may be modified to account for the individual circumstances associated with a particular structure. For example, for loads of short duration, such as snow, somewhat higher stresses may be used than with structures on which the full design load may be expected to continue for long periods. On the other hand, adverse circumstances such as certain treatments of wood, exposure to long-continued high temperature or long-term use at high moisture content may dictate modification to lower design stresses. Design stresses and their modification are presented in design and other handbooks.

Structural design is characteristically based on the use of allowable stresses. Increased attention, however, is being given to the development of reliability-based techniques for application to wood structures.

2. *Applications*

Typical examples of the wide range of applications of wood are discussed here to indicate some of the possibilities, but a complete review is not attempted.

2.1 Foundations

Wood piles have long been used to support buildings, bridges and waterfront structures. Because of the conditions to which most piles are exposed in service and the cost and other problems incident upon replacement, most piles are treated to provide protection against decay and termite attack and, when used in salt water, against attack by marine organisms (see *Preservative-Treated Wood*).

Round wood poles or square wood posts set in the ground and supported on a footing are used with supporting beams to provide the substructure for buildings, frequently in basementless structures. In this type of application, the posts extend far enough above ground to minimize the possibility of decay in the beams. In a similar type of construction, the poles or posts are longer and form not only the foundation but part of the farming of the building as well, the other structural members being fastened directly or indirectly to the poles or posts. The latter type of construction is especially suitable on sloping sites, since a minimum of disturbance to the soil is required—far less than would be the case for a typical foundation. The integral foundation–framing type of construction has advantages where the structure is subject to high wind and wave action, as in seacoast structures in a hurricane area.

A wall-type foundation system, commonly called the all-weather wood foundation, is made from pressure-treated dimension lumber and plywood. Essentially a wood frame wall, the panels are placed on a wood footing over a gravel bed. This type of foundation, widely accepted in North America, has a number of advantages. It can, for example, be installed during weather that precludes the pouring of a concrete foundation, hence its name. In addition, the installation time is less than for a conventional foundation and the wall may be insulated, providing a more comfortable basement than one with concrete walls.

2.2 Light Frame Construction

In North America and possibly to a lesser degree elsewhere in the world, the majority of residences and many farm and light industrial buildings are of light frame construction. The framing members are usually of dimension lumber 2 in. (nominal) (5 cm) in thickness by widths varying from 4 in. (10 cm) to ~12 in. (30 cm) depending upon the type of load applied. Typically, light frame structures have been "stick built," that is, made of individual pieces assembled on the site. Increasingly, however, prefabricated units in the form either of panels or three-dimensional modules are being utilized. In the past, light frame construction has made only limited use of engineering analytical methods; this is changing as the cost and availability of materials become ever more critical and as the requirements of building codes are refined.

Floor loads are carried by joists spanning from the exterior walls to a central girder or load-carrying partition. The width of the joists depends upon the span but is generally in the range 8–12 in. (20–30 cm). Roof loads are commonly carried by rafters resting on the exterior walls and abutting at a ridge board. Vertical loads are carried by studs normally 2 × 4 in. (5 × 10 cm) in cross section. Increasingly, the common rafter is being replaced by a truss (frequently called a trussed rafter), with the various truss components connected by metal plates attached to the members by teeth punched from the plate or by nails driven through the plate (see *Joints with Mechanical Fastenings*).

Flat trusses with similar jointing methods are becoming more common to support floors, instead of joists. A new technology termed the truss-framed system has recently been introduced. It consists essentially of a trussed rafter and a floor truss connected by wall studs framed into both the trusses, the whole constituting a rigid frame requiring no load-carrying framing in the interior. Not only does this construction use less structural lumber than conventional light framing, but all framing can be done with a single size, such as 2 × 4 in. (5 × 10 cm), rather than the more costly wide dimension lumber needed for solid joists and rafters. Fabrication in a plant rather than on the site permits better control of quality.

2.3 Panelized Roof Systems

A variety of panelized roof systems are available. One common in California uses a basic unit consisting of a 4 × 8 ft (1.2 × 2.4 m) panel of plywood to which are attached stiffeners (commonly 2 × 4 in. dimension lumber). These basic units are attached to purlins which span between primary supporting beams. Although the basic panels can be put into place individually after purlins are erected, efficiency can be gained by assembling the panels and purlins on the ground, lifting the whole unit into place by machine and installing the purlins into hangers previously attached to the primary beams. This system promotes safety in that only one person is required to be on the roof to set panels and there is always a roof surface on which to stand; walking on beams or purlins is not necessary.

2.4 Heavy Timber Structures

Heavy timber construction is a type included in nearly all major building codes. As mentioned in Sect. 1.2, its use is based on the inherent fire resistance of timber members of large cross section. Codes specify the minimum dimensions for various members in order to ensure the desired resistance to fire. Heavy timber construction was at one time based solely on the use of solid timbers. These are becoming increasingly difficult and costly to obtain, so glued laminated members are now more commonly used.

2.5 Beams and Columns

Beams and columns made from solid-sawn timbers are typically straight members of rectangular cross section. Their cross section and length are limited by the size of the log from which they are cut. Thus beam spans and column heights are similarly limited. The advent of glued laminated structural timbers (glulam) has largely removed size limitations in terms of both cross section and length. Thus glulam beams of 100 ft (30 m) span or more and of 5–7 ft (1.5–2 m) depth are becoming more common. Greater spans may be achieved through the use of cantilevered or continuous beams. Further, beam form is no longer necessarily straight and rectangular. Glulam beams may be straight with single or double taper, or they may be pitched and tapered with a curved lower face over part of the span. Columns may be tapered or straight and may be either solid or spaced.

2.6 Curved Members

The advent of glulam has removed also the limits on curvature which had existed for so long. Glulam arches may be (and are) used to support both roof and bridge structures. Although both two-hinged and three-hinged arches are feasible with glulam, two-hinged arches are less common because of difficulties in treatment and transportation. The curved forms possible are almost limitless, but one of the most common is the Gothic type so frequently found in church architecture. Gothic arch spans are commonly up to 80 ft (25 m).

Arches may also be used in domes for athletics or other facilities requiring large clear floor space. A typical form employs large arches radiating from a compression ring at the top of a dome to a tension ring at the base. Domes of this type (termed radial rib) have been constructed with clear spans up to 350 ft (105 m).

Even greater clear spans have been achieved with domes framed with wood members, largely laminated, to produce the needed curvature. Framing configurations differ and special connections between members are required. Two principal framing methods are employed, termed the Varax and the Triax domes; they represent patented systems. Structural analysis of such domes requires complex computer programs. A triangulated dome housing an athletics facility in Portland, Oregon has a clear span in excess of 500 ft (150 m).

2.7 Open Web Trusses

Open web roof and floor trusses offer a number of advantages including ease of erection because of light weight and the use of the open space for ductwork, plumbing, electrical lines and lines for fire sprinkler systems. One such open web system widely available uses steel tubing for the web members and machine stress-rated lumber for the top and bottom chords. Open web roof systems may span up to ~120 ft (35 m) and floor systems up to ~50 ft (15 m).

2.8 Plywood Components

Plywood is frequently used as roof, floor and wall sheathing in structures where the primary loads are carried by the framing. Not only does the plywood serve the normal sheathing function, but because of its high resistance to shear loads it may be designed to act as a diaphragm that will enhance the resistance of the structure to wind and earthquake loads. Stressed-skin panels, consisting of framing members with plywood bonded to one or both faces of the framing, provide efficient elements for carrying roof, floor or wall loads. Sandwich panels involving plywood bonded to a continuous lightweight core such as a foam or honeycomb are also efficient structural elements for roof, floor or wall. Stressed-skin or other types of panels may be used in folded plate roof systems designed to act as diaphragms. The plates are designed as deep intersecting beams acting together and supported at the ends.

In recent years, plywood nail-glued to joists has been used to provide semicomposite T-beam action resulting in reduced deflection and less squeaking of floors. Composite action is ignored in designing for strength. The special construction adhesives are generally suitable for on-site use. I-beams with plywood webs are used as structural members. Most have dimension lumber for flanges, but one manufacturer uses a specially made laminated material built up from veneer bonded with adhesive.

2.9 Wood Shells

Shell roofs of several forms may be constructed from layers of wood fastened either with nails or with adhesives. Only limited numbers of such structures have been made of wood.

2.10 Bridges

Wood has had wide application in bridge construction and has given long service; there are wood bridges more than 100 years old in North America and even older in Europe. The secret of long service is protective design to avoid entry of water (a tight deck with a water-resistant wearing surface, for instance), coupled with preservative treatment.

Bridges of a number of types have been built with wood. The type for a particular situation depends, of course, on such factors as the span, the type of traffic, the intended service life and aesthetics. Wood bridges are, for example, particularly suited to use in rural locations because their appearance can easily be made compatible with the surroundings.

The trestle-type bridge is common, particularly on railroads. In this type, girders or stringers with an appropriate deck system are supported on a substructure of piles or on a frame bent.

Girder bridges may be built with either solid sawn or glued laminated girders supporting a deck system. Solid sawn girders are, of course, limited in length and cross-sectional dimensions to the dimensions which can be cut from available timber; the maximum span for highway loads is ~30 ft (9 m). This limitation is largely eliminated if glued laminated members are used (see *Glued Laminated Timber*). Laminated girders up to 84 in. (2.1 m) in depth and 150 ft (45 m) long have been used. Deck systems vary, but nailed laminated decks of dimension lumber are common. These decks tend to become loose with changes in moisture content, with resulting water leakage and potential decay in the girders. Glued laminated deck sections obviate this difficulty. Various methods of attachment and joining, including special brackets and clips and dowels to transmit shear at the ends of the panels, have been developed and are in use.

A number of arch-supported bridges have been built. This type may be the most economical, depending upon the site conditions, because of savings in substructure. It does, however, require substantial height between foundations and roadway, and the site must be capable of resisting horizontal thrust in the abutments. A three-level interchange on an interstate highway near Mount Rushmore, South Dakota, involved, in the top bridge, an arch spanning 155 ft (47 m).

Truss-type bridges are becoming less common. They are practical for spans up to ~250 ft (75 m). Parallel-chord and bowstring trusses are the most common. The bowstring type may be the most economical because of the lower stresses in the web members. The advent of glued laminated members offers the possibility of continuous chords, reducing some of the connection problems found with shorter members. Improved fasteners can develop strong, rigid, concentric joints.

The Canadian Forest Service has developed designs for inverted kingpost bridges in which the stringers are designed to carry only the dead load while cables prestressed over the post carry the live load. Kingpost designs have been used for spans up to 125 ft (38 m) and queenpost trusses to 160 ft (50 m).

Suspension bridges are unusual, although the US Forest Service developed designs for spans up to 400 ft (120 m) in the 1930s. A suspension bridge in Switzerland is reported to span more than 540 ft (165 m).

Composite timber–concrete bridge decks have been built in numbers. The T-beam type consists of timber stringers forming the stem of the T and a concrete slab forming the flange or crossbar. Provision must be made for resisting horizontal shear, for example by cutting notches in the top of the timber, and uplift must be resisted by such means as driving mechanical fasteners into the top of the stringer. Composite slabs consist of 2 in. (50 mm) lumber nail-laminated with the wide surfaces vertical. The concrete is cast monolithically on top of the timber. Again, provision must be made to resist shear and uplift.

3. Wood-Based Materials

Wood is the base for a wide variety materials for use in many different applications. Some of those most likely to find use in construction are discussed briefly in this section. Many of these materials are discussed in detail in other articles (see *Lumber: Types and Grades*; *Lumber: Behavior Under Load*; *Plywood*; *Hardboard and Insulation Board*; *Particleboard and Dry-Process Fiberboard*; *Glued Laminated Timber*; *Laminated Veneer Lumber*; *Structural-Use Panels*). Depending upon the hazards which may be expected in service, these materials may have to be treated to resist fungal attack, marine borer attack, or to improve resistance to fire.

Softwood lumber is most commonly used to provide the structural framework of buildings in the form of dimension lumber or of timbers. In such applications as joists, girders, posts, studs and rafters, it is used as it comes from the sawmill or planing mill. In the form of trusses, dimension lumber may be used in relatively short lengths connected by metal truss plates. In addition, it may form the chords of I-beams or box beams in combination with a panel material such as plywood. It may, in addition, be used as the stringers in stressed-skin panels having faces of plywood or another panel material. In all such structural uses, it is preferable that the lumber be stress-graded to ensure sufficient strength and stiffness to support service loads. In most cases, also, it should be at a relatively low moisture content. Lumber may also serve a protective function as house siding in various forms. In such applications, appearance and dimensional stability are important characteristics.

Panel products such as plywood, fiberboard and hardboard may serve structural or other functions in buildings. As mentioned earlier, the high shear resistance of plywood leads it into floor and wall sheathing, particularly in structures expected to be subjected to high wind or earthquake loads and, because of its availability in large sheets, into use as the facings of stressed-skin panels. Fiberboards generally are used to provide insulation, but some forms are used as wall sheathing in light-frame buildings. Hardboards find a great deal of use as a base (or underlayment) for tile or linoleum floorings.

A product which is commonly termed laminated veneer lumber but which is marketed under a number of trade names is made from layers of veneer bonded together with the grain of each layer in the same direction as in glulam (see *Laminated Veneer Lumber*). Laminated veneer lumber (LVL) could serve any of the functions of sawn lumber, but is generally limited to special applications such as truss chords, scaffold planks and ladder rails.

Wood in round form finds use as piling foundations, as poles in pole frame buildings and as utility structures, either as single poles or as pole frames. In most uses, treatment to resist decay or other hazards is common.

See also: Weathering

Bibliography

American Institute of Timber Construction 1985 *Timber Construction Manual*, 3rd edn. Wiley, New York

American Institute of Timber Construction 1984 *Standard Specifications for Structural Glued Laminated Timber of Softwood Species*, AITC Standard 117–84, *Manufacturing*; AITC Standard 117–84, *Design*. American Institute of Timber Construction, Englewood, Colorado

American Society of Civil Engineers 1975 *Wood Structures: A Design Guide and Commentary*. American Society of Civil Engineers, New York

Freas A D 1971 Wood products and their use in construction. *Unasylva* 25 (101–03): 53–68

Freas A D (ed.) 1982 *Evaluation, Maintenance and Upgrading of Wood Structures.* American Society of Civil Engineers, New York

Freas A D, Moody R C, Soltis L A (eds.) 1987 *Wood: Engineering Design Concepts*, Vol. IV: Clark C Heritage Memorial Series on Wood. Pennsylvania State University, University Park, Pennsylvania

National Forest Products Association 1986a *National Design Specification for Wood Construction.* National Forest Products Association, Washington, DC

National Forest Products Association 1986b *Design Values for Wood Construction*, A supplement to the 1986 edition of National Design Specification, National Forest Products Association, Washington, DC

US Department of Agriculture 1987 *Wood Handbook: Wood as an Engineering Material*, Agriculture Handbook No. 72. US Department of Agriculture. Washington, DC

A. D. Freas
[Madison, Wisconsin, USA]

C

Cellulose: Chemistry and Technology

Cellulose as a material is a marvel of nature which has been and is of immense importance to mankind. It is the world's most abundant organic material. Cellulose incorporates about 40% of the carbon in plants and its total amount in the vegetable world is over 10^{11} t. With the growing shortage of petroleum products, research and development related to cellulose as a raw material for producing high-performance polymeric products and value-added chemicals have been prolific over the last twenty years.

Cellulose is a polydisperse linear syndiotactic natural polymer. It can be found mostly in wood and cottonseed hairs. The basic monomeric unit of cellulose is D-glucose, which links successively through a glycosidic bond in the beta configuration between carbon 1 and carbon 4 of adjacent units to form long-chain 1,4-β-glucans (Fig. 1). Each β-D-glucopyranose unit within a cellulose chain has three hydroxyl groups, namely two secondary and one primary hydroxyl groups. Intermolecular and intramolecular hydrogen bonding during biosynthesis has a profound effect on the morphology and reactivity of the cellulose chains. Intramolecular hydrogen bonding between adjacent anhydroglucose rings enhances the linear integrity of the polymer chain. Such intramolecular hydrogen bonding not only affects chain rigidity, but also the reactivity of the hydroxyl groups, particularly of the C-3 hydroxyl groups which hydrogen-bond strongly to the ring oxygens on adjacent anhydroglucose units. The regular ribbon-like cellulose chains allow an efficient, closely packed arrangement, permitting intermolecular hydrogen bonding to yield a fibrillar tertiary structure of high lateral order, i.e., high crystallinity. The loosely packed molecules remain as the amorphous region of cellulose macromolecules. Crystalline and amorphous regions make up 55–75% and 25–45%, respectively, in the cellulose fiber. The hydroxyl groups located in the amorphous regions are highly accessible and readily reactive in all chemical reactions, whereas those in crystalline regions are not readily accessible to reactant molecules and are completely inaccessible to some. The highly crystalline nature of cellulose provides strength and stiffness that make cellulose an excellent structural material. On the other hand, this unique property may hamper its reactivity for chemical modification and its solubility in industrial solvents. This might be one of the reasons that interest in cellulose-based technologies waned significantly after the 1950s as more melt-processible and solution-processible synthetic polymers became commercially available.

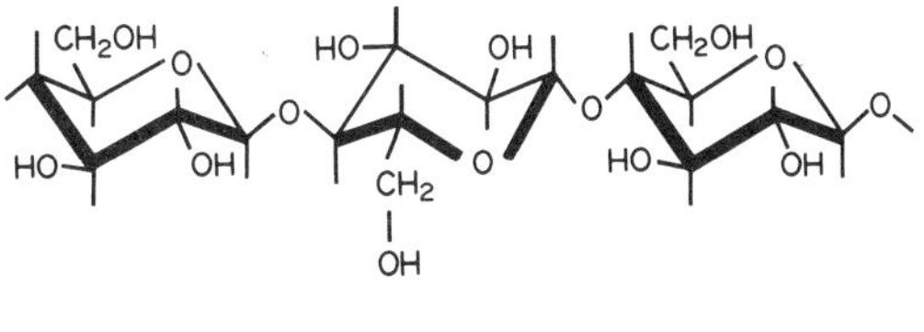

Figure 1
Cellulose structure

1. Swelling and Dissolution

Because of its strong intermolecular hydrogen bonding, and in spite of the presence of many polar hydroxyl groups, cellulose is practically insoluble in water or common organic solvents. In order to be dissolved, the hydrogen bonds must be destroyed with a suitable solvent. Cellulose solvents are commonly classified into four different systems, in which: (a) cellulose acts as a base, (b) cellulose acts as an acid, (c) cellulose forms a complex with metal ions, and (d) cellulose transforms into dissolvable intermediates which can be regenerated.

As cellulose is a hydrophilic polymer, it absorbs water readily and swells. However, the swelling is limited to the amorphous region of the fiber, i.e., intercrystalline swelling, as it is counteracted by the strong hydrogen-bond network of the crystallites. Stronger swelling agents that will achieve intracrystalline swelling are thus necessary if cellulose is to be completely swollen. For instance, it can be swollen in concentrated alkali solution by forming an alkali cellulose, which serves as an important intermediate for the production of cellulose derivatives. Liquid ammonia can swell cellulose markedly. It not only expands the crystal lattice, but also transforms the crystalline regions into an essentially amorphous configuration. Cellulose can also be swollen in highly polar solutions which ultimately results in dissolution. These solutions are made up of aqueous mineral acids, aqueous solutions of zinc chloride, calcium thiocyanate, aqueous alkalies, quaternary ammonium hydroxide and metal ammonia hydroxide complexes such as copper ammonia hydroxide. These solvents are able to break both the intermolecular hydrogen bonding network and the van der Waals forces between the surfaces of the ribbon-like cellulose chains, and subsequently penetrate into the crystalline areas. At the same time, however, all of these solvents will also degrade cellulose, particularly when oxygen is present. For homogeneous hydrolysis purposes, cellulose is dissolved (or degraded) in hydrogen fluoride, trifluoroacetic acid, 44% hydrochloric acid, 72% sulfuric acid,

68% nitric acid, 60% perchloric acid and 85% phosphoric acid. Regenerated fibers from the cupriethylenediamine hydroxide system were used for commercial production of cuprammonium rayon but the process is no longer competitive with the xanthate process. The regenerated cellulose or viscose rayon from the xanthate process is prepared by pretreating cellulose with sodium hydroxide to produce alkali cellulose which then reacts readily with carbon disulfide at room temperature to form an alkali-soluble cellulose xanthate. The viscose cellulose xanthate solution is forced through spinnerets and the filaments produced are coagulated and converted to cellulose by immersion in an aqueous solution of sulfuric acid and sodium sulfate. The reaction scheme for the production of cellulose xanthate is shown below:

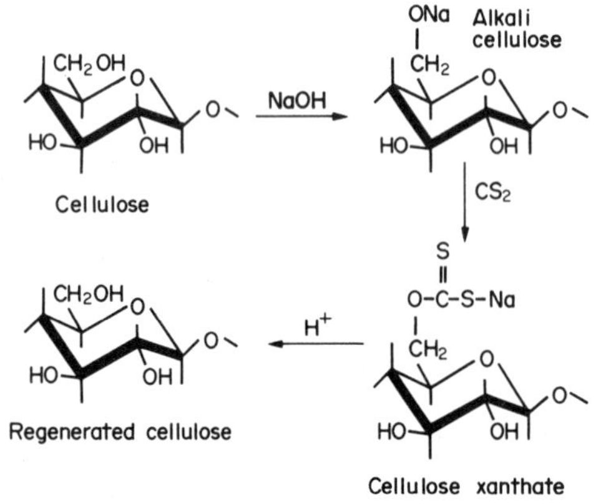

Cellulose carbamate can be prepared by heating cellulose with urea at 135 °C. The product is stable in the dry state, and is soluble in cold aqueous sodium hydroxide. Pure cellulose can be regenerated by alkaline hydrolysis of the carbamate group, as shown below:

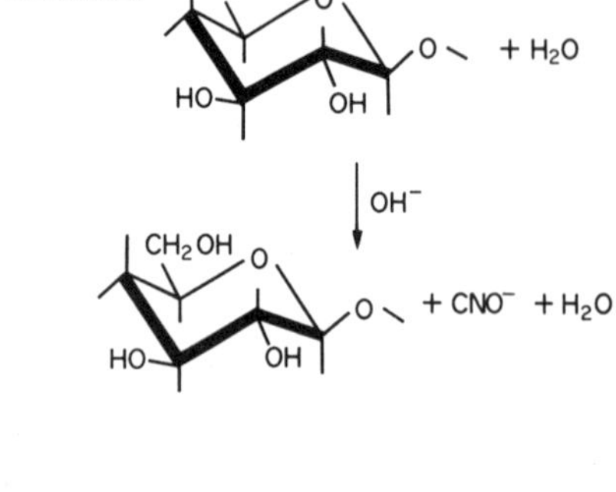

Rayon has been widely used in textiles and clothing. It can be made stronger than cotton and has been popularly used in tire cord.

Because of growing use, and in order to improve the competitive position of cellulose in relation to nonrenewable materials derived from petrochemicals, research and development of less-degrading organic solvents have made impressive strides since the 1970s.

The use of a combination of ammonium thiocyanate and ammonia achieves dissolution of cellulose. It was found that cellulose can best be dissolved in 72.1% NH_4SCN, 26.5% NH_3 and 1.4% H_2O. No chemical reaction takes place in this process. The use of lithium chloride (10%) and dimethylacetamide (DMAc) has proven to dissolve 16% of cellulose with a degree of polymerization (DP) of 550, and 4% with DP of 1700. The critical technique for this system is to activate cellulose with either water or liquid ammonia followed by DMAc exchange.

Cyclic tertiary amine oxides, such as *N*-methylpiperidine-*N*-oxide and *N*-methylpyrrolidine-*N*-oxide have been reported to be excellent solvents. With these agents, a typical solution was one in which 10% cellulose had been dissolved in the presence of 15.5% water at 100–105 °C. This system does not involve formation of cellulose derivatives.

Several systems that will dissolve cellulose do involve formation of cellulose derivatives. The use of dinitrogen tetroxide or nitroxyl chloride and dimethylformamide or dimethyl sulfoxide will dissolve cellulose by replacing hydroxyl groups with nitrite esters, which can be hydrolyzed readily in water to regenerate cellulose. Likewise, the use of dimethyl sulfoxide and paraformaldehyde is to replace hydroxyl groups with hydroxymethyl groups. This derivative is soluble in the solvent and can be regenerated in protic solvents.

All of these new solvent systems seem to have less degradative effect on cellulose. Regenerated cellulose can be made from these solutions. Owing to the use of new solvents, more useful products from homogeneous cellulose solutions can be anticipated.

2. *Cellulose Derivatives*

The intrinsic insolubility of cellulose motivated early workers to explore new derivatives based on reactions of the hydroxyl groups. The reactivity of the three hydroxyl groups at C-2, C-3, and C-6 on the D-anhydroglucopyranose unit offers a variety of possibilities for making useful derivatives. The properties of the derivatives depend heavily on the type, distribution, and uniformity of the substitution groups. The average number of hydroxyl groups replaced by the substituents is the degree of substitution (DS), the maximum being three. When side-chain formation is possible, molar substitution (MS) is used to denote the length of side chain, and the value can exceed 3.

Properties that are most strongly affected by DS and MS are the solubility and plasticity. Derivatives of a low DS are often more sensitive to water and may even be dispersive in water. In derivatives with a high DS of nonpolar substituents, the water solubility and the absorption of water is decreased, and the solubility in organic solvents is increased. Moreover, with an increase in DS or MS by nonpolar groups, the plasticity is increased. Many cellulose derivatives have reached commercial success and are playing an important role in various industries.

Cellulose derivatives can be made by either etherification, esterification, cross-linking or graft-copolymerization reactions.

2.1 Esters

Since cellulose possesses primary and secondary hydroxyl groups, it undergoes esterification with inorganic and organic acids in the presence of a dehydrating agent or by reaction with acid chlorides. The reaction requires the absence of water for completion, as it is a reversible reaction. The general reaction scheme can be illustrated as follows:

$$\text{Cellulose} + \text{Acid} \xrightarrow{\text{Catalyst}} \text{Cellulose ester} + H_2O$$

The resulting esters have entirely different physical and chemical properties from the original cellulose and are soluble in a wide range of solvents. Over 100 cellulose esters have been developed, and only the major ones which have important industrial applications will be described here.

Cellulose nitrate, erroneously called nitrocellulose, has a special place in the history books because it was the first synthetic polymer and the one on which the present-day plastics industry is founded. Hence, this derivative was responsible for many changes in industrial and military technology. Gun cotton replaced black powder as a propellant and the synthetic plastic, celluloid, initiated to a large extent the molding and fabrication of plastics. Cellulose nitrate is prepared by reacting cellulose with nitric acid in the presence of sulfuric acid and water. The degree of substitution is modified by varying the concentration of these three components as well as reaction conditions. The general reaction scheme is shown below:

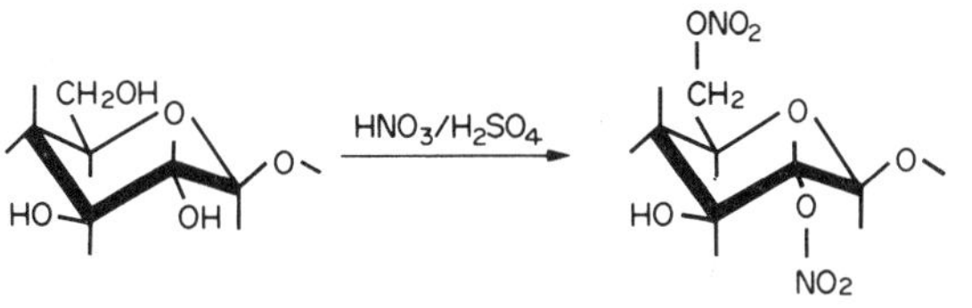

Cellulose   Cellulose nitrate

Today cellulose nitrate is still an important chemical for making propellants and explosives, as well as an ingredient for making lacquers, adhesives, plastics (celluloid) and a binder for printing inks and paints. Cellulose nitrates with different DP and DS have different solubility and mechanical properties for different applications (see Table 1).

Cellulose acetate is an important organic ester of cellulose. It is prepared by pretreating pulp with acetic acid or acetylating pulp directly with an excess of acetic anhydride, acetic acid and a catalyst to the triacetate stage and then saponifying to the diacetate. The general reaction scheme is shown below:

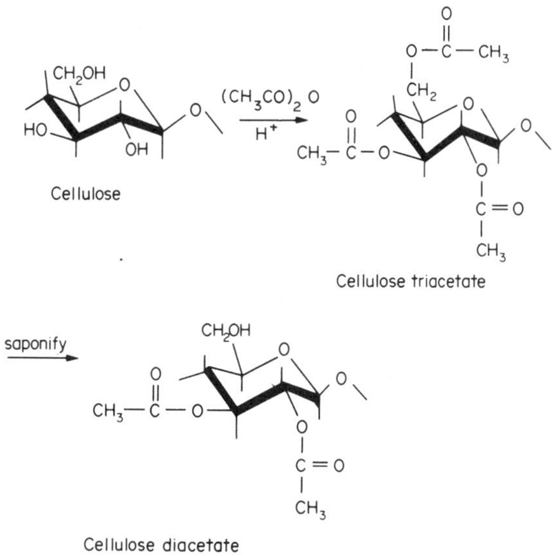

Table 1
Properties of cellulose nitrate

Nitrogen content (%)	Degree of substitution	Common solvent	Field of application
10.5–11.1	1.8–2.0	Ethanol	Plastics, lacquers
11.2–12.3	2.0–2.4	Methanol, esters, acetone	Lacquers, adhesives, binders
12.4–13.5	2.4–2.8	Acetone	Explosives, propellants

Cellulose acetate has the greatest commercial value for making fibers, cigarette filters, safety glass, photographic film and transparent sheeting. The properties of typical cellulose acetates are summarized in Table 2. The acetic acid content and acetyl content corresponding to the DS are also included. Other cellulose esters with commercial values are cellulose propionate, cellulose butyrates and cellulose esters of higher aliphatic acids such as stearic, lauric and palmitic acids.

2.2 Ethers

Cellulose ethers are made by the reaction of cellulose with aqueous sodium hydroxide and then with an alkylhalide, such as methyl and ethyl chloride. Cellulose ethers comprise methylcellulose, ethylcellulose, hydroxyethylcellulose, hydroxypropylcellulose and their derivatives. Most of the cellulose ethers with low DS are soluble in water, and those with higher DS are soluble in alkali solution or in organic solvents (see Table 3). Cellulose ethers have gained their position on the market due to their multifunctional effects. They exhibit useful properties of thickening, thermal gelation, surfactancy, film formation and adhesion. Those characteristics earn them application in areas such as foods, cosmetics, paints, construction, pharmaceuticals, tobacco products, agriculture, adhesives, textiles and paper. In pharmaceutical applications, cellulose ethers such as methyl cellulose and ethyl cellulose have been widely used as a bulking agent for treating various intestinal ailments as in diabetic diets, as ingredients for ointments and lotions, and as a commonly used binder for tablets. In food applications, cellulose ethers have been used in bakery products, as thickener for canned-fruit pie fillings and in frozen foods. They also have been used in adhesive and cement formulations.

The reaction of cellulose with sodium monochloroacetate can produce sodium carboxymethylcellulose (CMC), a water-soluble anionic linear polymer. Sodium CMC is probably used in more varied applications worldwide than any other water-soluble polymer known today. A major application of CMC is in detergent systems to prevent soil redeposition after the soil has been removed from fabrics by synthetic detergent. It is widely used in the petroleum industry as a water-soluble colloid in drilling-fluid systems. It is also used as an additive for the paper industry, a sizing agent for the textile industry, and has many applications in cosmetics and pharmaceuticals.

2.3 Cross-Linking

The properties of cellulose can be improved by cross-linking reactions, which have been utilized particularly in the textile, paper and cellulose-derivative industries.

Table 2
Properties of cellulose acetate

Acetic acid yield (%)	Acetyl content (%)	Degree of substitution	Common solvents	Field of application
52–54	37.3–38.7	2.2–2.3	Acetone, ethyl acetate	Plastics
54–56	38.7–40.1	2.3–2.4	Acetone, ethyl acetate	Fiber, film
56–58	40.1–41.6	2.4–2.6	Acetone, ethyl acetate	Plastics
61–62.5	43.7–44.8	2.9–3.0	Chloroform	Fiber, film, foil

Table 3
Solubility of cellulose ethers

Cellulose ether	Solubility/degree of substitution			Organic solvent
	Cold 4–8% NaOH	4–8% NaOH	Cold water	
Methylcellulose	0.1–0.4	0.4–0.6	1.3–2.6	2.5–3
Ethylcellulose		0.5–0.7	0.8–1.3	2.3–2.6
Hydroxyethylcellulose	0.1–0.4	0.5	0.5–1.0	
Ethylmethylcellulose			1.0–1.3	
Hydroxyethylmethyl cellulose			1.5–2.0	
Na-carboxymethyl cellulose	0.1–0.4	0.5	0.5–1.2	
Cyanoethylcellulose				2.0
Benzylcellulose				1.8–2.0

Covalent and ionic cross-linking agents have been used for improving dimensional stability, and dry and wet strength of paper. Covalent bonding is usually achieved by the formation of ester linkages by reaction of cellulose with a polycarboxylic acid and by formation of imine linkages by reaction of polyamines with oxidized cellulose. Urea-formaldehyde, melamine-formaldehyde and the polyamide-amine polymers have been used to form water-resistant bonds between fibers. Cellulose and cellulose derivatives can also be cross-linked with suitable agents to make superabsorbents; and to improve their fluid absorbency, fluid retention and dry and wet resilience.

Durable-press and crease-resistance (e.g., wash and wear) properties of cellulose textiles have been improved by use of cross-linking agents. Dimethylolethyleneurea, bis(methoxymethyl)uron, dimethylolpropyleneurea, dimethylol monocarbamates and glyoxal monourein derivatives are the important agents used.

2.4 Grafted Copolymers

Cellulose esters and cellulose ethers are prepared by substitution of cellulose hydroxyl groups with short-chain reagents. Cellulose can also be modified by introducing long-chain polymer(s) onto its main chain. The products are mostly grafted copolymers and in some cases block copolymer can also be made. A large number of different methods of grafting have been developed. By far the greatest effort has been via free-radical vinyl-polymerization routes. The general reactions are shown below:

Initiation

$$\mathrm{In} \longrightarrow \mathrm{In}\cdot$$

$$\mathrm{In}\cdot + \mathrm{RH} \longrightarrow \mathrm{R}\cdot + \mathrm{In}$$

$$\mathrm{R}\cdot + \mathrm{M} \longrightarrow \mathrm{RM}\cdot$$

Propagation

$$\mathrm{RM}\cdot + \mathrm{M} \longrightarrow \mathrm{RM}_2^{\cdot}$$

$$\mathrm{RM}\cdot + n\mathrm{M} \longrightarrow \mathrm{R(M)}_{n+1}^{\cdot}$$

Termination

$$\mathrm{R}\cdot + \mathrm{R}\cdot \longrightarrow \mathrm{R{-}R}$$

$$\mathrm{R}\cdot + \mathrm{RM}\cdot \longrightarrow \mathrm{R{-}MR}$$

$$\mathrm{R}\cdot + \mathrm{RM}_2\cdot \longrightarrow \mathrm{R{-}M_2{-}R}$$

where In, RH and M are initiator, cellulose and monomer, respectively.

Vinyl monomers that can be grafted onto cellulose include acrylic acid, acrylonitrile, methyl methacrylate, and many others. Grafted copolymers of cellulose have shown improved strength, surface properties, adhesion, dye receptivity and resistance to chemical, microbial and photochemical degradation and abrasion.

3. Degradation Reactions

Like any other polymer, cellulose is susceptible to degradation. Under the action of acids, alkalis, heat, radiation, ultraviolet light and mechanical stress, cellulose is degraded by a variety of mechanisms.

3.1 Acid Degradation

The glycosidic linkage in cellulose is sensitive to acid-catalyzed hydrolysis. The degree of sensitivity varies according to the accessibility (amorphous or crystalline region), concentration and type of acid, and temperature. The reaction involves rapid formation of an intermediate complex between the glycosidic oxygen and a proton; this is followed by the slow, rate-determining scission of the glycosidic bond adjacent to C-1. In the presence of excess water the reaction is first order, although in many cases of cellulose hydrolysis, the data can be fitted equally well to a zero-order plot. Hydrolysis products with a wide range of DP may be obtained. If the process is carried to completion, the final product will be D-glucose. Hence, saccharification of cellulose to provide D-glucose for fermentative production of ethanol has taken on greater significance in recent years. Acid hydrolysis may proceed as a homogeneous process with strong acids, which dissolves the substrate, or as a heterogeneous process with weaker acids. The rate and extent of acid hydrolysis depend on the amount of amorphous portion present, because swelling, acid penetration and actual reaction occur preferentially in the amorphous region.

3.2 Alkali Degradation

Cellulose is relatively stable toward alkali at temperatures below about 170°C. However, the rate of degradation under oxygen and the loss in DP are dependent on the number of aldehydic groups present. This means that degradation occurs in a stepwise (or endwise) fashion from the reducing end of the cellulose chain (so-called peeling reaction). Chemical attack that leads to an increase in carbonyl group causes cellulose to become more susceptible to alkali. When cellulose is heated with sodium hydroxide at temperatures above 170°C, a considerable reduction in DP due to random scission of glycosidic bonds takes place. Consequently, the newly formed reducing end-groups are also subject to the stepwise degradation that enhance weight loss of cellulose. When cellulose is heated in aqueous alkali above 250°C in an autoclave, low-molecular-weight substances such as formic, acetic, glycolic and lactic acids are generally formed. Alkaline degradation in commercial processes is usually undesirable. This reaction is responsible for most of the cellulose loss that occurs in wood-pulp digestion and refining under alkaline conditions.

3.3 Thermal Degradation

Thermal degradation or pyrolysis of cellulose proceeds through a gradual degradation, decomposition and charring on heating at temperatures below 250°C. It also affects mechanical properties of cellulose. There may be changes in dimension, papermaking proper-

Table 4
Various wood saccharification processes

	Dilute acid	Concentrated acid	Fermentation	Hydrogen fluoride
Glucose yield (%)	50	85–90	50	90–95
Acid consumption	medium	high	none	very low
Enzyme and chemical requirements	moderate	high	high	low

ties, softening, elastic modulus and torsional properties. Rapid pyrolysis of cellulose takes place at temperatures above 250 °C and gives levoglucosan and its thermolysis products, including various quantities of char, tar and volatile products. Industrial pyrolysis of cellulose to various fuels and chemicals including charcoal, levoglucosenone, pyroligneous acid, acetic acid, methanol, terpentine and tars have been reported (see *Thermal Degradation*).

3.4 Mechanical Degradation

Under mechanical stress, the mechanical energy supplied by compression and shear forces in milling and cutting is sufficient to change the topochemistry and reactivity of cellulose, and to cause homolytic chain scission. Mechanoradical formation, transformation and decay reactions seem to occur continuously during processes involving the absorption of mechanical energy, and their consequences are likely to concern the properties of disaggregation products. Stress-induced reactions in cellulose result in the disaggregation of fiber bundles, shortening of fiber length, i.e., loss of DP, increased accessibility and decreased crystallinity. It has been proposed based on esr studies that the primary cleavage of bonds takes place between C-2 and C-3; the relocalization of energy and the stabilization of free radicals consequently lead to the cleavage of glycosidic bonds. The result of this stress reaction process is the reduction of chain length, i.e., DP.

3.5 Photodegradation

Even pure cellulose is sensitive to ultraviolet light, but relatively stable when irradiated with light of wavelength greater than 340 nm. During photoirradiation, free radicals are generated as the reactive intermediates. Under the influence of solar irradiation, chain scission (i.e., depolymerization) takes place. With the irradiation of light greater than 280 nm, dehydrogenation takes place in addition to chain scission. With light of wavelength greater than 254 nm, dehydroxymethylation also takes place. When cellulose is contaminated with impurities such as metal ions and dyes, it undergoes degradation at a faster rate. Degradation results are similar to those at >254 nm, and dehydroxylation is also observed.

4. Chemicals from Cellulose

Many conversion technologies are available to generate low-molecular-weight chemicals from cellulose, ideally from waste paper and waste wood. The conversion of cellulose to glucose is usually considered to be the first step in the potential large-scale utilization of cellulose. This can be done by saccharification. In essence, there are four approaches to cellulose saccharification: (a) use highly concentrated mineral acids at ambient or low temperature, (b) use diluted acid concentration applied at high temperature and corresponding high pressure, (c) fermentation processes, and (d) hydrogen fluoride saccharification (Table 4). Fragmentation may also be done with alkaline treatment of cellulose, and significant advances have been made in the utilization of enzymes or enzyme extracts from fungi. The glucose produced can be fermented to ethanol with high yield using commercial processes, and may be the basis for a chemical industry. The usual route proposed is to dehydrate the ethanol to ethylene. It is also possible to oxidize ethanol to acetaldehyde or acetic acid and to use the ethanol directly to produce ethyl esters and butadiene. Cellulose also can undergo microbial fermentation to produce single-cell protein, a possible animal feed material.

See also: Cellulose: Nature and Applications; Chemical Composition

Bibliography

Bikales N M, Segal L (eds.) 1971 *Cellulose and Cellulose Derivatives*, Vol. 5, Pts. 4–5. Wiley-Interscience, New York
Fengel D, Wegener G 1984 *Wood: Chemistry, Ultrastructure, Reactions.* de Gruyter, Berlin
Hon D N-S 1988 Cellulose: A wonder material with promising future. *Polym. News* 13: 134–40
Zeronian S H, Nevell T P (eds.) 1985 *Cellulose Chemistry and Its Applications.* Ellis Horwood, Chichester

D. N.-S. Hon
[Clemson University, Clemson, South Carolina, USA]

Cellulose: Nature and Applications

Cellulose is the most abundant organic material on earth, comprising the principal component of plants. Over 100 billion metric tons grow annually. This article discusses the composition and structure, properties, sources and applications of cellulose.

1. Composition and Structure

Cellulose is a long-chain polymer of β-D-glucose in the pyranose form linked together by 1,4′glycosidic bonds to form cellobiose residues which are the repeating units in the cellulose chain (see Fig. 1).

As a result of the β configuration the alternate glucose units are rotated through 180° as can be seen from the figure. This structure strongly favors the organization of the individual cellulose chains in bundles with crystalline order held together by hydrogen bonds. X-ray diffraction studies over the past 75 years have established that the small crystallites are oriented parallel to the fiber axis, and have a monoclinic unit cell with dimensions of about 10.3 Å along the fiber axis (the cellobiose repeating unit), 8.2 Å and 7.9 Å with an angle of about 96°. These cells containing cellobiose residues of two chains are considered an adequate approximation of the structure of native cellulose, although some uncertainty still exists about fine structure details.

Cellulose is polymorphic, and the structure of native cellulose (cellulose I) is changed by strong alkali as in mercerization or regeneration to cellulose II, which is also monoclinic with the same fiber axis repeating unit of 10.3 Å, but with other dimensions of 8.0 Å and 9.1 Å and angle of 117°. Other polymorphic forms of cellulose have been reported as cellulose III and cellulose IV. However, there is considerable opinion that these are not true polymorphic forms, but are rather disordered versions or mixtures of cellulose I and II.

Cellulose consists not only of these highly ordered crystalline regions, but contains disordered or amorphous regions as well. These different regions do not have clearly defined boundaries, but rather blend into each other. Cellulose samples of different origin and history contain differing relative amounts of ordered and disordered material. This degree of crystallinity decreases in the order cotton > wood pulp > mercerized cellulose > regenerated cellulose.

The chain length of cellulose in solutions can be determined by various techniques such as osmotic pressure, light scattering, ultracentrifugation and viscosity measurements. Not unexpectedly, celluloses of different origins and treatment histories show considerable variation. Values for the degree of polymerization (DP) can range from less than 1000 for regenerated cellulose to 7–10 000 for wood pulp and as high as 15 000 for cotton.

The macrostructure of cellulosic materials as defined by light microscopy is specific for the plant material concerned, e.g., cotton, softwoods, hardwoods, grasses. However, the native celluloses regardless of origin are made up of fibrils of only a few tenths of a micrometer in diameter, near the limit of resolution of the light microscope. Electron microscopy has shown the fibrils to be composed of still finer microfibrils. These range from 100 to 300 Å in diameter, and thus may be presumed to contain several hundred cellulose chains.

Closely associated with cellulose in the plant cell walls are other carbohydrate polymers known collectively as hemicelluloses. They differ from cellulose in that they consist for the most part of sugars other than glucose, both pentoses and hexoses, are usually branched and have much lower degrees of polymerization, from less than 100 to about 200 sugar units. In softwoods hemicellulose mannan is the principal constituent, in hardwoods xylan, while in grasses arabinan occurs as well.

2. Properties

The most remarkable property of cellulose is its insolubility in water despite the fact that it is a polymer of glucose. Hydrogen bonding between cellulose chains is so intense that water cannot disrupt it by complexing with the hydroxyl groups. Other reagents, however, such as concentrated strong acids and bases, concentrated salt solutions and metal complexes do exert a solvent effect, especially on fractions of low molecular weight. Hemicelluloses are also water insoluble, but can be dissolved in alkali, permitting their separation from the total carbohydrate fraction, called holocellulose, leaving essentially pure or α-cellulose behind.

However, the presence of three hydroxyl groups on each anhydroglucose residue in the cellulose chain does make cellulose very hygroscopic, and it readily adsorbs and desorbs water with changes in relative humidity. Many applications of cellulosic materials

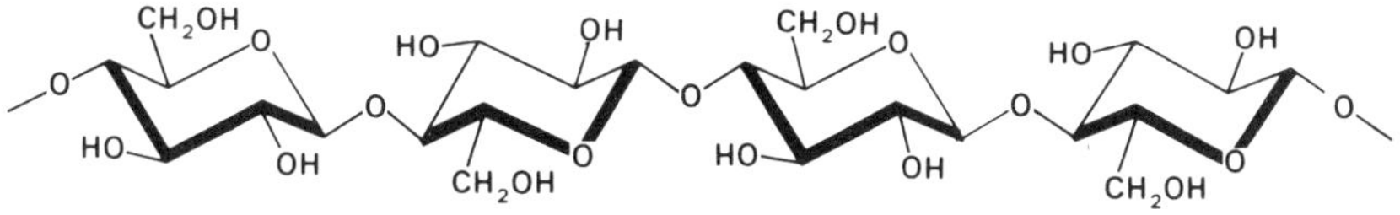

Figure 1
Portion of cellulose chain. Ring hydrogens have been omitted for clarity

are influenced by this property, both beneficially and detrimentally.

The mechanical properties of cellulose, cellulosic materials and cellulose derivatives are high in relation to their weight. These desirable strength properties coupled with the abundance of cellulosic materials have led to their wide industrial application.

Chemical properties of cellulose are influenced by its structure and functional groups. Reagents which interact with the hydroxyl groups must first penetrate the fibrils, so accessibility or availability of the hydroxyl groups is an important factor in all cellulose reactions. Cellulose derivatives may be prepared by esterification, etherification, xanthation and grafting.

Cellulose esters of strong inorganic acids such as nitric, sulfuric and phosphoric may be prepared by direct reaction of the acids with cellulose to give the respective esters. Other strong acids such as perchloric acid and the halogen acids do not esterify cellulose. Formic acid is the only organic acid which can directly esterify cellulose to any appreciable extent. However, the anhydrides and acid chlorides of other organic acids can yield high degrees of substitution in the presence of suitable acidic catalysts (with anhydrides) and bases (with acid chlorides).

Ethers of cellulose are formed in the presence of alkali with organic halides, sulfates and alkene oxides. Some important ethers are methyl cellulose, ethyl cellulose, carboxymethylcellulose, hydroxyethylcellulose and cyanoethyl cellulose.

Although cellulose xanthate is the ester of dithiocarbonic acid it is prepared as in etherification by the reaction of alkali cellulose with carbon disulfide. The resultant derivative is water soluble and can be regenerated to cellulose by coagulation in an acid medium.

Graft copolymers in which chains of monomer units of a different kind are attached to a backbone of cellulose can be prepared by reaction with ethylene oxide or β-propiolactone, but more generally by allowing vinyl monomers to polymerize on active free radical or ionic sites on the cellulose. The initiating sites may be formed by peroxides, irradiation, oxidation or mechanical shear. Properties of cellulose graft polymers can be controlled to enhance or decrease hygroscopicity, solubility and thermoplasticity.

Some reactions of cellulose involve degradation which can be desirable in controlling the viscosity and solubility of cellulose derivatives or producing sugars, or can be undesirable when the mechanical properties are adversely affected. Hydrolysis involves the scission of the acetal links between glucose units by acid with reduction of degree of polymerization (DP), and, when carried to completion, production of glucose. An initial rapid degradation to a levelling-off DP corresponds to hydrolysis of the amorphous cellulose. The hydrolysis rate then decreases to a new lower value as the crystalline regions are degraded.

Oxidative degradation of cellulose, especially by oxygen in the presence of alkali, involves a free radical mechanism which is catalyzed by transition metals (cobalt, iron, manganese). Solution viscosity is decreased indicating reduction of DP, and the introduction of aldehyde and carboxyl groups is significant. The exposure of cellulose to ionizing radiation also generates free radicals which can become involved in further oxidative degradation.

Since cellulose is a sugar polymer it serves as food for many microorganisms, and the degradation of cellulosic materials by the enzymes secreted by these organisms is of great importance. Enzymatic degradation is similar to hydrolytic degradation by acids, but is more localized because of the much larger size of the enzymes. This biodegradability of cellulosic materials is an intrinsic part of the carbon cycle, causes billions of dollars of damage to structures and fabrics, and may yet serve to convert abundant cellulose to chemicals and fuels.

3. Sources

The most abundant source of cellulose is from woody plants in which 40–50% of the plant material consists of cellulose. So plentiful a material should be relatively inexpensive, and in fact raw wood of low quality is cheap. The price of cellulose, however, depends on its purity and the extent to which it has been separated from the lignin and hemicelluloses with which it is naturally associated. Highly purified cellulose for use in the preparation of cellulose derivatives is known as chemical pulp or dissolving pulp, and results from extensive bleaching and extraction beyond that customary for ordinary paper pulps.

Seed hairs, of which the most important is cotton also consist of cellulose. Of all the commercial raw materials for cellulose, the cotton fiber contains the lowest percentage of noncellulosic material, 3–15% making purification simpler than for other cellulosic materials. The longer lint fibers are used in textile applications, while the shorter fibers removed from the seed are known as linters which are used for the production of chemical cellulose.

Bast fibers, the long fibers of the inner bark of various plants, include such cellulosic fibers as flax hemp, jute and ramie. Many cellulosic fibers are obtained from the leaves of plants. Since cellulose is the major constituent of the cell walls of all plants grasses, straws and agricultural residues have been extensively used as a source of cellulose.

4. Applications

Most of the applications of cellulosic materials are in the form of cotton, paper and wood. Regenerated cellulose products and cellulose derivatives which are based on chemical cellulose have a wide range of uses.

Regenerated cellulose refers to products which may have undergone chemical modification during processing to bring about changes in shape or form, bu which consist of cellulose in the final product. Parchment paper prepared by the gelatinization of cellulose with concentrated sulfuric acid and subsequent washing as well as vulcanized fiber in which strong zinc chloride solutions are used to swell the cellulose are examples. Viscose rayon, fibers prepared by spinning cellulose xanthate into an acid bath, and cellophane, the analogous film product, are also regenerated cellulose.

In describing the applications of cellulose derivatives it is more convenient to categorize them by the use rather than the specific cellulose derivative involved. Important fiber applications in addition to viscose rayon use cellulose acetate and cellulose triacetate. Earlier cellulosic rayons such as cellulose nitrate and cuprammonium rayon, a regenerated cellulose, are no longer produced.

Protective coatings in the form of lacquers, textile and paper coatings and varnishes use cellulose nitrate, cellulose acetate, the acetate propionate, acetate butyrate and ethyl-cellulose either as the main film-forming material or in combination with synthetic resins. Films and foils, principally of cellulose acetate, are used in photographic applications and packaging.

The first synthetic plastic, celluloid, was made from cellulose nitrate. Organic cellulose esters, chiefly cellulose acetate, but also mixed esters with higher acids for water resistance, are important in molding applications. Cellulose ethers find application as gums, viscosity modifiers and thickening agents. Nitrocellulose is still used as an explosive.

Newer applications of cellulose derivatives are based on graft polymers with hydrophilic monomers which show greatly enhanced water-holding capacity, and ionic substituents which are useful in ion-exchange applications.

See also: Cellulose: Chemistry and Technology

Bibliography

Atalla R H (ed.) 1987 *The Structures of Cellulose*, ACS Symposium Series No. 340. American Chemical Society, Washington, DC

Bikales N M, Segal L (eds.) 1971 *Cellulose and Cellulose Derivatives*, Vols. 4, 5. Wiley-Interscience, New York

Immergut E H 1963 *Cellulose*. In: Browning B L (ed.) *The Chemistry of Wood*. Interscience, New York, Chap. 4

Ott E, Spurlin H M, Graffin M W (eds.) 1954 *Cellulose and Cellulose Derivatives*, Vols. I–III. Wiley-Interscience, New York

I. S. Goldstein
[North Carolina State University, Raleigh, North Carolina, USA]

Chemical Composition

Wood is an extremely heterogeneous material, and it is not surprising that its chemical composition, anatomy and physical properties vary within wide limits. The wood of the conifers (softwood) differs in its chemistry from that of the arboreal angiosperms (hardwood), and among the latter there are differences between those native to the tropics and to the temperate zones. Chemical variations in cell-wall composition between species are rare but not unknown. Within each tree, the roots, stem and branches have a different chemical composition, and in the stemwood there are variations with the height above the ground and with the distance from the pith. Juvenile wood differs from mature wood, sapwood from heartwood, earlywood from latewood, and normal wood from reaction wood. The chemical composition of the wall is not the same for the tracheids and the ray cells in the softwoods, and this also applies to the fibers, vessels and ray cells in the hardwoods. The middle lamella (intercellular region), the primary wall and the secondary wall do not have the same composition. Within the secondary wall there are differences between the S_1, S_2 and S_3 layers. Despite all these variations it is, however, still possible to consider the overall chemical composition of wood (see *Macroscopic Anatomy*; *Ultrastructure*).

Wood is a composite material consisting of three major polymers—namely cellulose, hemicelluloses and lignin—which serve as skeletal, matrix and encrusting substances, respectively. The so-called extractives (extraneous components) are usually low-molecular-weight, extracellular compounds. Inorganic constituents (ash) generally amount to no more than 0.1–0.5% of the wood. The general chemical composition of normal and reaction woods of a typical softwood and a typical hardwood species is shown in Table 1.

1. Cellulose

Cellulose is the most abundant organic chemical on earth. It is estimated that over 100 billion tonnes of cellulose are produced every year in nature, where it is found in all land plants. Wherever it occurs, cellulose is present in a fibrillar form. Normal softwoods and hardwoods contain on the average $42 \pm 2\%$ cellulose. Cellulose is a 1,4-linked glucan, consisting of β-D-glucopyranose residues in the chair conformation, linked together by glycosidic bonds between C-1 in one unit and C-4 in the next to form long, linear chains (**1**). Every glucose residue is turned over 180° with respect to its neighbors.

The cellulose present in the secondary wall of wood cells has a degree of polymerization (DP) of 10 000. There are indications that this cellulose is monodisperse, that is, it consists of chain molecules of the same size. Primary cell walls contain 25% of a polydisperse cellulose with a DP of 3000. Well-known solvents for cellulose are the aqueous cupriethylenediamine hydroxide (CED) and triethylenediamine cadmium hydroxide (cadoxen). Dimethyl sulfoxide containing formaldehyde is a new and promising

Table 1
Chemical composition of normal and reaction woods in a typical softwood and hardwood species

Component	Eastern white pine (*Pinus strobus*)		Paper (white) birch (*Betula papyrifera*)	
	Normal wood %	Compression wood %	Normal wood %	Tension wood %
Cellulose	42	30	42	50
1,3-Glucan		3		
Glucomannan			3	2
Acetylgalactoglucomannan	18	9		
1,4-Galactan		10		
1,4; 1,6-Galactan				8
Acetyl-4-*O*-methylglucuronoxylan			30	22
Arabino-4-*O*-methylglucuronoxylan	11	9		
Lignin	28	38	24	17
Other hemicelluloses, pectin, ash	1	1	1	1

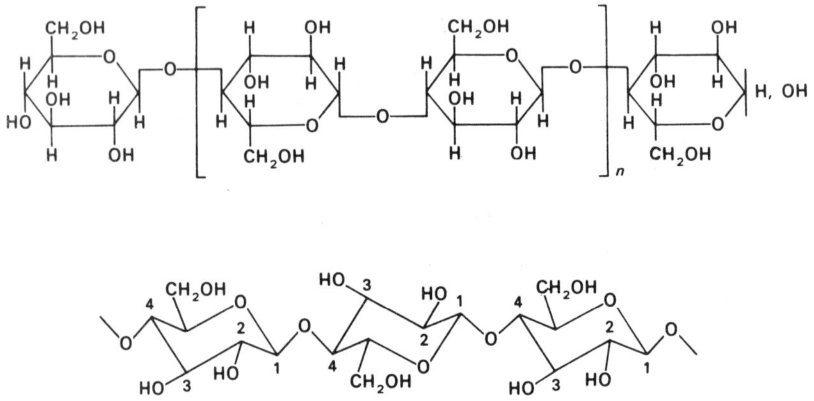

(1)

solvent. Cellulose swells but does not dissolve in aqueous sodium hydroxide. Each glucose residue contains on average three hydroxyl groups, which can be esterified or etherified to yield cellulose derivatives, many of them of commercial importance. Viscose rayon is cellulose that has been brought into solution as a xanthate (dithiocarbonate) derivative and regenerated in a fibrous form.

All cellulose fibers are partly crystalline—to the extent of 50–60% in wood. Cellulose can occur in two major and two minor crystallographic forms. In cellulose I, the modification in which cellulose is found throughout nature, the chains are all oriented in the same direction (parallel). They are bound together by strong hydrogen bonds in one of the two transverse directions, but by only weak forces in the second. In addition, there are hydrogen bonds between adjacent glucose residues within each chain. Crystalline, native cellulose has a rigid structure, which is both a chain lattice and a layer lattice, impermeable to water. Cellulose II is formed when the lattice of cellulose I is destroyed either by swelling with strong alkali or by dissolution. It is the thermodynamically more stable of the two modifications. In cellulose II, the chains are antiparallel, and there are hydrogen bonds between the chains not only within each layer but also between the layers.

Electron microscopy of untreated cellulose fibers reveals the presence of microfibrils of indefinite length and with a width of 1–2 nm in primary and 10–20 nm in secondary walls. The parallel arrangement of the chains in cellulose I makes it impossible for these chains to be folded in native cellulose.

The cellulose microfibrils present in the S_2 layer of tracheids and fibers of normal softwoods and hardwoods are oriented at an angle of 10–30° to the longitudinal axis. In the tracheids of juvenile and compression woods, the microfibril angle is 30–50°, a fact that is largely responsible for the high longitudinal shrinkage of these woods. In the gelatinous layer of tension wood fibers, which consists entirely of cellulose, the microfibrils are oriented almost parallel with the fiber axis.

2. *Hemicelluloses*

The hemicelluloses are linear polysaccharides of moderate size that are invariably associated with cellulose and lignin in plant cell walls. Pectin is a large, acidic and branched polysaccharide with 1,4-linked α-D-galacturonic acid residues as a major constituent. In wood it is present only in the primary wall. Some of the hemicelluloses can be isolated directly from wood by extraction with aqueous alkali, while others require prior removal of the lignin. Hemicelluloses do not occur as microfibrils in wood, and are more readily degraded by acids than is cellulose.

The predominant hemicellulose in hardwood is an acidic xylan. The xylan content can reach 35%, but in most species it is $25 \pm 5\%$. Hardwood xylans consist of a main chain of 1,4-linked β-D-xylopyranose residues, some of which carry a single, terminal 4-*O*-methyl-α-D-glucuronic acid unit attached to C-2. These side chains are distributed at random along the xylan backbone. Most hardwood species have on the average one side chain per ten xylose residues. In addition, there are on the average seven acetyl groups per ten xylose units attached to either C-2 or C-3 (**2**). Recently, it has been found that an L-rhamnose and a D-galacturonic acid residue occur near the end of the main chain. Hardwood xylans have a DP of 200 and are amorphous in their native state, although they can be induced to crystallize after the acetyl and some of the acid side chains have been removed. There is strong evidence that the xylan chains are oriented parallel to the chains of cellulose.

In addition to xylan, hardwoods contain less than 5% of a glucomannan, composed of 1,4-linked β-D-glucopyranose and β-D-mannopyranose residues. The glucose to mannose ratio is generally 1:2. The size of this polysaccharide is unknown.

The major hemicellulose (20%) in almost all softwoods is a family of galactoglucomannans, consisting of a main chain of 1,4-linked β-D-glucopyranose and β-D-mannopyranose residues, some of which carry a single residue of α-D-galactopyranose attached to C-6. The ratio of glucose to mannose is always about 1:3, but the ratio of galactose to glucose varies from 0.1:1 to 1:1. The former polysaccharide, which is only soluble in strong alkali, is often referred to as "glucomannan." Some of the hydroxyl groups in the main chain are esterified with acetic acid (**3**).

Softwoods also contain about 10% of an acidic xylan, which has the same main chain of 1,4-linked β-D-xylopyranose residues as that in the hardwoods, and the same type of side chains of 1,2-linked 4-*O*-methyl-α-D-glucuronic acid residues. There are, however, twice as many acid side chains as in the hardwood xylans—that is, one per five xylose units. Instead of acetyl groups, the softwood xylan contains one α-L-arabinofuranose residue per seven xylose units, attached to C-3 of the xylan backbone (**4**). The hemicelluloses in softwoods are isolated by extraction of delignified wood with aqueous alkali. Xylan and galactoglucomannan are separated from each other by forming an insoluble complex of the latter with barium hydroxide. The size of the native softwood hemicelluloses is not known, but their DP is at least 150.

Heartwood of larch trees contains 5–30% of an arabinogalactan, which can be regarded as both a hemicellulose and an extractive. Unlike most wood polysaccharides, this water-soluble polymer is not present in the cell wall, but occurs in the lumen of the tracheids and ray cells, from which it can be extracted with water. Larch arabinogalactan has a highly branched structure with a framework of 1,3-linked β-D-

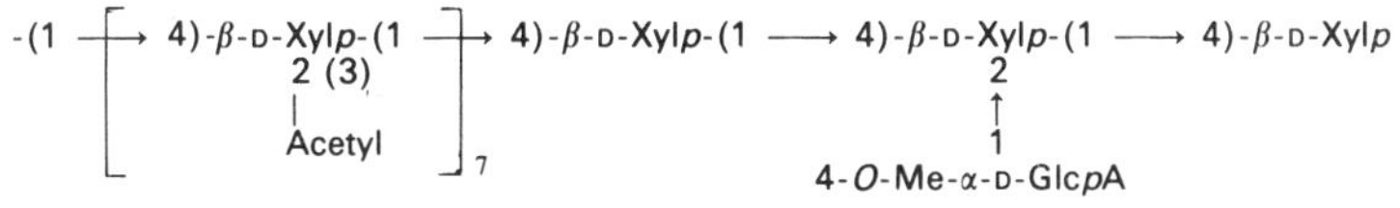

(**2**)

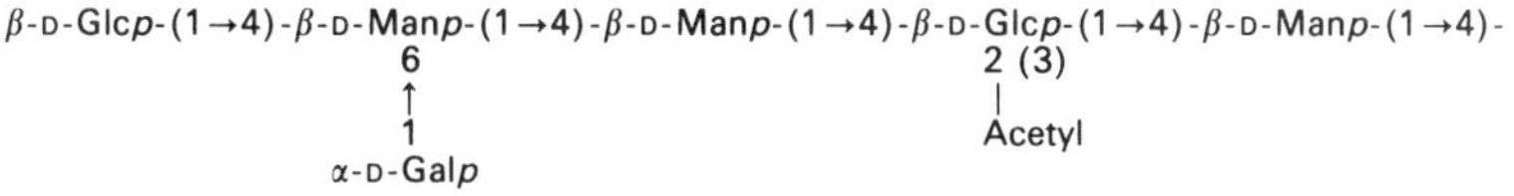

(**3**)

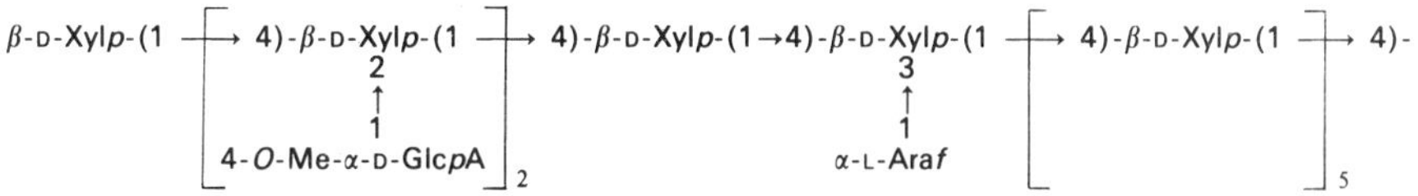

(**4**)

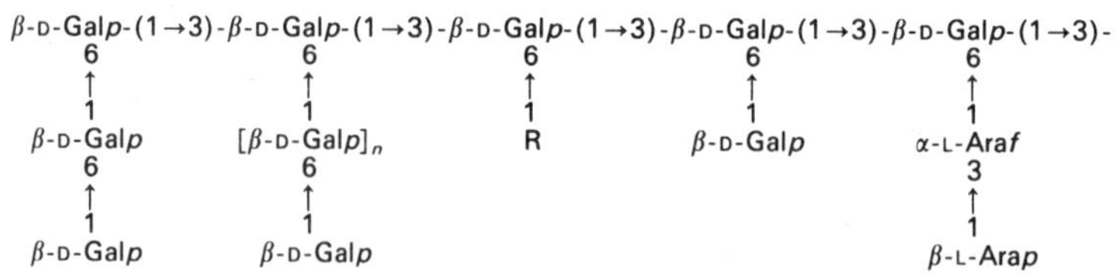

R = α-L-arabinofuranose or β-D-glucuronic acid

(5)

galactopyranose residues, all of which are substituted at C-6 with different types of side chains. Some of these consist of single residues of β-D-galactopyranose, β-D-glucuronic acid, or α-L-arabinofuranose, while others are 1,6-linked chains of β-D-galactopyranose residues of variable size. There are also side chains consisting of one β-L-arabinopyranose and one α-L-arabinofuranose residue (**5**). The major fraction of larch arabinogalactan has a molecular weight of 70 000, while a lesser part has a molecular weight of 12 000. Because of its highly branched nature, arabinogalactan forms aqueous solutions with a very low viscosity. It is one of the few wood hemicelluloses that has found industrial use in its own right.

3. *Lignin*

Lignin is a three-dimensional polymer composed of phenylpropane units that encrusts the intercellular space and the cell wall after the polysaccharides have been formed. Its function in wood is to cement the cells together and to impart strength to their wall. Most softwoods contain 30±4% lignin, and hardwoods of the temperate zone 25±3%. Lignin is formed in wood by a dehydrogenation–polymerization of *p*-coumaryl (**6**a), coniferyl (**6**b), and sinapyl alcohols (**6**c), all in the *trans* configuration. Lignin is best isolated by extraction with organic solvents of finely ground wood, preferably after pretreatment with polysaccharide-degrading enzymes. Almost all softwoods contain a guiacyl lignin with only one methoxyl group, while hardwoods have a guiacylsyringyl lignin with one or two methoxyl groups.

The following functional groups are present in 100 phenylpropane units of a spruce lignin: methoxyl (90 units), phenolic hydroxyl (20 units), phenolic ether (80 units), aliphatic hydroxyl (90 units), benzyl alcohol-ether (40 units) and carbonyl (20 units). About two thirds of the interunit linkages in lignin are carbon–oxygen and the remaining one third are carbon–carbon bonds. The major interunit linkage types are the phenylcoumaran (**7**; 10%), the pinoresinol (**8**; 10%), the arylglycerol β-aryl ether (**9**; 50%), and the diphenyl (**10**; 10%). There are also many other kinds of interunit bonds. The lignin present in hardwoods is similar to that in softwoods, but there are greater variations between different species. The major difference is the presence of two methoxyl groups in

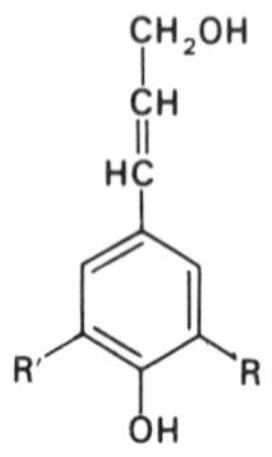

(a) R = H, R′ = H

(b) R = OCH_3, R′ = H

(c) R = OCH_3, R′ = OCH_3

(6)

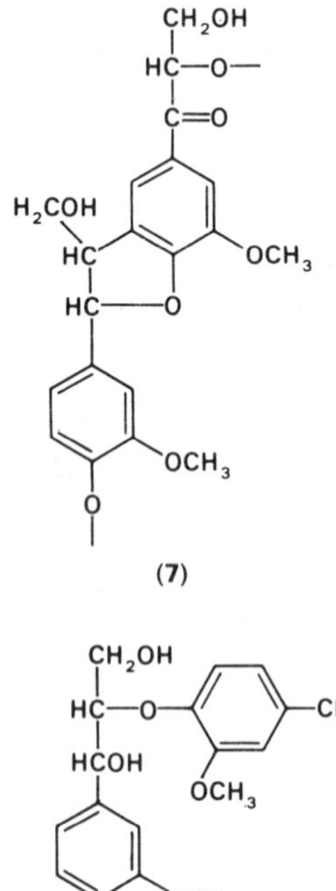

(7)

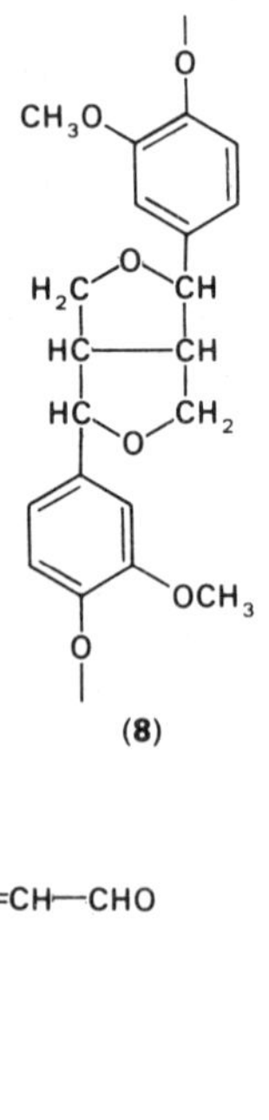

(8)

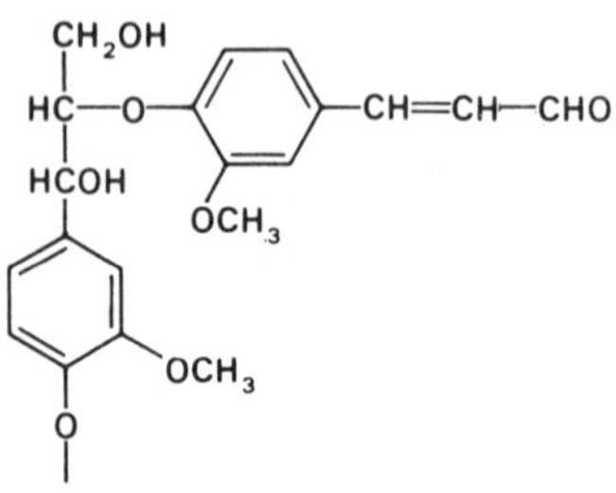

(9)

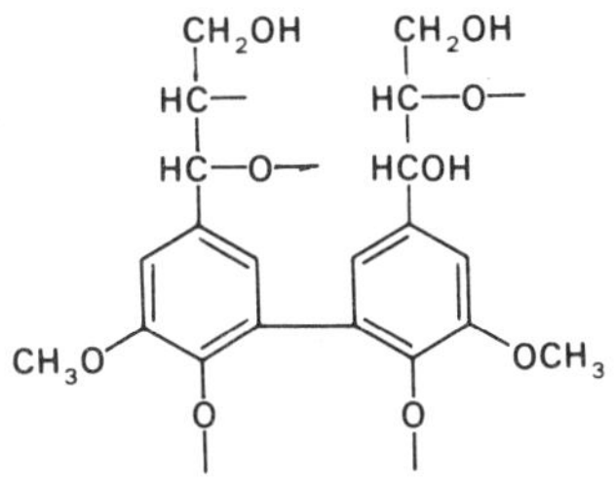

(10)

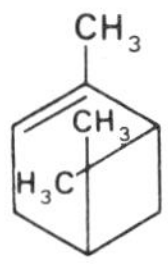

(11)

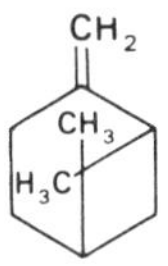

(12)

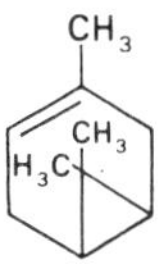

(13)

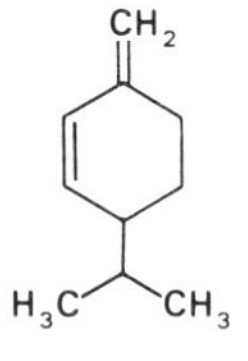

(14)

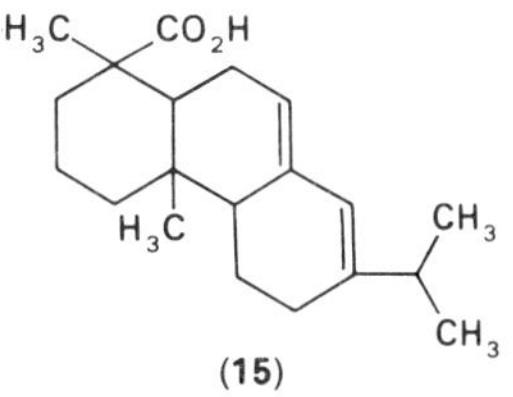

(15)

some of the phenylpropane units. Compression wood and wood of the monocotyledons have a lignin with a high content of unmethylated phenylpropane units. Tension wood has a normal hardwood lignin.

Lignin is the only cell wall component with an ultraviolet absorption. A guaiacyl lignin absorbs much more strongly at 280 nm than a guaiacylsyringyl lignin. The molecular weight of lignin is indeterminate. Lignosulfonates obtained in the sulfite pulping process can have a molecular weight of 10^6 or higher. While cellulose imparts tensile strength to wood, lignin is partly responsible for its compressive strength. It also offers a certain protection against microbial attack. Although some valuable chemicals are produced from lignin, such as lignosulfonates, vanillin and dimethyl sulfoxide, by far the largest part of the lignin removed from wood in the pulping processes is burned.

4. *Extractives*

The vast majority of the extractives in wood are of a low molecular weight and are located outside the cell wall. A few of them are reserve food materials, and others are protective agents, but most of them seem to serve no specific function. Many of them are located only in the heartwood. The large number of different extractives in wood can be classified as terpenes, resin acids, tannins, polyphenols, lignans, tropolones, fats, waxes and carbohydrates. The nature of its extractives is frequently typical of a tree species. Pines, for example, can be classified on the basis of either their monoterpenes or their heartwood polyphenols.

The monoterpenes, which are especially abundant in the pines, consist of two isoprene units, α- and β-Pinene (**11, 12**), Δ-3-carene (**13**), and β-phellandrene (**14**) are four of the most common. Resin acids, e.g., abietic acid (**15**), are diterpenes and consist of four isoprene units. Oleoresin, which exudes from a conifer stem wound, is composed of one part of terpenes and two to three parts of resin acids. It is formed in the resin canals and in the ray cells. Tall oil, an important byproduct in the kraft pulping industry, is a mixture of resin and fatty acids. Turpentine consists of various terpenes. If paraquat, a herbicide, is injected into a pine stem, there is a spectacular increase in the formation of oleoresin.

Tannins occur in both wood and bark. Gallotannins are esters of glucose with gallic acid. Commercial tannins, obtained from trees such as acacia or quebracho are of the condensed, polymeric type, and are based on a monomer such as catechin (**16**). Polyphenols comprise a variety of phenolic compounds. Pinosylvin and its monomethyl ether (**17**) are found in the heartwood of all pines, and taxifolin, a flavanone, (**18**) in Douglas fir and larches. Lignans are present, generally in small amounts, in both softwoods and hardwoods. They consist of two phenylpropane units. One lignan, conidendrin (**19**), was at one time recovered industrially. Cupressales trees, such as western red cedar, contain small amounts of tropolones in their heartwood—e.g., γ-thujaplicin (**20**). These seven-membered aromatic compounds are extremely toxic to microorganisms, and are responsible for the high decay resistance of cedar wood.

5. *Distribution of Polysaccharides*

As is evident from Table 1, compression wood contains considerably less cellulose and galactoglucomannan, while tension wood has a higher cellulose content than normal wood. Both reaction woods have a galactan not found in normal wood. In addition, there is a 1,3-linked glucan, called laricinan, in compression wood. Juvenile softwood has a lower cellulose and galactoglucomannan and a higher xylan content than mature wood. There is more xylan and less galactoglucomannan in earlywood than in latewood. Unlike the tracheids, the ray cells in softwoods contain twice as much xylan as galactoglucomannan, and in hardwoods these cells have a higher xylan content than the fibers and vessels.

Primary cell walls are characterized not only by the occurrence of a unique type of cellulose but also by the presence of polysaccharides not present in the secondary wall, such as arabinogalactan, galactoxyloglucan and pectin. It is believed that in the secondary wall the cellulose microfibrils are coated with hemi-

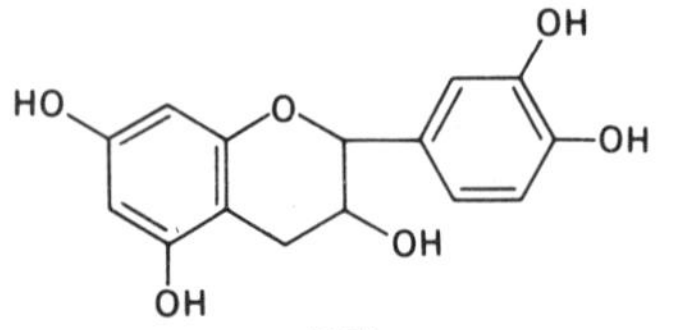

(16)

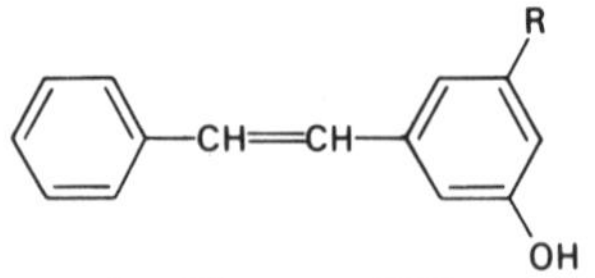

R = OH or OCH_3

(17)

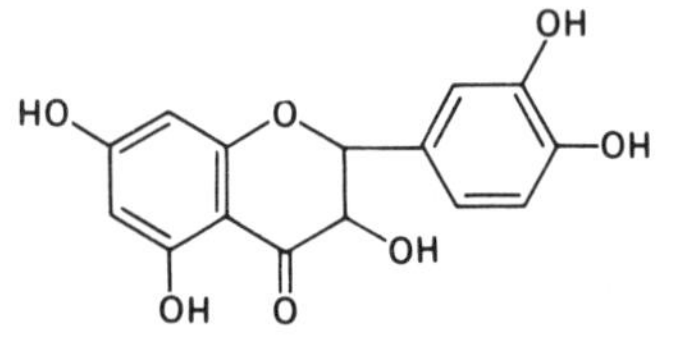

(18)

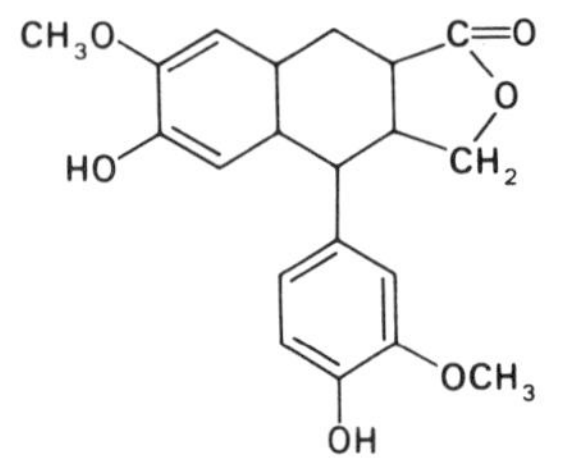

(19)

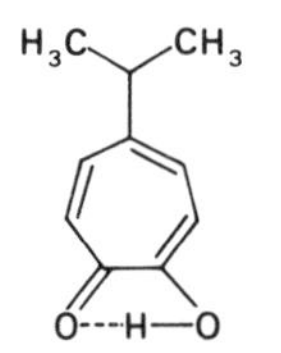

(20)

celluloses and surrounded by a matrix of hemicelluloses and lignin. There are indications that much of the galactoglucomannan occurs in the S_2 layer, and that S_3 has the highest concentration of xylan. Other results suggest that the outer portion of S_1 and S_2 have a high content of hemicelluloses. The galactan in compression wood is located in S_1 and the outer part of S_2. The pit membrane consists largely of pectin. The torus in bordered pits is composed either of pectin, or, in the Pinaceae, of pectin and cellulose. It is not lignified.

6. *Distribution of Lignin*

Lignin can be examined in wood under a light microscope after staining with reagents such as phloroglucinol or safranin-fast green. After treatment with permanganate, it can also be observed in an electron microscope. In recent years, quantitative data have been obtained with the aid of ultraviolet microscopy. The most recent technique entails application of energy-dispersive x-ray analysis.

The low lignin content of tension wood is entirely due to the presence of the gelatinous layer. The amount of lignin present per fiber is the same as in normal hardwood. Juvenile wood and earlywood have a higher lignin content than mature wood and latewood, respectively. The ray cells in both softwoods and hardwoods contain considerably more lignin than the tracheids or fibers. Hardwood vessels are also highly lignified. Of the total lignin present in normal softwoods and hardwoods, 20–25% occurs in the intercellular region and primary wall, while 75–80% is present in the cell wall. The concentration of lignin varies from 50% to 100% in the middle lamella–primary wall, and is 20–25% in the secondary wall. The true middle lamella at the cell corners in conifer latewood consists of lignin. In some conifer species, the S_1, S_2 and S_3 layers are lignified to the same extent, but in others the S_3 layer seems to have a higher lignin content. In opposite wood (wood formed opposite to compression wood), this layer is highly lignified. The lignin present in the secondary wall of conifer tracheids contains twice as many phenolic hydroxyl groups as that in the middle lamella–primary wall. The tracheids in compression wood have a unique distribution of lignin. Only 10% of the total lignin is located in the intercellular region. The lignin content of S_1 and the inner part of S_2 is 30–40%. The concentration of lignin in the outer portion of S_2, the so-called S_2(L) layer, is as high as 70%, or the same as in the middle lamella.

The secondary wall of fibers and ray cells in hardwoods contains a syringyl lignin, whereas the vessels contain a guaiacyl lignin. The lignin in the vessel middle lamella is also of the guaiacyl type, while the middle lamella of the fibers has a guaiacyl-syringyl lignin. Obviously, each individual wood cell controls the lignification of its wall.

See also: Cellulose: Chemistry and Technology; Constituents of Wood: Physical Nature and Structural Function

Bibliography

Adler E 1977 Lignin chemistry—past, present and future. *Wood Sci. Technol.* 11: 169–218

Atalla R H (ed.) 1987 *The Structures of Cellulose*, ACS Symposium Series No. 340. American Chemical Society, Washington, DC

Fengel D, Wegener G 1984 *Wood: Chemistry, Ultrastructure, Reactions.* de Gruyter, Berlin

Higuchi T (ed.) 1985 *Biosynthesis and Biodegradation of Wood Components.* Academic Press, Orlando

Hillis W E 1987 *Heartwood and Tree Exudates.* Springer, Heidelberg

Loewus F A, Runeckles V C (eds.) 1977 *The Structure, Biosynthesis and Degradation of wood.* Plenum, New York

Nevell T P, Zeronian S H (eds.) 1985 *Cellulose Chemistry and Its Applications.* Ellis Horwood, Chichester

Rowell R M (ed.) 1984 *The Chemistry of Solid Wood*, ACS Advances in Chemistry Series 207. American Chemical Society, Washington, DC

Sarkanen K V, Ludwig C H (eds.) 1971 *Lignins: Occurrence, Formation, Structure and Reactions.* Wiley-Interscience, New York
Sjöström E 1981 *Wood Chemistry: Fundamentals and Applications.* Academic Press, New York
Timell T E 1986 *Compression Wood in Gymnosperms.* Springer, Heidelberg
Young R A, Rowell R M (eds.) 1986 *Cellulose: Structure, Modification and Hydrolysis.* Wiley, New York

T. E. Timell
[State University of New York, Syracuse, New York, USA]

Chemically Modified Wood

Chemically modified wood is wood which has been treated by chemical processes to change the biological, chemical, mechanical or physical properties of the natural wood. This article primarily discusses modifications involving chemical reactions with the wood components and their effect on wood properties, as well as selected nonreactive treatments.

1. Reactive Modification of Wood

By far the most abundant functional group in wood is the hydroxyl group, and it follows that reactive chemical modification of wood depends principally on reagents which react readily with these groups. The distribution, accessibility and reactivity of these hydroxyl groups vary throughout the wood depending on the component (cellulose, hemicellulose or lignin), degree of order (crystalline or amorphous regions) and type of hydroxyl group (primary, secondary or phenolic). These variations influence the distribution of the reaction products in the modified wood.

Virtually every type of reagent capable of reacting with hydroxyl groups has been applied to the chemical modification of wood. These include acid anhydrides, inorganic acid esters, acid chlorides, chloroethers, aldehydes, lactones, reactive vinyl compounds, epoxides and isocyanates. Both monofunctional and difunctional reagents have been studied.

Since the modified wood must still possess the basic desirable properties of untreated wood, the reagents must be capable of reaction with wood under relatively mild conditions. Extremes of acid or alkaline conditions or temperature which might degrade the wood must be avoided. Penetration is favored by small polar molecules. With reagents which form a by-product in the reaction with hydroxyl groups an additional processing step for its removal may be necessary. Among the most successful modifying agents are acetic anhydride, acrylonitrile, epichlorohydrin, formaldehyde, methyl isocyanate, β-propiolactone and propylene oxide.

2. Properties of Chemically Modified Wood

The changes in wood properties which are most apparent after chemical modification are an increase in resistance to biological deterioration and an improvement in dimensional stability towards variations in humidity. Under the most favorable modification conditions, these improvements are achieved with no sacrifice of mechanical properties or with even a slight increase in strength. However, there are unavoidable increases in weight and cost.

Chemical modification of the hydroxyl groups in wood affects wood properties through two mechanisms. One relates to the increased steric influence of the larger substituent compared with that of the small hydroxyl hydrogen. The other involves the reduced hydrogen-bonding capability of the modified wood and its resultant lower hygroscopicity. Both mechanisms contribute to the observed improvements in properties.

Biological deterioration of wood, whether by fungi or by the symbiotic protozoa in termites, depends on enzymatic hydrolysis of the polymeric wood carbohydrates to simple sugars. Enzymatic action is extremely stereospecific, so the introduction of substituents on the hydroxyl groups prevents hydrolysis. The substitution of one hydroxyl per anhydroglucose unit in cellulose provides biological resistance. Decay fungi also need a high moisture content in the wood to thrive. Reduction of wood hygroscopicity through chemical modification also retards biological deterioration.

In contrast to conventional treatments of wood to impart biological resistance, chemical modification of wood provides treatments which are effective, nontoxic and nonleachable. As such they are certainly more environmentally acceptable. However, the high degree of modification required for effectiveness results in a rather expensive treatment.

Dimensional stabilization of wood by chemical modification takes advantage of the larger size of the substituent groups through a mechanism known as bulking. The larger volume of these groups occupies space in the swollen cell wall and prevents the cell wall from shrinking in response to loss of moisture. In effect, the wood has been preswollen and is used in this condition. Furthermore, the substituent groups are less hygroscopic, so less water is adsorbed at any specific humidity level. The stabilizing groups are permanently attached to the wood structure.

Chemical modification of wood can also provide stabilization by mechanical restraint at the molecular level by the formation of cross-links between cellulose chains or microfibrils which prevent the spreading and swelling of the cellulose structure by water. The most effective cross-linking agent is formaldehyde which can provide a high degree of dimensional stability (90%) at a weight gain of only 7% compared with bulking agents which require 20–30% weight gain for

70–75% stabilization. However, cross-linking with formaldehyde causes serious embrittlement of the wood as well as loss of abrasion resistance.

Chemical modification without cross-linking provides dimensional stability with no loss in toughness. In contrast, polymer impregnation often causes severe embrittlement because the disordered regions of the cellulose chains are embedded in a rigid matrix which restricts chain motion and permits failure at low levels of absorbed energy. In chemical modification the cellulose chains are still free to absorb energy by vibration without chain failure occurring at low impact stresses.

Since the strength properties of wood are higher at low moisture contents the lower equilibrium moisture content of the less hygroscopic chemically modified wood at any given humidity results in higher strength values. For example, chemically modified pine wood does not fail first under compressive stress in the upper part of the beam during bending strength determinations like untreated pine, but rather in tension at the lower surface, thus allowing higher values of fiber stress at the proportional limit and of rupture modulus. The lower moisture content of the modified wood provides sufficient increase in crushing strength to bring about this change in mode of failure.

In a specific example of chemical modification, wood acetylated with uncatalyzed acetic anhydride to a weight gain of about 20% shows reductions of swelling in high humidity of 70–80%. Acetylated pine becomes resistant to wood-destroying fungi and termites at a weight gain of about 18%. Acetylated wood distorts and checks less than untreated wood during weathering. The impact strength of acetylated wood is not decreased by the treatment, whereas wet compressive strength is doubled. Hardness and work to proportional limit are materially increased. The elastic modulus remains unchanged. Density is increased and equilibrium moisture content is reduced.

3. *Nonreactive Chemical Modification*

Chemicals which do not react with the wood components can also be used to bring about changes in wood properties. The bulking mechanism for dimensional stabilization is operative for any chemical which penetrates the cell wall. Water-soluble inorganic salts and organic compounds, although effective bulking agents, are hygroscopic and cause finishing problems and reduction in strength. Polyethylene glycols are very effective water-soluble bulking agents, and at higher molecular weights (~ 1000) are somewhat less hygroscopic than the small molecules. They can be applied to green wood by diffusion and used to stabilize gunstocks, carvings and turnings.

The use of bulking agents which are solvent-soluble but water-insoluble can avoid the hygroscopicity problem entirely. Cellosolve (ethylene glycol monoethyl ether) can be used as an intermediary solvent because it is soluble in water in all proportions; it is also used as a solvent for waxes, oils and resins. Replacement of the water in swollen wood by cellosolve followed by replacement of the cellosolve with molten waxes or oils affords a high degree of dimensional stability. However, the process is very slow.

Plasticization of wood can be accomplished by chemical treatment, most notably by the use of ammonia in gaseous or liquid form. The use of amines, amides and urea permits permanent plasticization by retention of the chemical or temporary plasticization followed by its removal.

Some natural woods are self-lubricating, such as lignum vitae which was long used for marine bearings. More common species of wood may be made self-lubricating by impregnation with waxes or lubricating oils. Internal lubricity improves machinability. Cutting properties of cedar pencil stock are improved by impregnation with paraffin wax or polyethylene glycol.

The resistance of wood to degradation by various chemicals such as acids and alkalis can be enhanced by preventing access of the reagents to the wood fibers by means of barriers. Filling the pores of wood with resistant materials such as waxes, tars, asphalt and sulfur improves chemical resistance. Still higher resistance to acids, even in the presence of organic solvents or at elevated temperature can be obtained by polymerization of thermosetting phenol–formaldehyde resins within the cell wall. This treatment does not resist alkali. Formation within the wood of an acid- and alkali-resistant resin by polymerization of furfuryl alcohol provides broad-spectrum chemical resistance to everything but oxidizing agents.

See also: Preservative-Treated Wood; Wood–Polymer Composites

Bibliography

Goldstein I S, Loos W E 1973 Special treatments. In: Nicholas D D (ed.) 1973 *Wood Deterioration and Its Prevention by Preservative Treatments*, Vol. 1. Syracuse University Press, New York, pp. 341–71

Rowell R M 1983 Chemical Modification of Wood. *For. Prod. Abstr.* 6(12): 363–82

I. S. Goldstein
[North Carolina State University, Raleigh, North Carolina, USA]

Chemicals and Liquid Fuels from Wood

The use of woody plants as a raw material for chemical applications has a long history in such embodiments as naval stores from pine trees, tanning agents from bark, rubber from latex and methanol from wood distillation. However, the chemical industry as we know it today has evolved from the use of cheap and abundant fossil fuels—first coal, and more recently petroleum and natural gas. Now that the cost of the fossil fuels has increased and concerns about their future availability and ultimate depletion are being raised, there has been a resurgence of interest in the use of renewable resources, chiefly wood, as the carbon source for the synthetic organic materials such as fibers, adhesives, plastics, coatings, rubbers, etc., so vital in our economy.

Chemicals are obviously materials, especially when used as synthetic organic polymers. However, in an energy economy based on carbon, chemicals and fuels are interdependent. The same chemical may serve either as a material or as fuel. In this article the use of chemicals as fuels is restricted to liquid fuels. It should be noted at the outset, however, that whereas there is more than enough annual growth of wood to meet our chemical needs, current consumption of liquid fuels far exceeds our capacity to produce them from all forms of biomass.

Any fuel or chemical derivable from fossil fuels can alternatively be derived from wood. There are no inherent barriers to the use of wood for these purposes. All the constraints are relative in comparison to the cost or ease of processing of the various raw materials.

1. Composition of Wood

Fortunately for the prospect of converting wood into chemicals, the general composition of wood is quite uniform, consisting principally of cellulose, hemicelluloses and lignin (see *Chemical Composition*). These cell-wall polymers comprise as much as 95% of the material, and provide a small number of antecedents for most of the chemicals attainable. This greatly simplifies the entire field of chemical conversion. Other components which are more variable in structure or quantity are extractives (the soluble compounds which are often species or genus specific) and bark.

Cellulose comprises about 50% of the wood. It is a long-chain polymer of β-D-glucose linked together by 1,4′-glycosidic bonds. For conversion into chemicals the important features of cellulose are that it is a glucose polymer, and that its highly ordered crystalline structure limits the accessibility of reagents and enzymes.

Hemicelluloses comprise about 25% of the wood. They are structural polysaccharides which consist of pentose and hexose sugars, are usually branched and have much lower degrees of polymerization than cellulose. Hemicelluloses show differences among species. In hardwoods pentosans (xylose polymers) predominate, while in softwoods the most abundant are hexosans (mannose polymers).

Lignin comprises the remaining 25% of the wood. It serves as a cement between the wood fibers, a stiffening agent within the fibers, and a barrier to enzymatic degradation of the cell wall. Lignins are three-dimensioned network polymers of phenylpropane units. Unlike the carbohydrate polymers they are not readily broken down into their precursors because of resistant ether and carbon–carbon bonds. Lignins from hardwoods have higher methoxyl contents than those from softwoods. The aromatic and phenolic character of lignin is important in chemical conversion.

Extractives may comprise up to 15% of the wood, but generally do not exceed 5%. They include a large variety of compounds such as volatile oils, terpenes, fatty acids, hydrocarbons, sterols, tannins, flavonoids, etc. Bark also contains large quantities of extractives. The familiar wood cell-wall polymers are present in lower quantity. Other bark components are suberin (long-chain hydroxy acid esters) and phenolic acids.

2. Historical Chemical Products from Wood

Extractives from pine trees provided the raw material for the naval stores industry, the oldest chemical industry in North America. These and other past or present chemical products are listed in Table 1. Most represent by-products of other processes. For example, the gum naval stores industry has been replaced by the recovery of oleoresins from the kraft pulping process. Turpentine is recovered from digester relief gases, while the fatty acids and resin acids are recovered from pulping liquors in the form of tall oil. About a million tonnes per year are recovered.

Phenolic acids extracted from bark are used as extenders for synthetic resin adhesives. Bark waxes can be used for general wax applications. High-purity chemical cellulose from the pulping of wood is the starting material for rayon, cellophane, cellulose esters for fiber, film and molding applications and cellulose ethers for use as gums.

Table 1
Historical chemical products from wood

Bark products	*Naval stores*
Phenolic acids	Pine oil
Waxes	Rosin
	Turpentine
Cellulose	Tall oil
Rayon	Tall oil fatty acids
Cellophane	Tall oil rosin
Cellulose esters	
Cellulose ethers	*Pyrolysis products*
	Charcoal
Extractives	Methanol
Rubber	Acetone
Tannins	Acetic acid
Lignin products	*Wood sugars*
Alkali lignin	Glucose
Lignosulfonates	Xylose
Dimethyl sulfide	Mannose
Dimethyl sulfoxide	Ethanol
Vanillin	Yeast

Natural rubber latex was for many years the only source of rubber, and natural rubber is still preferred for many applications. Tannins extracted from bark and heartwood of various species was traditionally used in leather processing. Pulping liquors also yield lignosulfonates (dispersants, adhesives, etc.) and alkali lignin (extender, stabilizer, reinforcing agent). Vanillin is produced by oxidation of lignosulfonates, and dimethyl sulfoxide by oxidation of dimethyl sulfide from kraft pulping.

Volatile products of wood pyrolysis such as acetic acid, methanol and acetone are no longer obtained from that source. However, wood hydrolysis to simple sugars is commercially practised in the USSR and Brazil. Acid pulping liquors also contain sugars which can be fermented to ethanol, yielding also yeast.

3. Potential Future Chemical Products from Wood

Those historical uses of chemicals from wood in Table 1 may be extended to related opportunities in by-product and extractives utilization. However, the greatest potential for chemicals from wood lies in the conversion of the cell-wall components. This source of raw material far exceeds the extractive components or the chemical by-products in volume.

Three mechanisms can be identified by which chemicals from wood may replace those from fossil sources. These are direct substitution of natural polymers such as rubber and cellulose derivatives for synthetic polymers, conversion of wood into the same intermediates which are now derived from fossil sources, and production of different but equally useful intermediates which are more readily obtained from wood.

The chemical industry has developed by the use of those chemicals which are readily obtainable from fossil sources. Even though it may be more difficult and expensive to produce the same chemicals from a renewable resource such as wood, there is a considerable advantage in doing so, since the existing interrelated structure and physical plant of the chemical process and polymer industries can remain intact using the same basic intermediates. The only modification would be the raw material source. On the other hand, a whole new industry for chemical and polymer production can develop from intermediates more readily obtainable from wood.

Two approaches to the chemical conversion of wood are possible. The processes useful in converting the cell-wall polymers in their natural mixed state to chemicals are inherently nonselective and are the same as those useful for the conversion of coal. Both gasification and pyrolysis are applicable to all carbonaceous materials. Alternatively, in contrast to the drastic, high-temperature nonselective processes, selective processing of the individual wood components can provide substantial yields of a wide variety of chemicals.

3.1 Nonselective Processes

Gasification of wood is its conversion by thermal reactions in the presence of controlled amounts of oxidizing agents to provide a gaseous phase containing principally carbon monoxide, hydrogen, water, carbon dioxide and methane or other hydrocarbons. The oxidizing agent may be oxygen, water, carbon dioxide or a mixture of these. Since gasification is generally carried out at temperatures of at least 550 °C, where the reaction rates are so fast that heat and mass transfer become controlling, gasifier design is critical.

The gasification of wood compares favorably with the gasification of coal as far as overall thermal efficiency is concerned (60–80%). Furthermore, wood gasification offers the following advantages over coal: (a) substantially lower oxygen requirements, (b) essentially no steam requirements, (c) lower cost for changing hydrogen/carbon monoxide ratios (which are already higher in wood gas), and (d) lower desulfurization costs. Offsetting these advantages are the economies of scale available in coal gasification systems not available in wood gasification because of procurement limitations.

The gas produced may be used directly for energy (see *Energy Generation from Wood*), or for the synthesis of organic compounds usable as liquid fuels or chemicals. By enrichment in hydrogen a synthesis gas is formed which is suitable for the production of ammonia, methanol, methane, or the higher hydrocarbons obtainable by the Fischer–Tropsch synthesis. Methanol may be further processed to formaldehyde, so important in adhesive technology. Even without specific processing of the gas to chemicals, significant quantities of chemicals are generated during the gasification step. For example, the following yields have been obtained: ethylene 5.2%; acetylene 0.8%; propylene 0.5%; benzene 1.8%; and toluene 0.3%. Higher acetylene and ethylene yields (up to 15%) have been obtained at 2000–2500 °C in an electric arc.

Pyrolysis of wood is its conversion by thermal reactions in the absence of added oxidizing agents to provide a volatile phase and a solid char. The volatile phase can be further separated into condensible liquids and noncondensible gases. Pyrolysis is carried out at lower temperatures than gasification, which passes through a pyrolytic stage before completion.

Depending on process type and temperature, oil yields can range from 1 to 40% and char yields from 10 to 40%. By manipulation of only three variables—temperature, heating rate, and gas residence time—the relative proportions of gaseous, liquid and solid products can be varied. Low temperatures favor liquids and char, low heating rates favor gas and char, and short gas residence time favors liquids.

The condensed liquids from wood pyrolysis consist of the so-called pyroligneous acid, which is an aqueous acidic layer, and the heavier wood tar from which the aqueous phase has been decanted. Both phases contain many compounds, with the tar much more complex since its components have not been selectively extracted on the basis of water solubility.

Major water-soluble components include methanol, acetic acid and acetone, with smaller quantities of other carboxylic acids, aldehydes, ketones, alcohols and esters. The yields of even the major components are low, with methanol (once produced commercially by pyrolysis) yield only 2% and acetic acid 3–7%.

The wood tar contains at least 50 identified phenolic compounds, which account for up to 60% of the tar. There are also higher acids, aldehydes, ketones, esters, furans and hydrocarbons. Tar yields in conventional pyrolysis can range from 4 to 12%.

The low yields of such volatile compounds as methanol, acetic acid and acetone from pyrolysis make it unlikely that they would ever regain commercial importance. They can be used internally for fuel in the carbonization process. Phenols from the pyrolysis oils have potential value, but even though some applications have been found as pitches, antioxidants, binders, etc., the complicated nature of the mixtures obtained in pyrolysis has made higher-value applications difficult. The corrosivity encountered in pyrolysis from acetic and formic acids can be removed to yield a high-energy liquid fuel suitable for use in high-efficiency boilers designed for fuel oil.

Direct liquifaction of wood to oil suitable for use as a boiler fuel may be achieved in 50% yield by reaction with carbon monoxide at 400 °C and 3×10^7 N m^{-2}. Hydrogenation of wood also affords high yields of liquid products.

Table 2 lists the chemical products obtainable from nonselective processing of whole wood.

3.2 Selective Processes

In selective processing the individual cell-wall components are either first separated from each other before further conversion to chemicals or fuels, or the components are sequentially removed from the polymeric composite we call wood. Hydrolysis in organic solvents, autohydrolysis and steam explosion are examples of the first type. The products in each case are an aqueous solution of hemicellulose sugars, an alkaline or organic solvent solution of lignin and a cellulose residue. In the second type the readily hydrolyzed hemicelluloses are first removed from the wood and converted into simple sugars under mild conditions that leave the cellulose and lignin essentially unaffected. Subsequent hydrolysis of the cellulose provides almost pure glucose and a lignin residue that can be further processed to phenols and other aromatic compounds.

Table 2
Chemical products from nonselective processing of whole wood

Synthesis gas ($CO + H_2$)
Ammonia
Methanol
Formaldehyde
Methane
Aliphatic hydrocarbons
Charcoal
Wood oils

Whether the cellulose is first isolated as a residue or removed to leave a lignin residue, its ultimate fate in selective processing of wood is hydrolysis to glucose. This may be effected by dilute or concentrated acids or by enzymes. Each of these methods has its advantages and disadvantages. With pure amorphous cellulose, enzymatic hydrolysis to glucose can be quantitative in several hours. However, with the crystalline and lignified cellulose usually available as a substrate the yields of glucose are lower and reaction times longer. Dilute acid hydrolysis at higher temperatures takes place in seconds (15 s at 240°C), but yields are only about 50%. Concentrated acids can provide quantitative yields of glucose in minutes, but acid recovery costs are high.

Once a glucose solution is available from the hydrolysis of cellulose, its further conversion to useful chemicals or fuels is effected chiefly by fermentation, although under acidic conditions first hydroxymethylfurfural and then levulinic acid are formed. In addition to ethanol, fermentation of glucose can yield such other organic chemicals as acetic, butyric, citric and lactic acids, acetone, butanol, glycerine and isopropanol. Lactic acid may be further processed into acrylic acid. The greatest potential for useful products from glucose, however, is through the familiar fermentation to ethanol. Alternatively to its use in motor fuels, ethanol could be converted to ethylene, the most important organic chemical of commerce, which can then be further processed to a multitude of chemicals and synthetic polymers. Oxygenated aliphatic chemicals are of special interest in glucose utilization, because they do not suffer the loss in mass which accompanies removal of oxygen to form hydrocarbons. By oxidation of ethanol to acetaldehyde an intermediate is obtained for the further production of acetic acid and acetic anhydride, acrylonitrile, butadiene and vinyl acetate.

Hemicellulose hydrolysis is very facile, yielding principally mannose from softwoods and xylose from hardwoods. Mannose will probably not become an important wood hydrolysis product since softwoods will generally find their way to higher-value applications in the form of lumber and pulp. That which is formed, however, may be readily processed along with glucose to yield the same products. Fermentation of xylose to ethanol is difficult, but ongoing research is showing promising improvements. Chemical conversion of xylose to xylitol by reduction and to furfural under acidic conditions can be carried out in high yield. Furfural has many potential applications as a chemical intermediate.

Under various hydrogenation and hydrogenolysis conditions lignin can be converted into phenols in yields of up to 50%, and these mixtures of alkylated and polyhydroxyphenols can be dealkylated and dehydroxylated to give simple phenol in about 20% and benzene in about 14% yield. Etherification yields aryl ethers which are useful as octane enhancers in motor fuels. Depending on their origin the lignin residues or solutions may be modified, etherified, esterified and condensed to change their physical properties and form useful adhesives and resins.

Table 3 lists some of the chemical products derivable from the selective processing of the principal wood components.

4. Economic and Other Considerations

The technical feasibility of converting wood into a variety of useful chemicals has been long established, and where economically feasible in a free market or desired in a controlled economy, chemicals from wood have been produced on a commercial scale.

One economic principle which warrants special emphasis for a multicomponent raw material like wood is the need to maximize the yield of products for greatest economic efficiency. Early schemes for conversion of wood to chemicals considered only single products produced from a portion of the wood, and consequently carried a higher raw material cost as an economic burden. The raw material cost of a cellulose-derived chemical such as ethanol, for example, can double when it is a single product, compared to its cost when coproducts are derived from the remaining wood components, the hemicelluloses and lignin. Yet such single-product plants have been designed and operated, and their failure to operate profitably (except in times of national emergency) has prejudiced many against the whole concept of chemicals from wood. Even established industries such as the petroleum industry could not thrive on a single product.

A number of integrated schemes for converting each of the components of wood into chemicals have been suggested. In addition to those directed at production of chemicals only, others include coproduction of fibers, food and fuels.

The most important considerations affecting the ultimate large-scale conversion of wood into chemicals and at the same time the most imponderable are the cost and availability of the fossil fuels, especially petroleum and natural gas for petrochemicals production. At some time depletion of these resources will lead to higher prices and make renewable raw materials for chemicals like wood economically attractive. Exactly when costs of petrochemicals will converge with and eventually exceed the cost of chemicals from wood cannot be predicted at this time.

Capital equipment costs in the conversion of wood to chemicals are greater than in conventional petrochemical processing. Reasons include the greater difficulty in storing and handling solids compared to liquids and the lower bulk density of wood. Wood and coal conversion facilities costs are comparable. On the positive side, individual plant size for wood conversion would be smaller. The need to procure a wood supply within a reasonable hauling radius would lead to plants of relatively modest size widely dispersed throughout the resource base. The lower investment needs could favorably influence decisions on incremental or replacement plant capacity.

The classical economic principles of supply, demand and profitability are not the only important factors in determining a change in a resource base. Political and social decisions may also have an important role as has been shown by the historical shift from wood to coal and then to oil and gas as primary energy resources.

Inertial barriers to resource base replacement must also be considered. Liquids and gases are easier to handle than solids, and the organic chemicals industry is accustomed to handling a liquid raw material by tanker or pipelines. The prospect of collecting a solid raw material over an area of several thousand square miles is foreign and uncomfortable. On the other hand, the forest products industries which regularly assemble even larger quantities of wood in their operations are equally uncomfortable with the idea of producing chemicals.

Our present reliance on fossil hydrocarbons for organic chemicals cannot continue any longer than the availability of this finite, depletable resource. Even though the future of chemicals from wood cannot be predicted with certainty, they will make an important contribution to the problems of organic chemicals supply that will be caused by the depletion of fossil hydrocarbons.

Utilization of wood for this application that is insensitive to species, size and fiber quality and that may equal or exceed in volume the traditional uses of wood will inevitably bring about major changes in how and where we grow our wood supplies, and how we allocate them to various end uses.

Table 3
Chemical products from selective processing of wood components

Cellulose	*Hemicellulose*
Glucose	Mannose
Ethanol	Xylose
Ethylene	Ethanol
Ethylene derivatives	Furfural
Butadiene	Xylitol
Oxygenated aliphatic chemicals	
Hydroxymethylfurfural	*Lignin*
Levulinic acid	Phenols
Sorbitol	Aromatic hydrocarbons
Glycols	Aryl ethers
	Adhesives
	Resins

Bibliography

Goldstein I S (ed.) 1981 *Organic Chemicals From Biomass.* CRC Press, Boca Raton, Florida

I. S. Goldstein
[North Carolina State University, Raleigh, North Carolina, USA]

Coconut Wood

Wood or xylem makes up "the principal strengthening and water-conducting tissues of stems and roots" (International Association of Wood Anatomists 1964). It is characterized by the presence of tracheary elements which are composed of vessel members and tracheids, and parenchyma which are soft tissues that store and distribute elaborated food materials.

Based on the above description, the stem of the coconut palm (*Cocos nucifera* Linn.) can be classified as wood. However, many of its features differ from those of conventional wood, i.e., the wood of gymnosperms (conifers or softwoods) and dicotyledons (broadleaved trees or hardwoods). Like other palms, coconut is a monocotyledon.

1. Structure and Stem Anatomy Affecting Utilization

The palms are more closely related to the hardwoods than to the softwoods. For example, some of the structural features of palms and hardwoods such as the sap-conducting tissues are similar. Despite these similarities, there are still a number of important differences between conventional wood and coconut palm stem wood (Table 1). These differences underline the fact that coconut wood cannot be used indiscriminately as a substitute for conventional wood. It must be sorted (graded) for efficient utilization, based on a knowledge of the various properties of different parts of the coconut stem.

Coconut stems commonly reach a diameter of around 30 cm. Observations show that for a given strain of coconut, minor variations in diameter from one stem to another or between locations are due to the growing conditions for individual stems during the early stages of their life.

Tomlinson (1961) described succinctly the general anatomy of a palm stem:

> The central cylinder is abruptly demarcated from the cortex by a wide peripheral sclerotic zone made up of congested vascular bundles separated from each other by narrow layers of parenchyma. Since each vascular bundle has a massive radially extended fibrous sheath external to the phloem and the ground parenchyma becomes sclerotic, this zone forms the main mechanical support of the palm stem.

In the lower portion of the stem, the dark color of the vascular bundles gives longitudinal surfaces a unique "quill-like" appearance and therein lies the decorative value of palm wood.

The important elements of a palm stem aside from the cortical tissue, according to Kaul (1960), as seen in cross section are as follows.

(a) Fibrovascular bundles (or vascular bundles) consist of fibrous and vascular portions. The first is sclerenchymatous, varying greatly in size and form; the second is made up of the phloem and xylem. The phloem lies between the fibrous and xylem tissues. The xylem is sheathed by parenchyma cells containing either large pores or both large and small pores.

Table 1
Main differences between conventional wood and coconut palm stem wood (adapted from McQuire 1979)

Conventional wood from softwood and hardwood trees	Coconut palm stem "wood"
Trees have secondary thickening, so the stem increases in diameter with age and may also produce annual rings	No true secondary thickening and no annual rings
Elongated cells give strength to the wood (fibers or tracheids) uniformly and continuously throughout the stem	Fibers are grouped together in distinct and isolated vascular bundles and distributed fiber bundles
Most trees produce heartwood in the stem as they grow older and larger. This heartwood is generally more durable than sapwood	No heartwood. The center of the stem is mostly composed of soft parenchymatous tissues
Wood contains ray cells which run from the surface towards the center of the stem	No ray cells
Remains of branches within the wood form knots	No branches and knots
Bark is completely separated from the wood by other tissues (cambium) and is relatively easy to remove	No cambium and "bark" is very difficult to remove from the wood

(b) Fibrous strands or bundles consist of purely fibrous elements. They are normally found in the cortex and less commonly in the central cylinder (e.g., in the dermal zone and even in the central region). They are usually very small although they can be large and circular.

(c) Ground tissue is parenchymatous and is one of the most important tissues in the stem anatomy of palms. Definitive forms of ground tissue vary from being compact, slightly loose to lacunar and sometimes very spongy. Starch grains are abundant and restricted to this tissue. It is hard and quite dense in texture in the dermal and subdermal zones but quite soft, loose and easily indented with a fingernail in the central zone.

One of the most interesting aspects of the palm stem is the common association of silica-containing cells called stegmata, which are more-or-less spherical in coconut. Calcium oxalate crystals also occur usually in the form of raphide clusters in distinct raphide sacs. Stegmata and raphides cause dulling of sawteeth during sawing.

The vascular bundles are readily recognizable in cross-sectional view to the naked eye due to the dark-colored fibrous sheath. They are closely spaced in the dermal and subdermal zones and widely scattered in the central zone. In the dermal zone, they range from 60–100 or more per square centimeter and are smaller than those in the subdermal and central zones. In the subdermal zone, they are fewer in number and even fewer in the central zone (i.e., about 8–12 per cm^2). They vary in shape, i.e., kidney-shaped or angular to circular, with the circular shape most apparent in the central zone.

Sclerenchyma fibers are thick-walled to very thick-walled (0.015–0.022 mm) and very long (2.3–2.9 mm). The phloem tissue is light colored, undivided or of a single strand and visible only through a hand lens or barely visible to the naked eye.

The xylem contains either one or more large pores (averaging 140–260 μm wide) and small pores. The small pores average 23–80 μm in width and are numerous in the central zone. Vessel elements are very long (average 2.5–2.8 mm) and usually with scalariform pitting.

2. Physical Properties

The more important physical properties of coconut lumber are moisture content, specific gravity, density and shrinkage (Table 2).

The newly-felled (green) coconut trunk contains a considerable amount of sap or water. The moisture content varies considerably along the height from butt to top and over the cross section from the core to the outer portion near the bark. In the core, the moisture content of the butt and top ends reaches up to 550% and 300%, respectively. In the outer portion, the top end has a higher moisture content than the butt end.

Most conventional timber species exhibit density gradients from the center of the stem towards the bark and from the butt of the trunk towards the top. The gradients, however, are not pronounced except in early-formed (juvenile) wood compared with later-formed (mature) wood. With coconut stems, the gradients are much more pronounced (Table 2). The density of the coconut wood decreases from butt to the top and increases from the center of the stem towards the bark.

In wood, the greatest shrinkage is in the tangential direction while longitudinal shrinkage is the least. For coconut lumber, radial and tangential shrinkages appear to be about the same (Table 2). Longitudinal shrinkage can be considered negligible for both. Unlike conventional wood, the transverse shrinkage of coconut wood appears to be independent of specific gravity, since material from the core and the dermal zones have very similar shrinkage values while their specific gravity values differ by a factor of more than 2.

Table 2
Physical properties of green coconut lumber from the Philippines

Property	Core (soft inner portion)	Mixed core and dermal zone	Dermal zone (hard outer layer)
Moisture content (%)	329	173	101
Basic specific gravity (based on oven-dry weight and green volume)	0.260	0.420	0.592
Density at the indicated moisture content ($kg\ m^{-3}$)	1115	1147	1190
Shrinkage, green to oven-dry condition (%)			
radial	6.1	5.7	6.0
tangential	5.7	6.0	6.7
volumetric	11.5	11.4	12.3

3. Mechanical Properties

Data on mechanical properties of green coconut lumber are shown in Table 3. As would be expected on the basis of specific gravity differences, the strength and stiffness of core wood is much less than the wood of the dermal zone. The static bending properties of coconut wood are not as high as conventional wood of equal specific gravity. For instance, the expected modulus of rupture (bending strength) of material of specific gravity equal to that of core wood would be 27.6 MPa for softwoods (coniferous wood) and 25.5 MPa for hardwoods (broadleaved wood). These values are 16% and 8% higher, respectively, than the value of 23.7 MPa for coconut core wood. Similarly, softwood and hardwood of equal specific gravity would be expected to have a modulus of rupture greater than that of coconut wood of the dermal zone by 26% and 24%, respectively.

In compression parallel and perpendicular to grain and in the closely related property of hardness, coconut wood has some values that are higher and others that are lower than would be expected in conventional wood of equal specific gravity. For instance, the maximum crushing strength parallel to grain of coconut core wood falls between the values that would be expected for softwoods and hardwoods of equal specific gravity. The only instance where coconut wood has substantially higher values is at fiber stress at proportional limit in compression perpendicular to grain of core wood, which is about twice the value for corresponding conventional wood.

Table 3
Mechanical properties of green coconut lumber from the Philippines

Property[a]	Core	Dermal zone
Static bending:		
modulus of rupture (MPa)	23.7	51.7
fiber stress at elastic limit (MPa)	14.1	30.4
modulus of elasticity (GPa)	3.0	7.2
Compression parallel to grain:		
maximum crushing strength (MPa)	12.1	28.8
fiber stress at elastic limit (MPa)	7.2	16.6
modulus of elasticity (GPa)	4.7	10.7
Compression perpendicular to grain:		
fiber stress at elastic limit (MPa)	1.8	3.8
Shear parallel to grain:		
maximum shearing stress (MPa)	2.2	5.2
Hardness:		
side grain (kN)	1.3	5.1
end grain (kN)	1.1	4.8
Toughness (Joule per specimen)	17.2	31.7

[a] All tests followed ASTM Standards D143-52 (reapproved 1972)

4. Processing of Coconut Wood

The utilization of wood from mature coconut palm stem involves several stages of processing. These include sawmilling of the stem (logs), seasoning of sawn wood, machining and preservative treatment. Research and development activities have been conducted in the Philippines to generate benchmark information and to establish a comprehensive range of technologies on coconut wood utilization.

4.1 Saw-Milling

Saw-milling is the process of cutting or milling coconut stem into boards or lumber by using one of the following items of equipment or tools: (a) bandsaws, (b) circular saws, (c) power chainsaws, (d) two-man ripsaws, and (e) axes and bolos. Each of these has its own characteristics and specific applicability. For large-scale sawing operations, the bandsaw and circular saw are most applicable. The others are applicable in small-scale operations and sawing can be done in the coconut plantation.

When coconut stem is sawn with a 1300 mm bandsaw using a stellite-tipped blade having a pitch of 4.45 cm, tooth angle of 45°, clearance angle of 20°, sharpness angle of 25° and gullet depth of 1.5 cm, the input rate can be up to 1.06 m^3 of coconut wood per hour. Using high-speed alternately-swaged-teeth steel blade with a similar tooth configuration could result in an input of 0.47 m^3 per hour. Lumber recovery for both types of saw blades can be up to 49.0%. Sawing time for an average diameter stem of 22 cm and 3 m long would take an average of 13.30 minutes for the stellite-tipped saw blade and 29.85 minutes for the alternately-swaged-teeth blade.

A circular saw 112 cm in diameter with a stellite, inserted tooth can cut 50 coconut stems 30.5 cm in diameter and 4.8 m long in 8 hours. Lumber recovery is 34.2–36.7%. Saw teeth must be sharpened after cutting 25 coconut stems.

Using a power chainsaw, a team of 3 chainsaw operators can produce 7.81 m^3 of coconut lumber in 8 hours. Lumber recovery (hard and partly hard portions) is 35%–38%.

4.2 Seasoning

Drying of sawn coconut lumber in the Philippines can be done by air or kiln drying. Lumber 25–50 mm thick can be air-dried to a moisture content of 15.5–19% in less than 4 months. Drying in a steam-heated kiln to 10% moisture content takes about 4–5 days for 25 mm green lumber and about 14 days for 50 mm lumber.

4.3 Machining

In the manufacture of coconut wood products, the most common operations employed are planing, turning, boring, shaping and sanding.

To obtain good surface quality of coconut wood products, the following are recommended for machining operations: for planing, preferably use 30° rake angle at 1.6 mm and 0.8 mm depth of cut at a cutter-head speed of 3000–4000 rpm; for shaping, use 2 knives at a spindle speed of 7000–10 000 rpm; for turning, preferably use a spindle speed of 1500–2500 rpm; for boring and mortising, preferably use 1700 to 2000 rpm.

In machining coconut wood, particularly the hard outer (dermal) portion, carbide-tipped blades or cutters must be used. Ordinary high-speed steel cutters dull easily, owing to the relatively thick-walled fibrovascular bundles and also the presence of silica containing cells called stegmata.

There is a difference in power consumption in machining the "soft" and "hard" portions of the coconut wood, with the soft portion requiring less electrical energy. This is primarily attributed to the difference in density of the two portions.

4.4 Preservation

Coconut wood is subject to deterioration by termites, beetles and fungi. When conditions favor the attack of these organisms, the material would last only for a limited number of months or years, depending on the portion of the stem where the wood was taken.

The low natural durability of coconut wood (especially the soft portion) can be overcome by proper application of standard chemical wood preservatives, employing established treatment processes. Preservative treatment is the key to effective utilization of wood products.

Treatment of coconut wood (hard portion) measuring 128 mm × 152 mm × 6 m with a moisture content of 35–65% subjected to 6 hours of hot bath using coal tar creosote and 10–12 hours cold bath can result in an average creosote retention of 272 kg m^{-3}. For solid round coconut stem measuring 30.5 cm in diameter and 6 m in length with a moisture content of 35–120% subjected to 10 hours hot bath and 12–15 hours cooling, an average creosote retention of 115–173 kg m^{-3} can be obtained. This method when properly applied and when used with the right kind of preservative can provide adequate protection to coconut wood against the attack of wood-destroying organisms. The cost of treatment is comparatively lower than for the pressure method. The hot and cold soaking method is applicable in treating electric power and communication poles, fence posts and other farm timbers, bridge timbers and other construction materials (see *Preservative-Treated Wood*).

5. *Uses of Coconut Wood*

Coconut wood can be used for diverse purposes. If properly treated, it can be used for fence posts, electric and telecommunication poles, cross arms, small-span bridge timbers, road guardrails and sign boards.

For housing, coconut wood can be utilized for posts, beams, rafters, trusses, purlins, studs, ceiling joists and hangers, fascia boards, tongue and groove flooring, siding, ceiling and insulation boards, shingles, floor tiles (parquet), door and window jambs, door frames and panels, window frames and blades, louvers, stair braces, stringers and steps, handrails, balusters and railings, finished lumber, cabinet frames, forms and scaffoldings.

Coconut wood can also be used for household utensils, interior decoration, bookshelves, picture frames, canes, night sticks, novelty items, tool handles, furniture, wooden shoes, boxes and crates. It can even be used for vehicle and truck bodies, boat plankings, firewood and charcoal.

Bibliography

Committee for Forestry 1985 *The Philippines Recommendations for Coconut Timber Utilization*, PCARRD Tech. Bull. Series No. 60. Philippine Council for Agriculture & Resources Research & Development (PCARRD), Los Banos, College, Laguna, Philippines

International Association of Wood Anatomists 1964 *Multilingual Glossary of Terms Used in Wood Anatomy*. Committee on Nomenclature, International Association of Wood Anatomists, c/o Rijksherbarium, Leiden, p. 45.

Kaul K W 1960 The anatomy of the stem of palms and the problem of the artificial genus *Palmoxylon* Schenk. National Botanic Gardens, Lucknow, India

Killman W 1983 Some physical properties of the coconut palm stem. *Wood Sci. Technol.* 17(3): 167–85

McQuire A J 1979 Anatomical and morphological features of the coconut stem in relation to its utilization as an alternative wood source. *Proc. Coconut Wood—1979.* Philippine Coconut Authority, Quezon City, Philippines, pp. 1–8

Mosteiro A P 1978 Utilization of coconut palm timber: Its economic significance in some countries in the tropics. *FORPRIDE Dig.* 7(1): 44–57

Mosteiro A P 1980 The properties, uses and maintenance of coconut palm timber as a building material. *FORPRIDE Dig.* 9(3,4): 46–55

Mosteiro A P 1981 Preservation of coconut palm wood for villagers in the tropics. *NSDB Technol. J.* 6(2): 35–42

Tomlinson P B 1961 *Anatomy of the Monocotyledons II. Palmae.* Clarendon Press, Oxford

J. O. Siopongco, J. P. Rojo, A. P. Mosteiro
and J. E. Rocafort
[Forest Products Research and Development
Institute, College, Laguna,
Republic of The Philippines]

Constituents of Wood: Physical Nature and Structural Function

For many applications of wood and wood-based materials, it is helpful to understand what structural role each of the major chemical constituents plays in the development of strength and stiffness and how these relate to the fibrous, cellular structure of the wood. As with industrial multiphase materials, some of the components in wood cell walls act as a structural framework (reinforcement) and others act as the bulking and stiffening matrix in which the reinforcement is embedded.

1. The Occupation of Space in a Unit of Wood Cell Wall

Figure 1 is a schematic illustration of a typical small unit of space (12 nm × 43 nm cube) in one of the layers (e.g., the S2 layer) of the wall of a wood fiber (or other cell). Six microfibrils are shown, each with cross-sectional dimensions of 10 nm × 3.5 nm. The microfibrils contain the crystalline polymer of β-D-glucose residues (cellulose), with or without crystal-order defects. Some microfibrils appear to be quite pure in their composition; others may have occasional altered linkages or nonglucose monomer substitutions.

Figure 1 also shows extractable short hemicellulose chains between the microfibrils. The hemicelluloses also exhibit some degree of orientation but have considerably less crystallinity than the cellulose. Other substances (lignin, proteins, inorganic matter, air and especially water) occupy the remainder of the 12 nm × 43 nm space. The architecture of the woody plant cell wall then consists of many of these spaces. This general scheme appears to hold for the entire spectrum of lignified cellulosic plants.

The overall picture of a unit-volume wood cell wall at the supramolecular level is as follows.

(a) A crystalline, filamentous, solid-phase material (the microfibril) composed nearly exclusively of cellulose I, which is impenetrable to water, accounts for approximately 42% of the dry solid mass, but a somewhat lower proportion of the total volume under normal ambient conditions because of the presence of air and water.

(b) Two interpenetrating solid-phase systems, one composed of an extensively branched, three-dimensional, amorphous polymer (lignin), and the other composed of a complex of relatively linear (but nevertheless partially branched), partially paracrystalline polymers of a variety of molecular sizes and solubilities (hemicelluloses), account for most of the remaining solid fraction and create a matrix of polymer materials in which the filamentous microfibrils are embedded.

(c) A third very delicate and tenuous interpenetrating system of solid inorganic matter exists, in addition to the lignin and hemicelluloses, which remains as a recognizable structure when wood is carefully ashed. It comprises only a small proportion, of the order of 1%, of the dry mass of wood.

(d) A fourth and fifth interpenetrating system can be ascribed to the presence of water and air within the cell-wall structure. Both water and air are always present, but their proportions vary with the dryness of the ambient environment. If little water is present, as in the case of wood in a low-humidity environment, the water only exists in the form of (i) water of constitution and (ii) adsorbed water, wherein the H_2O molecules are hydrogen-bound to the surfaces of the carbohydrate and protein molecules present. This binding may be monomolecular or polymolecular. If the wood is wet, or subject to extreme conditions of humidity, there may also be free water present, which penetrates and swells the spaces between microfibrils. Air, or in general, gas occupies spaces within wood not occupied by solid or liquid substances.

(e) Substances such as fats, starches, resins, gums, organic crystals and other extractives, and proteins are usually not regular components of the spaces or unit volumes in wood cell walls but exist in specific locations as inclusions or depositions; except for the proteins and some extractives, they are typically found only in cell lumens and intercellular spaces and are thus not considered as intrinsic molecular components of wood fibers and other cells in wood.

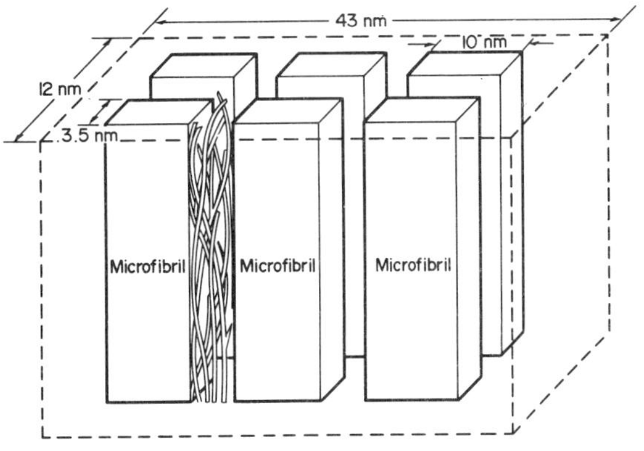

Figure 1
Diagrammatic representation of a small element of wood cell wall. Six microfibrils are present. Between two of the microfibrils, bent and branched rods representing hemicellulose chains are shown. Hemicelluloses, as well as the other substances present in wood, occupy the spaces between microfibrils throughout the cell wall (after Mark 1980)

2. Structural Functions of the Constituents

A unit volume of wood cell wall of the size order shown in Fig. 1 can be described as a filamentary composite (see Sect. 1). Since the fiber (or other cell) wall is built up of layers of such filament-reinforced matter, it is also accurate to describe it as a laminated composite.

Based on experimental evidence (direct and indirect) and deduction from experience with analogous industrial materials and structures, structural functions can be assigned to each of the components found in a typical unit volume of fiber wall. As mentioned earlier, certain components act as the framework (reinforcement) and others act as the stiffening matrix for the reinforcement. In fibers and other plant cell walls, the microfibril provides the structural reinforcement. Since the microfibril is principally if not exclusively cellulose I, the mechanical properties of that crystalline substance are used identically for the properties of the microfibril.

The crystalline arrangement of cellulose I within the microfibril is shown in Fig. 2. The total number of cellulose chains that pass through a given cross section of a microfibril vary according to the size of the microfibril, but it is always greater than 20 and may be an order of magnitude greater than that. In Fig. 2a, a small array of five cellulose chains is shown as viewed along the axis of the chain. The crystalline unit cell, not to be confused with the wood cell, is monoclinic ($a \neq b \neq c$, $\alpha = \beta = 90°$, $\gamma \neq 90°$), and contains disaccharide segments of two chains. The unit cell dimensions are $a = 0.778$ nm, $b = 0.820$ nm, $c = 1.034$ nm (chain axis) and $\gamma = 96.5°$. Computations have been made for several physical and mechanical properties of this solid-state microfibrillar entity.

Of all the properties, the set of elastic constants applicable to a given structural solid material are generally the most important. These elastic constants may be measured physically in laboratory tests on materials that are available in the form of large specimens. For microcrystalline materials such as cellulose I, these elastic constants may be derived on the basis of known crystal structure and the energetics of bond stretching and bond-angle change within that structure. Specifically, the sum of the energies associated with deforming each bond length and bond angle within the unit cell are equated to the overall elastic strain energy of the unit cell itself. The unit cell undergoes a change in length in the direction of the chain axis so that the repeating unit length L (= 1.034 nm) experiences an incremental displacement $dL = Le_3$, where e_3 is the strain in the direction of the chain axis. The axial deformation Le_3 is related to the deformations of all the primary and secondary bond lengths and angles within the unit cell. An apportionment of this displacement is made (vectorially) on the basis of the orientations of each bond length and angle. For a prescribed dL it is assumed that the bond-deformation energies are minimized. The method yields an inverse gross spring-force constant $1/K$ for the 1.034 nm repeating unit as the sum of terms containing $1/K_i$ values for all the bonds therein. Then, the energy in straining the unit cell volume is equated to the overall bond-deformation energy by

$$W = \tfrac{1}{2} C_{33} e_{33}^2 \Delta V = \tfrac{1}{2} K \, dL^2 \quad (1)$$

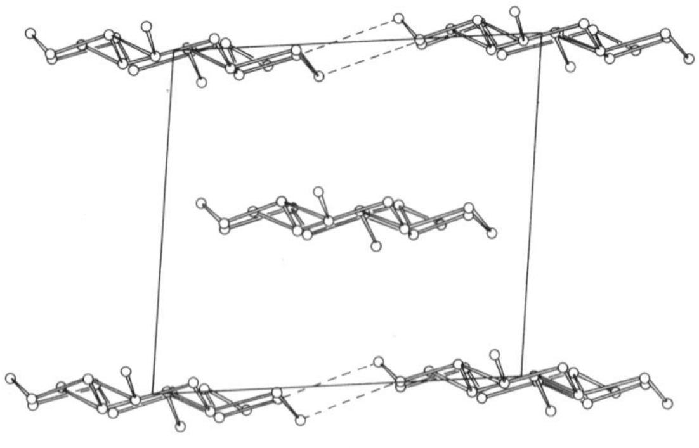

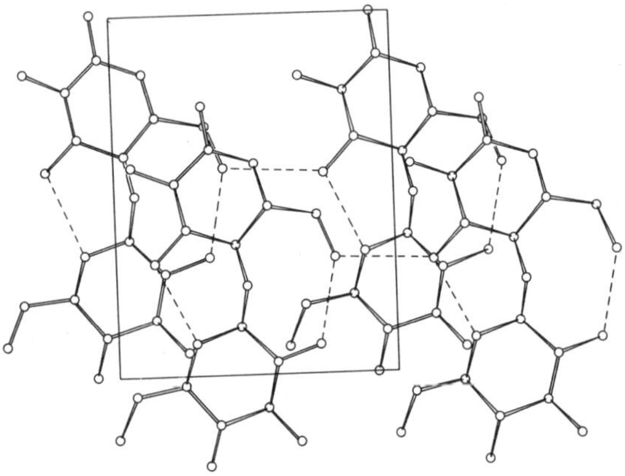

Figure 2
Projections of the unit cell of ramie cellulose. Dashed lines represent hydrogen bonds. (a) An a–b projection. The crystal repeat unit contains one center chain and parts of 4 corner chains. (b) A b–c projection; the rectangle shows the position of the unit cell. Two corner chain segments are shown; they run the length of the rectangle. Two similar chain segments, representing center chains, are displaced downward by $\frac{1}{4}c$, or 0.2585 nm (after Woodcock and Sarko 1980. © American Chemical Society, Washington, DC. Reproduced with permission)

where C_{33} is the elastic stiffness constant in the chain axis direction and ΔV is the unit cell volume. Solution of Eqn. (1) yields $C_{33} = KL^2/\Delta V$. It is possible to derive a full set of elastic constants for crystalline cellulose by this method. Table 1 gives a set of technical elastic constants for cellulose I.

From the standpoint of physical/mechanical behavior, the matrix phase in a wood or other cellulosic cell wall consists of the relatively unoriented or amorphous, short-chained or branched, polymers of various molecular species (i.e., lignin and hemicellulose and any other polymers such as pectopolyuronides that might be present in a given layer), plus all of the tiny voids and the gases and adsorbed water associated with them—in other words, everything else that surrounds the microfibrils in their local environment. Note that this is a broader definition of "matrix" than is used by biologists and chemists. Since the distribution of components such as the hemicelluloses, lignin and void space is nonuniform, there is point-to-point variation in the matrix elastic constants. Regions of dense molecular aggregation have high local moduli of elasticity and rigidity, and the more open regions where air and/or water is present in the cell-wall voids have small, perhaps insignificant, moduli. The moduli overall take on some sort of average values, but deformation takes place preferentially in the regions of lower aggregate macromolecular packing density.

As properties of the noncellulose molecular species and/or water content change, the matrix constants change, whereas undegraded microfibrillar cellulose I is considered invariable, at least within constant-temperature conditions.

The matrix is collectively taken to be isotropic. Thus, a single value is calculated for the modulus of elasticity, shear modulus and Poisson's ratio for each physical-state condition. Typical constants of this type representing very rigid, average, and very compliant matrices are given in Table 2.

Table 1
Engineering elastic constants for cellulose I: Young's modulus E, shear modulus G and Poisson's ratio ν

Elastic constant	(GPa)	Elastic constant	($\times 10^{-3}$)
E_{11}	16.4	ν_{12}	−0.141
E_{22}	25.2	ν_{21}	−0.0921
E_{33}	246.4	ν_{32}	33.6
G_{12}	2.58	ν_{23}	3.44
G_{23}	0.240	ν_{31}	41.0
G_{31}	0.173	ν_{13}	2.74

Table 2
Engineering elastic constants for the matrix surrounding the cellulosic microfibrils in wood cell walls

	Elastic constant		
Assumption	E (GPa)	G (GPa)	ν
High stiffness	7.0	2.7	0.30
Average stiffness	1.8	0.7	0.30
High compliance	0.2	0.08	0.30

3. *Properties of the Composite*

Once the structural functions of the various chemical constituents in the wood cell wall are determined, a variety of modelling systems can be employed to determine the physical and chemical properties of the composite unit volume element of the fiber wall.

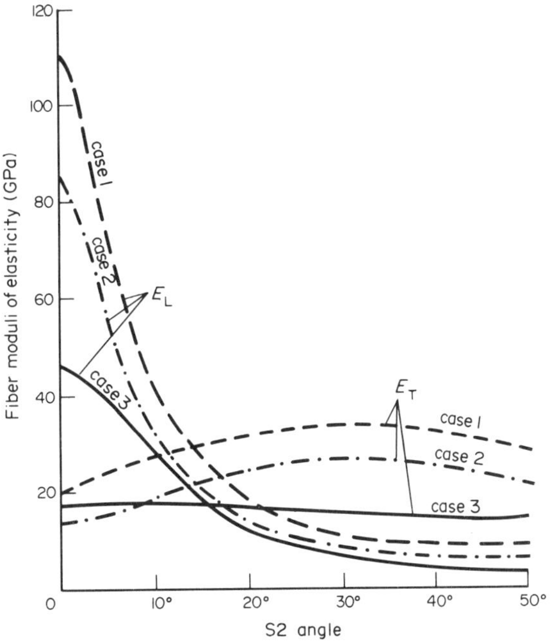

Figure 3
Curves of theoretical variation of wood-fiber elastic moduli versus S2 angle in the direction of the fiber axis (E_L) and in the transverse circumferential direction (E_T). The three sets of curves represent extreme variations in the theoretical elastic constants employed for the framework (cellulose) of the cell wall. Curves extend to 50°, about the maximum S2 angle ever observed in a cellulosic fiber (courtesy Empire State Paper Research Institute)

Orthotropic elasticity theory has been used to determine mechanical elastic constants for wood and wood-derived fibers, using both two-dimensional and three-dimensional models. These have often been layered systems representing the layers of the cell wall. Anisotropic dielectric constants have been determined for wood cells from consideration of the dielectric characteristics of their components, in similar fashion. Another set of models has been developed to predict the swelling and shrinkage behavior of wood on the basis of molecular/microfibrillar modelling.

Figure 3 shows the axial and transverse moduli of elasticity for the wood-fiber cell wall as a function of (a) the elastic constants of the cellulosic microfibril (parametric curves) and (b) variation of the S2 angle of the fiber (abscissa) with respect to the fiber axis. It may be concluded, on the basis of these models and on laboratory experiments on wood and fibers, that:

(a) cellulose itself *in the microfibril* is a linear elastic substance—the major portion of strain is immediately elastic and proportional to stress;

(b) the stiffness of the framework (structural polysaccharide, principally cellulose) is so great that it vastly predominates over the effects of cellular geometry and nonstructural molecular components when wood is tested parallel to the grain or fibers are loaded in tension.

See also: Chemical Composition; Ultrastructure

Bibliography

Barrett J D, Schniewind A P, Taylor R L 1972 Theoretical shrinkage model for wood cell walls. *Wood Sci.* 4: 178–92

Gillis P P, Mark R E 1973 Analysis of shrinkage, swelling and twisting of pulp fibers. *Cellulose Chem. Technol.* 7: 209–34

Mark R E 1967 *Cell Wall Mechanics of Tracheids.* Yale University Press, New Haven, Connecticut

Mark R E 1972 Mechanical behavior of the molecular components of fibers. In: Jayne B A (ed.) 1972 *Theory and Design of Wood and Fiber Composite Materials.* Syracuse University Press, Syracuse, New York, pp. 49–82

Mark R E 1980 Molecular and cell wall structure of wood. *J Educ. Modules Mater. Sci. Eng.* 2: 251–308

Mark R E 1984 Fiber structure. In: Mark R E (ed.) 1984 *Handbook of Physical and Mechanical Testing of Paper and Paperboard.* Dekker, New York, pp. 445–84

Mark R E, Gillis P P 1983 Mechanical properties of fibers. In: Mark R E (ed.) 1983 *Handbook of Physical and Mechanical Testing of Paper and Paperboard.* Dekker, New York, pp. 409–95

Norimoto M, Hayashi S, Yamada T 1978 Anisotropy of dielectric constant in coniferous wood. *Holzforschung* 32: 167–72

Preston R D 1974 *Physical Biology of Plant Cell Walls.* Chapman and Hall, London

Salmén L 1986 The cell wall as a composite structure. In: Bristow J A, Kolseth P (eds.) 1986 *Paper Structure and Properties.* Dekker, New York

Salmén L, Kolseth P, de Ruvo A 1985 Modeling the softening behavior of wood fibers. *J. Pulp Pap. Sci.* 11(4): J102–7

Woodcock C, Sarko A 1980 Packing analysis of carbohydrates in polysaccharides, Part 11, Molecular and crystal structure of native ramie cellulose. *Macromolecules* 13: 1183–87

R. E. Mark
[State University of New York, Syracuse, New York, USA]

Cork

Cork is a natural material with a remarkable combination of properties. It is light yet resilient; it is an outstanding insulator for heat and sound; it has exceptional qualities for damping vibration; has a high coefficient of friction; and is impervious to liquids, chemically stable and fire resistant. These attributes have made cork commercially attractive for well over 2000 years.

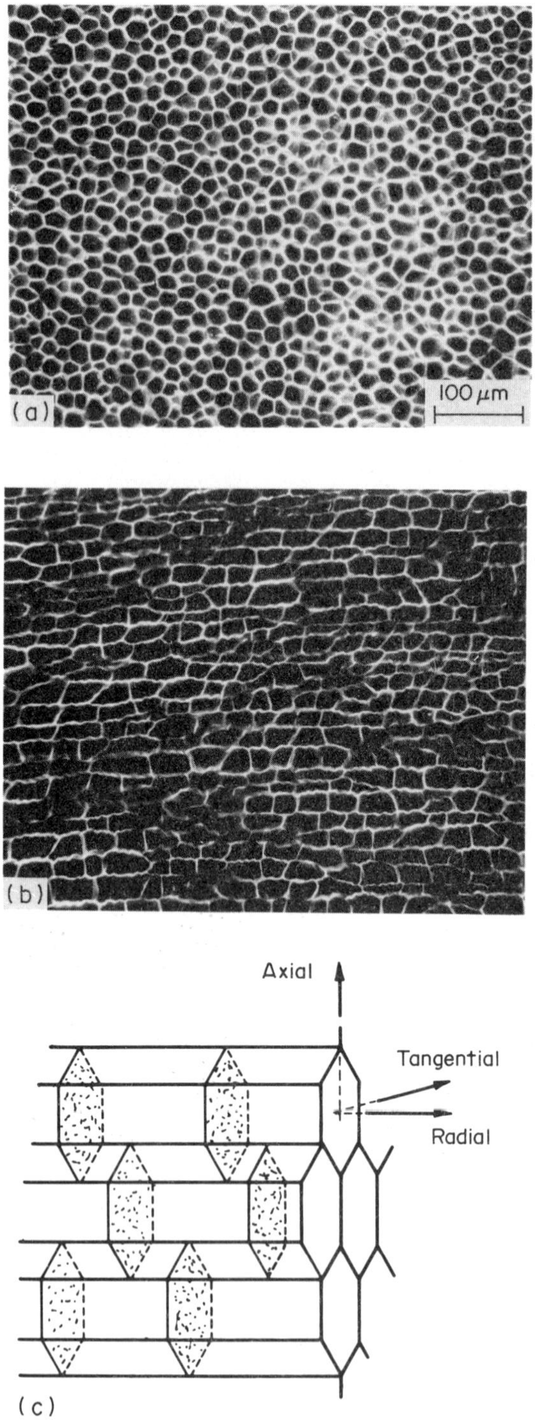

Figure 1
Cells in cork: (a) a section normal to the radial direction of the tree, (b) a section normal to the axial direction of the tree (i.e., the direction parallel to the trunk). The cells (c) are hexagonal prisms

1. Origins, Chemistry and Structure

Commercial cork is the outer bark of *Quercus suber* L., the cork oak. The tree, first described by Pliny in AD 77, is cultivated for cork production in Portugal, Spain, Algeria, France and parts of California. As Pliny says, the cork oak is relatively small and unattractive; its only useful product is its bark which is extremely thick and which, when cut, grows again.

Cork occupies a special place in the history of microscopy and of plant anatomy. When Robert Hooke perfected his microscope around 1660, one of the first materials he examined was cork. What he saw led him to identify the basic unit of plant and biological structure, which he called "the cell." His book, *Micrographia* (Hooke 1664) contains careful drawings of cork cells showing their hexagonal-prismatic shape, rather like cells in a bee's honeycomb. Subsequent studies by Lewis (1928), Natividade (1938), Eames and MacDaniels (1951), Gibson et al. (1981), Ford (1982) and Pereira et al. (1987) confirm Hooke's picture, and add to it the observation that the cell walls themselves are corrugated like the bellows of a concertina (Fig. 1), a structure which gives the cork extra resilience. The cells themselves are considerably smaller than those in most foamed polymers: the hexagonal prisms are between 10 and 40 μm in height and measure between 10 and 15 μm across the base, with a cell-wall thickness of between 1 and 1.5 μm. There are between 4×10^4 and 2×10^5 cells per cubic millimeter.

The cell walls of cork are made up of lignin and cellulose, with a thick secondary wall of suberin and waxes (Eames and MacDaniels 1951, Sitte 1962, Esau 1965, Zimmerman and Brown 1971, Pereira 1982). All trees have a thin layer of cork in their bark. *Quercus suber* is unique in that, at maturity, the cork forms a layer several centimeters thick around the trunk of the tree.

Table 1
Average properties of cork

Density	120–240 kg m^{-3}
Young's modulus	4–60 MN m^{-3}
Shear modulus	2–30 MN m^{-2}
Poisson's ratio	0.15–0.22
Collapse strength (compression)	0.5–2.5 MN m^{-2}
Fracture strength (tension)	0.8–3.0 MN m^{-2}
Loss coefficient	0.1–0.3
Coefficient of friction	0.2–0.4
Thermal conductivity	0.025–0.028 J mK^{-1}

Table 2
Anisotropic properties of cork (density 170 kg m^{-3})

Young's modulus E_1	20 ± 7 MN m^{-2}
Young's modulus E_2, E_3	13 ± 5 MN m^{-2}
Shear moduli G_{12}, G_{21}, G_{13}, G_{31}	2.5 ± 1 MN m^{-2}
Shear moduli G_{23}, G_{32}	4.3 ± 1.5 MN m^{-2}
Poisson's ratio ν_{12}, ν_{13}, ν_{21}, ν_{31}	0.05 ± 0.05
Poisson's ratio ν_{23}, ν_{32}	0.5 ± 0.05
Collapse strength, compression, σ_1^*	0.8 ± 0.2 MN m^{-2}
Collapse strength, compression, σ_2^*, σ_3^*	0.7 ± 0.2 MN m^{-2}

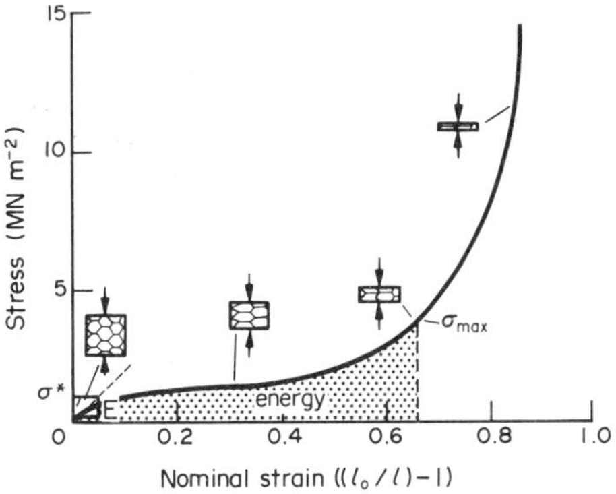

Figure 2
A compression stress–strain curve for cork, showing the modulus E and the collapse stress σ^*

2. Properties

Many of the properties of cork derive from its cellular structure (Gibson et al. 1981, Gibson and Ashby 1988). When cork is loaded in compression or tension, the cell walls bend, and this bending allows large elastic change of shape. For this reason, the elastic moduli are low (Tables 1, 2)—roughly one hundred times lower than those of the solid of which the cell walls are made. When compressed beyond the linear-elastic limit (Fig. 2) the cell walls buckle, giving "resilience," meaning a larger nonlinear elastic deformation. The prismatic shape of the cells makes cork anisotropic. Young's modulus E_1 measured in a direction parallel to the prism axis (the "radial" direction of Fig. 1) is almost twice as great as E_2 and E_3, measured in directions normal to this axis (the "axial" and "tangential" directions); the shear moduli, too, are anisotropic. Most anisotropic of all is Poisson's ratio, which, for compression parallel to the prism axis, is almost zero. This means that compression in this direction produces little or no lateral spreading—a property useful in gaskets, and in corks which must seal particularly tightly. The wide range of values for a given property derives from the range of density of natural cork (which depends on growing conditions and the age of the tree) and on its moisture content.

Cork is a particularly "lossy" material, meaning that, when deformed and then released, a considerable fraction of the work of deformation is dissipated. This fraction is measured by the loss coefficient (Table 1); for cork it has a value between 0.1 and 0.3, giving the materials superior vibration and acoustic-damping properties. The high coefficient of friction, too, derives from the ability to dissipate energy. When an object slides or rolls on a cork surface, the cells deform and then recover again as the slider passes; the energy loss gives friction.

The cellular structure gives cork a particularly low thermal conductivity. Heat is conducted through a cellular solid by conduction through the cell walls, by conduction through the gas within the cells and by convection of this gas (Gibson and Ashby 1988). The cells in cork are sufficiently small that convection is suppressed completely, and the thin cell walls contribute little to heat transfer; the thermal conductivity is reduced to a value only slightly above that of the gas contained within the cells themselves.

Cork is remarkable for its chemical and biological stability. The presence of suberin, waxes and tannin give the material great resistance to biological attack and to degradation by chemicals. For this reason, it can be kept in contact with foods and liquids (such as wine) almost indefinitely without damage. And for reasons which are not fully understood, cork is fire-resistant: it smoulders rather than burns.

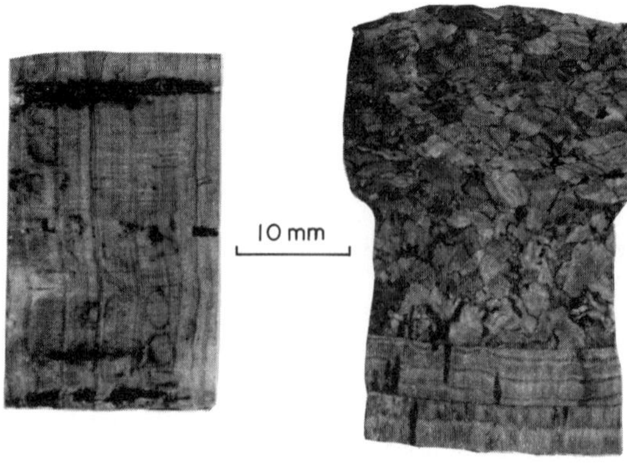

Figure 3
Ordinary corks (a) have their axis at right angles to the prism axis to prevent leakage through the lenticels (dark lines). A laminated cork (b) makes better use of the symmetry of the cork structure

3. *Uses*

For at least 2000 years, cork has been used for—among other things—"floats for fishing nets, and bungs for bottles, and also to make the soles for women's winter shoes" (Pliny, AD 77). Few materials have such a long history or have survived so well the competition from man-made substitutes. We now examine briefly how the special structure of cork has suited it so well to its uses.

3.1 Bungs for Bottles and Gaskets for Woodwind Instruments

Connoisseurs of wine agree that there is no substitute for corks made of cork. Plastic corks are hard to insert and remove, they do not always give a good seal, and they may contaminate the wine. Cork corks are inert and seal well for as long as the wine need be kept. The excellence of the seal derives from the elastic properties of the cork. It has a low Young's modulus E and, more importantly, it also has a low bulk modulus K. Solid rubber and solid polymers above their glass transition temperature have a low E but a large K, and it is this that makes them hard to insert into a bottle and gives a poor seal when they are in place.

One might expect that the best seal would be obtained by cutting the axis of the cork parallel to the prism axis of the cork cells; the circular symmetry of the cork and of its properties are then matched. This is correct: the best seal is obtained by cork cut in this way. But natural cork contains lenticels—tubular channels that connect the outer surface of the bark to the inner surface, allowing oxygen into, and CO_2 out from, the new cells that grow there. The lenticels lie parallel to the prism axis (the radial direction): a cork cut parallel to this axis will leak. This is why almost all commercial corks are cut with the prism axis (and the lenticels) at right angles to the axis of the bung (Fig. 3(a)).

A way out of this problem is shown in Fig. 3(b). The base of the cork, where sealing is most critical, is made of two disks cut with the prism axis (and lenticels) parallel to the axis of the bung itself. Leakage is prevented by gluing the two disks together so that the lenticels do not connect. The cork, when forced into the bottle, is then compressed (radially) in the plane in which it is isotropic, exerting a uniform pressure on the inside of the neck of the bottle.

Cork makes good gaskets for the same reason that it makes good bungs: it accommodates large elastic distortion and volume change, and its closed cells are impervious to water and oil. Thin sheets of cork are used, for instance, for the joints of woodwind and brass instruments. The sheet is always cut with prism axis (and lenticels) normal to its plane. The sheet is then isotropic in its plane, and this may be the reason for cutting it so. But it seems more likely that it is cut like this because the Poisson ratio for compression down the prism axis is zero. Thus when the joints of the instrument are mated, there is no tendency for the sheet to spread in its plane and wrinkle.

3.2 Floor Covering and the Soles of Shoes

Manufacturers who sell cork flooring claim that it retains its friction even when polished or covered with soap, and experiments (Gibson et al. 1981) confirm this.

Friction between a shoe and a cork floor has two origins. One is adhesion: atomic bonds form between the two contacting surfaces, and work must be done to break them if the shoe slides. Between a hard slider and a tiled or stone floor this is the only source of friction; and since it is a surface effect, it is destroyed by a film of polish or soap. The other source of friction is due to anelastic loss. When a rough slider moves on a cork floor, the bumps on the slider deform the cork. If cork were perfectly elastic no net work would be done: the work done in deforming the cork would be recovered as the slider moves on. But if the cork has a high loss coefficient (as it does) then it is like riding a bicycle through sand: the work done in deforming the material ahead of the slider is not recovered as the slider passes on, and a large coefficient of friction appears. This anelastic loss is the main source of friction when rough surfaces slide on cork; and since it depends on processes taking place below the surface, not on it, it is not affected by films of polish or soap. The same thing happens when a cylinder or sphere rolls on cork, which therefore shows a high coefficient of rolling friction.

3.3 Packaging and Energy Absorption

Many of the uses of cork depend on its capacity to absorb energy. Cork is attractive for the soles of shoes and flooring because, as well as having good frictional properties, it is resilient under foot, absorbing the shocks of walking. It makes good packaging because it compresses on impact, limiting the stresses to which the contents of the package are exposed. It is used as handles of tools to insulate the hand from the impact loads applied to the tool. In each of these applications it is essential that the stresses generated by the impact are kept low, but that considerable energy is absorbed.

Cellular materials are particularly good at this. The stress–strain curve for cork (Fig. 2) shows that the collapse strength of the cells (Table 1) is low, so that the peak stress during impact is limited. Large compressive strains are possible, absorbing energy as the cells progressively collapse. In this regard, its structure and properties resemble polystyrene foam, which has replaced cork (because it is cheap) in many packaging applications.

3.4 Thermal Insulation

Trees are thought to surround themselves with cork to prevent loss of water in the hotter seasons of the year. The two properties involved—low thermal conductivity and low permeability to water—make it an excellent material for insulation against cold and damp (the hermit caves of southern Portugal, for example, are lined with cork). It is for these reasons that crates and boxes are sometimes lined with cork. And the cork tip of a cigarette must appeal to the smoker because it insulates (a little) and prevents the tobacco getting moist.

3.5 Indentation and Bulletin Boards

Cellular materials densify when they are compressed or indented. So when a sharp object, like a drawing pin, is stuck into cork, the deformation is very localized. A layer of cork cells, occupying a thickness of only about one quarter of the diameter of the indenter, collapses, suffering a large strain. The volume of the indenter is taken up by the collapse of the cells so that no long-range deformation is necessary. For this reason the force needed to push the indenter in is small. Since the deformation is (nonlinear) elastic, the hole closes up when the pin is removed.

Bibliography

Eames A J, MacDaniels L H 1951 *An Introduction to Plant Anatomy*. McGraw-Hill, London

Emilia Rosa M, Fortes M A 1988 Stress relaxation and stress of cork. *J. Mater. Sci.* 23: 35–42

Esau L 1965 *Plant Anatomy*. Wiley, New York, p. 340

Ford B J 1982 The origins of plant anatomy: Leeuwenhoek's cork sections examined. *IAWA Bull.* 3: 7–10

Gibson L J, Ashby M F 1988 *Cellular Solids: Structure and Properties*. Pergamon, Oxford, Chap. 12

Gibson L J, Easterling K E, Ashby M F 1981 The structure and mechanics of cork. *Proc. R. Soc. London Ser. A* 377: 99–117

Hooke R 1664 *Micrographia*. Royal Society, London, pp. 112–21

Lewis P T 1928 The typical shape of polyhedral cells in vegetable parenchyma. *Science (N.Y.)* 68: 635–41

Natividade J V 1938 What is cork? *Bol. Junta Nac. da Cortica (Lisboa)* 1: 13–21

Pereira H, Emilia Rosa M, Fortes M A 1987 The cellular structure of cork from Quercus suber L. *IAWA Bull.* 8: 213–18

Sitte P 1962 Zum Feinbau der Suberinschichten in Flaschenkork. *Protoplasma* 54: 55–559

Zimmerman M H, Brown C L 1971 *Trees Structure and Function*. Springer, Berlin, p. 88

M. F. Ashby
[University of Cambridge, Cambridge, UK]

D

Decay During Use

Decay of wood during use is a serious problem, causing significant losses in strength in stages of decay so early as to be difficult to diagnose even microscopically. The key to preventing decay in wooden buildings is the exclusion of water. Although emphasis in this article is placed on experience with wood in structures, the principles of wood decay are equally applicable to any wood in use.

1. Causal Fungi and their Requirements for Growth

Decay of wood is caused by fungi which are members of the group of higher fungi called Basidiomycetes related to the common edible mushrooms. These fungi are primitive plants which lack chlorophyll and therefore cannot manufacture their own food, relying instead on the organic matter produced by plants possessing chlorophyll such as the wood produced by trees. The decay fungi, as scavengers, play an important role in the forest, returning organic matter to the soil, returning nutrients for reuse by other plants, preventing buildup of material which would make access for wildlife impossible and reducing the fire hazard produced by combustible material on the forest floor. However, the same natural process occurring in wood in structures is considered undesirable and so preventive measures are necessary.

Decay fungi have four basic requirements for growth and production of decay in wood: air (they are aerobic organisms), water, a favorable temperature and a food source. The food is of course the wood itself. This food may be made unavailable to the fungus by poisoning it with a preservative, but normal use of wood in above-ground portions of structures does not usually involve such treatment. Decay fungi have a wider tolerance to temperature than people who occupy wooden structures. Freezing temperatures will stop the progress of decay but will not kill the decay fungi. Temperatures must usually exceed about 60 °C before thermal death occurs. Decay fungi require so little oxygen that it is usually not possible to control decay by limiting this factor in structures; water submersion or deep burial, however, excludes oxygen sufficiently that wood in such conditions is not attacked by decay fungi. Thus, the remaining factor—water—is the key to controlling decay in buildings. Throughout history a major goal of the design and construction of structures has been to shed water effectively. If this is done, and no water leaks are allowed, there is no reason why above-ground portions of wooden buildings should decay. Wood is a hygroscopic material and takes up moisture from the surrounding air, but the moisture content attained by wood in 100% relative humidity is not sufficient to support decay; a source of liquid water such as rain, plumbing leaks or ground contact is required.

There are other fungi which inhabit wood, namely those which produce mold and blue stain, but they do not cause the extensive loss in wood strength which results from the action of decay fungi. These fungi live principally on storage materials in the wood, such as starch, and therefore do not break down the wood cell walls as do decay fungi. Since they have pigmented cell walls or spores, their main effect on wood is discoloration. However, although they tend to grow faster, they have generally the same requirements for growth as decay fungi and therefore their presence may indicate that the wood has been exposed to conditions conducive to decay.

2. Sources of Inoculum

Any living cells of a fungus, when transferred to a new location favorable for growth, have the potential for reproducing the fungus in this new site. Decay fungi produce spores which act like seeds of higher plants in that they are capable of starting a new growth of fungus in a new area where conditions are favorable. The fungi produce great numbers of such spores light enough to be carried by the winds, thereby assuring that fungi are present wherever conditions conducive to decay exist. Moreover, the careless manner in which wood products are often handled tends to ensure that wood in structures is infected before being placed in service.

Wood may be shipped and stored in the green condition (see *Hygroscopicity and Water Sorption*), thereby allowing time for it to become infected. Wood products may be delivered to the job site and set on the soil; since decay fungi live in the soil this provides an opportunity for them to move into the wood before it is placed in the structure. If the wood is allowed to dry and remains dry throughout the life of the structure, such preinfection may be of no consequence. However, if conditions later become conducive to decay, the necessary fungi may already be present.

3. Types of Decay

There are two major categories of decay produced by Basidiomycetes: brown rot and white rot. The brown-rot fungi decompose only the carbohydrate fraction of wood, leaving the lignin modified but not metabolized (see *Chemical Composition*). The residual lignin is darker in color than the cellulose, leading to the brown coloration of the rot, and has very little strength, leading to the excessive shrinkage and cubic checking pattern typical of brown-rotted wood.

White-rot fungi decompose all the major components of wood and presumably may lead to 100% loss in cell wall material. Some of the white-rot fungi appear to attack lignin preferentially in the early stages of decay, while others, called simultaneous rots by some workers, decompose both the lignin and the carbohydrates at a rate approximately proportional to their original relative concentration in the wood.

A third category of severe wood deterioration—soft rot—is not caused by Basidiomycetes and therefore should not be included under the term decay. Nevertheless, soft rot can cause such serious decomposition of wood and loss of strength in certain situations that it must be considered in any treatise concerned with the effects of wood decomposition. Soft rot is caused by fungi of the mold and stain group (Ascomycetes and Fungi Imperfecti). It is not known whether fungi which normally inhabit wood as mold or stain producers are also capable of producing soft rot under the right conditions, or if only special strains of these organisms are capable of producing soft rot when present. In any event, soft rot occurs in environmental conditions under which growth of decay fungi would not be expected, such as in preservative-treated wood, and very wet or water-submerged wood. The hyphae of soft-rot fungi grow within the S2 layer of the secondary wall of wood cells, producing cavities parallel to the microfibrils making up that layer (see *Ultrastructure*). After removal of the S2 layer there is so little left of the cell walls that the softened wood either sloughs off or fails. Soft-rot fungi produce deterioration similar to that of brown-rot fungi in that they primarily remove the carbohydrates from the cellulose-rich S2 cell wall layer, but, unlike brown-rot fungi, they metabolize some lignin as well.

Some bacteria are capable of decomposing wood also, but their attack is usually very slow and occurs only in conditions of high moisture, such as saturation or submersion. Because many of these bacteria can grow under anaerobic conditions, their attack is often the sole cause of degradation in submerged or buried wood. Bacterial attack would not be expected in buildings.

4. Importance of Water

As stated above, the key to decay in wood structures is water. Wood kept dry will never decay. Furthermore, since the major goal of structural design and construction is the shedding of water, keeping wood dry in a structure should be the easiest, cheapest and most effective method of assuring that the wood does not decay. Building leaks, design features which lead to water infiltration, features which tend to trap water and leaks in or condensation on interior plumbing should be carefully avoided to preserve the life of a wooden structure.

5. Microscopic and Chemical Effects of Decay

Not all decay fungi attack wood in the same way, and a given fungus may produce different effects in different woods. Therefore, not all the features discussed below will necessarily be found in any given piece of decayed wood. However, the following microscopic features may frequently be required for diagnosing the presence of early stages of decay; the more they are present, the more confidence may be placed in the diagnosis.

A fungus can be thought of as being like a ball of cotton with filaments running in all directions. The filaments that make up the body of a fungus are called hyphae. Individual hyphae are too small to be seen easily without a microscope, but massed together they form what is known as a mycelium, which can readily be seen on the wood surface. The hyphae of decay fungi may contain very distinctive loop-like structures called clamp connections. Only Basidiomycetes have them, and the only Basidiomycetes which would be present in wood would be decay fungi. Therefore, if microscopic examination reveals hyphae with clamp connections, then the fungus is a decay fungus. Unfortunately, many decay fungi do not produce clamp connections and some which produce them in culture do not produce them in wood. Clamp connections, then, are a valuable diagnostic tool if present, but their absence supplies no useful diagnostic information.

The hyphae of decay fungi may invade wood in early stages by growth through ray cells and large longitudinal elements, such as resin ducts or vessels. This is true of mold and stain fungi too, but unlike these fungi the decay fungi normally leave these elements and ramify throughout the wood. In some cases decay-fungal hyphae may be most numerous in the strength-providing longitudinal elements (fibers or tracheids). Penetration may be through pits or by means of boreholes dissolved through the wood cell walls by the enzymes from the hyphae. The boreholes of decay fungi are usually numerous. They enlarge to several times the hyphal diameter, tend to run in series (serial boreholes) through a number of wood cell walls in a relatively straight line (as seen in radial section) and bear little relationship to pits. The boreholes produced by mold and stain fungi are infrequent and tend to remain at the diameter of the penetrating hypha, resulting in a "necked-down" appearance to the large hypha penetrating a small hole. They tend to be solitary, with the penetrating hypha changing direction after penetration, and penetration is frequently through pits with no borehole formation at all. The hyphae of soft-rot fungi may produce serial boreholes, but they normally consist of large, frequently pigmented hyphae which "neck-down" considerably to pass through the wall and are therefore not too difficult to differentiate from those of decay fungi.

Early stages of attack on the cell walls may be characterized by separations within or between cells, and shallow depressions or gouges at the wall–lumen interface. In later stages of decay, white-rotted wood cell walls may appear thinned in a uniform fashion from cell to cell, whereas brown-rotted wood cells appear shrunken and distorted on a very irregular basis from cell to cell. Soft rot results in cavities within the S2 layer of the secondary wall which parallel the microfibrils in this layer and tend to be diamond-shaped or elongated with pointed ends, making soft rot easily distinguishable from brown rot or white rot under the microscope.

White-rot fungi decompose all the major cell-wall components and some remove them at rates approximately proportional to the amounts present in sound wood, thereby leading to the observed progressive thinning of the cell walls.

Brown-rot fungi decompose primarily the carbohydrate fraction of the wood. Lignin is altered by their attack, but little of it is removed. In early stages of brown rot the wood constituents may be depolymerized more rapidly than the decomposition products are metabolized by the fungus, leading to a buildup of low-molecular-weight molecules which constitute mass but provide no strength.

6. *Effects of Decay on Strength Properties*

Resistance to impact loading is the mechanical property of wood most sensitive to the presence of decay. Toughness, for example, may be largely destroyed at a stage of decay so early that it may be hard to detect even microscopically (Table 1). Those strength properties relied upon in the use of wood as a beam, a common mode of service, are also severely diminished in very early stages of decay. In such early decay stages there is little difference noted between brown rot and white rot with regard to the magnitude of strength losses. However, as decay advances, further effects on strength are much more severe in brown rot than in white rot.

Table 1
Expected strength loss at an early stage of decay (5–10% weight loss) in brown-rotted softwoods (after Wilcox 1978)

Strength property	Expected strength loss (% of strength of sound wood)
Toughness	⩾80
Impact bending	80
Static bending (MOR and MOE)	70
Compression prependicular to grain	60
Tension parallel to grain	60
Compression parallel to grain	45
Shear	20
Hardness	20

Because of the drastic reduction in most strength properties caused by decay at such an early stage as to be difficult to diagnose, it is easy to see that it is considerably safer to prevent decay in the first place than to rely upon ability to detect and diagnose it accurately in sufficient time to effect a remedy before failure occurs.

7. *Natural Decay Resistance*

Some woods have what is called natural decay resistance, that is, their heartwoods are infiltrated with chemicals at the time of heartwood formation which make that wood unpalatable to decay fungi. Through both laboratory testing and field performance, various woods which appear in the US commercial market have been rated according to the natural decay resistance of their heartwoods (Tables 2 and 3). Decay-susceptible wood will perform adequately in structures

Table 2
Relative grouping of some US domestic woods according to heartwood decay susceptibility based upon laboratory tests and and field performance (after US Department of Agriculture 1987)

Resistant or very resistant	Moderately resistant	Slightly resistant or nonresistant
Baldcypress (old growth)[a]	Baldcypress (young growth)[a]	Alder
Catalpa	Douglas fir	Ashes
Cedars	Honeylocust	Aspens
Cherry (black)	Larch (western)	Basswood
Chestnut	Oak (swamp chestnut)	Beech
Cypress (Arizona)	Pine	Birches
Junipers	Eastern white[a]	Buckeye
Locust (black)[b]	Southern	Butternut
Mesquite	Longleaf[a]	Cottonwood
Mulberry (red)[b]	Slash[a]	Elms
Oak	Tamarack	Hackberry
Bur		Hemlocks
Chestnut		Hickories
Gambel		Magnolia
Oregon white		Maples
Post		Oak (red and black species)
White		Pines (other than longleaf, slash and eastern white)
Osage orange[b]		Poplars
Redwood		Spruces
Sassafras		Sweetgum
Walnut (black)		True firs (western and eastern)
Yew (Pacific)[b]		Willows
		Yellow poplar

[a] The southern and eastern pines and baldcypress are now largely second growth with a large proportion of sapwood. Consequently, substantial quantities of heartwood lumber of these species are not available
[b] These woods have exceptionally high decay resistance

Table 3
Relative grouping of some woods imported into the USA according to heartwood decay susceptibility based upon laboratory tests and field performance (after US Department of Agriculture 1987)

Resistant or very resistant	Moderately resistant	Slightly resistant or nonresistant
Angelique	Andiroba[a]	Balsa
Apamate	Apitong[a]	Banak
Brazilian rosewood	Avodire	Cativo
Caribbean pine	Capirona	Ceiba
Courbaril	European walnut	Jelutong
Encino	Gola	Limba
Goncalo alves	Khaya	Lupuna
Greenheart	Laurel	Mahogany, Phillippine: Mayapis
Guijo	Mahogany, Phillippine: Almon	Mahogany, Phillippine: White lauan
Iroko	Mahogany, Phillippine: Bagtikan	Obeche
Jarrah	Mahogany, Phillippine: Red lauan	Parana pine
Kapur	Mahogany, Phillippine: Tanguile	Ramin
Karri	Ocote pine	Sande
Kokrodua (Afrormosia)	Palosapis	Virola
Lapacho	Sapele	
Lignum vitae		
Mahogany, American		
Meranti[a]		
Peroba de campos		
Primavera		
Santa Maria		
Spanish cedar		
Teak		

[a] More than one species included, some of which may vary in resistance from that indicated

if kept dry. Moderately decay-resistant wood may perform adequately, even if occasionally wetted but then allowed to dry out. Highly decay-resistant wood may perform adequately in above-ground exposure even if allowed to remain wet for substantial periods of time. Most woods, even if highly decay-resistant, will provide only limited service in contact with the soil, which is the most hazardous terrestrial exposure to which wood can be put. To provide adequate service in ground contact, most woods must be pressure-treated with an effective preservative.

8. *Detailing for Shedding Water*

The first part of a building to intercept rainwater is the roof. Flat roofs have become popular in some parts of the world, no doubt primarily for economic reasons, but they are dangerous for a wooden structure in that the performance of the entire structure depends solely upon the effectiveness of the membrane placed on top of the flat roof. One pinhole or small crack in this barrier may jeopardize the safety of the entire structure. It is better to provide a slope and drainage so that the natural force of gravity is working for rather than against the building and assists in the shedding of water. Roof overhangs are useful in removing the drip line from the rest of the structure and in protecting the walls and wall penetrations from direct contact with rainwater. Gutters and downspouts are also important since they collect water from the roof and dispose of it safely away from the structure rather than allowing it to drip, splash or run down various portions. If no roof overhang is provided, the flashing and detailing at the intersection between the roof and the vertical walls becomes critical. Flashings and moisture barriers behind exterior wall coverings should be designed and assembled so that, at each joint, the tendency is to force water out, away from the structure. It is surprising how many structural failures due to decay in wooden buildings can be traced to careless reversal of the flashing or shingling of moisture barriers so that water coming in contact with these structures is drawn into the interior of the wall rather than being shed. Careful attention to such apparently minor features may be major factors in attaining adequate service from a wooden building.

Being a hygroscopic material, wood will adjust its moisture content to the humidity of the air around it. This means that if the air becomes more moist, the wood will increase in moisture content, resulting in swelling. If the surrounding air becomes drier, the wood will lose moisture and shrink. Wood undergoes normal fluctuation in size and shape with the annual change in seasons and the concomitant differences in relative humidity. Such changes must be taken into account in the design and construction of wooden buildings. If wood is put into a structure at a higher moisture content than it is expected to reach when at equilibrium with its surroundings, shrinkage must be expected, and this overall reduction in the size of each member must be taken into account during construction. On the other hand, dry wood products which will be expected to take on moisture in service must be provided with sufficient gaps between them to allow for the resultant swelling. Both wood shrinkage and swelling with the crushing which can accompany it may result in gaps which can lead to water infiltration.

9. *Ventilation*

Most processes associated with habitation of structures, such as kitchen, bathroom and laundryroom operations, may introduce significant amounts of water vapor into the living space of the structure which must be vented to the outside. Furthermore, moisture leaving the soil on which a wooden structure is built may condense on wood surfaces or on plumbing in contact with wood and provide sufficient water to induce decay problems. The solution to both these problems is adequate ventilation.

All activities within the living space which release large amounts of water vapor should be vented mechanically (using fans if necessary) to the outside to prevent the necessity for such water to make its way through the walls or ceiling of the structure. Substructure areas where the soil is likely to be moist should have sufficient cross ventilation supplied through or above the foundation to keep the relative humidity of the air in the subspace below the dew point at all seasons of the year. If the soil is so wet that the amount of ventilation required becomes excessive, a vapor barrier should be placed over the soil surface in the sub-area to prevent moisture from leaving the soil.

10. Hazards of Ground Contact

Soil is the most hazardous terrestrial environment in which wood can be placed. Soil is usually moist and, if so, has the capacity of supplying moisture to the wood at a perfect rate to be conducive to decay. In addition, decay fungi live in the soil on cellulosic residue such as wood scraps, roots, stumps and other plant debris. Therefore, placing wood in contact with the soil may provide both an actively growing decay fungus and the one missing factor necessary to produce decay—water. In many parts of the world, ground contact has the additional hazard of exposing wood to direct entry by subterranean termites (see *Deterioration by Insects and Other Animals During Use*).

11. Insulation and Vapor Barriers

As a result of the need to reduce heating cost, great emphasis is currently being placed on providing adequate insulation in structures. In some cases the addition of insulation to existing wooden structures is advocated. Some insulations, such as the closed-cell plastic foams, may also act as effective vapor barriers. Other insulation materials, like fiberglass placed on a metal foil, have vapor barriers incorporated in them to improve their performance as insulation materials. Both types of materials when added to wooden structures may be extremely dangerous if they impede the natural flow of moisture vapor out of the living area of the structure or change the location of the dew point in such a way as to cause condensation in walls or ceilings. At present many construction and design professionals are inadequately informed in this area. Such insulation practices, if not based on adequate knowledge and theory, though well-meaning in their intent, may end up costing more energy than they save in terms of replacement of wood which they have made vulnerable to rot.

See also: Building with Wood; Preservative-Treated Wood; Radiation Effects; Weathering

Bibliography

Eslyn W E, Clark J W 1979 *Wood Bridges: Decay Inspection and Control*, USDA Agriculture Handbook No. 557. US Government Printing Office, Washington, DC

Graham R D, Helsing G G 1979 *Wood Pole Maintenance Manual: Inspection and Supplemental Treatment of Douglas-Fir and Western Redcedar Poles*, Research Bulletin 24. Oregon State University, Corvallis, Oregon

Meyer R W, Kellogg R M (eds.) 1982 *Structural Uses of Wood in Adverse Environments*. Van Nostrand Reinhold, New York

Nicholas D D (ed.) 1973 *Wood Deterioration and its Prevention by Preservative Treatments*. Syracuse University Press, New York

Rosenberg A F, Wilcox W W 1982 How to keep your award-winning building from rotting. *Wood Fiber* 14: 70–84

Scheffer T C, Verrall A F 1973 *Principles for Protecting Wood Buildings from Decay*, USDA Forest Service Research Paper FPL 190. US Forest Products Laboratory, Madison, Wisconsin

US Department of Agriculture 1987 *Wood Handbook: Wood as an Engineering Material*, USDA Agriculture Handbook No. 72. US Government Printing Office, Washington, DC

Wilcox W W 1978 Review of literature on the effects of early stages of decay on wood strength. *Wood Fiber* 9: 252–57

W. W. Wilcox

[University of California, Berkeley, California, USA]

Deformation Under Load

The two outstanding features of the deformation behavior of wood under load are that it is highly anisotropic and, due to the biological nature of wood, very variable. Furthermore, the deformation of wood is time dependent even under normal ambient conditions. For many applications, however, the time-dependent deformation constitutes such a small fraction of the total deformation that the former may be safely neglected. Wood is then treated, to a first approximation, as a linearly elastic material. In this article wood will also be considered as a viscoelastic material.

The nature of the deformation behavior of wood can be clarified by reference to the glass transition temperature, T_g, of its components. Lignin and hemicellulose as wholly or largely amorphous polymers and the amorphous portion of cellulose (see *Chemical Composition*) all exhibit glass transition temperatures. In the dry state, values of T_g are about 230 °C for cellulose, 180 °C for lignin and 200 °C for hemicellulose. At room temperature, wood is therefore clearly in a glassy state. Adsorbed moisture, however, acts as a plasticizer and lowers T_g. Lignin, being less hygroscopic, is not affected as much as cellulose and hemicellulose. The T_g of the latter two falls below room temperature at the fiber saturation point, so that wet wood will have some components that are in a rubbery state.

1. Wood as a Linearly Elastic Material

Figure 1 shows what might be considered typical stress–strain curves for wood in tension and compression. They are for wood loaded in a direction parallel to the grain, in other words, parallel to the long axis of the majority of the cells (see *Macroscopic Anatomy*). Both curves have an initial portion which is more or less rectilinear, followed by a curvilinear portion to the point of maximum stress. The stress at the junction of the rectilinear and curvilinear portions is referred to as the fiber stress at proportional limit. The slope of the stress–strain curve below the proportional limit represents the modulus of elasticity parallel to the grain. Although it has been suggested that the modulus of elasticity is slightly higher in tension than in compression, the difference, if it exists at all, is so small that it can be neglected for all practical purposes.

Beyond the proportional limit the deformation becomes nonlinear, and the degree to which this takes place is much greater in compression than in tension. In tension the inelastic deformation is restricted to molecular mechanisms which are not well understood. Since wood is a porous material, deformation in compression can additionally take place by crinkling and folding of cell walls and localized buckling into the cell cavities. The inelastic deformation of wood is largely permanent, but wood does not exhibit plasticity in the same sense as do, for instance, metals. Figure 1 also shows that the strength is greater in tension than in compression (see *Strength*).

At any point in wood there are three mutually orthogonal planes of structural symmetry defined by three orthogonal directions, as shown in Fig. 2. The directions are longitudinal (L), which is parallel to the axis of the tree trunk and to the majority of wood cells; tangential (T), which is tangential to the annual or growth rings; and radial (R), which is orthogonal to the other two. Cylindrical anisotropy therefore seems to be the most suitable model for wood. However, since most pieces of wood are cut from positions away from the center of the tree trunk, it is customary to neglect the curvature of the growth rings and to treat wood as an orthotropic material. The matrix of elastic compliances (S_{ijkl}) for orthotropic materials is:

$$\begin{bmatrix} S_{1111} & S_{1122} & S_{1133} & 0 & 0 & 0 \\ S_{2211} & S_{2222} & S_{2233} & 0 & 0 & 0 \\ S_{3311} & S_{3322} & S_{3333} & 0 & 0 & 0 \\ 0 & 0 & 0 & 4S_{2323} & 0 & 0 \\ 0 & 0 & 0 & 0 & 4S_{1313} & 0 \\ 0 & 0 & 0 & 0 & 0 & 4S_{1212} \end{bmatrix}$$

Since the matrix is symmetric, we have nine independent elastic constants. Equating the x_1 and L directions, the x_2 and T directions and the x_3 and R directions, we can express the compliances in terms of three Young's moduli, E_L, E_R and E_T, three shear moduli, G_{LT}, G_{LR} and G_{TR} and six Poisson ratios, ν_{LT}, ν_{TL}, ν_{LR}, ν_{RL}, ν_{TR} and ν_{RT}. Thus, $S_{1111}=1/E_L \ldots$, $S_{1122}=-\nu_{TL}/E_T \ldots$, and $4S_{1212}=1/G_{LT} \ldots$. The first subscript in the Poisson ratios indicates the direction of

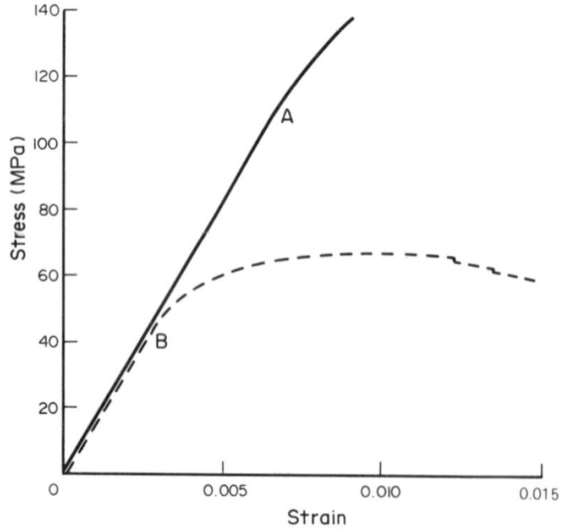

Figure 1
Stress–strain curves in tension and compression parallel to grain for air-dry samples of Bishop pine (*Pinus muricata*):——tension;- - -compression; A, B proportional limits in tension and compression, respectively

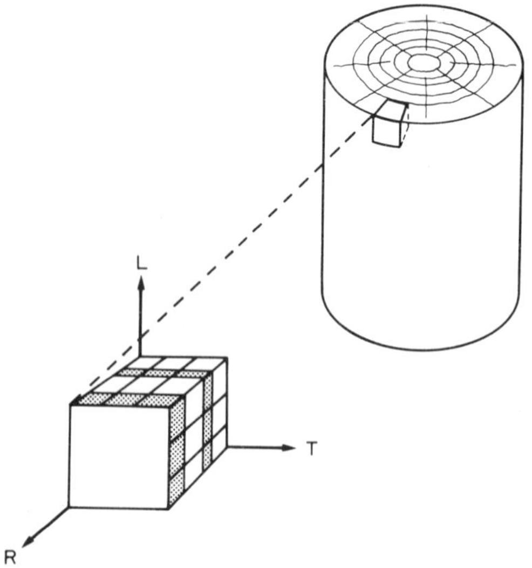

Figure 2
Wood element showing principal structural axes: longitudinal (L), radial (R) and tangential (T), together with the tree trunk section from which it was taken

applied stress and the second the direction of lateral expansion or contraction.

Some typical values for the elastic constants are shown in Table 1. Note that E_L is an order of magnitude larger than either E_T or E_R; compared with this difference, that between the two latter values, although substantial of itself, is relatively small, which explains why it is possible to neglect the curvature of the growth rings and consider wood to be orthotropic. Note also that some Poisson ratios are greater than 0.5, which is not contradictory as long as they are complemented by appropriately smaller ratios in other directions. Some of the values are less than 0.05 and very difficult to measure. They are therefore often calculated from the symmetry relations: $\nu_{LT}/E_L = \nu_{TL}/E_T$, $\nu_{LR}/E_L = \nu_{RL}/E_R$ and $\nu_{TR}/E_T = \nu_{RT}/E_R$.

The matrix of compliances given above applies only when the principal directions of structural symmetry coincide with the coordinate axes. If they do not, some or all of the zero terms in the original matrix may become nonzero in the transformed matrix, so that for instance normal stresses give rise to shear strains and, conversely, shear stresses give rise to normal strains. In lumber such an "off-angle" condition is known as cross grain, meaning that the grain or fiber direction is not parallel to the geometric long axis of the piece. This also leads to a coupling between bending and torsion, that is, a cross-grained piece of lumber tends to twist as well as bend if subjected to bending moments, or bends as well as twists if subjected to torsional moments.

2. *Factors Influencing the Elastic Constants of Wood*

The elastic moduli, both Young's moduli and shear moduli, are subject to various influencing factors, much as are the strength properties of wood (see *Strength*). First there is a difference between species, which can amount to an order of magnitude, as shown by balsa and birch in Table 1. A major part of the species difference can be explained on the basis of wood density, which is a measure of porosity and thus of the amount of wood substance present per unit volume (see *Density and Porosity*). The Poisson ratios, on the other hand, are relatively unaffected by density.

Above the fiber saturation point (about 28%) (see *Hygroscopicity and Water Sorption*) moisture content has no effect on elastic properties, but as moisture content decreases below this level, the elastic moduli increase. The Young's modulus E_L parallel to the grain increases about 30% when wood initially above the fiber saturation point is dried to 12% moisture content. The Young's moduli E_T and E_R and all three shear moduli are more sensitive to moisture content than is E_L, whereas particular Poisson ratios may either increase or decrease with decreases in moisture content.

Increases in temperature result in decreases in the elastic moduli. The effect depends on moisture content, dry wood being less sensitive than wood of high moisture content. In the range from -100 to $+100\,°C$ the relationship between the elastic moduli and temperature is linear. Taking the room temperature values as a base, E_L decreases (increases) by 0.1–0.7% for each 1 °C increase (decrease) in temperature, depending on moisture content. For the two other Young's moduli and the shear moduli, the effect of temperature is more pronounced, ranging from 0.5 to 1.5% per 1 °C, again depending on moisture content. The effect of temperature on Poisson ratios is not known.

As a natural product of biological origin, even clear, straight-grained wood of the same species is variable in its properties. For the modulus of elasticity E_L as determined from bending tests, the coefficient of variation (standard deviation expressed as a percentage of the mean) is about 20%. Assuming a normal distribution, this means that about 68% of all values are within 20% of the mean.

In most practical applications it is a matter not of clear, straight-grained wood but of lumber which has a number of growth characteristics or defects that affect strength and stiffness. Since lumber is usually used in the form of linear elements, such as beams or columns, the effective modulus of elasticity E_{eff} along the axis of the piece is of much more concern than any of the other elastic constants. The two main defects of interest are cross grain and knots. Figure 3 shows the variation of E with grain angle and illustrates clearly that small amounts of cross grain (small angles of deviation) can lead to substantial reductions in E_{eff}. Knots similarly lead to reductions in stiffness. Although the knot itself may be harder and denser than

Table 1
Elastic constants of some woods[a]

Species	Density ($kg\,m^{-3}$)	E_L (MPa)	E_T (Mpa)	E_R (MPa)	G_{LT} (MPa)	G_{LR} (MPa)	G_{TR} (MPa)	ν_{LT}	ν_{TL}	ν_{LR}	ν_{RL}	ν_{TR}	ν_{RT}
Balsa	100	2440	38	114	85	124	14	0.49	0.01	0.23	0.02	0.24	0.66
Birch	620	16300	620	1110	910	1180	190	0.43	0.02	0.49	0.03	0.38	0.78
Scots pine	550	16300	570	1100	680	1160	66	0.51	0.02	0.42	0.04	0.31	0.68
Sitka spruce	390	11600	500	900	720	750	39	0.47	0.02	0.37	0.03	0.25	0.43

[a] Hearmon (1948)

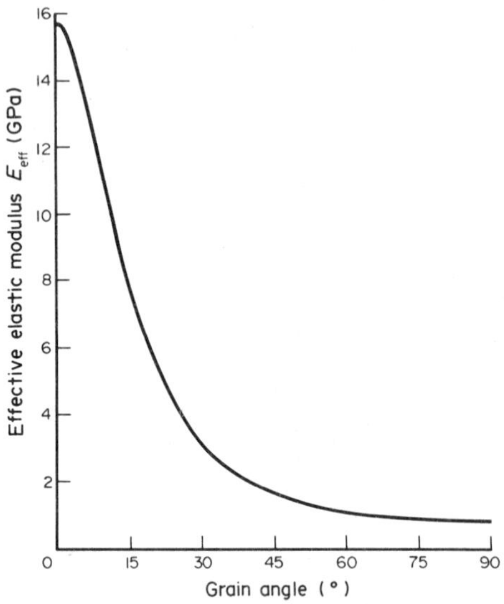

Figure 3
Effective modulus of elasticity as a function of grain angle for air-dry Douglas fir (*Pseudotsuga menziesii*) (calculated from data in Hearmon 1948)

the surrounding wood, the directions of its fibers are more or less perpendicular to the long axis of the piece of lumber. Furthermore, grain distortions around the knot have a detrimental effect. Reductions in E_{eff} are proportional to the severity of cross grain and the size of knots. These variations in stiffness are in turn correlated with strength, so that measurements of the axial modulus of elasticity of lumber can be used in the nondestructive evaluation of structural lumber grades (see *Nondestructive Evaluation of Wood and Wood Products*).

3. *Wood as a Viscoelastic Material*

While the linearly elastic model is adequate for many applications, wood under load does in fact undergo time-dependent deformation in addition to the instantaneously elastic deformation. The additional deformation or creep occurs at a rate which decreases with time, as shown in the curves of Fig. 4. If stresses are low, the deformation can reach virtual equilibrium after a period of loading of 2–3 years. At high stresses, however, a point may be reached where an inflection appears in the creep curve and the creep rate increases again. Once the creep rate does increase again, failure is inevitable unless the load is removed. This process of delayed failure or static fatigue is known as the duration-of-load effect in wood (see *Strength*).

The creep deformation can be divided into two parts, one of which is recoverable after unloading and the other is not. They are also referred to as the delayed elastic response and the flow, respectively. Recovery of creep is the inverse process, being rapid at first and proceeding at a decreasing rate, asymptotically approaching its final value. Both forms of creep can occur at all stress levels.

Within limits, wood can be considered a linear viscoelastic material to which the Boltzmann superposition principle applies. One limiting factor is stress level, which in dry wood at room temperature generally should not exceed 60% of the short-term static strength. At elevated temperature and moisture content, nonlinear behavior may become evident at lower stress levels. Linear behavior persists at higher stress levels in tension than in compression. Another limiting factor is time, because the flow is only approximately linear for short times and occurs at a decreasing rate which virtually vanishes after a few years of loading.

Creep under constant load and steady conditions can be represented by equations of the type

$$\varepsilon_t = \varepsilon_0 + mt^n \tag{1}$$

where ε_t is the total creep strain, ε_0 is the initial strain, t is time and m and n are constants. For limited timescales it is also possible to use the equations of the Burgers model (Kelvin and Maxwell models in series):

$$\varepsilon_t = \sigma_0\{J_0 + J_\infty[1 - \exp(-t/\tau)] + t/\eta\} \tag{2}$$

where σ_0 is the applied stress and J_0, J_∞, τ and η are parameters of the spring and dashpot elements of the model. The three terms in braces represent the instantaneous elastic, delayed elastic and flow components, respectively. For an extended timescale, the model can be generalized by adding terms of delayed elastic response with different retardation times τ_i.

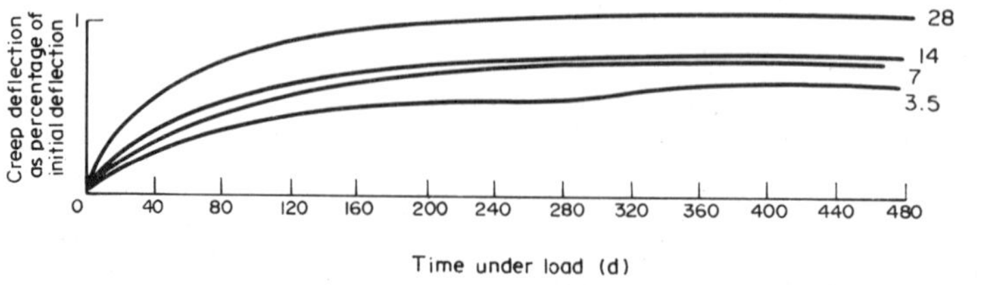

Figure 4
Creep of air-dry beams of mountain ash (*Eucalyptus regnans*) at 40 °C (after Kingston 1962)

Creep deformation is often expressed as a fraction of the initial—presumably instantaneously elastic—deformation. As a general rule of thumb, fractional creep of up to 1.0 can be expected with long periods of loading under moderate conditions, that is, the additional time-dependent deformation can reach about the same order of magnitude as the initial elastic deformation. Fractional creep tends to be greater across the grain than parallel to the grain and greater in compression than in tension.

Creep in wood can be greatly increased if there is simultaneous moisture sorption, as for instance in a green beam drying under load. Even more pronounced effects are obtained if there is moisture cycling, as illustrated in Fig. 5. It should be pointed out, however, that the results in Fig. 5 are based on beams only 2 × 2 mm in cross section; in larger members the effect of humidity cycling is much reduced, since the resulting changes in moisture content tend to be confined to the surface layers.

During dynamic loading (vibrational loading), the resulting deformation of the viscoelastic material is subject to damping. In forced vibration this manifests itself by a phase difference between stress and strain, and in free vibration by a gradual decay of amplitude. In either case the damping capacity can be expressed in terms of the logarithmic decrement, which in the case of free vibration is defined as the natural logarithm of the ratio of two successive amplitudes. The value of the logarithmic decrement depends on grain direction, moisture content, temperature and frequency. For dry wood at room temperature it is in the region of 0.02–0.05.

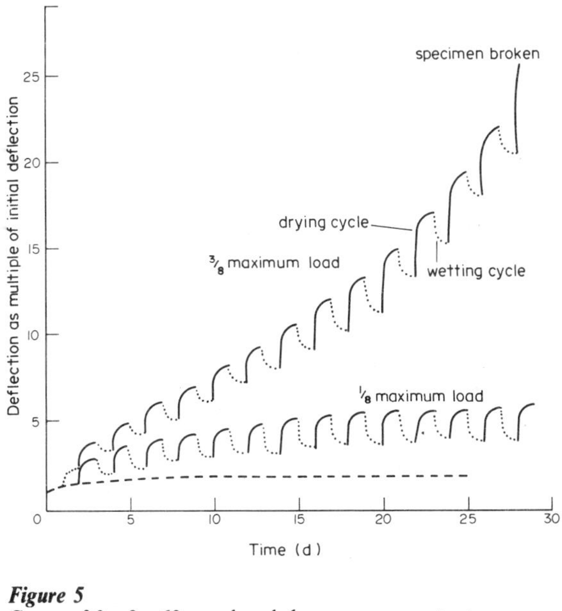

Figure 5
Creep of 2 × 2 × 60 mm beech beams: --- constant conditions ($\frac{3}{8}$ maximum load, 93% relative humidity); and alternative drying (——) and wetting (· · · ·) cycles (after Hearmon and Paton 1964)

See also: Lumber: Behavior Under Load; Stringed Instruments: Wood Selection

Bibliography

Back E L, Salmén N L 1982 Glass transitions of wood components hold implications for molding and pulping processes. *Tappi* 65 (7): 107–10

Bodig J, Goodman J R 1973 Prediction of elastic parameters for wood. *Wood Sci.* 5: 249–64

Bodig J, Jayne B A 1982 *Mechanics of Wood and Wood Composites.* Van Nostrand Reinhold, New York

Dinwoodie J M 1981 *Timber: Its Nature and Behavior.* Van Nostrand Reinhold, New York

Hearmon R F S 1948 *The Elasticity of Wood and Plywood,* Forest Products Research Special Report No. 7. Department of Scientific and Industrial Research, London

Hearmon R F S, Paton J M 1964 Moisture content changes and creep of wood. *For. Prod. J.* 14: 357–59

Jayne B A (ed.) 1972 *Theory and Design of Wood and Fiber Composite Materials.* Syracuse University Press, New York

Kelley S S, Rials T G, Glasser W G 1987 Relaxation behavior of the amorphous components of wood. *J. Mater. Sci.* 22: 617–24

Kingston R S T 1962 Creep, relaxation and failure of wood. *Res. Appl. Ind.* 15: 164–70

Kollmann F F P, Côté W A Jr 1968 *Principles of Wood Science and Technology,* Vol. 1. Springer, Berlin

Schniewind A P 1968 Recent progress in the study of the rheology of wood. *Wood Sci. Technol.* 2: 188–206

Schniewind A P 1981 Mechanical behavior and properties of wood. In: Wangaard F F (ed.) 1981 *Wood: Its Structure and Properties,* Educational Modules for Materials Science and Engineering Project. Pennsylvania State University, University Park, Pennsylvania

Schniewind A P, Barrett J D 1972 Wood as a linear orthotropic viscoelastic material. *Wood Sci. Technol.* 6: 43–57

US Forest Products Laboratory 1987 *Wood Handbook: Wood as an Engineering Material,* USDA Handbook No. 72. US Government Printing Office, Washington, DC

A. P. Schniewind
[University of California, Berkeley, California, USA]

Density and Porosity

Wood is a porous material. Its porosity exists either in the form of cell cavities, the interconnecting pit system, resin ducts and intercellular spaces, or as voids within cell walls. Since the density of the cell wall material is relatively constant, it follows that the density (or relative density) of wood is a measure of its gross porosity. Gross porosity has such an important influence on mechanical and other physical properties of wood that it is the most useful index of the suitability of wood for many end-product uses.

1. Density and Relative Density

Weight and volume of a body would seem to be among the easiest physical characteristics to measure. In reality, for a porous material such as wood, the measurement of volume is extremely difficult, particularly so in the determination of cell wall and wood substance densities. For most materials, density is defined as the mass per unit volume at specified conditions. Wood density, however, is usually based on the moisture-free or oven-dry weight, and on the moisture-free volume, the fully swollen or green volume or the so-called "volume at test," which can be at any specified moisture content between the moisture-free condition and the fiber saturation point. (Volume at 12% moisture content is frequently used.)

The density of wood is affected by the density of the wood structure, the moisture content of the material (which affects both volume and weight) and the amount of mineral and extraneous materials present. Comparison between species is facilitated somewhat by using relative density, defined as the ratio of the density of the material to the density of pure water at 4 °C. Relative density is always based on a wood density in which the numerator is the moisture-free weight, thus eliminating variability caused by moisture.

2. Relative Density as a Measure of Porosity

The relative density of wood substance is about 1.5, which means that a piece of dry wood with a relative density of 0.5 contains about one volume part of substance to two parts of air. Relative density is a measure of gross porosity, but does not provide information on the nature of the pores or the distribution of pore sizes. The latter may be more relevant when dealing with processes involving the flow of liquids through wood, diffusion rates or the availability of surface area to molecules.

Optical and electron microscopy are useful in determining pore size and geometry, but volume determinations are difficult. Optical microscopy is limited in its usefulness to pores with diameters greater than 5 μm, while with electron microscopy there are problems of nonrepresentative sampling. One of the most useful techniques is mercury porisometry, which has permitted the measurement of pores ranging in size from 100 μm to 3 nm.

3. Relative Density of Wood Substance and Constituents

Dry cell wall densities have been reported as varying from 1.44 to 1.50 $g\,cm^{-3}$. These variations may result from differences in either the proportion, densities or arrangement of the basic cell wall substances, or in the amount of voids within the wall. The void volume of the dry cell wall does not exceed 5%, but varies from wood to wood and accounts for a considerable amount of the variability in cell wall density. The relative degree of order of the cellulosic constituent, expressed as a crystallinity index, has been found to be highly correlated to the variability in cell wall density. Densities of the constituent substances range from 1.35 $g\,cm^{-3}$ for softwood lignin to 1.67 $g\,cm^{-3}$ for softwood hemicellulose, and the main constituent, α-cellulose, has a density of 1.52 $g\,cm^{-3}$. Wood substance densities measured in situ have been found to range from 1.497 to 1.529 $g\,cm^{-3}$ (see *Chemical Composition*).

The void volume, expressed as a fractional part V of the total volume of wood, is given by

$$V = 1 - \rho_1[(1/\rho_w) + (M_b/\rho_b) + (M_f/\rho_f)] \qquad (1)$$

where ρ_1 is the bulk relative density (volume based on current moisture content), ρ_w is the relative density of the cell wall substance, M_b is the bound water (water held within the cell wall) content in grams per gram of dry wood, M_f is the free-water (water held within the cell cavities) content similarly expressed, ρ_b is the average relative density of the bound water (1.014) and ρ_f is the relative density of water.

If Eqn. (1) is set equal to zero, the maximum moisture content of totally saturated wood can be calculated.

4. Effects of Moisture Content on Density

In use, wood may vary in moisture content from not more than a few percent (e.g., in the heated interior of buildings in cold climates) to the virtually fully saturated condition (as might be found in parts of cooling towers). When lumber weights must be calculated, the density value of interest must include the effect of both weight and volume.

In research, density is frequently reported on a moisture-free weight and volume basis. As moisture content increases from the dry condition up to the fiber saturation point (the moisture content at which the cell walls are fully saturated and the cell cavities free of water), the weight increases and, as a result of swelling, so does the volume (see *Hygroscopicity and Water Sorption*). These factors compensate for each other to some extent when the density being determined includes the weight of the water. For all measurements of relative density where the weight component is the moisture-free weight, the increasing moisture content results in a volumetric swelling and a decrease in sample density. Since swelling does not take place with increasing moisture content above the fiber saturation point, the relative density of a sample will remain constant at higher moisture contents. Obviously, where weight of the water is included in the density measurement, density will continue to increase.

For moisture contents of 0–30%, the foregoing relationship may be expressed mathematically as follows:

$$D_1 = D_0 \frac{1+M_1}{(1+0.84D_0M_1)} \tag{2}$$

where D_1 is the density of the wood at the moisture content M_1, D_0 is the density at the moisture-free condition and 0.84 is an empirical factor applying to all species.

When moisture content adjustments are made to relative density values, the weight component remains constant at the moisture-free condition and Eqn. (2) takes the form

$$\rho_1 = \rho_0/(1+0.84\rho_0 M_1) \tag{3}$$

where ρ_1 is the relative density at the moisture content M_1, and ρ_0 is the relative density at the moisture-free condition.

5. *Effect of Extractive Content on Relative Density*

Wood contains varying amounts of extraneous materials. These materials may be contained either within the cell walls or the cell lumens, but in either case are not considered an essential structural part of the cell wall. They may be present from trace amounts up to more than 20 wt% of the wood, depending on species. Heartwood generally contains larger quantities than sapwood (see *Chemical Composition*).

6. *Variation in Density*

The relative density of wood is extremely variable. Balsa may have a value as small as 0.1, while some tropical hardwoods such as lignum vitae can be as large as 1.3. Most commercially important woods in North America have dry-volume relative densities of 0.3–0.8.

Woods with green-volume-basis relative densities of 0.36 or less are considered to be light; 0.36–0.50 to be moderately light to moderately heavy; more than 0.50, heavy.

A number of commercially important North American woods and their average densities are listed in Table 1. Relative density values within a species commonly have a coefficient of variation of about

Table 1
Average relative densities of some commercially important North American species

Softwood species	Relative density	Hardwood species	Relative density
Western red cedar	0.31	Quaking aspen	0.35
Coast Douglas fir	0.45	American basswood	0.32
Western hemlock	0.42	Paper birch	0.48
Eastern white pine	0.34	Shagbark hickory	0.64
Slash pine	0.54	Sugar maple	0.56
White spruce	0.37	White oak	0.60

[a] Green-volume basis

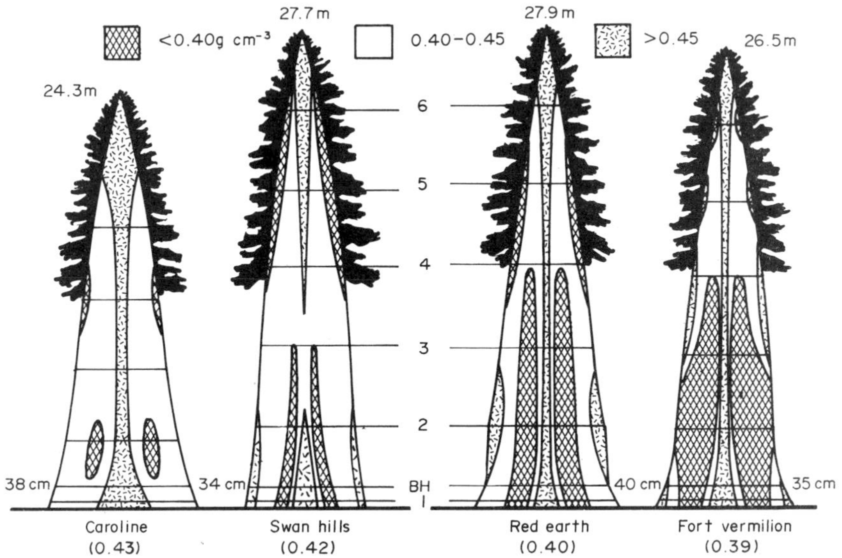

Figure 1
Diagrammatic representation of the average within-tree variation in white spruce trees from four sites in the Canadian boreal forest region (after Jozsa 1988)

10%. As a product of biological growth, the anatomical characteristics of wood which result in this variability are affected by a host of environmental factors, such as temperature, precipitation and wind. In addition, such factors as age and position of the material in the stem are important.

Some insight into the variability in relative density within a tree stem and how those patterns can vary between trees can be gained from Fig. 1. A great deal of variability in relative density exists between trees of the same species. Studies have generally found this to be a highly to moderately heritable trait so the potential for genetic control through tree improvement programs is good.

However great the variations in density may be within a single species, a greater variation exists for many species within a single annual ring. In some species, earlywood may have a relative density of less than 0.2, whereas that of the latewood within the same ring may exceed 0.8 (Fig. 2); however, in other species the difference between earlywood and latewood density may be insignificant.

See also: Constituents of Wood: Physical Nature and Structural Function; Hygroscopicity and Water Sorption; Macroscopic Anatomy; Shrinking and Swelling; Ultrastructure

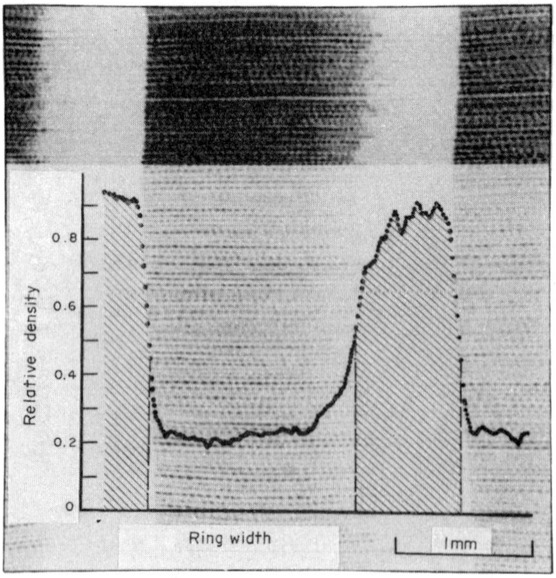

Figure 2
Intra-ring density profile for an annual ring of Douglas-fir superimposed on a negative radiographic image from which the density profile was determined (after Parker et al. 1980)

Bibliography

Elliott G K 1970 *Wood Density in Conifers*, Commonwealth Forest Bureau Technical Communication No. 8. Commonwealth Agricultural Bureaux, Oxford

Haygreen J G, Bowyer J L 1982 *Forest Products and Wood Science—An Introduction.* Iowa State University Press, Ames, Iowa

Jozsa L A 1988 The impact of climate change and variability on wood quality. In: MacIver D C, Street R B, Auclair A N 1988 *Proc. Symp./Workshop Forest Climate* '86. Atmospheric Environment Service, Downsview, Ontario

Kellogg R M, Wangaard F F 1969 Variations in the cell-wall density of wood. *Wood Fiber* 1: 180–204

Kollmann F F P, Côté W A Jr 1968 *Principles of Wood Science and Technology*, Vol. 1, *Solid Wood.* Springer, Heidelberg

Megraw P A 1985 *Wood Quality Factors in Loblolly Pine.* Tappi Press, Atlanta, Georgia

Panshin A J, de Zeeuw C 1980 *Textbook of Wood Technology: Structure, Identification, Properties and Uses of the Commercial Woods of the United States and Canada.* McGraw-Hill, New York

Parker M L, Bruce R D, Jozsa L A 1980 *X-ray Densitometry of Wood at the W.F.P.L.*, Forintek Canada Corporation Technical Report No. 10

Paul B H 1963 *The Application of Silviculture in Controlling the Specific Gravity of Wood*, USDA Technical Bulletin No. 1288. US Government Printing Office, Washington, DC

Stamm A J 1964 *Wood and Cellulose Science.* Ronald Press, New York

US Forest Products Laboratory 1987 *Wood Handbook: Wood as an Engineering Material*, USDA Handbook No. 72. US Government Printing Office, Washington, DC

R. M. Kellogg
[Forintek Canada Corporation, Vancouver, British Columbia, Canada]

Design with Wood

Wood is a remarkable design material. It has generally high strength-to-weight ratios, requires only limited amounts of energy in preparation for end use and can be made more durable by various treatments. It is aesthetically pleasing, comes from a renewable resource and its countless successful applications testify to its recognized value. In many places in the world, wood is the main structural material used in residential and other light-frame construction. Heavy construction now makes increased use of highly refined laminated timber products that are available in a wide variety of sizes and shapes, and are suitable for all types of structures.

The design of main load-bearing systems is usually covered by law and requires certification of engineering competence on the part of the designer. However, this is not always necessary for the design of other objects or systems. The design process must be comprehensive and cover all attributes desired in the finished product. The primary consideration is the expected lifetime of the product. A wooden toy may be designed to lead a short rigorous life, whereas a building is designed to last for many years. There are other considerations if the design object is a structure such as a building. For example, it must be able to carry the expected loads, provide adequate fire safety, provide the desired interior environment, and resist all expected agents of degradation during the expected life. The structure may need to meet acoustic, vibrational or other special requirements, all of which must be met within the constraints of economy. Nonstructural wood systems share many of these demands and will probably have further special requirements such as surface finish performance.

1. Design Considerations

Although all design can be recognized as a combination of art and science, this is particularly true with wood design because the available scientific knowledge is frequently incomplete, and it is therefore necessary to resort to knowledge accumulated through years of experience. For example, to know the materials the designer must develop a subjective familiarity with wood products including accurate mental pictures of the species, its appearance and the range of variation that can be expected. Commercial wood-product specifications must be clearly understood since some designs may require special sorting or manufacture or both, and this can narrow sources of availability and increase cost.

Details of successful and unsuccessful construction must be absorbed from observation and the experience of others who can point out important features. Some aspects of the design may be quite demanding on wood characteristics such as freedom from warp, swelling or shrinkage, and the occurrence of defects such as knots. The designer must be aware of these matters when defining material specifications because a particular species or species group may be desirable or certain grades most suitable. All of this is necessary because wood is a variable biological product and continues to behave like a living material during service, changing size according to environmental conditions. Even within a species and grade, wood products will vary significantly in appearance, and in physical, chemical and biological properties.

As a natural material, wood is subject to attack by natural forces; these must be arrested to give the required life span to the design. With the proper knowledge, long-term service can be readily obtained from wood. There is considerable overlap between structural and nonstructural design, with possible change of priorities in factors relating to safety and appearance.

2. Moisture and Temperature Considerations

In service all wood contains some moisture, the amount varying with the temperature and humidity of the environment. Moisture-content changes can cause shrinkage or swelling, and this must be anticipated in the design. In fine wood products, the moisture content must be controlled from the time of manufacture of parts through fabrication and after the wood product is put in service. Many surface finishes retard moisture flow in and out of the wood, thus offering a degree of protection against environmental change in service. Special treatment can stabilize wood against shrinkage and swelling, but it is expensive. The control of moisture in construction wood products is not usually as stringent, but serious problems can occur. Undue restraint on a wood structural member experiencing a moisture change can cause failure in the material. Thus close attention must, for example, be given to structural connector spacing, particularly in large members, to preclude the development of tensile failures perpendicular to grain owing to shrinkage restraint of the connectors. Constrained wood parts may buckle or create other unsightly defects in response to a moisture increase caused by a normal environmental change.

Wood engineering manuals and design specifications offer rules and advice on adjusting design stresses and elastic moduli for common end-use situations. Design for particular conditions of moisture and temperature should be undertaken with great care. When conditions of high stress are added to unusual moisture–temperature conditions, experienced advice is mandatory and, if this is not available, tests should be carried out in a simulated environment. The influences of cycling moisture content between high and low levels and/or cycling temperature on wood strength are not well covered by available research but indications are that these cycles can cause strength degradation.

Temperature as an isolated variable also influences the strength of wood, with strength decreasing as temperature increases. This effect is more pronounced at higher wood moisture contents. In the range of temperatures normal to the human environment, engineering adjustments of permissible stresses are generally omitted. If temperatures exceed 65 °C for extended periods or are extremely low, it is necessary to consult the literature available on the subject.

Higher wood moisture contents short of full saturation provide opportunities for growth of fungi which destroy the material. Decay will stop when the moisture content falls, but will begin again with the return of higher moisture. In building construction, ground proximity or contact, leaks, rainwater standing on horizontal exposed surfaces, and condensation are the primary offenders in creating moisture levels conducive to fungal growth (see *Decay During Use*).

Where there is any chance of intermittent moisture accumulation, dual priority in wood design must be to keep the wood dry while also providing backup ventilation. Insulated walls, attic spaces and crawl spaces are examples of danger zones in buildings that must be ventilated. Horizontal, weather-exposed wood surfaces are to be avoided because water can more easily spill over into poorly ventilated cracks in addition to saturating the horizontal surface. Effective drainage on the exterior of the structure is necessary for prevention of decay. Some structural situations, such as salt storage buildings and certain types of animal housing, produce high-moisture environments hostile to wood, thus necessitating preservative treatment.

Vapor barriers are commonly used in conjunction with insulation in walls, floors and ceilings to control condensation. Condensation can reduce the effectiveness of the insulation and can cause decay and paint problems. No vapor barrier can be expected to be perfect and allowance for ventilation to remove trapped condensation is necessary in good design.

Wood is often exposed to the weather for architectural or other reasons, but this is a questionable practice in many geographical locations. If successful local examples exist, they should be checked out as to species of wood, fastening and geometry of design. Weathering effects from alternate wetting and sunlight combine to create surface degradation and other potential cosmetic defects (see *Weathering*).

Although dry wood will last indefinitely, a stable temperature–humidity environment is even more conducive to long satisfactory service. Wood that remains completely submerged in water will not decay, although animal life such as marine borers may attack and destroy the wood.

3. *Thermal and Electrical Properties*

Wood is usually classified as a thermal insulator (see *Thermal Properties*) but does not perform as well as commercial insulation since its insulation qualities decrease with increasing moisture content and are lower in woods of higher density. Wood is also classified as an electrical insulator (see *Electrical Properties*); its electrical resistance changes rapidly with increasing moisture content within the normal service range (i.e., near oven-dry to 20% moisture content). At higher moisture values resistance loss is smaller and erratic. In the event that thermal or electrical considerations must be included in the design, available literature should be consulted with regard to the specific situation.

4. *Insect Protection*

Insects can attack wood (see *Deterioration by Insects and Other Animals During Use*) and this fact must be considered in both the design and the selection of materials. Information concerning the local incidence of insect problems and their control must be appraised early in the design procedure. The many areas of the world where wood is a historic building material indicate that local insect problems can be either avoided or economically controlled. In other areas, insect threats can be very serious and may be the major consideration of design.

5. *Fire Considerations*

The fire safety of wood construction is a matter of concern that is important but sometimes overrated. Uncontrolled contents of buildings and combustible decorative materials frequently cause fire risks that overshadow safety features incorporated into the basic structure. In the USA sprinklered wood construction is widely accepted and provides for a reduction in risk from the total fire load including contents. The experimental designer must, however, always keep fire in mind as highly efficient structures using slender or thin wood parts may not be practical because of potential fire problems. Fire tests and qualification of new construction systems can be very expensive, and hence proposed new constructions that cannot be generically related to already approved construction should be developed with this potential cost in mind (see *Fire and Wood*).

6. *Chemical Treatment*

Wood can be chemically treated to change its properties in response to special environment needs (e.g., to resist decay and insects, to retard burning, or to resist the effects of moisture (see *Preservative-Treated Wood*)). The designer should be careful in selecting treated wood, taking into consideration problems of handling, possible toxicity of treatment materials and the likelihood of interaction with other components of design such as corrosion of fasteners.

7. *Acoustic and Vibrational Considerations*

Acoustics and vibration in wood-building construction have been the subject of increased research in recent years, with some results becoming regulatory in the USA. Acoustic and vibrational problems are, of course, not confined to wood but occur with all structural materials. As a material, wood has a favorable vibration damping coefficient, and its acoustic performance varies according to the application (see *Acoustic Properties*). The methods of study underlying current acoustic/vibration recommendations for construction have, of necessity, been mainly empirical. This is because the total wall, floor, ceiling or building represents a complex unit that is generally intractable from an analytical standpoint. The available design information on acoustics relates to specific constructions including fastener details. New constructions could require testing for code approval.

8. Strength–Load Duration

The strength properties of wood and connections are time dependent at constant moisture and temperature, subject to complex relationships. Contemporary engineering methodology for dealing with this aspect of design is relatively simple because the present base of research findings is limited. Simplified factors tied to available research and backed by experience are used to adjust allowable stresses and elastic moduli for design purposes. Many common structural designs use prescribed loads that have a relatively low probability of occurrence and a limited period of expected duration at full design load. This situation, quite typical in ordinary building design, is relatively straightforward and backed by an extensive successful history. In such applications, tabulated allowable mechanical properties can be adjusted for the expected load duration and compared with design-load stresses with good assurance of satisfactory performance over the design lifetime. The load duration adjustments used in such a case are specified by building codes and can be found in wood engineering handbooks.

9. Importance of Structural Connections

If established practices of connection design as found in the handbooks and manuals are followed, and if loading and environment are not unusual, tabulated design values for connectors can be expected to yield good results. Large joints must be carefully designed, keeping the need for allowable wood movement in mind and giving special attention to the possibility of development of high tensile stresses perpendicular to the grain as a result of shrinkage. Shear stresses within heavy timber joints can also become critical and are given careful attention by the experienced engineer.

Adhesive connections are a natural choice for wood assembly (see *Adhesives and Adhesion*), but practical considerations frequently preclude their use. Many wood adhesives require high-quality machining and fitting of parts. Most have rigid temperature requirements and all require considerable expertise and discipline in their application. For structural purposes, adhesive usage is confined, with few exceptions, to prefabrication of parts. The difficulty in inspecting finished adhesive connections requires good assembly quality control along with subsequent destructive test sampling. The possibility of joint stresses arising from moisture- or temperature-related size changes must also be kept in mind. If the changes occur slowly, some relief can be expected as a result of stress relaxation. Finally, the designer must clearly understand the costs of suitable adhesive joining and balance these against the resultant gain in product performance.

Mechanical connectors such as nails, bolts, screws, shear plates, timber connector rings and similar traditional devices have a long service record. With the exception of timber connector rings, completed joints fabricated with these mechanical devices are relatively easy to inspect and simple field assembly is usually possible under quite adverse field conditions. The toothed metal-plate connector, although not well suited to field assembly, is a relatively new and efficient device that has found wide acceptance in light-frame building component construction. It has also found more recent use as a connector for upholstered furniture frames. This newer connector, available in a variety of sizes, thicknesses and tooth patterns, has given rise to a significant industry in the USA that produces the plates and provides extensive engineering design services using sophisticated computer systems. These services, including the required practical expertise, are readily available in most countries (see *Joints with Mechanical Fastenings*).

Advances in metal plate and other new connectors, coupled with the development of computer systems for complex analyses, have fostered the development of new, longer-span structural components consisting of smaller lumber parts. Much of this progress has been made possible through the use of computer-based techniques of structural analysis that emerged in the 1960s and have since been adapted into improved design techniques for trusses, frames and composite beams. Special programs can treat floor and wall systems whereas others deal with furniture. These tools are available to the designer but do not eliminate the requirement for the consideration of the important fundamental aspects of wood use. The computer system only analyzes a model; the model is created by the designer and is thereby limited by the designer's insight and knowledge of materials, connections, structure and loads. As an example, the modelling can accommodate the behavior of a partially rigid mechanical wood connection but the user of the system must supply the quantitative inputs that describe the partial rigidity.

See also: Building with Wood

Bibliography

American Institute of Timber Construction 1985 *Timber Construction Manual*, 3rd edn. Wiley, New York

Avery-Phares 1980 *The Woodbook*. Avery-Phares, San Francisco, California

Hoyle R J Jr, Woeste F E 1988 *Wood Technology in the Design of Structures*, 5th edn. Iowa State University Press, Ames, Iowa

Percival D H, Suddarth S K 1981 Structural wood systems. *Educ. Modules Mater. Sci. Eng.* 3: 291–311

US Forest Products Laboratory 1987 *Wood Handbook: Wood as an Engineering Material*, USDA Handbook No. 72. US Government Printing Office, Washington, DC

S. K. Suddarth
[Purdue University, West Lafayette, Indiana, USA]

Deterioration by Insects and Other Animals During Use

Insects and certain other animals can damage wood by attacking living trees, invading freshly cut logs and unseasoned wood, and infesting partially seasoned or fully seasoned wood. This article is primarily concerned with those groups of insects that initiate or continue attack in seasoned wood in use. Other animals which degrade wood in use will also be considered.

Because some of the insects that initially invade only unseasoned wood sometimes survive the drying, milling, and manufacturing or construction processes, they may continue to damage the seasoned wood. As the insects in this category cannot reinvade seasoned wood, greatest concern lies with identifying evidence of their damage in wood in use. Such evidence, if not properly interpreted, might lead to an unfounded expectation that the wood will continue to deteriorate.

If logs are stored or rafted in brackish water, or if wood is placed in use in a marine environment, there are several types of animals, referred to collectively as marine borers, which cause deterioration. Wooden poles, chemically treated or otherwise, are attacked by several species of birds known as woodpeckers.

Greatest emphasis in this article is placed on the recognition of damage in wood and the assessment of its significance. In those instances where the organism will continue to attack the wood in use, the principles of prevention and control procedures are briefly discussed.

The economic losses associated with the degradation of wood in use by insects and animals amount to many millions of dollars each year. Reasonable estimates indicate that subterranean termites alone cause losses of over 750 million dollars each year in the USA (Mauldin 1982), marine borers probably account for twice this amount. Woodpeckers cause considerable, though unestimated, damage to wooden utility poles in most parts of the USA. Since the usual methods of chemical treatment of the poles do not prevent attack, woodpeckers present a special problem.

1. Termites

Termites are typical insects in their general characteristics, but differ from many insects in that they are social: they live together in groups or "colonies." There are different forms (castes) of termites in the colonies, each of which perform different functions. The most conspicuous termites are the reproductives which emerge from mature colonies in large flights at least once a year. The time and conditions for flights vary with the species. Their function is to pair off, mate and establish new colonies. Flying termites are often confused with flying ants and it is important to distinguish the two kinds of insects (Fig. 1). Winged termites vary in color from nearly black to yellowish-brown. Most are up to 14 mm long, including the folded wings, which are slightly opaque to milky white or gray.

The most numerous termites in a colony are wingless workers and nymphs (immature stages). They are seen only when a nest or a piece of infested wood is broken open. Workers are up to 8 mm long, soft-bodied and almost white in color. The functions of workers, and older nymphs, are to gather food, enlarge nests, and care for the reproductives, soldiers, very young nymphs, and eggs.

There is a soldier caste (also wingless) in most of the wood-feeding species. Soldiers resemble the workers except that they have a highly modified head, often with very large jaws (mandibles) to protect the colony from natural enemies such as ants. They are fed and groomed by the workers. The soldiers in some species exude a sticky substance from a head gland which acts as a protective agent.

Thousands of termite species are found in the tropical and temperate regions (from the equator to between 45 and 50° north and south latitudes), with the greatest number occurring in the tropics. Their importance as destroyers of wood in use varies with climate and wood species. Highly populated areas in

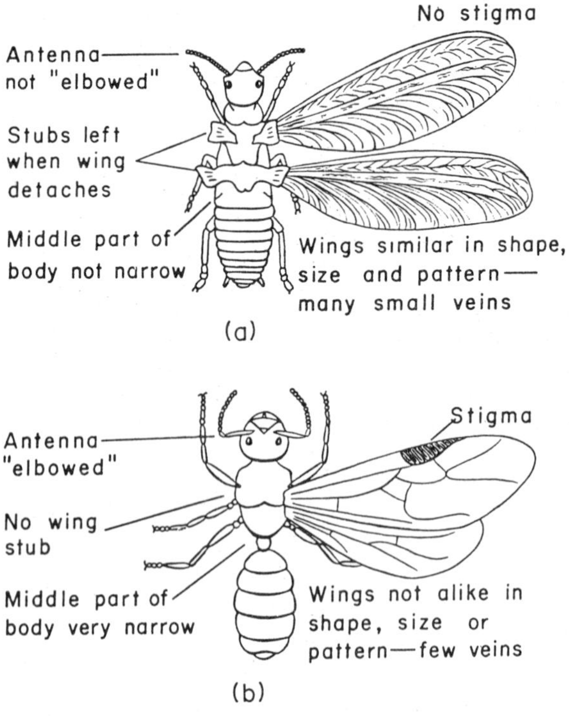

Figure 1
Comparison of (a) winged termite with (b) winged ant (Courtesy US Department of Agriculture)

temperate climates suffer larger economic losses than less-populated areas in warmer climates and greater termite activity.

Termites may be broadly classified as: subterranean termites, which usually nest in the ground and require a constant supply of moisture; drywood termites, which live in seasoned wood without access to moisture other than that in the wood; dampwood termites, which nest in constantly wet wood, usually in soil contact; and harvester termites, which forage for grass and other vegetation which they use to grow their fungus food in their nests. Subterranean and drywood types are of greatest importance as destroyers of wood in use the world over. These two kinds will be discussed briefly. More detailed information on termite biology may be found in Krishna and Weesner (1969, 1970).

By far the greatest amount of insect-related damage to wood is done by termites. The subterranean termites cause the largest proportion of this damage, particularly in temperate climates. The potential for damage varies with the number of buildings and the length and severity of cold temperatures. Drywood termites, the second most important type, occur throughout the tropics and into the coastal areas of warm, temperate regions.

Termite colony foundation begins with the pairing of a single male and female. They break off their wings and then excavate a small cell in wood at an appropriate site, which varies according to their need for moisture. The original pair cares for the nymphs that hatch from the first batch of eggs. As the nymphs mature, they take over the care and feeding of the reproductives and the newly laid eggs and younger nymphs. During the first year several soldiers are produced. The reproductives continue to produce offspring for many years, and the colony growth varies with the termite species, suitability of the site and availability of food. Several years after a colony is initiated it will begin producing winged reproductives which leave each year. Only a minute fraction of these reproductives leaving the colony survive long enough to find a mate and establish a new colony.

Subterranean termites will eat any substance which contains cellulose. Wood, paper and vegetable fibers are their usual fare. They also penetrate, but do not eat, plastics, rubber and some metal which is buried in soil. The cost of subterranean-termite damage to wood and wood products will continue to rise with the increase in numbers of dwellings and the cost of replacing damaged wood. Repairing damage accounts for about one-third of the total annual loss. Chemical treatments to prevent and control termite infestations make up the remainder (Koch 1972).

Since subterranean termites are so common, it is important to be aware of the conditions that lead to termite attack and to be able to recognize the damage or evidence of activity before it becomes serious (Ebeling 1975, Mallis 1982a, Moore 1986). Most species of subterranean termites in temperate climates swarm during the day. These winged forms are often the only evidence that is conspicuous to occupants of infested buildings. Even if no one is present when the flight occurs, the termites' wings are often found in large numbers, particularly at a light source such as a window. Accumulations of discarded wings are typical of termites but not of ants.

More careful examination is necessary to locate subterranean-termite activity and damage in wood. Whenever they leave the protection of the soil, subterranean termites build mud shelter tubes over the surfaces which separate the soil and the wood above, such as foundation walls. These shelter tubes provide protection from their natural enemies, particularly ants. The tubes are constructed of soil and wood particles held together by fecal material. Their color is that of the soil from which they emerge. Termites also construct mud shelters over the exterior of wood being consumed and fill in gaps between boards to seal out their enemies.

Wood damaged by subterranean termites may show few, if any, external signs of deterioration. Pounding or probing will reveal the hollowed-out portions of the wood and, when broken up, will reveal the live termite workers and soldiers, if the infestation is active (see Table 1).

Prevention of attack by subterranean termites involves good building practices and the use of insecticidal treatments. These good building practices are essentially those which also prevent wood decay—proper clearance, drainage and ventilation (see *Decay During Use*).

Preventive chemical treatments involve the use of pressure-treated wood, if the wood is in or near ground contact, and the creation of a chemical soil barrier at all points of possible entry from the soil onto foundations, utility pipes, piers and the like (Rambo 1985).

Treatment of existing infestations involves three basic steps. The first is mechanical alteration to render the structure, or its immediate surroundings, less favorable to termite occupation. The second step is soil treatment which is similar to a preventive treatment except for the added problem of gaining access to the soil which must be treated underneath concrete and other areas not readily accessible. The third step is the injection of chemical agents into void spaces in foundation walls, piers, and posts (Rambo 1985).

Drywood termites have much smaller colonies than subterranean termites. It is common, however, for several scattered colonies to exist in wooden structural members and furniture in a building (Ebeling 1975, Mallis 1982a, Moore 1986). The most commonly encountered evidence of drywood-termite presence is the flight of winged forms. This occurs at night with most drywood species. Although drywood termites make no conspicuous shelter tubes, they do drop fecal pellets from their workings which accumulate in piles beneath the attacked wood (Table 1). Probing and

Table 1
Characteristics of damage to wood in use caused by common wood-boring insects and marine borers

Surface holes	Frass (bore dust)	Galleries (tunnels)	Wood attacked		Reinfestation	Name of insect or marine borer
			Part and class	Condition		
None	mixture of soil and fecal material	partly concentric with annual rings; surfaces plastered with light-tan fecal material	sapwood and heartwood of softwoods and hardwoods	seasoning or seasoned	yes	subterranean termites
None	tiny, hard round-ended pellets with 6 longitudinal ridges; less than 1 mm long	irregular cavities not following annual rings; may contain pellets	sapwood and heartwood of softwoods and hardwoods	seasoned	yes	drywood termites
None or obscure	coarse, sawdust-like, contains insect fragments; below nest only	partly concentric with annual rings; surfaces clean and smooth; no frass present	sapwood and heartwood of softwoods and hardwoods	often damp and partly decayed; prefer soft-grained woods	yes	carpenter ants
Numerous, round, 1–2 mm diam	none	round, 1–2 mm diam, up to 12 mm long; numerous; in surface layer of wood	outer, exposed surface of softwoods and hardwoods	submerged in salt or brackish water	yes	gribble (*Limnoria* spp)
Obscure	none	rapidly enlarge from 1 mm on surface up to 25 mm diam; lined with chalky material	sapwood and heartwood of softwoods and hardwoods	submerged in salt or brackish water	yes	shipworm
Round, 10–12 mm diam	coarse, sawdustlike, below surface openings only	round, 10–12 mm diam, 15–18 cm long; primarily with grain of wood	sapwood and heartwood of softwoods	seasoned, soft-grained	yes	carpenter bees
Round, 5–6 mm diam	absent or excelsior-like	oval, up to 13 mm long diam	sapwood and heartwood of softwoods	unseasoned	no	roundheaded borers (sawyers)

Round, 4–5 mm diam	coarse, tightly packed	round, up to 5 mm diam	sapwood and heartwood of softwoods	unseasoned	no	horntails (wood wasps)
Round, 2.5–7 mm diam	fine to coarse powder; tightly packed, tends to stick together	round, 1.6–10 mm diam; numerous; random	sapwood of hardwoods primarily; minor in softwoods	seasoning and newly seasoned	rarely	bostrichid powderpost beetles
Round, 1.6–3 mm diam	fine powder with elongate pellets conspicuous; loosely packed	round, up to 3 mm diam; numerous; random	sapwood of softwoods and hardwoods; rarely in heartwood	seasoned	yes	anobiid powder-post beetles
Round, 0.8–1.6 mm diam	fine, flour-like; loose in tunnels	round, up to 1.6 mm diam; numerous; random	sapwood of ring- and diffuse-porous hardwoods only	newly seasoned with high starch content	yes	lyctid powder-post beetles
Round, 1.6-2.5 mm diam	coarse to fine powder, bark-colored; tightly packed in some tunnels	round, up to 2.5 mm diam	inner bark and surface of sapwood only in softwoods and hardwoods	unseasoned, under bark only	no	bark beetles
Round, 0.5–3 mm diam	none	round, same diam as holes on surface; across grain; walls darkly stained	sapwood and heartwood of softwoods and hardwoods	unseasoned logs and lumber	no	ambrosia beetles
Oval, 3–10 mm long diam	coarse to fibrous; some blunt pellets; loosely to tightly packed	oval, up to 12 mm long diam; size varies with species	sapwood and heartwood of softwoods and hardwoods	unseasoned logs and lumber	no	roundheaded borers (general)
Oval, 6–10 mm long diam	very fine powder and tiny pellets; tightly packed in tunnels; ripple marks on walls	oval, up to 10 mm long diam; numerous in outer sapwood	sapwood only of softwoods, primarily pine	seasoning to seasoned	yes	old-house borer
Oval, 3–13 mm long diam	coarse powder, some blunt pellets; tightly packed	flat oval up to 10 mm long diam; winding	sapwood and heartwood of softwoods and hardwoods	seasoning	no	flatheaded borers

pounding the wood will help to determine the extent of infestations.

Drywood termites are easily transported in infested furniture and other wooden articles, as well as lumber. Thus, they sometimes occur at sites far removed from their natural habitats. The colonies in colder climates survive only in heated buildings and their development is very slow.

Prevention of drywood termites in tropical and semitropical regions is all but impossible. Properly pressure-treated wood is free from attack, but any unpainted surface of untreated wood is a potential point of entry. The application during construction of a silica aerogel desiccant dust into wall voids and attic spaces has been shown to provide good protection of the coated surfaces (Ebeling 1975).

Once drywood termites have infested wood, several methods of control are possible. The most widely used system for structures is fumigation under gas-proof tarpaulins. Fumigation is a procedure to be employed by professionals only (Ebeling 1975, Mallis 1982a).

2. *Beetles*

Adult beetles are characterized by having their front wings modified into hardened wing covers. There are more kinds of beetle than any other kind of insect; fortunately, only a relatively few species are of importance in the destruction of wood in use. Some species of beetles begin their attack on wood in living trees, dying trees, and freshly cut logs. Others attack seasoning or seasoned wood and even decaying wood. Some of the beetles which begin their attack in dying trees, freshly cut logs, or seasoning wood survive the processing and air-drying of wood and complete their development after wood has been placed in use. They cause relatively minor structural damage and rarely require treatment. They do cause the downgrading of lumber and are thus of economic concern.

Only beetles which attack and reattack seasoning and seasoned wood are discussed. Those which initiate attack in wood prior to seasoning only are shown in Table 1.

Of those beetles which attack seasoning or seasoned wood, powderpost beetles are the most common. There are three families of powderpost beetles: Anobiidae, Bostrichidae and Lyctidae. Differences in their biology and habits make species in each family a slightly different problem. Their damage and the types of wood attacked are summarized in Table 1.

The powderpost beetles primarily attack sapwood. Anobiids also attack heartwood. Lyctids oviposit only in wood pores and thus only attack ring-porous or diffuse-porous hardwoods such as oak, ash, walnut and cherry. Anobiids and bostrichids attack both hardwoods and softwoods; bostrichids being important for only a few species of hardwoods.

The adult powderpost beetles emerge from round flight holes in the wood surface during the spring and summer. After mating, the females lay eggs on or in suitable wood and then die. Life cycles range from one year or less for lyctids and bostrichids to two or more years for the anobiids. Anobiids are the ones found in coniferous structural framing timbers in buildings.

Powderpost beetles are usually discovered when small piles of fine wood powder are observed on or below wood surfaces. The round adult flight holes are easily seen. In heavily damaged wood, the sapwood portion is reduced to a mass of powdered wood held in place by a thin outer wood surface—hence the name "powderpost." In larger-dimensioned lumber, the structural strength loss is often not significant. When hardwood trim and other small-dimensioned wood components are so damaged, they may need replacement.

Powderpost beetles are destroyed if wood is properly kiln-dried, but they can reinvade wood after drying. Careful removal of scrap hardwood from storage facilities will reduce attack by those species attracted to hardwoods. When wood has a finish on the surface, it is no longer attacked by powderpost beetles except in joints, cracks and the like. Good clearance and ventilation underneath buildings will reduce the likelihood of anobiid powderpost-beetle attack. Wood should be inspected before use to determine whether it is free of infestation. Infested wooden articles or stacked lumber can be successfully treated by vault fumigation, fumigation under a gas-proof tarpaulin, or heat sterilization (Ebeling 1975, Mallis 1982b, Moore 1986).

Hardwood or softwood components in structures cannot be treated successfully by surface application of insecticides if they are covered with a finish. Such infestations will slowly decline if the building is centrally heated and has no moisture problems. Unfinished wood surfaces in uninhabited portions of buildings can be treated by surface application of a residual insecticide.

The only other beetle of significance which attacks seasoned wood in use is one of the roundheaded borers, the old-house borer. They attack primarily pine sapwood, but also spruce or fir, as soon as the moisture content drops below the fiber saturation point; their infestation continues, one generation after another, until the moisture content drops below 10%. These beetles occur in the eastern half of the USA and in the temperate zones of Europe, Asia and Africa. The sapwood may be reduced to a powdery consistency with an outer thin shell of wood holding it in place (Table 1). Reducing moisture content of wood is a long-term means of reducing the attack, but chemical treatment is sometimes necessary. Unfinished wood can be treated by surface application of a residual insecticide, such as lindane. Extensive infestations and those in inaccessible portions of buildings are best treated by fumigation of the structure under a gas-proof tarpaulin (Ebeling 1975, Mallis 1982b, Moore 1986).

3. Carpenter Ants

Carpenter ants are black to reddish-brown wood-nesting insects up to 12 mm long. They are typical of all ants in their social habits. The colonies are initiated by individual mated females. These ants excavate nests from soft-grained or partially decayed wood. The wood that is removed in nest building is not consumed but ejected from the tunnels. The fibrous wood fragments are a tell-tale sign of the nest location (Table 1). Several years after a nest is begun it will consist of several thousand individual workers and immature forms in addition to the queen.

Carpenter ants nesting in buildings are usually found in porch posts, roof areas and foundation timbers near the soil. They also nest in wooden posts and poles. If they are not controlled, over long periods they can cause structural weakening of infested wood.

Control of carpenter ants is not difficult if the nests are located and insecticide dusts or sprays are directed into them. Treating where ants are generally seen is rarely successful. Drilling holes into the nest cavities, which are completely free of wood fragments, improves the penetration and effectiveness of the chemical treatment. Excessive moisture should be eliminated to reduce the wood's attraction for carpenter ants as well as to stop deterioration by decay fungi (Ebeling 1975, Moore 1986).

4. Carpenter Bees

There are several species of large bees, called carpenter bees, that make their nests in wood. They superficially resemble bumble bees in size and color but have abdomens that are shiny black instead of hair-covered. These bees do not live in colonies and are not aggressive to humans as are bumble bees.

The females make perfectly round nest entrance holes 12 mm in diameter in the bottom or vertical faces of wood in well-lit protected places. The wood is not consumed by the bees, and the coarse wood fibers excavated accumulate below the entrance holes (Table 1). Carpenter bees prefer soft-grained wood such as redwood or cedar but will also attack weathered pine. They do not attack wood with a good coat of paint. Stains or pressure-injected metallic-salt wood preservatives do not hinder their attack.

The nests are constructed, stocked with pollen, and eggs laid in late spring or early summer in temperate climates. The parents die before the new brood emerges. In warmer climates there may be year-round activity, but new bees emerging in midsummer in more moderate climates remain sexually immature until the following spring when mating and nest building resume. Carpenter bees tend to reoccupy old nesting sites and this can eventually lead to extensive damage.

Control of carpenter bees consists of applying a properly labelled insecticide dust or spray into the nests and sealing the holes with a hard-setting wood putty. Woodpeckers sometimes become a secondary problem when they discover the bee nests and damage wood in gaining access to the bee larvae inside the tunnels.

5. Marine Borers

With the exception of a few resistant tropical woods, any untreated wood placed into use or stored in salt or brackish water, throughout most of the world, is subject to damage by marine borers. The speed with which damage occurs depends upon climate, pollution, borer species, and other factors. Damage occurs most rapidly in warmer climates. Marine-borer-damaged wood is sometimes incorporated into manufactured products (particularly lumber) and might be confused with insect-damaged wood. On a worldwide basis, economic losses caused by marine borers amount to more than a billion US dollars a year (Bletchly 1967, Koch 1972, US Forest Products Laboratory 1987).

The most important marine borers with respect to wood damage are molluscan borers called shipworms (*Banksia* sp. and *Teredo* sp.) and crustacean borers of the genus *Limnoria*, called gribble. Both types of borer excavate wood for food as well as shelter. Shipworms are the most destructive. The immature forms enter the wood through pinhole-sized openings which never increase in size. As the shipworms grow in length and diameter, the tunnels become increasingly larger; up to 2.5 cm in diameter and 1 m or more long. The tunnels, which are excavated with a pair of small boring shells at the head end are lined with a white, shell-like deposit. They may eventually honeycomb the interior of infested wood with little exterior evidence of attack (Table 1).

The ant-sized gribbles, which resemble sowbugs, extend their small tunnels into the outer few centimeters of wood. The surface of infested wood becomes spongelike in appearance and erodes away, particularly at the waterline. As the tunnelled wood breaks away, the burrows are extended into new depths (Table 1).

Prevention of marine-borer attack requires dual treatment of exposed wood: heavy loadings of water-borne metallic preservatives followed by coal-tar creosote. The more thorough the treatment, the longer its effectiveness. Treated wood must be further protected from mechanical damage to the treated outer layer. Jacketing, including metal, concrete and plastic, has been employed to protect the creosoted surfaces. On an experimental basis, two fumigants were effective in preventing and controlling marine borers over a 20-month exposure period (Helsing et al. 1984).

6. Woodpeckers

Woodpeckers have been reported to seriously damage wooden utility poles in temperate zone areas of Europe, Asia and North America. Several genera in the family Picidae are injurious. The incidence of woodpecker attack and the cost of pole replacement varies over their total range. Thus, it is not possible to set a value on their economic impact (Rumsey 1970).

Woodpeckers make excavations up to 0.6 m in depth in poles for nests and roosts. There is sometimes only 3 cm of wood shell surrounding the cavity. This weakens the poles, which may subsequently break easily in storms. It is common for the birds to make many false starts or probe holes in the face of a pole before completing a nest. Some species appear to prefer nesting in poles to trees, probably because the poles are in strategic locations in open spaces.

Most poles used to support power and telephone lines are pressure-treated with chemicals which protect them from nearly all destructive plant and animal organisms. These chemicals do not deter woodpeckers. Some birds will nest in new poles with a surface residue of creosote so high that their nestlings do not survive. Many methods of preventing woodpecker damage have been tried. Research has been done with repellents, decoy poles, coatings and wrappings. Wrap-on coatings of heavy polyethylene are promising, but the best, though not infallible, protection comes from wrapping the pole with 5 cm by 5 cm hardware cloth from a point about 5 m above groundline to the top.

See also: Preservative-Treated Wood; Fumigation

Bibliography

Bletchly J D 1967 *Insect and Marine Borer Damage to Timber and Woodwork: Recognition, Prevention and Eradication.* HMSO, London

Ebeling W 1975 *Urban Entomology.* University of California, Richmond, pp. 128–216

Helsing G C, Graham R D, Newbill M A 1984 Effectiveness of fumigants against marine borers. *For. Prod. J.* 34(6): 61–63

Koch P 1972 *Utilization of Southern Pines*, Vol. 1: *The Raw Material.* US Department of Agriculture, Washington, DC, pp. 667–734

Krishna K, Weesner F M (eds.) 1969, 1970 *Biology of Termites*, Vols. 1, 2. Academic Press, New York

Mallis A 1982a Termites. *Handbook of Pest Control*, 6th edn. Franzak and Foster, Cleveland, Ohio, pp. 177–257

Mallis A 1982b Wood-boring, book-boring, and related beetles. *Handbook of Pest Control*, 6th edn. Franzak and Foster, Cleveland, Ohio, pp. 277–309

Mauldin J K 1982 The economic importance of termites in North America. In: Breed M D, Michener C D, Evans H E (eds.) 1982 *Biology of Social Insects: Proc. 9th Congr. Int. Union Study Social Insects.* Westview Press, Boulder, Colorado, pp. 138–41

Moore H B 1986 Pest management of wood-destroying organisms. In: Bennett G M, Owens J (eds.) 1986 *Advances in Urban Pest Management.* Van Nostrand Reinhold, New York, pp. 313–33

Rambo G W (ed.) 1985 *Approved Reference Procedures for Subterranean Termite Control.* National Pest Control Association, Vienna, Virginia

Rumsey R L 1970 Woodpecker attack on utility poles—a review. *For. Prod. J.* 20(11): 54–59

US Forest Products Laboratory 1987 Protection from organisms that degrade wood. *Wood Handbook: Wood as an Engineering Material.* US Government Printing Office, Washington, DC, Chap. 17, pp. 1–17

H. B. Moore Jr.
[North Carolina State University, Raleigh, North Carolina, USA]

Drying Processes

Solid wood and many other forest products are dried before being put into use for various reasons, but primarily for control of product moisture content. Lumber has been and continues to be dried in air or kilns or both. Most other wood products are dried using heated-chamber systems. Most mechanical and physical properties of wood depend upon final moisture content and hence change as wood is dried. Wood, particularly in lumber form, can be degraded by the drying process depending on the severity of drying conditions. This article discusses why and how lumber is dried, and the types of defects that can result.

1. Moisture Content

The wood of a felled tree contains varying amounts of water, depending upon the species, type of tissue and position within the tree. The moisture content of freshly sawn lumber (expressed as a percentage of oven-dry mass) can range from as low as 35% in the heartwood of Douglas-fir to as high as nearly 300% for old-growth redwood. When exposed to atmospheric conditions the wood slowly desorbs this moisture to its surroundings, ultimately reaching an equilibrium moisture content of 6–10% in an interior site and 12–15% for an exterior exposure (see *Hygroscopicity and Water Sorption*). The time required to reach this equilibrium value can be very long and may even be years in the case of large-sized timbers or logs.

2. Reasons for Moisture-Content Control

It is generally desirable that wood in service be dry, because dry wood is stronger (see *Strength*) and less susceptible to degradation (biological, thermal or chemical) than wet wood (see *Decay During Use*; *Thermal Degradation*). However, there is no benefit in drying wood to a moisture content other than the equilibrium moisture content it will reach in service, because there is no effective way of sealing moisture into or out of wood permanently. Furthermore, shrinkage and swelling are directly related to moisture content (see *Shrinking and Swelling*), so that dimensional change in service depends directly on the magnitude of moisture-content change the wood undergoes. If wood was not dried before being put into service, much greater dimensional change would ultimately occur. Ideally, then, wood should be dried to the equilibrium moisture content corresponding to service conditions before secondary manufacture and being placed in service. Thus, it can be seen that wood drying is not performed for the control of in-service properties, with the exception of initial dimensional change. Rather, it is carried out in order to reduce weight to minimize shipping costs (this applies mainly to lumber) and to prepare wood for subsequent manufacturing operations—wet wood cannot be glued effectively, treated well with preservatives, or painted.

3. Drying Methods for Lumber and Other Solid Wood Products

Hardwood or softwood lumber is generally either air dried or kiln dried. It is invariably kiln dried if the lumber is to be remanufactured in any way. Air drying has been used for thousands of years and as the name implies involves the simple practice of piling the lumber in open air and waiting for drying to proceed slowly. Exact techniques used vary with region, climate, species, lumber grade and the prevailing economic situation.

Air drying has several advantages over kiln drying. It requires no special equipment and thus it is simple and capital inexpensive. It only requires reasonably favorable weather conditions and land space. It can result in reduced degradation of difficult-to-dry species as the mild drying conditions and slow drying result in lower internal stress development in wooden boards. This is particularly true for thicker, refractory timbers. Air drying does, however, have many major disadvantages. Since weather (wind, temperature, relative humidity and precipitation) cannot be regulated, there is little control over the drying process. Air drying is much slower than kiln drying. Many drying conditions are such that the moisture content of the lumber cannot be reduced below about 12%, which is too high for many types of interior wood use.

3.1 Air Drying

In air drying, lumber is stacked into packages generally around 1.25 m wide by 1.25 m high by variable length. These units are then placed as many as four-high in the air yard. Each layer of lumber within the package is separated by stickers ranging from 12 to 40 mm in thickness, placed every 0.6–1.25 m perpendicular to the long length of the lumber. The use of stickers permits air to pass through the unit. The thinner the sticker the less the air flow, and hence the slower the drying rate.

The stacks of lumber are placed in the air yard in varying configurations to allow handling with forklifts and to obtain maximum wind movement and concentration per hectare.

The ends of the boards are sometimes coated with a water-resistant end coating to prevent excessive end drying that leads to end checks and splits. The top units of the stacks are often covered with metal, wood or plastic roof covers to retard too-rapid drying of the upper layers and to prevent penetration and absorption of rain or snow into the units.

Approximate times necessary to dry 25 mm thick lumber of various species are given in Table 1. The minimum time applies to lumber placed in the yard during the best drying weather, spring and summer. Maximum times are for lumber yarded during autumn and winter. Further information on air drying can be found in Rietz and Page (1971) and Wengert (1977).

3.2 Kiln Drying

With kiln drying, lumber is placed in a dry kiln. A modern-day dry kiln is nothing more than a large chamber in which the air movement, temperature and relative humidity are controlled. This control permits the drying rate to be both regulated and accelerated. The lumber can also be dried to any desired final moisture content.

The physical layout and primary components of a steam-heated dry kiln are shown in Fig. 1. Commercial kilns range in capacity from 24 m^3 to as large as 350 m^3. A dry kiln has five primary components:

(a) Housing—this is usually a prefabricated outer shell made of aluminum nowadays, although concrete structures are still common; brick or wooden designs are less common.

Table 1
Approximate time to air dry green 25 mm thick lumber to 20% moisture content (after Rietz and Page 1971)

Species	Air-dry time (d)
Softwoods	
coastal Douglas fir	20–200
western hemlock	60–200
ponderosa pine	15–150
loblolly pine	30–150
light-type sugar pine	15–90
sinker-type redwood	200–365
Engelmann spruce	20–120
Hardwoods	
red alder	20–180
black cottonwood	60–150
hickory	60–200
silver maple	30–120
northern white oak	80–250
southern red oak	100–300
sweetgum (heartwood)	70–300
black walnut	70–200
yellow poplar	40–150

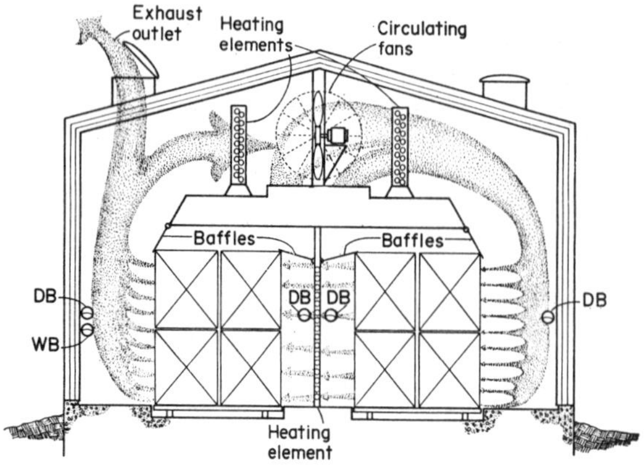

Figure 1
Cross-sectional view of a lumber dry kiln; DB = dry bulb, WB = wet bulb

(b) Heating—the lumber in a kiln is dried in air that has been heated by steam-heat-exchanger coils or by the addition of combustion gases from a wood residue, natural gas or oil burner. Conventional kilns use air temperatures from ambient up to 95 °C; high-temperature kilns, used when more-rapid drying can be tolerated, operate at temperatures above the boiling point of water up to about 115 °C.

(c) Humidity—this is controlled through two different mechanisms. When the humidity is too high, vents located on the roof open, enabling the hot moist kiln air to be exhausted. At the same time cool dry air is sucked into the kiln, thereby reducing the relative humidity. If the humidity is too low, steam is sprayed into the kiln to raise it to the desired level.

(d) Air circulation—the air or combustion gases present in the kiln serve several purposes. The primary purpose is to deliver heat to the lumber to evaporate the water in it and to carry away the evaporated moisture. Using constant static pressure in the side plenums, the gases assist in maintaining the uniform temperature and humidity conditions needed for quality control. Air velocities typically range from 1 to 3 $m\,s^{-1}$.

(e) Sensing-control—the conditions within the kiln are monitored and recorded using dry- and wet-bulb, liquid- or vapor-filled thermometers connected normally to a recorder–controller, which is set manually. Multiple dry-bulb thermometers are usually employed whereas a single wet-bulb sensor is standard practice (Fig. 1).

Drying in conventional kilns is accomplished using a series of temperature and relative-humidity conditions designed to be mildest when the lumber is the wettest. The series of combinations of desired air temperature and relative humidity is known as the kiln schedule. As the moisture content changes the schedule is made progressively more severe. Once all of the free water has been evaporated the lumber becomes less defect-prone and kiln conditions are further harshened to increase the rate of drying.

Softwood lumber is usually dried using a time schedule and the conditions are changed at predetermined times, usually every 12 or 24 hours. Hardwoods and thicker softwood lumber are dried using moisture-content schedules. These schedules are based on an actual measurement of the moisture content of boards within the charge being dried. Both types of schedule have largely been derived empirically by trial and error over a number of years.

Kiln-drying times for 25 mm thick lumber for a number of North American softwood and hardwood species are given in Table 2. The exact time required for a given type of material depends upon the properties of the material, the kiln and the schedule being used. Lower-grade softwood lumber is generally kiln dried from its original green condition until 95% of the charge is below 19% moisture content. The upper grades are dried to moisture contents between 6 and 10%. Hardwoods, and in particular thicker stock and species which are more difficult to dry, are first partially air dried and then kiln dried until the moisture content is between 6 and 10%.

Table 2
Approximate kiln-drying time for 25 mm thick lumber (after Rasmussen 1961)

Species	Time required to kiln dry 25 mm stock (d): 20 to 6% moisture content	Green to 6% moisture content
Softwoods		
coastal Douglas fir		2–4
western hemlock		3–5
ponderosa pine		3–6
loblolly pine		3–5
light-type sugar pine		3–4
sinker-type redwood	5–7	20–24
eastern, black, red, white spruce		4–6
Hardwoods		
red alder	3–5	6–10
cottonwood	4–8	8–12
hickory	4–12	7–15
red, silver (soft) maple	4–6	7–13
red oak	5–10	16–28
sweetgum (heartwood)	8–12	15–25
black walnut	5–8	10–16

4. Drying Defects

The drying of lumber can cause a wide variety of defects. These defects can be divided into three different classes based on their cause: (a) shrinkage, (b) fungal and (c) extractive.

4.1 Shrinkage-Related Defects

The fact that wood shrinks as it dries can lead to the development of several different drying defects. During the initial or early stages of drying, the outer shell of a board loses its moisture and once the fiber saturation point (see *Hygroscopicity and Water Sorption*) is reached the surface layer begins to shrink. The more massive inner core, however, is still above the fiber saturation point and has not started to shrink; in effect, the core restrains the outer shell from shrinking as much as it normally would. This differential shrinkage rate relative to thickness produces tensile stresses in the outer shell and compensating compression stresses in the inner core. The outer tensile stresses can be so large that they exceed the elastic limit in tension perpendicular to grain and a permanent set takes place. In some cases the stresses may be greater than maximum strength and a fissure or surface check results.

As drying progresses, the inner core ultimately begins to fall below the fiber saturation point and it shrinks. This is termed the second stage of drying. The tensile set developed during the first stage of drying has a major influence since it now restrains in turn the inner core from shrinking normally. This results in a stress reversal, the outer shell being under a compressive stress and the inner core being under a tensile stress. The compressive stresses on the surface can frequently close the surface checks that were easily visible during the first stage of drying, leaving the user with the impression that they are not present.

If the tensile stresses in the inner core are greater than the tensile strength perpendicular to grain, internal ruptures or honeycombing occurs. It cannot be seen by viewing a board's surface. Both honeycombing and surface and end checks usually originate along the wood rays since these are planes of weakness.

Once drying is completed, a board is still in a stressed condition, the outer shell under compression and the inner core under tension. This condition has, unfortunately, been termed case hardening. Such a condition presents no problem unless the board is remanufactured in such a way that the stresses are unbalanced relative to the thickness and width. If this is done, considerable distortion can occur. Fortunately, proper kiln drying can relieve this stressed condition. At the end of drying the charge is exposed to a short period of much higher relative-humidity conditions, thereby forcing a compressive set on the outer shell. If this final compressive set is roughly equal to the initial tensile set, all of the stresses are eliminated and the final lumber is stress free. This treatment is called conditioning. If the conditioning treatment is too long, stress reversal reoccurs and the lumber is permanently left in a reverse-case-hardened condition. This cannot be corrected.

End checks and end splits develop for exactly the same reasons and differ only with respect to differential moisture gradients in the longitudinal or lengthwise direction of the lumber.

Collapse of lumber is thought to be caused by capillary tension in those cell lumens that are fully saturated and/or by compressive stresses in the inner core that exceed the compressive strength. Collapse is a flattening or severe distortion of the cells. When severe, collapse appears as local depressions or corrugations on the surface.

Warp is one of the most common drying defects of wood. It is a distortion of the shape and form of a piece because of differences in radial, tangential and longitudinal shrinkage. Warp is greater the lower the final moisture content, because greater shrinkage will have taken place.

4.2 Fungus-Related Defects

Sap stain (blue stain), decay and molds can all develop in lumber as it waits to be dried or under certain drying conditions. The sapwood of most species is more susceptible to fungal attack than the heartwood with its higher extractive content. All of these defects generally occur early during drying when the wood is above the fiber saturation point and the fungi have the necessary requirements of food supply, free water, oxygen and acceptable wood temperature. They can be prevented by rapid air or kiln drying, particularly of the surface layer, or through treatment with a toxic antifungal chemical solution.

4.3 Extractive-Related Defects

The extractives in wood can cause unsightly discolorations on the wood surface as a result of either concentration or chemical change during drying. Examples of the most common problems are sticker stains where the extractive movement causes darkening along the outer edges of or under the stickers, leading to stripes crossing a board at intervals. Brown stain occurs on the five-needle pine species and results from an enzymatically induced polymerization of the extractives leading to the presence of light to very dark brown streaks and blotches on the surface. Resin bleeding occurs in some softwoods if the resin or pitch has not been exposed to a high enough temperature for a sufficiently long period for the volatile fractions to volatilize, thereby fixing the resin. These defects can generally be controlled through proper drying practice.

5. *Factors Affecting Drying Rate*

The drying rate of lumber is affected by both internal or material properties of the lumber itself and by external factors, that is, the kiln conditions. The movement of water through wood as a result of drying takes place as mass transport of liquid water and as diffusional transport of bound water and water vapor. The bound-water diffusion occurs in the cell walls and water-vapor diffusion through the void structures.

Except for highly permeable species, the greater part of moisture loss occurs through diffusion. Thus, those factors that affect wood's diffusional characteristics (see *Fluid Transport*) in general also control drying rate. As a general rule, and all other factors being equal, the drying rate decreases with increasing specific gravity and thickness. Since the diffusion coefficient of water through wood increases with increasing moisture content, the drying rate decreases as moisture content decreases and hence wood has a falling-rate-type drying curve. Increasing wood temperature raises the drying rate since the diffusion coefficient also increases with rising temperature.

The external factors that play a significant role are wet-bulb depression, air temperature and air velocity. Temperature, as noted above, affects the wood temperature, which in turn affects the diffusion coefficient. The wet-bulb temperature is the dominant factor because it represents the driving force leading to evaporation. Air velocity is of secondary importance once surface moisture has been evaporated although it affects the surface diffusion resistance and heat-transfer rate to the board surface.

In general, the drying rate of lumber is not limited by heat or mass transfer. Rather, the rate at which lumber can be dried is controlled by its propensity to defect development.

6. *Special Drying Methods*

Over the years a large number of alternative methods to air and conventional kiln drying have been tried. They include such methods as radiofrequency and microwave drying, boiling in oil, solvent seasoning, vapor drying and press drying. As noted previously, while often highly effective from a heat- or mass-transfer viewpoint these techniques generally had excessive defect development problems, and when slowed down to obtain acceptable defect levels lost their time-economics advantage.

Two more recent drying methods are, however, worth noting. The first of these is dehumidification drying. With this method, a well-insulated chamber similar to a dry kiln is employed. In contrast to heating via steam coils or combustion gases and venting of dry kilns, moisture is removed through the cooling coils of a heat pump. In this manner the kiln is heated using the latent heat of condensation from the compressor and small amounts of heat from supplemental electrical heating coils. Dehumidification is quite common in Europe and is being increasingly used in the USA particularly for hardwoods. Drying times are intermediate between air drying and conventional kiln drying. This is now a viable commercial method.

Solar drying does not yet have proven commercial acceptance but appears promising particularly for the less developed countries. In solar kilns the energy of the sun is either simply used directly via a greenhouse-type kiln design or in some cases combined with a thermal storage system to offset diurnal problems. Some solar kilns employ forced convection whereas other designs simply use free convection, which results in slower drying rates.

Both dehumidification and solar kilns have several distinct disadvantages. Neither has an effective system for adding humidity to the kiln environment if it is needed, and this represents a major problem in conditioning for stress reduction. They also both tend to operate at lower temperatures than conventional or high-temperature kilns and thus drying times are longer. Their major advantages are lower capital investment and operating costs.

7. *Drying of Other Forest Products*

Considerable volumes of veneer, particles, fiberboard and paper sheets are also dried. These materials present much less of a production problem because of their limited thickness. Their thinness precludes the development of steep moisture gradients that result in so many problems in the case of lumber. Although still highly variable in properties, the additional step of comminution somewhat reduces the piece-to-piece and within-piece variability that is also a major problem with the drying of solid wood products.

Paper is dried by initially dewatering the fibers, after they have been deposited on a mesh screen, by passing it through a series of paired nip rollers that mechanically press out most of the free water bringing the sheet to near the fiber saturation point. The paper sheet then passes through a series of heated dryer rolls utilizing conductive heat transfer. Various supplemental heat- and mass-transfer modifications to this basic approach have evolved over the years, including air impingement, pocket ventilation and microwave heating (see *Paper and Paperboard*).

Veneer is dried by conveying the thin sheets through a steam- or combustion-gas-heated chamber. Driers with the air flow being parallel to the veneer surface (across or longitudinal) have now largely been replaced by perpendicularly impinging jet driers. Modern veneer driers operate at temperatures between 125 and 275 °C with air velocities between 8 and 25 m s^{-1}. Veneer driers do not use humidification because of their high temperatures.

Particles are conventionally dried in rotating drum driers at temperatures as high as 500 °C, which is well above the combustion point of wood. The particles can be exposed to such extreme temperatures as long as there is sufficient evaporative cooling to leave the wood temperature below the combustion point. The particles are conveyed through the drier using high air velocities, the smaller lighter particles being passed through more rapidly, thereby preventing excessive thermal degradation.

Little information is available on the fundamental aspects of particle, flake or fiberboard drying.

See also: Plywood

Bibliography

Bachrich J L 1980 *Dry Kiln Handbook*. Simons International, Vancouver

Bramhall G, Wellwood R W 1976 *Kiln Drying of Western Canadian Lumber*, Western Forest Products Laboratory Report VP-X-159. Canadian Forest Service, Vancouver

Cech M Y, Pfaff F 1977 *Kiln Operator's Manual for Eastern Canada*, Eastern Forest Products Laboratory Report OPX192E. Canadian Forest Service, Ottawa, Ontario

Koch P 1985 *Utilization of Hardwoods Growing on Southern Pine Sites, Volume II, Processing*, USDA Forest Service Agriculture Handbook No. 605. US Government Printing Office, Washington, DC

Pratt G H 1986 *Timber Drying Manual*, 2nd edn. (revised by C H C Turner). Building Research Establishment, Department of the Environment, Princes Risborough Laboratory, Buckinghamshire

Rasmussen E F 1961 *Dry Kiln Operator's Manual*, USDA Forest Service Agricultural Handbook No. 188. US Government Printing Office, Washington, DC

Rietz R C 1978 *Storage of Lumber*, USDA Forest Service Agriculture Handbook No. 531. US Government Printing Office, Washington, DC

Rietz R C, Page R H 1971 *Air Drying of Lumber: A Guide to Industry Practices*, USDA Forest Service Agricultural Handbook No. 402. US Government Printing Office, Washington, DC

Wengert E M 1989 *Drying Lumber and Other Forest Products*. Springer, Berlin (in press)

Wengert E M, Grimm P, Lamb F, Muench J 1988 *Opportunities for Dehumidification Drying of Hardwood Lumber in Virginia*. Lumber Manufacturers' Association of Virginia, Sandston, Virginia

Wengert E M, Lamb F 1982 *A Predryer Symposium for Management*. Lumber Manufacturers' Association of Virginia, Sandston, Virginia

Wengert E M, Lamb F 1983 *Making Management Decisions in Lumber Drying*. Lumber Manufacturers' Association of Virginia, Sandston, Virginia

D. G. Arganbright
[University of Massachusetts,
Amherst, Massachusetts, USA]

E

Electrical Properties

The electrical properties of wood are complex because of its complex, hygroscopic structure. Wood consists of small crystal-like regions dispersed in a matrix of amorphous material. The amorphous material is hygroscopic, so wood contains moisture in proportion to the humidity of its environment. The moisture content of wood affects its electric properties much more than any other factor (see *Hygroscopicity and Water Sorption*; *Constituents of Wood: Physical Nature and Structural Function*).

The electrical properties considered here are conductivity, dielectric constant, loss tangent, dielectric strength and piezoelectric behavior.

1. Conductivity

Conductivity is a material property that determines the current density resulting from a given voltage gradient in the material. Its reciprocal is resistivity.

(*a*) *Moisture content.* The conductivity of wood varies enormously with its moisture content. Oven-dry wood has a conductivity of the order of $10^{-16}\,\mathrm{S\,m^{-1}}$, and at 30% moisture content, about 10^{-3}–$10^{-4}\,\mathrm{S\,m^{-1}}$. In this range, the relationship between conductivity and moisture is roughly logarithmic (Fig. 1). As the moisture content increases above 30%, the conductivity increases much more slowly, and at complete water saturation is usually no more than 50 times its value at 30%.

(*b*) *Chemical treatments.* Extraneous water-soluble electrolytes in wood greatly increase its electric conductivity provided the wood also contains about 9% or more moisture. If the moisture content is less than 9%, the electrolytes have no significant effect.

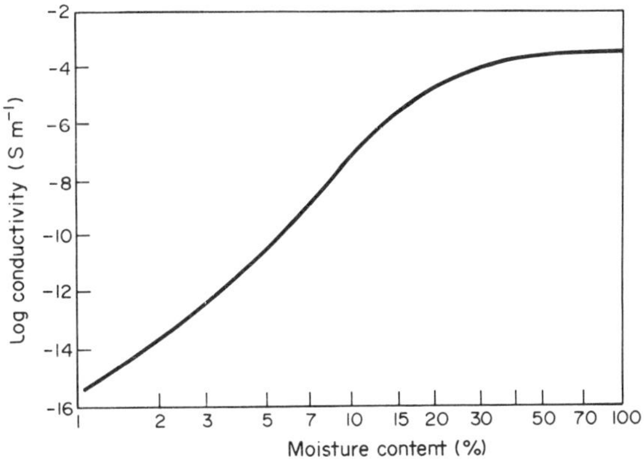

Figure 1
The dc conductivity of a typical wood specimen at room temperature as a function of its moisture content

(*c*) *Temperature.* The conductivity of wood increases with increasing temperature, with an interaction with moisture content; see Fig. 2.

(*d*) *Applied voltage.* The conductivity of wood usually increases as the applied voltage increases (Evershed effect). Data on this phenomenon are sparse, but it appears that the Evershed effect is significant only when the potential gradient is less than $15\,\mathrm{V\,mm^{-1}}$, and is negligible for alternating voltages of 60 Hz or greater.

(*e*) *Variation with time.* When the conductivity of wood is observed using metal electrodes, the apparent conductivity decreases with time after the electrodes are set in place; this decrease occurs whether or not the electrodes are energized. The cause is apparently due to polarization or chemical effects at the wood–metal interface. If contact with the wood is made through silver paint applied to the wood and permitted to dry, these time-related effects are much smaller.

(*f*) *Frequency.* Wood contains both bound and unbound charge carriers. The bound charge carriers cannot contribute to the dc conductance, but as they can be displaced somewhat from their bound positions, they can contribute to the ac conductance. The ac conductivity of wood increases as the frequency increases, but cannot exceed the conductivity that

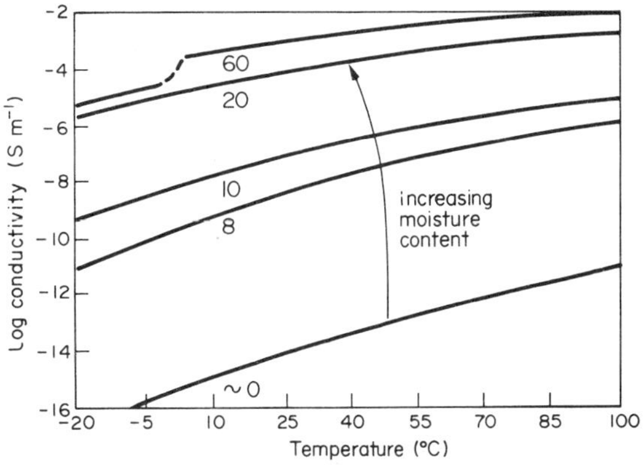

Figure 2
The dc conductivity of a typical wood specimen as a function of temperature, at various levels of moisture content. Data for approximately zero moisture content are based in part on extrapolation and therefore are not as reliable as other data

would exist if the bound charge carriers were, in fact, unbound.

(*g*) *Grain direction*. The conductivity parallel to the grain is about double that across the grain. There is no consistent significant difference between various directions across the grain.

(*h*) *Density*. There is a weak tendency for the conductivity to become greater as wood density increases.

2. Dielectric Constant

The dielectric constant of a material is a measure of the electric energy it will absorb from an electric field and store internally as polarization.

(*a*) *Moisture content*. The dielectric constant of wood varies greatly with moisture content, with a strong interaction with frequency (Fig. 3) and a lesser interaction with temperature.

At frequencies greater than about 1 MHz, the dielectric constant of moist wood appears to be essentially the sum of that of dry wood and of water. At these frequencies, the predominant mechanisms of polarization are induced dipole moment in the atoms and molecules of the moist wood, and alignment of the fixed dipole moments of water molecules when the moisture content is high enough that at least some of the water molecules are essentially unbound.

At lower frequencies, the dielectric constant of moist wood can become very large due to the contribution of fixed dipole and interfacial polarizations that are too slow to be effective at the higher frequencies.

The very-low-frequency dielectric constant of wet wood can be as large as 10^6 at room temperature, while typical corresponding values for dry wood are less than 10.

(*b*) *Frequency*. The dielectric constant of wood varies with frequency because polarization takes a finite time to occur. Some mechanisms of polarization occur faster than others and, as the frequency increases, the contributions of the slower polarizations fade out because they cannot follow the more rapidly alternating electric field. As a result, as the frequency increases, there is a general decrease in dielectric constant.

(*c*) *Temperature*. The dielectric constant of wood varies with temperature primarily because some of the polarizations, notably the fixed dipole and interfacial polarizations, are thermally activated. These predominate at low frequencies and moderate-to-high moisture contents, so the temperature effect is also greater under such conditions. At higher frequencies (1 MHz or more) or low moisture content (less than 8%), the temperature effect is relatively small (Fig. 4).

(*d*) *Grain direction*. The dielectric constant of wood is greater when the electric field is parallel to the grain than when it is perpendicular. The difference depends on other variables, but ranges from a factor of about 2 to 10 or more.

(*e*) *Density*. The dielectric constant of wood increases roughly in proportion to wood density.

(*f*) *Electrode design*. The apparent dielectric constant of wood is somewhat erratic and unstable when it is measured using metal electrodes in contact with the wood. Electrodes painted onto the specimen using silver paint generally give relatively stable and reproducible measurements.

3. Loss Tangent (Dissipation Factor)

When a material is polarized by an electric field, some of the polarization energy is dissipated as heat. The amount of energy lost in relation to the total polarization energy is expressed by the loss tangent or dissipation factor; other quantities also may be used.

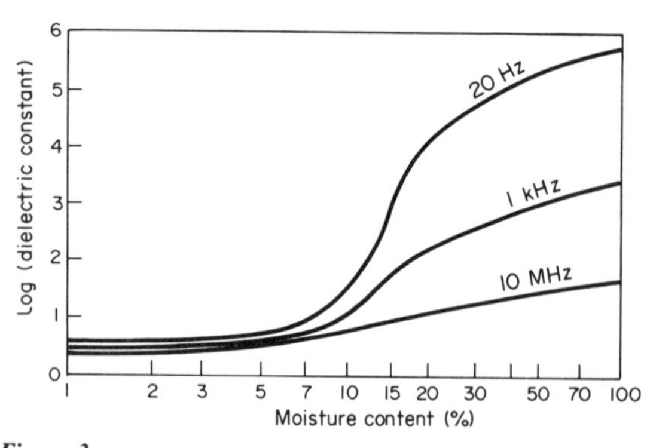

Figure 3
The dielectric constant of a typical wood specimen as a function of its moisture content, field parallel to the grain, room temperature, and at various frequencies

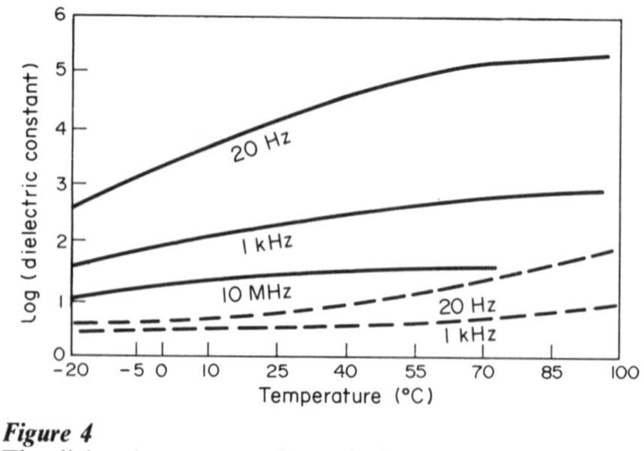

Figure 4
The dielectric constant of a typical wood specimen as a function of its temperature, field parallel to the grain, at various combinations of frequency and moisture content (—— 24%; --- 6%)

(*a*) *Moisture content.* Under most conditions, the loss tangent of wood increases as the moisture content increases, but under some combinations of frequency and temperature, the loss tangent goes through maximum or minimum values as the moisture content is increased (Fig. 5).

The loss tangent at room temperature can vary from about 0.01 for dry wood at about 100 Hz to about 100 for wet wood at middle audio frequencies (James 1975). At higher frequencies (about 10 MHz), the variation is much less, ranging from about 0.1 to 1.0 as the moisture content is increased from bone dry to soaking wet.

(*b*) *Frequency.* The variation in loss tangent of wood with frequency is quite complicated. As the frequency is varied from near zero to the microwave region, the loss tangent goes through maximum and minimum values, depending on the temperature and moisture content. The frequencies at which these extremes occur increase as either temperature or moisture content increases.

(*c*) *Temperature.* As with other variables, the loss tangent of wood may increase or decrease with increasing temperature, depending on the frequency and moisture content (Fig. 6).

At frequencies less than about 1 kHz, the loss tangent of wood that contains a substantial amount of free water may be decreased sharply as the temperature is reduced below freezing point. In particular, under these conditions the loss tangent of soaking wet wood may be much smaller than when the moisture content is in the 15–25% range.

(*d*) *Grain direction.* At some combinations of temperature, frequency and moisture content, the loss tangent of wood is larger when the electric field is parallel to the grain (longitudinal) than when it is perpendicular (radial or tangential) to the grain; at others, the reverse is true. There is a tendency for the longitudinal values to be the larger at lower moisture contents and higher frequencies, and vice versa.

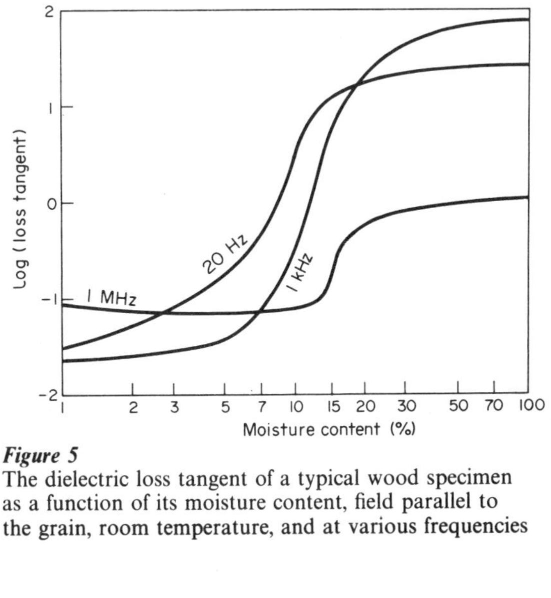

Figure 5
The dielectric loss tangent of a typical wood specimen as a function of its moisture content, field parallel to the grain, room temperature, and at various frequencies

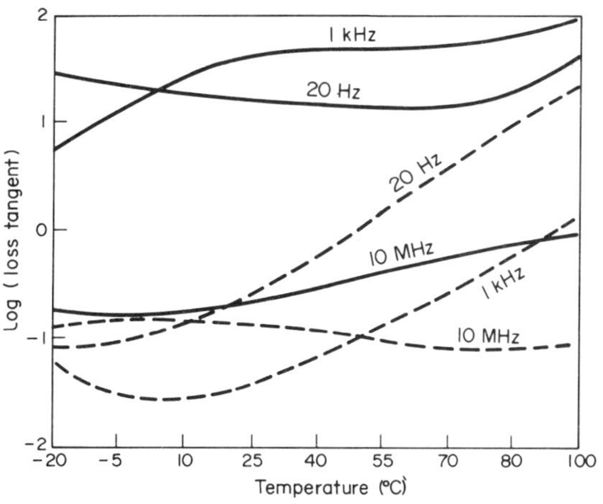

Figure 6
The dielectric loss tangent of a typical wood specimen as a function of its temperature, field parallel to the grain, and at various combinations of frequency and moisture content (—— 24%; --- 6%)

(*e*) *Other factors.* There is no consistent recognizable relationship between loss tangent and wood density. Electrode effects are similar to those discussed under conductivity and dielectric constant. Extraneous soluble salts or other electrolytes in wood greatly increase the loss tangent when sufficient moisture is present. This phenomenon is essentially related to conductivity.

There is some evidence that, at lower frequencies at least, the loss tangent increases as the intensity of the applied electric field increases. This phenomenon is consistent with, and may be related to, the Evershed effect (Nanassy 1972).

4. Dielectric Strength

The dielectric strength of a material is the minimum electric field that will cause, in a relatively short time, a catastrophic increase in current and usually local damage to the material.

The dielectric strength of dry wood, with the field perpendicular to the grain, is of the order of 10–20 kV mm^{-1}. At 15% moisture content it is about 1.0–2.0 kV mm^{-1}, and at 20% moisture content, about 0.5 kV mm^{-1}.

The dielectric strength of wood is typically 2–3 times larger for the field perpendicular to the grain than parallel to the grain. The logarithm of the dielectric strength decreases approximately linearly with increasing moisture content. There is a tendency for the apparent dielectric strength to be larger when the specimen thickness is reduced.

5. Piezoelectric Properties

The semicrystalline structure of wood causes it to have a small degree of piezoelectric activity, most readily observed as a transverse voltage gradient resulting from a longitudinal stress. The effect decreases as the wood moisture content increases, probably due largely to the leakage of charge around the crystallites. The effect increases with increasing wood density, increasing temperature and increasing degree of crystallinity of the wood. Maximum piezoelectric moduli are of the order of 10^{-13} C N^{-1}.

Bibliography

Brown J H, Davidson R W, Skaar C 1963 Mechanism of electrical conduction in wood. *For. Prod. J.* 13(10): 455–59

Galligan W L, Bertholf L D 1963 Piezoelectric effect in wood. *For. Prod. J.* 13(12): 517–24

James W L 1975 *Dielectric Properties of Wood and Hardboard; Variation with Temperature, Frequency, Moisture Content and Grain Orientation*, USDA Forest Service Research Paper FPL-245. Forest Products Laboratory, Madison, Wisconsin

Knuffel B, Pizzi A 1986 The piezoelectric effect in structural timber. *Holzforschung* 40(3): 157–62

Lin R T 1967a Review of the electrical properties of wood and cellulose. *For. Prod. J.* 17(7): 55–61

Lin R T 1967b Review of the dielectric properties of wood and cellulose. *For. Prod. J.* 17(7): 61–66

Maeda H, Fukada E 1987 Effect of bound water on piezoelectric, dielectric, and elastic properties of wood. *J. Appl. Polym. Sci.* 33(4): 1187–98

Nanassy A J 1972 Dielectric measurement of moist wood in a sealed system. *Wood Sci. Technol.* 6: 67–77

Peyskens E, De Pourq M, Stevens M, Schalck J 1984 Dielectric properties of softwood species at microwave frequencies. *Wood Sci. Technol.* 18(4): 267–80

Skaar C 1964 Some factors involved in the electrical determination of moisture gradients in wood. *For. Prod. J.* 14(6): 239–43

W. L. James
[Madison, Wisconsin, USA]

Energy Generation from Wood

Wood was the dominant pre-industrial thermal energy source for Western European and North American societies, and remains a dominant fuel in many African and Asian societies today (Tillman 1978). This form of biomass is used throughout the world as a primary or supplementary fuel for household heating and cooking, and it is also used in industrial applications, such as the raising of process heat and steam, the generation of electricity, and the cogeneration of both process energy and electricity. Wood is the dominant fuel source for the forest products industry in the USA, Canada and Scandinavia. It is also used as a fuel by electric utilities and independent power producers in the forested regions of North America. Current estimates of wood fuel consumption in the USA lie in the range 2.5×10^{12}–5×10^{12} MJ year^{-1} (2.5×10^{15}–5×10^{15} Btu year^{-1}) and estimates of worldwide wood fuel consumption can be placed at two to three times the US consumption.

The dominant means for generating energy from wood involves direct combustion, gasification being the primary alternative. Both of these types of systems involve similar physical and chemical reactions, and both result in the generation of process heat and electricity. Consequently, the dominant concerns in evaluating wood as a fuel involve the properties of the material, dominant combustion reactions, typical combustion systems, dominant gasification reactions and typical gasification systems.

1. Fuel Characteristics of Wood

Wood is the dominant biomass used in energy applications. It can be grown reasonably rapidly, and can be produced either as a residue from logging or wood product manufacturing activities. It can be stored "on the stump" in the forest and can be produced throughout the year. These availability characteristics distinguish wood from other forms of biomass such as agricultural crops (i.e., straw). Agricultural crops typically are grown annually and are harvested in a short season. Their use as fuel involves storage of significant quantities of material.

Wood can be produced on a year-round basis, eliminating the storage problems associated with crop materials. At the same time, wood fuels are grown in a dispersed or diffuse manner. Wood fuel production and use involves gathering/harvesting followed by material concentration at a single point. This is a critical distinction of wood from the fossil fuels, which are produced at a single location (i.e., a coal mine) and can then be distributed to end-point locations.

Chemically, wood is a partially oxygenated solid composed of cellulose, the hemicelluloses, lignin and extractives. It is highly volatile, with modest amounts of aromatic structures. Representative summative, proximate and ultimate analyses of softwoods and hardwoods are shown in Table 1, along with comparative values for a representative bituminous coal and a representative middle distillate oil. Also shown in Table 1 are the atomic hydrogen/carbon and atomic oxygen/carbon ratios of wood, bituminous coal and distillate oil. It should be recognized that these are representative values only. For example, the carbon content of softwoods can range from 48% to 58%, depending upon species, age of the wood, type of wood (i.e., compression wood) and other factors. It should be

Table 1
Chemical analyses of wood fuels with comparisons to bituminous coal and distillate oil (oven-dry basis)

	Fuel			
Analysis type	Softwood	Hardwood	Coal	#2 Oil
Summative				
% cellulose	42	45	N/A	N/A
% hemicelluloses	28	33	N/A	N/A
% lignin	30	22	N/A	N/A
Proximate				
% volatiles	75.2	78.1	35.4	N/A
% fixed carbon	23.0	19.0	56.2	N/A
% ash	1.7	2.5	8.4	N/A
Ultimate				
% carbon	53.0	51.0	74.6	87.2
% hydrogen	6.1	6.2	4.8	12.5
% oxygen	38.8	39.9	8.9	Nil
% nitrogen	0.1	0.2	1.5	0.02
% sulfur	Nil	Nil	1.8	0.3
% ash	1.7	2.5	8.4	Nil
Higher heating value (MJ kg^{-1})	20.4	19.7	31.1	44.9
Atomic ratios				
hydrogen/carbon	1.38	1.46	0.77	1.72
oxygen/carbon	0.55	0.59	0.09	0.0

Sources: Babcock and Wilcox 1977, Schwleger 1980, Tillman 1987 N/A: not applicable

noted, however, that wood is a very low-sulfur, low-nitrogen fuel with the consequence that it can be burned in an environmentally desirable manner.

Typically, wood has a so-called higher heating value HHV(d) (heating value of oven-dry wood) of about 19.5–23.7 MJ kg^{-1}. This variation is largely dependent on the ultimate analysis of the wood fuel. The higher heating value can be calculated using the following formula:

$$HHV(d) = 0.475C - 2.38 \qquad (1)$$

where C is percentage carbon. The correlation coefficient for this equation is 0.982 (Tillman 1980).

The values presented above relate to wood in the oven-dry state. In reality, wood moisture content ranges from 8% to 67% on a total-weight basis, or from 9% to 200% on a dry-weight basis. The two equations for calculating moisture content for any sample are shown below:

$$MC(t) = 100\ (\text{wt water})/(\text{wt wet wood}) \qquad (2)$$

$$MC(d) = 100\ (\text{wt water})/(\text{wt dry wood}) \qquad (3)$$

where MC(t) is moisture content, total-weight basis and MC(d) is moisture content, dry-weight basis. Variations in moisture content are a function of wood species, degree of wood processing and type of wood material. The value of MC(t) is typically 25–75% for bark, 30–60% for sawdust, 16–40% for planer shavings and 8–15% for sanderdust (Junge 1975, Resch 1978).

Physical properties of wood that are significant from an energy perspective include specific gravity, specific heat and thermal conductivity. Specific gravity, on an oven-dry basis, typically lies in the range 0.44–0.51. Typical values of specific heat (in J kg^{-1} K^{-1}) can be calculated as follows:

$$H_{sp}(d) = [1114 + 6.95(T - 273)] \qquad (4)$$

$$H_{sp}(t) = \frac{H_{sp}(d) + 10MC(d)}{1 + MC(d)/100} \qquad (5)$$

where H_{sp} (d) and H_{sp} (t) are specific heats of oven-dry wood and wet wood, respectively, and T is the temperature (K). Thermal conductivity λ (in W cm^{-1} K^{-1}) can be calculated using the following equations, depending on wood moisture content:

$$\lambda(MC(t) < 40\%) = (5.18 + 0.096MC(t))S + 0.57\,(1 - (S/1.46)) \qquad (6)$$

$$\lambda(MC(t) > 40\%) = (5.18 + 0.131MC(t))S + 0.57\,(1 - (S/1.46)) \qquad (7)$$

where S is the specific heat capacity of the dry wood (US Forest Service 1987). These chemical and physical properties largely govern how wood will function as a fuel for combustion, or as a feedstock for thermochemical gasification.

2. *Fundamentals of Wood Combustion and Gasification*

Chemically, wood combustion and gasification involve free-radical reaction sequences leading to the final products. Combustion is typically summarized as follows:

$$C + O_2 \rightarrow CO_2 + 33.9\ \text{MJ kg}^{-1} \qquad (8)$$

$$H_2 + 0.5\,O_2 \rightarrow H_2O + 142.9\ \text{MJ kg}^{-1} \qquad (9)$$

Gasification typically involves partial oxidation of the wood feedstock along the lines of the following reaction:

$$\text{wood} + O_2 \rightarrow CO + H_2 + CH_4 + CO_2 + H_2 + \text{others} \qquad (10)$$

The wood/oxygen mass ratio may vary between 2 and 3 kg air/kg dry wood.

The mechanisms associated with the conversion of wood into useful energy by either combustion or gasification are highly complex, and understood only in somewhat general terms. The sequence of reactions which begin either the process of combustion or gasification is as follows: (a) fuel particle heating and drying, (b) solid particle pyrolysis to volatiles and char, and (c) volatile compound pyrolysis. From there the chemical pathways diverge to some extent. The particle heating and drying reactions are physical phenomena, and are governed largely by Eqns. (5)–(7). Solid phase pyrolysis is then initiated as wood temperatures exceed 550 K. Hemicelluloses begin to pyrolyze first. Lignin tends to require the highest temperatures for pyrolysis. The products of solid phase pyrolysis include a host of condensible and noncondensible volatiles including CO_2, CO, CH_4, C_2H_6, CH_3OH, and a wide variety of other compounds. Radicals and fragments including OH, CH_3, and others may be produced directly by the pyrolysis process. The char generated is not completely fixed carbon, but remains an oxygenated compound. One empirical formula derived for such char is $C_{6.7}H_{3.3}O$ (Bradbury and Shafizadeh 1980). Volatile pyrolysis then provides for the dominant means of generating free radicals. Typical radical generation sequences can be postulated as follows:

$$C_2H_6+M\rightarrow 2CH_3+M \qquad (11)$$

$$C_2H_6+M\rightarrow C_2H_5+H+M \qquad (12)$$

$$CH_4+M\rightarrow CH_3+H+M \qquad (13)$$

$$CH_3+O_2+M\rightarrow CH_3O_2+M \qquad (14)$$

$$CH_3O_2\rightarrow CH_2O+OH \qquad (15)$$

Of particular importance are sequences generating hydroxy radicals since these are the most reactive species in the processes of wood combustion and gasification (Tillman 1987a, b). The subsequent reaction chains vary, depending on whether combustion or gasification is pursued. Combustion reactions proceed to the point where CO_2 and H_2O are formed. Combustion typically occurs in a significant excess of oxygen. Excess oxygen or air levels of 25–75% are common in modern wood-burning systems. Alternatively, excess oxygen levels can be expressed as equivalence ratios R_{eq}, defined as follows:

$$R_{eq}=(F\,A^{-1}\text{ actual})/(F\,A^{-1}\text{ stoichiometric}) \qquad (16)$$

where F is the amount of fuel available for combustion and A is the amount of atmospheric air for combustion. Typical equivalence ratios for wood combustion are 0.57–0.80. Therefore, product gases from wood combustion typically contain 79% N_2, 12–17% CO_2 and 5–9% O_2. Flame temperatures associated with wood combustion range from 1270 K to 2100 K, depending upon moisture content of the wood, level of excess oxygen, temperature of the air used in the combustor and heat content of the fuel. Adiabatic flame temperatures T_{ad} (in K) can be calculated as follows:

$$T_{ad}=650-10.1\,MC(\text{t})+1734\,R_{eq}+0.6\,(T_A-298) \qquad (17)$$

where T_A is the temperature of the combustion air (K). The constant, 650, varies modestly as a function of the higher heating value of the wood fuel (Tillman and Anderson 1983). Thermal efficiencies associated with wood combustion systems typically range from 65 to 80%, depending upon the condition of the wood and the combustion regime employed. Table 2 presents a summary heat and materials balance for a wood-fired combustor burning 50%-moisture wood and using 40% excess air (equivalence ratio = 0.71).

Table 2
Representative energy balance for a wood-fired combustor (conditions: hardwood fuel, 50% moisture content, 40% excess air, 460 K final stack temperature

Parameter	Percent
Energy inputs	
Wood	100
Air	0
Energy losses	
Unburned Carbon and Ash	0.9
Dry Stack Gas	7.1
Moisture in Stack Gas	22.8
Radiation & Mfg. Margin	1.5
Total of losses	32.3
Heat available to steam and blowdown	67.7

Gasification is designed to produce a useful fuel for subsequent combustion. Typical gasification reactions include the following:

$$C+H_2O\rightarrow CO+H_2 \qquad (18)$$

$$C+CO_2\rightarrow 2CO \qquad (19)$$

$$CO+H_2O\rightarrow H_2+CO_2 \qquad (20)$$

These yield the product gases that can then be utilized in producer or synthesis gas utilization systems. If air is used as the oxidant, then the gas typically contains some 6–7 MJ m^{-3} of chemical energy; and if pure oxygen is used in the reactor, the gas typically contains some 14–15.5 MJ m^{-3} of chemical energy. Gasifiers typically operate in the temperature regions of 1050–1280 K. Gasifier efficiencies are measured on a hot gas and cold gas basis. Hot gas efficiencies include the sensible heat in the product gas while cold gas efficiencies only include the chemical energy in the producer gas. Typical gasifier efficiencies are 82–92% and 70–80% for hot and cold gas, respectively.

3. *Wood Energy Utilization Systems*

Wood energy utilization systems include both direct combustion technologies and gasification technologies. Combustion technologies capable of handling wet wood fuel include pile burners, such as Dutch ovens or Deitrich cells, inclined grate systems, Wellons cells, Lamb–Cargate cells, and a wide variety of similar devices. Pile burners can be operated either as combustors or as gasifiers. Wet wood combustion systems also include spreader-stoker boilers. Spreader-stokers are the dominant technology for the burning of wood fuels. They may include fixed or travelling grate systems, air-swept stokers or mechanical paddle-wheel stokers, and other variations. The most common systems use chain-driven travelling grates and air-swept fuel stokers. Dry fuel systems are dominated by suspension burners. Cyclonic combustors also have been used for wood-fired systems.

Recent innovations in wood combustion systems have focused on the introduction of fluidized bed combustion. Fluid bed systems include conventional "bubbling bed" designs and the more recently developed circulating fluidized bed combustors. Fluidized bed systems burning wood fuel commonly co-fire with coal and/or refuse-derived fuel from municipal solid waste systems. These systems have been pioneered by such manufacturers as Lurgi, Ahlstrom and Gotaverken.

The largest-sized, direct-combustion-based wood-fired systems typically burn 80–100 t of fuel per hour. It is more common to encounter projects consuming 10–30 t of wood fuel per hour. Typical stand-alone power plants of very large sizes generate 40–50 MW of electricity with efficiencies measured as heat rates. Typical heat rates are in the region of 14.7 $MJ\,kWh^{-1}$. Typical cogeneration facilities have electricity outputs of 5–10 MW, and heat rates chargeable to power ranging from 6.5 to 10 $MJ\,kWh^{-1}$.

Gasification is most appropriate for small-scale systems. Gasifier designs may use counter-current flow of wood and product gas (including heat) for maximum heat and mass transfer. Alternatively, co-current gasifiers have been designed for dry fuels. Fluidized bed gasification has become a significant technical option in recent years. Gasifiers, followed by extensive gas clean-up systems, can be coupled to diesel or spark ignition engine generator systems. Such systems can be built as stand-alone power plants with heat rates of 14.2 $MJ\,kWh^{-1}$. Such systems also can be built as cogeneration devices if a use exists for low-grade process heat (i.e., hot water at 360–400 K). Maximum gasification systems are in the range of 100 $GJ\,h^{-1}$ fuel input.

Oxygen-blown wood gasification systems also have been proposed. These produce a synthesis gas that can be subsequently converted into liquid fuels, such as methanol or gasoline (see *Chemicals and Liquid Fuels from Wood*).

See also: Thermal Properties

Bibliography

Babcock and Wilcox. 1977 *Steam: Its Generation and Use.* Babcock and Wilcox, New York

Bradbury A G W, Shafizadeh F 1980a Chemisorption of oxygen on cellulose char. *Carbon* 18: 109–16

Bradbury A G W, Shafizadeh F 1980b Role of oxygen chemisorption in low-temperature ignition of cellulose. *Combust. Flame* 37: 85–89.

Junge D C 1975 *Boilers Fired with Wood and Bark Residues.* Forest Research Laboratory, Oregon State University, Corvallis, Oregon

Resch H 1978 Energy recovery from wood residues. Preprint from 8th World Forestry Congress, Jakarta, Indonesia

Schwleger B 1980 Power from wood. *Power* 124(2): S-1–S-32

Shafizadeh F 1982 Chemistry of pyrolysis and combustion of wood. In: *Progress in Biomass Conversion*, Vol. 3. Academic Press, New York, pp. 51–76

Tillman D A 1978 *Wood as an Energy Resource.* Academic Press, New York

Tillman D A 1980 Fuels from waste. In: *Kirk-Othmer Encyclopedia of Chemical Technology*, Vol. II, 3rd edn. Wiley, New York

Tillman D A 1987a Biomass combustion. In: Hall D O, Overend R P (eds.) 1987 *Biomass.* Wiley, London, pp. 203–9

Tillman D A 1987b Wood energy. In: *Encyclopedia of Physical Science and Technology*, Vol. 14. Academic Press, San Diego, pp. 641–55

Tillman D A, Anderson L L 1983 Computer modelling of wood combustion with emphasis on adiabatic flame temperature. *J. Appl. Polym. Sci. Appl. Polym. Symp.* 37: 761–74

US Forest Service 1987 *Wood Handbook.* US Government Printing Office, Washington, DC

D. A. Tillman
[Issaquah, Washington, USA]

F

Fire and Wood

The fire performance of wood depends on many variables. These include the nature of the wood specimen itself (density, moisture content, etc.), specific fire exposure, (flaming or smoldering) and geometry (i.e., thick or thin specimen).

Although combustible, wood has numerous fire-performance advantages when compared to many synthetic polymeric materials. These include:

(a) lower calorific value on a weight basis and lower rates of heat release in similar fire exposures;

(b) the presence of absorbed moisture, which tends to retard ignition;

(c) predictable charring/ablation rates; and

(d) minimal excess pyrolyzates (unburned fuel gases) which are produced when wood burns.

Extending these general properties to the use of wood in construction, the material has other advantages from a fire-safety standpoint. Amongst these are the fact that wood members are nonyielding in the early stages of a fire. This is in contrast to unprotected metal structural systems which yield early (and suddenly) in fully developed fires. In addition, wood members exhibit a predictable decrease in cross-sectional area during fires. This property, coupled with design philosophies (prescribed in building codes) which call for the use of repetitive wood members, provides a system with a substantial margin of safety in structural fire-performance applications. Finally, as compared to most synthetic polymers, absorbed moisture present in wood retards ignition of thick wood members, and the spread of fires from wood sources due to convection is minimal since excess pyrolyzates are rarely produced. Thus, physical barriers must usually be disrupted before wood-fueled fires will spread.

1. Fire Performance of Wood

Chemical composition, macrostructural details (e.g., grain orientation, density, moisture content) and microstructure all affect the fire performance of wood. Heat-transfer processes which occur during fires result in the development of steep thermal gradients in a burning member from exposed surface to unaffected core. At the same time, mass transport of moisture caused by combustion creates steep moisture-content gradients, ranging from zero moisture content at a burning surface to the pre-fire equilibrium moisture content at the central, unaffected core. Such moisture-content gradients are in turn accompanied by stress gradients and stress concentrations due to the shrinkage associated with dehydration. These lead to the formation of fissures which are familiar in burned wood.

In addition to the substantial moisture and stress gradients from burning wood surfaces to unaffected core areas, similar transitions and variations exist in the chemical processes occurring during fire exposure. Specifically, exothermic oxidation of carbonaceous chars created by pyrolysis predominate at or near burning wood surfaces. Simultaneously, the underlying char base undergoes complex thermal degradation reactions known collectively as pyrolysis. Successively lower levels of oxygen are present, moving from the wood surface to the base of this thermal-degradation zone. Eventually, a sharp boundary can be seen between thermally degraded wood and wood which is substantially unaffected and able to resist loads. Cracks through the char layer terminate in this zone as can be seen in Fig. 1.

Four broad temperature ranges have classically been associated with thermal degradation and fast combustion of wood in air. These are:

(a) ambient to 200 °C—generation of noncombustible gases such as CO_2, formic acid, acetic acid and H_2O occurs.

(b) 200 °C to 280 °C—endothermic heating and surface darkening occur. Gases listed above, as well as CO, continue to evolve.

(c) 280 °C to 500 °C—exothermic heating begins, accompanied by significant production of combustible fuel gases.

(d) above 500 °C—sustained gas-phase combustion of pyrolyzed gases occurs, and solid-state surface combustion consumes the charcoal residues remaining after fuel gases have been liberated. In this temperature range, the char front moves rapidly into the substrate.

Just as important as the temperature ranges listed above are time–temperature relationships. For example, although listed ignition temperatures of wood are usually at least 150 °C, long-term exposure of wood to temperatures in excess of 90 °C can result in thermal degradation conditions which have been observed to lead to ignition in the field. The ignition of such thermally degraded wood (sometimes referred to generically as pyrophoric carbon) may be a simple function of its extended exposure to moderately elevated temperatures, creating thermally active chars. It may also, however, be a function of low thermal inertia

Figure 1
Scanning electron micrograph of a fissure tip in the transition zone between whole and burned wood. The arrows indicate a longitudinal crack along the major grain direction caused by mechanical property differences in burned and unburned areas. The vertical crack originated at the initial wood surface

and conductivity (often expressed as the product "kpc") or the form of the wood subjected to heating (as with wood fiberboard and cellulose insulation which self-heat exothermically).

Chemically, wood is composed of three major polymers: cellulose, hemicellulose and lignin (see *Chemical Composition*), each of which have their own characteristic responses to thermal degradation conditions which are based on their polymer composition (i.e., chain order, crystallization, unsaturation, ring structure, etc.).

Two reaction pathways (each having characteristic reaction rates) exist when wood is subjected to high temperatures. These compete in the combustion process for available fuel and depend on composition, density and presence or absence of treatments (such as fire retardants) in the wood. One such pathway involves polymer cleavage or fission. This yields low-molecular-weight, gaseous pyrolyzate fuels which burn through gas-phase combustion. The second pathway involves dehydration of cellulose and yields chars and tars in the substrate. This pathway favors solid-state or glowing combustion. A shift in reaction rate of either of these competing reactions will change the character of the combustion and the char yields observed.

"Catalysts" such as fire retardants, metallic preservatives, fungal activity (or residue) or unusual mineral contents will alter the balance or ratio between these possible reaction pathways. Usually cell-wall polysaccharides (cellulose and hemicellulose) favor the cleavage, gas-phase combustion pathway. Lignins and extractive chemicals which contain a wide variety of cyclic rings and unsaturation follow pathways favoring solid-state combustion. Thermal analysis data show lignin to be the most stable with regard to oxidative degradation, followed by cellulose. Hemicellulose (coincidentally less ordered than cellulose) shows the least resistance to oxidative degradation and thus will participate in combustion earlier.

Microstructure, an expression of the cellular, three-dimensional nature of wood, affects the observed kinetics of wood combustion. For this reason, isolated individual chemical components of wood behave differently in a combustion environment than when they are intimately combined into the complex physical and chemical arrangement found in wood cells. For example, observed combustion reactions in wood vary from exothermic to endothermic, depending on the surface exposed to thermal radiation (i.e., end-grain vs. side-grain) or gross form (whole wood vs. wood meal). As a result of these structural effects, confusion has developed in reconciling research results from small, isolated specimens of wood with observations on larger members which possess intact wood structure. Data from the small, isolated specimens of specific, isolated chemcial groups (such as alpha-cellulose) indicate that combustion of wood follows first or second-order reaction kinetics. Conversely, results that take wood structure into account show zero-order kinetics, implying diffusion-controlled reaction mechanisms. In fact, diffusion control of combustion reactions in wood is consistent with both microstructural features in wood and char, and is also suggested by research linking moisture movement, pressure-driven flow and thermal conductivity effects with microstructural details.

Ignition depends on chemical composition, physical factors (especially thermal conductivity) and exposure geometry. Smoldering ignition, in which solid-state combustion predominates, occurs in wood-based materials of low thermal conductivity. This takes place when wood is exposed over time in situations where thermal inputs from the environment and heat produced by exothermic oxidative degradation are greater than heat losses from the material. In such cases, heat from the environment plus the exothermic heat is fed back into the substrate and increases decomposition rates. Eventually, smoldering combustion occurs if the available fuel is not consumed before actual ignition occurs. Such processes may also have biological origins (i.e., thermogenic bacteria). Examples are smoldering fires in raw material storage areas of particleboard or paper plants, improperly installed cellulose or fiberboard insulations, or fiberboards shipped in boxcars without being allowed to cool following manufacturing. In general the phenomena of smoldering ignition in wood substrates are less well understood than those of flaming ignition.

For flaming ignition to occur, wood substrates must be raised to temperatures sufficient for production of an adequate supply of fuel gases to allow sustained, gas-phase combustion. Assignment of a single, unique ignition temperature for wood (either piloted or non-piloted), however, is ill-advised. Variables such as specimen thickness, moisture content, rate of heating and mode of thermal input (i.e., radiative, convective or conductive) are all important and will affect the temperature required. Preheating and thermal history will also affect an observed "ignition temperature."

2. *Wood used in Construction*

Contemporary references occasionally refer to historical aspects of fire performance of wood in events like the Great Fire of London (1666), the Chicago Fire or other conflagrations. Following such events it has been common to see new fire safety design criteria expressed through building code changes in countries or areas where such events occur. Historically, the USA has had an abundance of wood available for construction. The advent of the textile industry in prerevolutionary New England led to a style of building known as "mill-type" construction. This involved wood posts and beams of large dimension used with floors of heavy timber or planks, three or more inches in thickness. Interestingly, such mill-type construction is still recognized today as a fire-resistant method of building in all US building codes.

Today wood joist, floor/ceiling assemblies and stud wall constructions which utilize a variety of surface finish materials afford predictable levels of fire performance for given design situations and building use.

3. *Fires Affecting Structures*

Fire performance of wood is related to the type of fire exposure which occurs. Structural fires can occur both within buildings and on a building's exterior. In both cases factors of importance are ignition source and form of surface materials present.

In exterior use, ignition sources are typically either (a) low intensity, such as grass or trash fires, or (b) high intensity, such as wild fires or fires caused by exposure to a fire of an adjoining structure. An exposure which involves both interior and exterior fire impact would occur when a fire, originating in the interior of a building, extends to siding materials via a window opening. At greatest risk in exterior fire exposures are wood roofing materials which have not been treated with durable fire retardants. Their impact on fire spread has been known for decades, especially in the western USA.

4. *Fires Within Structures/Compartment Fires*

Interior fires typically cause a greater structural impact than those originating outside a wood-framed structure and are also more common. Such incidents usually begin with ignition at one (often small) item, device or location. If such an incident is to pose a serious threat to the structure, the fire must grow to fully involve the room of origin which is often called a compartment. (Whether or not the fire does grow to this extent can be expressed through probability treatments which lend themselves to modelling. Fault-tree analyses may be conducted or state transition techniques applied. Such analytical methods assign probabilities in response to a series of "what-if" questions to determine the compound probability that a small fire will grow and spread in a given environment.)

Growth of the fire to where it threatens the room at large is divided into two regimes or growth periods. These are the pre-flashover and post-flashover periods. In the former, boundaries of the affected room (walls, doors, windows, etc.) are not threatened and will contain the small, growing pre-flashover fire. Occupants of such pre-flashover rooms will not usually be at risk during this period of fire growth, which is in contrast to the post-flashover period.

4.1 Pre-Flashover Fires

A substantial amount of literature exists on pre-flashover fires which has, as a common thread, the impact of finish materials, furnishings and room geometry on fire growth. Room size, ventilation, rapidity of flamespread over surfaces and ability of furnishings to sustain combustion and liberate heat after ignition are crucial variables. The performance of combustible finish materials such as wood panelling can be gauged by their rate of heat release as determined under various test conditions. Specific variables such as material thickness, sensitivity to ignition and moisture content are particularly important.

In compartment fires, completion of the flamespread state (which must be preceded by sustained ignition) signals the end of the pre-flashover stage. During this stage the fire will have spread from the point of localized ignition over combustible surfaces liberating heat and generating further gases for combustion throughout the compartment. Before this, the fire was essentially a two-dimensional phenomenon of limited intensity within a three-dimensional space. As the fire continues to grow hot gases accumulate in the upper levels of the compartment and radiative heat transfer increases.

In the pre-flashover stage, variables important to fire performance of wood are (a) susceptibility to ignition and (b) rate of flamespread following sustained ignition. Ease of ignition, though ill-defined from a quantitative standpoint, is an expression of the minimum flame or heat source required to ignite a sample of specified size and geometric configuration.

In terms of ease of ignition, two types of wood fuels classified as thermally thick or thermally thin are recognized. Plywoods less than 6 mm thick, for example, are typical of thermally thin fuels. Their thin cross section allows rapid attainment of thermal equilibrium with a potential ignition source and consequently ignition will occur readily. Conversely, a thermally thick fuel of the same wood species (such as a piece of dimension lumber) displays a substantial temperature gradient from exposed surface to the unexposed (back) face of the member when heated. For this reason, the thick member is much less susceptible to ignition from a similar source. Additionally, because of the effects of thermal inertia on thick wood fuels, substantial levels of radiative feedback from other burning members are necessary for sustained ignition. Thin fuels, in addition to showing easier ignition, will generally show much higher rates of flamespread in a given use-configuration than thick members of similar species.

From a practical standpoint, the use of thin plywood (because of its ease of ignition and high flamespread) is regulated in building by requiring that such materials be backed by a gypsum wallboard membrane, which essentially converts it to a thick fuel. Conversely, thicker layers of wood products (i.e., 15–16 mm), such as plywood or tongue-and-groove panelling, may be used without a backing over supporting framing because such materials do not show the ease of ignition or high rates of flamespread common to thin wood members.

4.2 Post-Flashover Fires

Eventually, the nature of a fire will change from the two-dimensional phenomenon discussed above to a three-dimensional phenomenon with volumetric involvement of the compartment. When this occurs, the affected room is said to go to "flashover" and the post-flashover fire stage begins. This stage of a fire is characterized by:

(a) the compartment becoming untenable for human life,

(b) gas temperatures in excess of 550 °C, and

(c) exposed combustible surfaces appearing to ignite more or less simultaneously.

Interestingly, the post-flashover process can be described using the kinetics of a constantly stirred tank reactor.

In the post-flashover fire, one of the properties that become important when considering the fire performance of wood is the ability to produce a char layer. This layer regulates the flow of heat and reactants into the pyrolysis zone as wood burns and will limit its rate of heat release in a post-flashover fire. The charring occurs at a rate of about 0.8 mm min^{-1} after an initial period of rapid burning. Research has shown that wood composites such as particleboard, thick plywood, laminated beams and lumber have charring rates similar to whole wood. Heat release, which is interrelated with charring, is a function of the thermal inertia of the material in a general sense. Thus, products such as plywood which have density and thermal conductivity similar to whole wood can be expected to show similar rates of heat release when exposed at identical radiant flux levels. Wood products which have substantially different materials properties (such as cellulose insulation or wood fiber sound-deadening board) can be expected to show charring rates and rates of heat release consistent with those differences.

5. Wood Composites

Forms of wood found in construction have expanded from timbers, lumber and decking to a wide variety of wood-based materials. Products such as plywood, laminated veneer lumber and more recently composites of plywood and particleboard (such as wafer board and strand board) are now commonly used. These products behave similarly to whole wood of equal dimensions with respect to fire performance, provided voids are not present and thermosetting adhesives are used in their fabrication.

Exceptions to these generalizations exist in several notable cases. For example, "comply" studs and joists composed of veneer faces over a particleboard core can be expected to show similar charring rates but reduced load-carrying ability if the veneer on a fire-exposed face is allowed to burned through since the veneer carries a disproportionately large share of the load. Likewise, when engineered wood structural elements with reduced cross-sectional areas are used (such as wood-composite I-beams or composite beams with repetitive metal tubing for web materials), load-bearing fire endurance can be expected to be reduced as compared with members of equal initial load-bearing capacity of either whole wood or laminated design. If an adequate thermal barrier is included in the design to protect such beams, they can be expected to function safely, however.

Other examples include truss designs where metal connectors transfer stresses between members. These cannot be expected to perform as well as the older style mill-type or post-and-beam constructions they replace. Similarly, nailed, metal truss plates can be expected to have poorer fire endurance than component member sizes alone would predict if a protective membrane is not included to reduce heat transfer to such plates in a fire incident. In these situations, field experience has shown that the metal connectors act as a heat sink, leading to early dehydration and charring at joints with subsequent early failures to sustain imposed loads.

6. *Fire Retardants*

Fire-retardants are applied to wood in two ways. One method involves surface coating to protect the underlying wood members. Such treatments increase time to ignition and reduce flamespread following ignition. Typically, such materials are of little benefit in a post-flashover fire, since they do little to increase fire endurance.

The second application is by pressure treatment. Most of these are based on mixtures of waterborne salts such as borates, sulfates and phosphates. Such treatments both reduce flamespread and produce wood elements which are accepted in the building codes as having improved fire endurance. Since they are water-soluble, however, such treatments are not suitable for exterior exposure. They must also be carefully formulated and/or applied to avoid corrosion of fasteners, pipes and associated metal in contact with treated wood. In addition, their appearance is not suitable for architectural uses.

The functioning of waterborne salts is complex, and a substantial literature exists describing how such chemicals act as retardants, reducing either gas-phase flaming combustion or increasing smoldering (glowing or solid-state) combustion. In most cases, it is believed that a catalysis occurs due to the presence of the fire retardant, which promotes char development through increased glowing (or solid-state) combustion. This decreases gas-phase combustion since the enhanced char layer reduces heat and oxygen transfer at the pyrolysis zone. Release of fuel gases formed in the pyrolysis zone (which would normally burn in the gas phase) is also reduced and is expressed by reduced flamespread values. In practice, formulations of such treatments must result in reduced flamespread without adverse smoldering tendencies, corrosive properties, or high levels of hygroscopicity.

In the past decade, a number of patented, exterior, durable fire-retardant formulations have been developed. These are based on pressure treatment of wood with water-soluble, thermosetting polymer systems which are heat cured following pressure treatment. Contents include melamines, urea formaldehydes, dicyandiamide, borates and other materials. Developed in both the USA and Canada, these are used mostly for fire-retardant treatment of wood roofing materials but they can also be applied to siding products which will have acceptable appearance for architectural uses.

Many other specialized wood products rely on fire retardants for their successful, safe utilization. Treatment of low-density fiberboard acoustical tiles, for example, drastically reduces their flamespread. Medium- and higher-density fiberboard products achieve low flamespread ratings by the inclusion of fire retardants such as aluminum trihydrate (which includes relatively large amounts of water of hydration) to reduce flammability. In both these cases, the fire retardant can be added to the furnish (fiber feedstock) to obtain integral treatment through the entire thickness of the product. Likewise, low flamespread particleboards are available for subsequent overlaying with fine veneers or high-pressure laminates for panelling or work surfaces.

Fire-retardant treated, loose-fill and spray-on insulations based on recycled newspaper or wood pulp are widely used in the USA. These systems are especially susceptible to smoldering ignition if not formulated properly because of their low thermal conductivity and high surface area. Use of proper fire-retarding and manufacturing techniques and proper installation practices are necessary for their safe use. Likewise, insulation boards based on wood fibers perform well but must be protected from exposure to heat sources such as torches or hot surfaces which can lead to either smoldering fires or rapid flamespread. While technology exists for fire-retardant treatment of such products little practical application has occurred.

See also: Building with Wood; Thermal Degradation

Bibliography

Atreya A 1983 Pyrolysis, ignition and fire spread on horizontal surfaces of wood. Ph.D. dissertation, Harvard University

Browne F L 1958 *Theories of the Combustion of Wood and its Control*, USDA Forest Products Laboratory Report 2136. USDA Forest Products Laboratory, Madison, Wisconsin

Goldstein I S 1973 Degradation and protection of wood from thermal attack. In: Nicholas D D (ed.) 1973 *Wood Deterioration and Its Prevention by Preservative Treatments.* Syracuse University Press, Syracuse, New York, pp. 307–39

Kanury A M 1972 Ignition of cellulosic solids—A review. *Fire Res. Abstr. Rev.* 14(1): 24–52

Meyer R W (ed.) 1977 Trends in fire protection. Symposium proceedings. *Wood Fiber* 9: 1 170

Parker W J 1985 Prediction of heat release rate of wood. *Proc. 1st Int. Fire Safety Symp.* Hemisphere, New York.

Parker W J 1988 Prediction of heat release rate of wood. Ph.D. dissertation, George Washington University (also available from University microfilms, Ann Arbor, Michigan)
Roberts A F 1971 Problems associated with the theoretical analysis of the burning of wood. *13th Symposium (Int.) Combustion*. The Combustion Institute, Pittsburgh, Pennsylvania, pp. 893–903
Zicherman J B, Williamson R B 1981 Microstructure of wood char, Part I: Whole wood. *Wood Sci. Technol.* 15: 237–49
Zicherman J B, Williamson R B 1982 Microstructure of wood char, Part II: Fire retardant treated wood. *Wood Sci. Technol.* 16: 19–34

J. B. Zicherman
[Berkeley, California, USA]

Fluid Transport

Two important forms of fluid transport through wood are bulk flow and diffusion. Bulk flow of fluids through the interconnected voids of the wood structure occurs under the influence of a static or capillary pressure gradient, and is sometimes referred to as momentum transfer because it can be attributed to a momentum–concentration gradient. Diffusion consists of two types: intergas diffusion which includes the transfer of water vapor through the air in the lumens of the cells, and bound-water diffusion which takes place within the wood cell walls. Some practical applications of bulk flow are the pressure treatment of wood with liquid preservatives and the impregnation of chips with pulping chemicals. Diffusion occurs during the drying of wood, in the migration of moisture through solid wood exterior walls; and in interior woodwork in response to seasonal changes in relative humidity. Diffusion may be accompanied by changes in wood moisture content and by swelling or shrinking.

Permeability is a measure of the ease with which fluids are transported through a porous solid by bulk flow. Porosity is the volume fraction of void space in a solid. It is clear that a solid must have some porosity to be permeable but the presence of high porosity is no guarantee that the body will have any permeability. A "closed-cell" porous structure could have zero permeability. In wood the lumens of the tracheid or fiber cells are interconnected by openings in the membranes of the bordered-pit pairs. In hardwoods, longitudinal flow can be greatly enhanced by the vessels, which in many cases have few constrictions. The porosities of dry woods may extend from a minimum of 0.30 for a very high-density tropical wood to a maximum of 0.96 for a very low-density wood such as balsa. Most softwoods used in construction have a porosity of approximately 0.65 corresponding to a density of approximately 500 $kg\,m^{-3}$. The range of permeabilities in woods is much wider than that of porosities and is of the order of one million to one for a given grain direction. There is very little correlation between porosity and permeability in woods. For example, the sapwood of the southern pines is usually relatively permeable whereas the heartwood of intermountain Douglas fir has extremely low permeability, even though these woods have a similar structure and porosity. Red oak has a very high permeability in the fiber direction due to its very large earlywood vessels despite the low porosity of this wood. Although porosity is no predictor of wood permeability, it is a measure of the maximum possible retention of a treating liquid.

1. Bulk Flow and Permeability

The most important path for flow in softwoods is from tracheid to tracheid through bordered-pit pairs which have small openings (radii 0.01 – 2 μm) in their membranes. In hardwoods the open vessels provide excellent longitudinal flow paths because there is little resistance at the ends of the cells and many vessels are uninterrupted for relatively long distances. There is extensive pitting between the fiber cells of hardwoods but the membranes apparently have no openings. This absence of openings does not prevent bulk flow, however, because the fiber lumens can be impregnated with liquids. It is probable that liquids flow through these membranes by diffusion.

Apart from the vessels of hardwoods, the condition, number and size of the openings in the pit of membranes of softwoods are the most important factors determining wood permeability and treatability (Siau 1984). When sapwood changes into heartwood, pit membrane openings may close because of pit aspiration, attributed to capillary forces resulting from the removal of free water, and pit incrustation by extractives and phenolic substances (Côté 1963). These conditions can decrease permeability very significantly, explaining the low heartwood permeability of most woods. Pit aspiration occurs only in softwoods since there are no tori in hardwood pits. Côté (1963) also describes the formation of tyloses, principally in the heartwood vessels of hardwoods. These very effectively inhibit or close off flow in the fiber direction. An example is white oak, which has extremely low permeability and is used for cooperage for this reason.

Pit aspiration in softwoods occurs not only during heartwood formation due to drying in the standing tree, but also in kiln and air drying of the sapwood. In latewood, the pits are smaller in diameter and the membranes are thicker, resulting in a structure which is mechanically more rigid than that in the earlywood. Because of this greater strength, many of the latewood pit membranes are able to resist the capillary forces and do not aspirate. Thus, the latewood of dried softwoods is usually much more permeable and treatable than the earlywood despite the higher density of the former. This is particularly true of Douglas fir and the southern pines. In hardwoods, on the other hand, this situation may be reversed because of larger earlywood vessels. Many of these phenomena have been discovered by microscopic examination of impregnated wood (Siau 1984).

The anisotropic nature of wood structure is a most important factor affecting relative permeability in the transverse and longitudinal directions. The tracheid cells in softwoods and fiber cells in hardwoods have a length–diameter ratio of approximately 100. The greatest resistance to flow occurs at the interconnecting pit openings in the cell walls. Therefore, the permeability decreases approximately in proportion to the number of walls traversed per unit area. On this basis, Comstock (1970) proposed a model in which the ratio of longitudinal to transverse permeability can vary from 10 000 to 40 000 in softwoods, depending upon the length of the overlap of the cell ends. In hardwoods there can be a ratio as high as one million to one due to the contribution of the vessels.

Several investigators have made permeability measurements of wood, mostly in the longitudinal direction, and have attempted to use it as a predictor of unsteady-state wood treatability with preservatives. In many cases the results correlate poorly, mainly because of exceptions to Darcy's law. This law is based upon the general steady-state law:

$$\text{flux} = \text{conductivity} \times \text{gradient} \tag{1}$$

In Darcy's law, the conductivity or permeability is assumed to be constant. Darcy's law for liquids may be stated as

$$k = QL/(A\Delta P) \tag{2}$$

where k is the permeability, Q is the flow rate, L is the length in the direction of flow, A is the cross-sectional area and ΔP is the pressure differential. In this case, Q/A is the flux and $\Delta P/L$ the pressure gradient.

When permeability is calculated from the Darcy equation, it may be defined as being numerically equal to the flow rate through a unit cube with unit pressure differential across two opposite faces. It is clearly a property of the anatomical structure of the wood and the viscosity of the fluid. When permeability is multiplied by viscosity, the more meaningful specific permeability is obtained. This is numerically equal to the flow rate of a fluid of unit viscosity through a unit cube with unit pressure differential. In this case, specific permeability is a function only of the wood structure and is assumed to be independent of the fluid:

$$K = k\eta \tag{3}$$

where K denotes specific permeability and η viscosity.

Most wood-science literature gives permeabilities in darcys, one darcy being equal to the product of permeability in $cm^3\ cm^{-1}\ atm^{-1}\ s^{-1}$ and viscosity in centipoise ($\eta_{water} = 1$ cp). To relate this to other more fundamental units,

$$1\ \text{darcy} = 9.87 \times 10^{-9}\ \text{cm}^2 = 0.987\ \mu\text{m}^2$$

Darcy's law and its assumptions and limitations are discussed by Muskat (1946) and Scheidegger (1974). The principal assumptions of Darcy's law are as follows.

(a) The flow is viscous and linear. Therefore, the velocity is directly proportional to the pressure differential.

(b) The fluid is homogeneous and incompressible.

(c) The porous medium is homogeneous.

(d) There is no interaction between the fluid and the substrate.

(e) Permeability is independent of length in the direction of flow.

The capillary structure of wood is extremely complex, particularly that of the hardwoods (Siau 1984). Wood therefore does not represent the uniform, circular, parallel-capillary model which would be described well by Darcy's law. The compressibility of liquids is negligible but, for gases, a term is added to the Darcy equation to account for compressibility. In the case of water there are strong interactions with the cell wall because hydrogen bonds form between the highly polar molecules and the sorption sites on the wood. Nicholas (1973) explains a lower resistance to flow of oils than of water of equal viscosity on this basis. Permeability generally decreases with specimen length, particularly in woods of low permeability (Siau 1984). This could be attributed to a progressive closing of flow passage as cells are reached having no openings to adjacent cells (Bramhall 1971). This is a significant departure from Darcy's law, making it difficult to predict the flow properties of large pieces of timber, such as those treated with preservatives, from permeability measurements made on small clear specimens.

There are serious experimental problems in the measurement of wood permeability with liquids, because of the presence of particulates which can clog pit openings and dissolved air which can cause air blockage. Kelso et al. (1963) describe the procedures which must be followed to obtain consistent results by careful deaeration and filtering of the liquid. Comstock (1967) has succeeded in showing that the specific permeability of wood measured with a nonpolar liquid is equal to that measured with a gas when the necessary precautions are followed.

The measurement of gas permeability avoids the main difficulties experienced in liquid measurements, and it is for this reason that most of the research in wood permeability has been done with air or other gases.

There are many exceptions to Darcy's law when gas flow is employed. One is due to the high compressibility of gases, for which a factor is applied to the Darcy equation. In addition to this, molecular slip flow

(Knudsen flow) occurs in the gaseous phase (Scheidegger 1974, Muskat 1946, Siau 1984). This is actually diffusion actuated by a static rather than a partial pressure gradient and it occurs when the radius of the opening is less than the mean free path of the molecules, resulting in collisions with the surface of the capillary. The molecular-slip flow term in the combined flow equation is proportional to the radius cubed whereas the viscous term, as defined by the Poiseuille equation for viscous flow, is proportional to the fourth power of the radius. The resulting Klinkenberg or Adzumi equations predict that the permeability increases linearly with the reciprocal of the average pressure. Typical plots are shown in Fig. 1(a) in which the permeability of membrane filters is plotted as a function of reciprocal average pressure. It follows from the equations that a mean effective radius can be calculated from the intercept divided by the slope. The smaller the radius, the greater the fraction of slip flow, corresponding to a greater relative slope. Calculations from these linear plots indicate a mean effective radius of 0.14 μm for the Millipore filter and 0.12 μm for the Sartorius filter. Wood specimens with very small pit openings and a highly inhomogeneous structure frequently produce curvilinear permeability–reciprocal-pressure relationships as depicted in Fig. 1(b) for longitudinal flow through basswood and maple. The curvilinear function has been interpreted by Petty (1970) as resulting from very high and very low permeability elements in series, such as a tracheid lumen followed by very small pit openings. By the application of curve-fitting techniques, the function may be analyzed into two straight lines whose slopes and intercepts make possible the calculation of a large and a small radius. Such an analysis of the basswood curve resulted in radii of 33 μm and 0.17 μm which could be interpreted as corresponding to vessel lumens and intervessel pit openings.

Another exception to Darcy's law is the nonlinear flow resulting from the kinetic-energy loss at the entrance to a capillary. This loss may be calculated from a term added to the Poiseuille equation for viscous flow. The equation is usually presented as

$$\Delta P = 8\eta LQ/(\pi r^4) + m\rho Q^2/(\pi^2 r^4) \tag{4}$$

where m is a constant (≈ 1.19), ρ is the density and r is the radius of capillary.

Siau and Petty (1979) have rewritten Eqn. (4) by substituting dimensionless flow velocity (Reynolds number) in place of one of the powers of Q in the second term. The Reynolds number Re can be written:

$$Re = 2\rho Q/(\pi r \eta) \tag{5}$$

Then, Eqn. (4) may be rearranged to

$$\Delta P = [8\eta LQ/(\pi r^4)][1 + 0.074 Re(r/L)] \tag{6}$$

Equation (6) shows that the nonlinear component of flow, due principally to kinetic energy, is a function of Reynolds number and the L/r ratio of the capillary

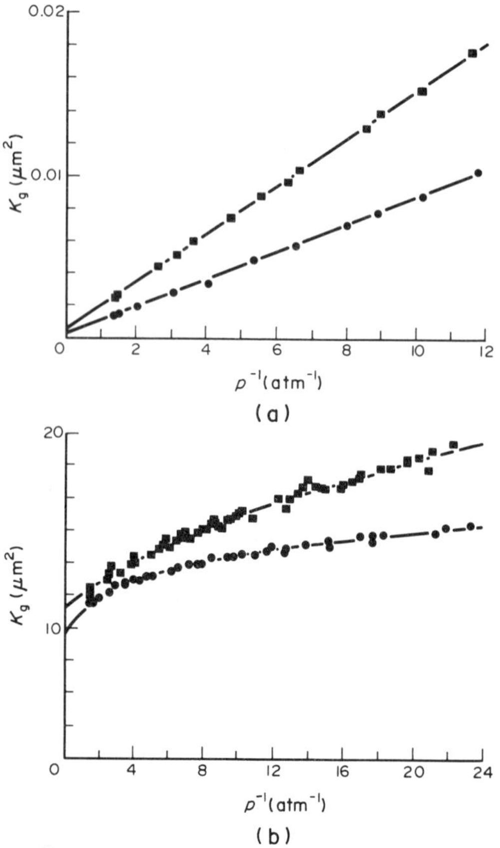

Figure 1
(a) Plots of permeability as a function of reciprocal average pressure for membrane filters according to the Klinkenberg equation: ■, Millipore VM filter, slope 0.00147 μm^2 atm; ●, Sartorius 11311 filter, slope 0.00084 μm^2 atm. (b) Plots of longitudinal (specific gas) permeability as a function of reciprocal pressure for specimens 4.5 cm long: ■, basswood; ●, maple; the curves can be analyzed into two linear Klinkenberg plots similar to those in (a) from which two radii can be calculated (Courtesy of *Wood and Fiber*)

and is independent of the radius alone. These facts have not been generally recognized. It is apparent from Eqn. (6) that there is a critical Reynolds number Re'' for the onset of nonlinear flow due to kinetic-energy loss which has a value approximately equal to the L/r ratio of the capillary. The critical Reynolds number for turbulence Re' is generally accepted as 2000, but since most capillaries have L/r much less than this, nonlinearity would be expected at relatively low values of Re. An interesting parallel between kinetic-energy and turbulence losses is that, in both cases, ΔP tends to be proportional to Q^2, making the two effects indistinguishable in many flow measurements. Siau and Petty (1979) have measured Re'' for short capillaries with L/r from 6 to 660 and with radii of 0.004–0.5 mm as experimental verification of Eqn.

(6). These results are directly applicable to wood structure in which tracheid lumens and vessels are interconnected by pit membrane openings which may be as short as 0.1 μm and with radii from 0.01 to 2 μm. Therefore, nonlinear flow would be expected with Reynolds numbers in the range 0.1–10. Scheidegger (1974) quotes several references in which *Re''* from 0.1 to 75 were observed in porous media.

In a recent investigation, Kuroda and Siau (1988) found that the longitudinal air permeability of the wood of loblolly pine, Douglas-fir and white spruce decreased with increasing flow rate. Since the maximum Reynolds number would occur at the pit openings between the tracheids, values of *Re''* were calculated from the air velocity at the onset of permeability decrease and from the radii of the pit openings as obtained from plots similar to Fig. 1(a). These values of *Re''* ranged from 0.4 to 1.6 which is much lower than the value of *Re'* corresponding to turbulence. They were in close agreement with values of *Re''* calculated from the L/r ratio of the openings using the method described by Siau and Petty (1979). No evidence of nonlinear flow was detected in hardwoods in which the longitudinal flow is essentially through long uninterrupted vessels having a high L/r ratio.

Nearly all permeability research has involved steady-state measurements, whereas it is unsteady-state flow which occurs during wood-preservation treatments and other liquid impregnations. There have been many attempts to predict unsteady-state behavior from steady-state measurements, but these have been only marginally successful. Siau (1976) has found that unsteady-state pressure versus time permeation curves of wood specimens have a very different shape from those calculated from the steady-state permeability using the diffusion differential equation. However, if two or three zones of widely differing permeabilities are assumed to be present within the wood, along with other variable parameters including the model of Petty (1970) for high-and low-permeability elements in series, it is possible to generate theoretical permeation curves which are in very close agreement with experimental curves. Therefore, if the extremely inhomogeneous structure of wood is taken into account, unsteady-state behavior could probably be predicted from steady-state permeability measurements.

At the present time, perhaps the greatest value of steady-state permeability research lies in the ability to use flow measurements to determine the effective radii and lengths of openings in the porous structure, thus making it a valuable tool in the study of the anatomical wood structure and its relationship to flow properties.

2. *Capillarity*

Surface-tension forces result from the imbalance of intermolecular forces at the surface of a liquid, and are therefore found only at liquid–gas interfaces. The magnitude of the capillary pressure at a circular interface may be calculated from the Jurin equation:

$$P_0 - P_1 = (2\gamma \cos\theta)/r \tag{7}$$

where P_0 is the pressure in the gas phase, P_1 the pressure in the liquid phase, θ the contact angle at a solid surface, γ the surface tension of liquid–gas interface.

It is clear from Eqn. (7) that the capillary pressure differential increases as the radius decreases, and that the term $(P_0 - P_1)$ is positive when $\theta < 90°$ and negative when $\theta > 90°$. The former case is true of liquids which wet the surface of a capillary, such as water in glass or wood. In the latter case there is no wetting of the solid surface such as with mercury in glass or wood. It is also a consequence of Eqn. (7) that, when $\theta < 90°$ and $P_0 = 1$ atm, a small radius (less than 1 μm) results in a negative value of P_1, known as capillary tension. This can explain the rise of the sap columns in tall trees to a much greater height than could be supported by osmotic root pressure or atmospheric pressure. It is necessary that the capillary radius at the openings of the mesophyll cells in the leaves be sufficiently small to provide the appropriate tension.

Eqn. (7) may be simplified in form assuming $\gamma = 0.072\ \mathrm{N\,m^{-1}}$ for water–air and $\theta = 0°$:

$$P_0 - P_1 = 1.46/r \tag{8}$$

where $(P_0 - P_1)$ is in atm and r in μm.

Equation (8) reveals that a radius of 0.1 μm gives $(P_0 - P_1)$ of 14.6 atm or P_1 of negative 13.6 atm, assuming $P_0 = 1$ atm. This tension is sufficient to support a column over 120 m high. When Eqn. (8) is applied to the practical problem of forcing minute air bubbles in a waterborne preservative solution through the pit openings between the cell lumens, which have radii from 2 down to 0.01 μm, pressures between 0.73 and 146 atm are required. These pressures may be much higher than those required to overcome the viscous forces described by Poiseuille's and Darcy's laws. They would be somewhat less with oily preservatives because of their lower surface tensions.

Capillary forces resulting from the removal of free water during drying can have a profound effect on wood properties. When the lumens are completely filled with water, concave menisci form at the exposed pit openings, causing tension which can result in collapse of the wood structure. This problem is most common in woods of low density and low permeability (Siau 1984). On the other hand, gradients of capillary pressure can aid in the removal of free water from partially filled lumens of softwoods due to decreasing radii of menisci in the tapered, overlapping ends of softwood tracheids (Spolek and Plumb 1981).

The aspiration of the membranes of bordered-pit pairs in softwoods is also a consequence of capillary tension during removal of free water. When the interface recedes to the membrane surface, the menisci are concave away from the dry side. Atmospheric pressure forces the membrane against the aperture on the wet side, where it is held in position after drying by hydrogen-bonding forces.

Wood posts or poles which are buried in the ground attain an equilibrium moisture content with the surrounding soil, principally due to capillary forces within the porous structures of both the soil and the wood. This manifestation of wick action is discussed by Baines and Levy (1979).

Mercury porosimetry is an application of Eqn. (7) in which mercury is forced into a porous body to measure the size and distribution of pore sizes. Stayton and Hart (1965) have applied this to wood by assuming a surface tension of 0.435 N m^{-1} and a wetting angle of 130°. When these values are substituted into Eqn. (7) the simplified relationship for mercury is

$$P_1 - P_0 = 6.12/r \tag{9}$$

With Eqn. (9), the radius corresponding to a measured volume of injected mercury may be calculated to determine the volumetric pore-size distribution.

3. *Diffusion*

Diffusion has generally been described by Fick's law, similar in form to Darcy's law, except that a concentration gradient is used:

$$J = -D\,dc/dx \tag{10}$$

where J is the flux (kg m^{-2} s^{-1}), D the diffusion coefficient (m^2 s^{-1}) and dc/dx the concentration gradient (kg m^{-4}).

Skaar (1954) has shown that Fick's law may also be expressed in terms of moisture-content or partial-vapor-pressure gradients (with appropriate modifications of the coefficient) and with identical results assuming isothermal conditions. Chemical or water potential has also been proposed as a driving force (by Siau 1983) and, again, identical results should be obtained under isothermal conditions. Diffusion coefficient for bound-water diffusion through the cell wall have been measured by Stamm (1959, 1964) over a wide range of temperatures and moisture contents. These coefficients can be related to concentration, partial vapor pressure or water potential. Stamm's values can be approximated by the following equation (Skaar and Siau 1981):

$$D_{BT} = 7 \times 10^{-6} \exp(-E_b/RT) \tag{11}$$

where D_{BT} is the transverse bound-water diffusion coefficient (m^2 s^{-1}), $E_b = 38\,500 - 290$ M(J mol^{-1}), M the moisture content (%), R the universal gas constant (8.31 J mol^{-1} K^{-1}) and T the temperature (K).

Equation (11) indicates that bound-water diffusion through the cell wall is activated by both temperature and moisture content. Indeed, Stamm (1964) has found a linear relationship between $\ln D_{BT}$ and $1/T$ in accordance with an equation similar to the Arrhenius equation. Thus, it becomes apparent that Fick's law is not obeyed for bound-water diffusion because the coefficient has extreme variability with both temperature and moisture content. Fick's law can then be regarded as a convenient relationship between flux and gradient for the purpose of calculation of a coefficient or a flux under specific conditions.

The coefficient of intergas diffusion of water vapor through the air in the lumens (D_V) is much higher than that for bound-water diffusion and is only mildly temperature-activated, being proportional to $T^{1.75}$. With a relatively constant coefficient, intergas diffusion essentially obeys Fick's law.

When wood is considered as a composite of cell-wall substance and air-filled lumens, the low resistance of the lumens becomes negligible, especially at low moisture contents, in comparison with the resistance of the crosswalls, when moisture diffuses through wood in the transverse direction. Siau (1984) proposes a model in which the transverse diffusion coefficient of gross wood (D_T) can be calculated from the value for the cell wall (D_{BT}):

$$D_T = D_{BT}/[(1 - v_a)(1 - v_a^2)] \tag{12}$$

where v_a is porosity.

Equation (12) indicates direct proportionality between the diffusion coefficient of wood and that of cell-wall substance and a decreasing coefficient with increasing wood density attributable to the increasing thickness of the crosswalls.

When bound water moves in the fiber direction, the diffusivity of the cell wall (D_{BL}) is approximately 2.5 times D_{BT}, but the much higher conductivity of the air in the lumens (D_V) may become the principal conductive parallel path. When the diffusivity for integras diffusion of water vapor in air (D_a) is converted to a basis of cell-wall substance and then related to the moisture content of wood so that it can be applied to the composite structure of wood, it is designated as D_V. Its value increases rapidly with temperature because of increasing wood vapor pressure with temperature, and it decreases as wood moisture content increases as a result of the increasing slope of the sorption isotherm with increasing moisture content. Thus D_{BT} increases with moisture content while D_V decreases. Similar to D_{BT}, D_T increases with moisture content as explained above, while values of D_L, calculated for the cell-wall model, decrease with moisture content due to the predominance of the high lumen conductivity (Siau 1984).

As bound-water diffusion is apparently a temperature-activated process it should follow that a temperature gradient will actuate bound-water diffusion in addition to that resulting from moisture-content difference. Skaar and Siau (1981) have proposed a steady-state version of Fick's equation to account for temperature effects in nonisothermal bound-water movement. This equation would be expected to apply in the transverse direction only where the cell-wall resistance dominates. The equation has the form

$$J = -K_{\mathrm{M}}\left[\frac{E_{\mathrm{b}}}{T}\left(\frac{M}{RT - \partial E_{\mathrm{b}}/\partial M}\right)\frac{dT}{dX} + \frac{dM}{dX}\right] \quad (13)$$

where K_{M} is the coefficient based on a gradient of moisture content ($\mathrm{kg\,m^{-1}\,s^{-1}\,\%^{-1}}$) and X(m) is the distance in the direction of the flux.

An equation of similar form based on a gradient of chemical or water potential was proposed by Siau (1983):

$$J = -K_{\mathrm{M}}\left[\frac{dM}{dX} + \frac{h(\partial M/\partial h)_T}{RT} \times \left(\frac{\partial \mu_1^0}{\partial T} + RT\frac{\partial \ln h}{\partial T} + R\ln h\right)\frac{dT}{dX}\right] \quad (14)$$

where h is the relative vapor pressure and μ_1^0 is the chemical potential of saturated vapor in equilibrium with a level surface of pure water ($\mathrm{J\,mol^{-1}}$).

It is probable that Eqns. (13) and (14) are equivalent. To compare their characterization of nonisothermal diffusion it is necessary to measure E_{b} and K_{M} over the appropriate range of M and T and the sorption isotherm over the range of T and h. Such an investigation was conducted by Avramidis and Siau (1987a). This was followed by 43 nonisothermal experiments on western white pinewood (Avramidis et al. 1987, Avramidis and Siau 1987b) in which opposite gradients of moisture content and temperature were maintained in the radial direction. It was demonstrated that diffusion could proceed against a moisture-content gradient by maintaining an opposing thermal gradient of sufficient magnitude. Analysis of the results using Eqns. (13) and (14) indicated that they both described the diffusion relatively accurately, particularly under conditions where balanced temperature and moisture-content gradients produced a net flux of zero.

Bound-water diffusion may result from gradients of static pressure, capillary tension, electrical potential and water potential. Diffusion could therefore explain the passage of liquids through the apparently continuous pit membranes of hardwoods. A differential of capillary tension could cause the flow of sap as a result of the transpiration occurring in the leaf. Palin and Petty (1981) have measured the permeability of the cell wall by utilizing the osmotic pressure of a solution rather than static pressure. In this case the movement of water through the cell wall could be regarded as bound-water diffusion, although the data were analyzed by use of the Darcy equation. Indeed, the fact that several alternative gradients can cause diffusion indicates that bulk flow, bound-water diffusion and integras diffusion usually occur simultaneously when there is a flux through the composite wood structure.

See also: Hygroscopicity and Water Sorption

Bibliography

Avramidis S, Kuroda N, Siau J F 1987 Experiments in nonisothermal diffusion of moisture in wood. Part II. *Wood Fiber Sci.* 19: 407–13

Avramidis S, Siau J F 1987a An investigation of the external and internal resistance to moisture diffusion in wood. *Wood Sci. Technol.* 21: 249–56

Avramidis S, Siau J F 1987b Experiments in nonisothermal diffusion of moisture in wood. Part 3. *Wood Sci. Technol.* 21: 329–334

Baines E F, Levy J F 1979 Movement of water through wood. *Inst. Wood Sci.* 8(3): 109–13

Bramhall G 1971 The validity of Darcy's law in the axial penetration of wood. *Wood Sci. Technol.* 5: 121–34

Comstock G L 1967 Longitudinal permeability of wood to gases and nonswelling liquids. *For. Prod. J.* 17(10): 41–46

Comstock G L 1970 Directional permeability of softwoods. *Wood Fiber* 1(4): 283–89

Côté W A Jr 1963 Structural factors affecting the permeability of wood. *J. Polym. Sci. Part C* 2: 231–42

Kelso W C Jr, Gertjejansen R O, Hossfeld R L 1963 *The Effect of Air Blockage upon the Permeability of Woods to Liquids*, Technical Bulletin 242. University of Minnesota Agricultural Experiment Station, Minneapolis, Minnesota

Kuroda N, Siau J F 1988 Evidence of nonlinear flow in softwoods from permeability measurements. *Wood Fiber Sci.* 20: 162–69

Muskat M 1946 *The Flow of Homogeneous Fluids through Porous Media.* Edwards, Ann Arbor, Michigan

Nicholas D D (ed.) 1973 *Wood Deterioration and Its Prevention by Preservative Treatments*, Vol. 2. Syracuse University Press, Syracuse, New York, pp. 322–24

Palin M A, Petty J A 1981 Permeability to water of the cell wall material of spruce heartwood. *Wood Sci. Technol.* 15: 161–69

Petty J A 1970 Permeability and structure of the wood of Sitka spruce. *Proc. R. Soc. London, Ser. B* 175: 149–66

Scheidegger A E 1974 *The Physics of Flow through Porous Media*, 3rd edn. University of Toronto Press, Toronto

Siau J F 1976 A model for unsteady-state gas flow in longitudinal direction of wood. *Wood Sci. Technol.* 10: 149–53

Siau J F 1983 Chemical potential as a driving force for nonisothermal moisture movement in wood. *Wood Sci. Technol.* 17: 101–5

Siau J F 1984 *Transport Processes in Wood.* Springer, Heidelberg

Siau J F, Petty J A 1979 Corrections for capillaries used in permeability measurements of wood. *Wood Sci. Technol.* 13: 179–85

Skaar C 1954 Analysis of methods for determining the coefficient of moisture diffusion in wood. *J. For. Prod. Res. Soc.* 4: 403–10

Skaar C, Siau J F 1981 Thermal diffusion of bound water in wood. *Wood Sci. Technol.* 15: 105–12

Spolek G A, Plumb O A 1981 Capillary pressure in softwoods. *Wood Sci. Technol.* 15: 189–99

Stamm A J 1959 Bound-water diffusion into wood in the fiber direction. *For. Prod. J.* 9(1): 27–32

Stamm A J 1964 *Wood and Cellulose Science.* Ronald, New York

Stayton C L, Hart C A 1965 Determining pore-size distribution in softwoods with a mercury porosimeter. *For. Prod. J.* 15: 435–40

J. F. Siau
[Keene, New York, USA]

Forming and Bending

Giving shape to wood by forming and bending is a technique thousands of years old. Today, forming is used to transform economical straight wood into simply-curved or complex three-dimensional forms. Applications of the technique include the manufacture of furniture, motor vehicles, ships, airplanes and sports equipment, as well as musical instruments and works of art.

Normal forming techniques involve selection and preparation of stock, softening, forming and setting. Softening is omitted when thin laminations are bent and glued together to produce curved shapes (see *Glued Laminated Timber*). Since bending by laminating is a relatively simple process, and since the most common method of softening in forming is by steaming, this article is primarily concerned with steam bending. It is also possible to apply forming techniques to particleboard that has been specially formulated with surface layers containing thermoplastic binders. These boards can be postformed by the application of dry heat (Lang et al., 1983).

Forming has many advantages: it is simple and fast, requires little energy and reduces material waste. Forming and bending do, however, require high levels of expertise and appropriate equipment and facilities. Undesirable side effects can include reduced stability, an increase in the swelling capacity and weakening of certain strength properties.

Permanent form alterations of single pieces are accomplished by the introduction of minute compression failures into the structure. Species showing certain discontinuities in their anatomical structure are prized for their good bendability. Examples of useful discontinuities are ring-porous vessel patterns (black locust, chestnut), broad rays (beech, sycamore), banded parenchyma (elm, sapele), radial series of vessels (jelutong), or combinations of these (oak, ash, afzelia, mansonia). Species without these features or those with relatively high lignin content, which impedes plasticization, have only limited formability. This is the case with most softwoods (particularly compression wood) and many exotic species.

1. Material Selection

For industrial forming, species with good bending qualities are generally chosen; these include hickory, birch, maple, walnut, mahogany and sweetgum, in addition to those named above.

Only straight-grained clear material can be used in production since growth characteristics (e.g., pith, knots and cross grain) and defects (e.g., surface checks and decay) lead to stress concentrations, and thus production defects. However, bending by laminating can be done with any species and with ordinary lumber.

2. Preparation of Stock

The blanks are first dried to the proper moisture content, that is, 8–12% for bending by laminating and 15–20% for steam bending. They are then cut to the desired dimensions with some allowance for the volume lost during forming. The surface is planed to reduce the tendency to crack. For the same reason, mortises and holes are cut only after forming.

Green wood is often flexible enough for bending. Industrially this is rarely done, since it involves long drying times and equipment tie-up, and results in high wastage caused by shrinkage and checking. Wood is therefore plasticized by steaming and boiling in water. This sharply increases compressibility, so that lengthwise-oriented cells tend to buckle and the cell cavities tend to collapse. Plasticization also reduces the tensile strength across the grain, as seen by the buildup of checks and delamination.

In practice, softening is usually performed with saturated steam at about 100 °C and vapor pressures of 1.1–1.3 bar. Since completely softened material easily checks during bending, wood is only partially plasticized. For each millimeter of material thickness, dry wood requires about two minutes to plasticize, and moist wood about one minute.

Wood can also be softened by other means (e.g., with urea which makes it thermoplastic, or with ammonia which gives better results than steaming). However, owing to high investment costs, these methods have not found much acceptance.

3. Forming

In the production of laminated wood, cold bending is used to make glued laminated timber or curved furniture parts from veneer. Cold-formed objects must be glued or mechanically held.

Solid wood is normally steam bent. Severe curvatures require forming with a supporting strap, a steel band placed on the convex side of the piece. This limits the tensile strain and causes most of the strain to shift to the compression side, where it is taken up by minute compression failures. Temperature, applied force and forming time are varied according to species, blank dimensions and the desired radius of curvature. Under expert control, wastage is less than 1%.

In addition, wood can be formed by densification, and partly by molding. In these processes, the cells are crushed and wedged together under pressure and heat to densities of up to 1400 kg m^{-3}. The mechanical properties can be significantly improved while the swelling capacity diminishes. These types of products are mainly used in specialized applications.

4. Setting

Forming gives rise to mechanical stresses in the blank. Therefore, the blanks must be fixed in the new form until the stresses have dissipated. Drying at 60–90 °C, cooling in a fixed form and storage in a climatically controlled chamber for 2–4 weeks, finally lead to form stability.

The curvatures of steam-bent wood are not completely stable, and tend to straighten out with temperature and humidity changes. In practice, a water-repellent surface coating is applied, or some mechanical restraint is introduced to assist in maintaining the shape.

See also: Macroscopic Anatomy

Bibliography

Bariska M, Schuerch C 1977 Wood softening and forming with ammonia. In: Goldstein I S (ed.) 1977 *Wood Technology: Chemical Aspects*, ACS Symposium Series No. 43. American Chemical Society, Washington, DC, pp. 326–47

Darzinsh T A 1976 *Modification of Wood.* Indian National Scientific Documentation Centre, New Delhi

Lang K, Scheithauer M, Göllner R, Möller A 1983 Herstellung, Verarbeitung und Anwendung nachformbarer Holzpartikelwerkstoffe. *Holztechnologie* 24: 215–20

Peck E C 1957 *Bending Solid Wood to Form*, USDA Agricultural Handbook No. 125. US Government Printing Office, Washington, DC

Stevens W C, Turner N 1970 *Wood Bending Handbook.* HMSO, London

US Forest Products Laboratory 1987 *Wood Handbook: Wood as an Engineering Material*, USDA Agricultural Handbook No. 72 (revised edn.). US Government Printing Office, Washington, DC

M. Bariska
[University of Stellenbosch, Stellenbosch, South Africa]

Fumigation

Damage from decay and insect attack is a major problem of wood in service. Even extremely dry wood can be vulnerable to attack by certain insects, and wood is often used in places where it cannot be kept dry. Decay fungi and insects can colonize the wood and rapidly reduce its service life. Certain chemicals have commonly been used to control surface deterioration, but they generally cannot penetrate beyond the surface to control internal decay. However, in the late 1960s it was found that even though wood is impermeable to liquids, common agricultural fumigants can migrate through it. This discovery opened new avenues for controlling the deterioration of wood.

1. Fumigants

Fumigants used to protect wood fall into two broad categories: those that rapidly eliminate established fungi or insects, usually within 48 hours, but do not remain in the wood for long periods; and those that move more slowly through the wood but remain for 3–17 years (Table 1). The short-term chemicals are used to kill insects in buildings and to eliminate plant pathogens from logs destined for export, while the long-term chemicals are used to protect wood from decay for 10 or more years.

All fumigants are toxic chemicals that can be hazardous to humans when misapplied. Fumigant use is regulated in many states and countries. Before using any fumigants, it is advisable to consult the appropriate government authorities for restrictions that may apply in a particular area.

1.1 Short-Term Fumigation

Short-term fumigant treatments generally employ highly volatile chemicals that rapidly penetrate the wood, eliminate insects and decay fungi, and escape into the atmosphere. They provide rapid control but no long-term protection against reinvasion. Short-term fumigation is typically applied to small wood pieces, such as historic wood artifacts in museums, to eliminate insect pests. Because these pieces often go unobserved for long periods, powderpost beetle infestations of the dry wood can cause substantial damage. Short-term fumigation is also effective against insects and fungi in logs destined for export.

Occasionally the need arises to fumigate large structures to eliminate internal insect pests, such as old house borers or drywood termites. This space fumigation, as it is called, is commonly used in agricultural storage buildings to eliminate insect infestations in grain.

The two chemicals most commonly used for short-term fumigation are methyl bromide and sulfuryl fluoride (Vikane). Both these compounds are colorless, odorless gases. They are often mixed with low levels of chloropicrin, which serves as an indicator gas. Methyl bromide is used to eliminate the oak wilt fungus (*Ceratocystis fagacearum*) from oak logs prior to export. Sulfuryl fluoride is most often used to eliminate drywood termite nests in wood structures.

Short-term fumigation is done under an airtight seal. Small objects are treated in airtight bags. For space fumigation, the area to be treated is cleared and sealed off with tarpaulins. A fixed amount of chemical is then pumped under pressure into the treatment space. A series of fans, strategically located inside the structure, mixes the chemical, which remains in the building for 20–24 hours. Since both chemicals are highly toxic, it is imperative that personnel performing the treatment wear respirators. After treatment the building is ventilated, and may then be safely entered.

The chemical dosage and length of treatment depend on the area to be treated, the type of seal used to contain the chemical, the temperature, and the wind speed, if any. These factors are used to express a dosage in grams of fumigant per hour, which is converted to grams of fumigant per 1000 liters times

Table 1
Characteristics of currently registered wood fumigants or their active ingredients

Chemical	Trade names	Persistence in wood	Application method	Target	Uses
Trichloronitromethane	chloropicrin Pic-clor Timber-Fume	>18 years	internal	fungi	poles piles timbers
Methylisothiocyanate	MIT	>9 years	internal	fungi	—[a]
MIT + chlorinated C_3 hydrocarbons	Vorlex Di-Trapex	>18 years	internal	fungi insects	poles piles
Methyl bromide	Bromo-Gas Meth-o-Gas Celfume	<48 hours	space	insects fungi	logs
Sodium *n*-methyl-dithiocarbamate	Vapam, Wood-Fume, Metam Sodium	<10 years	internal	fungi	poles piles chips
Sulfuryl fluoride	Vikane	<48 hours	space	insects	buildings furniture

[a] Not currently registered, but has shown great promise

the duration of treatment. Fortunately, several convenient slide rules, or Fumiguides, are available, which may be used to calculate these values quickly.

Short-term treatments usually do not alter the wood's appearance. However, when fumigating historic wood, it is advisable to expose a small section before treating the whole object, to ensure that the treatment does not damage the wood's color or finish.

1.2 Long-Term Fumigation

For some wood members it is expedient to provide slower, longer-term control of decay. This type of protection is particularly needed for large-dimension timber or roundwood that is decaying internally in areas that initial preservative treatments cannot reach. For such long-term protection, the chemical in its liquid form is poured into steep, angled holes drilled into the wood. These holes are then plugged with tight-fitting wooden dowels. The liquid chemical volatilizes and migrates through the wood, eliminating established decay fungi up to 4 m from the point of application. Fumigants applied in this way can arrest decay within 1 year and protect the wood from reinvasion for up to 17 years (Fig. 1).

Three currently registered fumigants are commonly used for long-term protection of wood. They are Vapam (32.1% sodium *n*-methyldithiocarbamate), Vorlex (20% methylisothiocyanate in chlorinated C_3 hydrocarbons) and chloropicrin (trichloronitromethane). Of these three, Vapam is the most commonly used and the easiest to handle. Vapam decomposes slowly in the presence of organic matter to release a mixture of 14 different compounds, including carbon

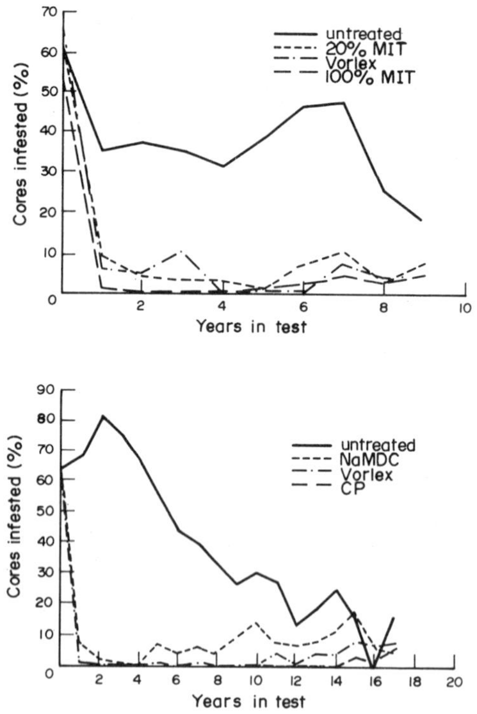

Figure 1
Ability of Vapam (NaMDC), Vorlex, chloropicrin (CP) and methylisothiocyanate (MIT) to eliminate and prevent Basidiomycete colonization of Douglas-fir poles, as measured by culturing core samples removed from each pole at yearly intervals

disulfide, carbonyl sulfide and methylisothiocyanate (MIT). MIT is believed to be the main fungitoxic component of Vapam, but recent studies suggest that other compounds may also provide some long-term protection. Of the three chemicals, Vapam remains in the wood for the shortest time—its volatile fungitoxic compounds cannot be detected 2 years after treatment. Yet poles treated with Vapam have remained free of decay fungi for up to 10 years.

One drawback of treatment with Vapam is its relatively low concentration of active ingredient per liter of formulated chemical. Vapam is a 32.1% solution that decomposes to produce MIT at a 40% conversion rate. As a result, only 12–16% of the applied chemical is released as MIT. Moreover, in many structures to be treated, the decay is so far advanced that the number of treatment holes must be limited. For such cases, chloropicrin and Vorlex, which have relatively high levels of active ingredient, may provide both more rapid control and better long-term control; residues of chloropicrin and Vorlex have been shown to persist in wood for more than 18 years. However, both these chemicals are highly volatile and highly toxic. They are best applied to wood located in areas where there is ample ventilation and where direct human contact is unlikely to occur—for example, utility poles, pier pilings and bridge timbers.

Tests indicate that most wood species can be treated with fumigants, but that chemicals will move faster through some woods than through others. For example, fumigant movement is slower through lodgepole pine than through southern pine; thus, southern pine will likely require more frequent retreatments than lodgepole pine. Any species of wood to be fumigated should be tested for fumigant movement first, if this has not already been done.

2. *Fumigant Applications*

Fumigants have been most widely used to control internal decay in utility poles. They have also helped to preserve laminated building timbers, marine pilings and bridge timbers. They have slowed the attack of marine borers for up to 3 years; further studies are now underway on full-sized marine pilings. Fumigants have also successfully controlled root rots in living trees and freshly cut stumps. Vapam is used to eliminate the pine wood nematode from wood chips destined for export.

In summary, fumigants can provide long-term protection against decay in most large poles and timbers. The only fumigant application that does not work well by itself is the remedial treatment of wood in contact with the ground. Such treatment alone does not keep subterranean termites and soft-rot fungi from attacking the wood's surface. Consequently, fumigant treatment in these cases is recommended only if some type of preservative wrap is applied at the same time.

3. *The Future of Wood Fumigation*

The fumigation of wood is a relatively recent method for arresting fungal and insect attack. Its growing popularity parallels an increasing demand, prompted by environmental concerns, for safer chemicals. Several research projects have been undertaken to learn more about how fumigants function in wood and what their effect is on the environment, with the aim of developing safer formulations and improved methods of application without sacrificing effectiveness.

Two chemical formulations have recently been identified as safer than those now used, yet still effective for long-term remedial fumigation of wood. They are solid MIT and Mylone. MIT is a crystalline solid which sublimes at room temperature to rapidly eliminate decay fungi established in the wood. It remains in sufficient concentrations to limit reinvasion for up to 10 years. This chemical must be encapsulated for safe use; however, studies indicate that encapsulation does not alter its effectiveness. Several field tests of MIT are underway. Mylone is a crystalline powder which slowly decomposes to produce MIT and several other compounds. The natural slow rate of decomposition can be altered by adding certain buffers. High pH (basic) buffers markedly accelerate decomposition, while low pH (acidic) buffers inhibit it. These buffers may be useful for selectively treating wood in varying stages of decay. For example, Mylone with a high pH buffer might be applied to actively decaying wood, while the same chemical without a buffer might be applied to protect a sound piece of wood. Another potentially safer fumigant is Vapam in its solid formulation. In its sodium salt form, Vapam produces a powder that, when hydrated, rapidly eliminates fungi established in wood. Studies are now underway to better develop this formulation for use in the field.

Because it is effective in lengthening the service life of wood, fumigation is increasingly a part of routine wood maintenance programs. While fumigants should never replace the careful design and construction practices that help ensure long service life, they improve the performance of wood used under adverse conditions.

See also: Decay During Use; Deterioration by Insects and Other Animals During Use; Preservative-Treated Wood

Bibliography

Graham R D 1973 Preventing and stopping internal decay of Douglas fir poles. *Holzforschung* 27: 168–73

Kinn D N, Springer E L 1985 Using sodium N-methyldithiocarbamate to exterminate the pine wood nematode in wood chips. *Tappi* 68(12): 88

Liese W, Knigge H, Rütze M 1981 Fumigation experiments with methyl bromide on oak wood. *Mater. Organ.* 16(4): 265–80

Morrell J J, Corden M E 1986 Controlling wood deterioration with fumigants: a review. *For. Prod. J.* 36 (10): 26–34

Morrell J J, Smith S M, Newbill M A, Graham R D 1986 Reducing internal and external decay of untreated Douglas-fir-poles: A field test. *For. Prod. J.* 36(4): 47–52
Ruddick J N R 1984 Fumigant movement in Canadian wood species. *Int. Res. Group Wood Preserv.* IRG/WP/3296. Stockholm, Sweden
Thies W G 1984 Laminated root rot: The quest for control. *J. For.* 82: 345–56
Young E D 1972 Short exposures in structural fumigation with Vikane fumigant. *Down to Earth* 27(4): 6–7
Zahora A R, Corden M E 1985 Gelatin encapsulation of methylisothiocyanate for control of wood-decay fungi. *For. Prod. J.* 35(7/8): 64–69

J. J. Morrell
[Oregon State University, Corvallis, Oregon, USA]

Future Availability

Projections of the future availability of wood necessarily depend on assumptions. These assumptions are conditioned by developments that have led to current conditions for availability of wood and on judgements as to what developments there may be in the future. In this article, key developments that have affected the availability of wood over the past two decades are reviewed as a prelude to presentation of judgements about future prospects. As will become apparent, prospects for future availability of wood vary markedly, depending on the country or region and product being considered, especially the contrast between industrialized and developing countries. A discussion of current inventories of forest land and wood is given elsewhere (see *Resources of Timber Worldwide*).

1. Evolution of the Current Situation

1.1 Industrialized Countries

Worldwide consumption of industrial roundwood (wood used for lumber, plywood, pulp and other end products) increased by more than 68% between 1955 and 1985 (Food and Agriculture Organization 1986). This increased consumption was centered mostly in the industrialized areas of North America, Japan and Western Europe.

The increase in consumption was attributable in part to rising demands brought about through expansionary economic policies in the industrialized countries, a lowering of barriers to trade and relatively inexpensive sources of energy and other raw materials. The increase in consumption in the industrialized countries was also attributable in part to the availability of previously untapped timber resources in Canada, the western USA, Siberia and Southeast Asia. Production of industrial roundwood also increased in the southern USA and Western Europe. Most of the increased output in the industrialized countries consisted of softwood species used in the manufacture of lumber, plywood and pulp. For example, in the USA between 1955 and 1985, production of industrial roundwood of softwood species increased by 28% and of hardwood species by 40%.

Japan was the major exception to increased output of timber products in the industrialized countries. Harvest in Japan declined from 53×10^6 m^3 in 1967 to a low of 32×10^6 m^3 in 1983. While domestic harvest declined, demand increased rapidly. Imported timber products accounted for about two-thirds of Japanese consumption of timber products in the 1970s, up from one-third in the early 1960s.

With the major exception of tropical hardwood logs, trade in timber products during the 1960s and 1970s increased primarily among the industrialized countries. For example, in the early 1960s, North America, Japan and Europe (including the Soviet Union) accounted for 90% of the value of world exports of forest products and 82% of the value of world imports of these products. In the mid-1980s the industrialized countries accounted for 87% of exports and 81% of imports. The value of world trade expanded sevenfold during this period.

In the industrialized countries, rising demands for timber products resulted in substantial increases in prices for timber products, especially softwoods, in the early 1970s. These price increases were responsible in part for developments in technology and industry practices that had the effect of stretching timber supplies. Examples include the substitution of plywood for lumber in the 1960s and 1970s, substitution of particleboard and other fiberboards for plywood, and pulping of hardwoods.

In the importing countries, relatively few barriers affected trade in chips, logs and other raw materials, but various types of restriction were imposed on lumber, plywood and other processed products. In the 1970s, various types of pressures developed in exporting countries concerning the propriety of exporting raw materials, especially logs.

In the industrialized countries, forest management generally became more intensive and utilization improved in response to rising prices. General awareness of the environmental effects of forestry practices and noncommodity forest outputs affected land use, especially that of public land.

1.2 Developing Countries

For many of the developing countries, the post-World-War-II period was one of profound social and economic change. Living standards in some countries increased, but remained low in others. In general, food and other necessities took priority over timber products in patterns of consumption and plans for development. There was a wide range of attitudes toward the role of forests and forestry among the developing countries. These included the exploitation of timber reserves, the shifting of agriculture with the effect of destroying both forest and the land, and relatively intensive afforestation and reforestation with plantations of introduced species, as in Brazil and Chile.

Individual countries participated in the growth of world trade in timber products, primarily through the export of tropical hardwood logs to industrialized countries. This pattern was especially evident in shipments from the Philippines, Malaysia and Indonesia to Japan.

Especially in the late 1970s, concern developed over the exploitation of reserves of tropical hardwoods. Although few data are available to document either the extent of remaining reserves or the rate of depletion, a general consensus exists that current rates of harvest cannot be maintained indefinitely, especially for preferred species (Sommer 1976). The possibility of depletion of timber inventories has given momentum to efforts to implement restrictions on log exports from several countries of Southeast Asia that would affect primarily shipments to Japan, Taiwan, South Korea and Singapore.

The end of relatively inexpensive energy in 1974 precipitated what had been a slowly developing crisis in wood for fuel in the developing countries. The jump in prices for petroleum-based fuels in the 1970s all but placed them out of reach of people in many countries. The result has been the virtual elimination of trees from around cities in some countries, especially in Africa (Arnold and Jongma 1977).

The World Bank (1978) and other international organizations have proposed several actions in response to the fuelwood crisis, including establishment of energy plantations accessible to population centers and the mixing of production of agricultural products with the production of wood for fuel.

Under the auspices of the United Nations, the developed and developing countries have discussed possibilities for cooperation in reforestation and other management of tropical forests after harvest. Much remains to be learned, however, about silvicultural and other management practices for tropical hardwoods. Also, tropical hardwood timber grown under management with relatively short rotations will probably result in commercial timber with strength and other properties that are different from the properties of the old-growth timber currently being harvested.

2. *The Future*

2.1 Industrialized Countries

After World War II, the years from 1973 to 1975 may well have been a turning point in the economies of the industrialized countries. The period of relatively inexpensive energy and other raw materials came to an end. A series of world economic cycles began, increasing in intensity and duration following the relative economic tranquility of the previous two decades. Alternative scenarios could be drawn for future rates of economic growth. If a consensus exists, it is probably that continued gains can be expected in real economic growth in the industrialized countries, albeit at slower rates than in the 1960s and early 1970s. Also, general expectations are for continued growth in demand for major timber products—higher for pulp, paper and other fiber-based products than for solid lumber and plywood.

The situation for supplies to meet growth in demand will probably differ markedly over the coming two decades compared with the previous two. In the future, any expansion of output in Canada is likely to be at slower rates than in the past (Reed 1987), production of softwood lumber, plywood or both may decline on the west coast of the USA (Haynes and Adams 1979), and expansion of output in Scandinavia will probably be limited (Food and Agriculture Organization and Economic Commission for Europe 1986). This contrasts with the general expansion of supplies from these areas in the 1960s and early 1970s.

Further expansion of supplies in the industrialized countries will depend in part on the intensification of timber management. Because of the economic importance of forestry to the Canadian economy, intensification of forest management in Canada has the support of government, industry and labor. Some potential exists for increased harvest of timber in western Europe, but the expectation is for growing dependence on offshore sources for timber products.

The government of Japan expects that the demand for lumber and plywood will grow slowly, if at all, and that the demand for pulp and paper products will increase over the coming decades (Japanese Ministry of Agriculture, Forestry and Fisheries 1980). Although the potential exists for significant expansion of Japanese production after the year 2000, the consensus seems to be that a significant share of Japanese needs for timber products will continue to be met by imports.

A report by the US Department of Agriculture Forest Service (1980) suggests the potential for significant expansion of timber supplies if programs could be developed that would encourage owners of small woodlands to take advantage of economic opportunities to intensify management of forest lands.

Projections of supplies and demands for timber products in the industrialized countries generally assume that historical linkages with the developing countries will continue. This generally means the expectation that exports of tropical hardwood logs, veneer or both and of lumber and plywood will continue to be available from the developing countries. Exports of timber products from the developed to the developing countries are generally not considered significant prospects for growth of markets.

Plantations of exotic species in New Zealand, Australia and other areas of the world will reach maturity in the year 2000 and beyond. These plantations have the potential to affect traditional patterns of world trade in timber products. Realization of the potential, however, requires resolution of problems related to costs other than wood, market acceptance of end products and distance from final markets.

2.2 Developing Countries

Several issues bear on the prospects for future availability of wood in the developing countries—wood for fuel, the conflict between forestry and agriculture for land use, forest management and depletion of timber inventories, and development of processing facilities to meet domestic demands as well as to meet goals related to employment and economic development.

Over the past decade, nearly every developing country with a timber resource has proposed to construct the capacity to process its timber. In general, these proposals have two intents: first, to stimulate economic development through employment and construction of the transportation infrastructure necessary to manage the timber resource; and second, to substitute domestically produced timber products for similar items that otherwise would have to be imported. Among the problems to be faced in these ventures are availability of capital, a trained labor force and equipment. In addition, problems in the manufacture of pulp and paper are associated with procuring chemicals and other additives necessary to the manufacturing process.

Developing countries vary in their ability to implement development plans centered on forestry and timber processing. Joint ventures have been a common mode of development of timber resources. For example, joint ventures have been one of the responses of Japanese investors to the tightening of log-export restrictions by countries in Southeast Asia.

Increased construction of processing capacity in the developing countries with the intent of substituting domestically produced items for imported ones will limit markets for the products of developed countries. This capacity may also lead to increased trade among the developing countries.

2.3 Final Outlook

To the year 2000, little basis exists for expecting significant change in the availability of wood, especially in the industrialized countries. Prices will continue to increase in the industrialized countries, leading to the substitution of one product for another and acting as incentive for improved utilization and intensified timber management. Major world trade flows in timber products will continue to be among the industrialized countries and among Pacific Rim countries.

In the developing countries to the year 2000, the problems of availability of wood for fuel and shifting patterns of agriculture may be insurmountable in some areas.

Beyond the year 2000, almost any scenario for the future availability of wood can sound plausible. The price of large saw-timber-sized trees will probably continue to increase, while prices of smaller, more abundant trees may decline. Innovations in fiber-processing technology, however, may tend to blur the distinction between solid and fiber-based products in the industrialized countries. Probably more processing of timber will be in developing countries with a resource base and in countries such as New Zealand with a resource based on exotic plantations. Mounting population pressures will exacerbate the problems of wood for fuel and shifting agriculture, however, and these problems may exist well into the next century.

See also: Industries Based on Wood in the USA: Economic Structure in a Worldwide Context; Resources of Timber Worldwide; Trade, Prices and Consumption of Timber in the USA

Bibliography

Allan L S 1979 Marketing of New Zealand's forestry products *N. Z. J. For.* 24(2): 214–16

Arnold J E M, Jongma J 1977 Fuelwood and charcoal in developing countries: An economic survey. *Unasylva* 29(118): 2–9

Food and Agriculture Organization 1979 *Yearbook of Forest Products, 1977.* Food and Agriculture Organization, Rome

Food and Agriculture Organization 1986 *Yearbook of Forest Products, 1985.* Food and Agriculture Organization, Rome

Food and Agriculture Organization and Economic Commission for Europe 1986 *European Timber Trends and Prospects to the Year 2000 and Beyond.* United Nations, New York.

Haynes R W, Adams D M 1979 Possible changes in regional forest products output and consumption during the next 50 years. *For. Prod. J.* 29(10): 75–80

Japanese Ministry of Agriculture, Forestry and Fisheries 1980 *Basic Plan for Japan's Forest Resources and Long-Range Demand and Supply Projections for Important Forest Products.* Japanese Ministry of Agriculture, Forestry and Fisheries, Tokyo

Levack H H 1979 Future national wood supply. *N. Z. J. For.* 24(2): 159–71

North R N, Solecki J J 1977 The Soviet forest products industry: Its present and potential exports. *Can. Slavonic Pap.* 19(3): 281–311

Reed F L C 1987 Forest policy and forest management in Canada, closing the gap between promise and performance. *Proc. Conf. Int. Forest Policy and Practice.* Lewis and Clark College, Portland, Oregon

Sommer A 1976 Attempt at an assessment of the world's tropical moist forests. *Unasylva* 28(112–113): 5–25

US Department of Agriculture Forest Service 1980 *An Analysis of the Timber Situation in the United States 1952–2030,* Review Draft. US Department of Agriculture, Washington, DC

World Bank 1978 *Forestry: Sector Policy Paper.* World Bank, Washington, DC

D. R. Darr
[US Forest Service, Arlington, Virginia, USA]

G

Glued Joints

Glued joints are used in wood construction to increase the size of available materials and also for product assembly. Natural adhesives have been used for furniture and other nonstructural products throughout most of recorded history (US Forest Products Laboratory 1981). Development of durable synthetic resin-based adhesives prior to World War II allowed the use of adhesives in demanding structural products and applications.

Three types of joints are readily defined with respect to the grain directions of the members joined together: edge-to-edge grain, end-to-end grain and end-to-side grain (where edge grain, also called side grain, refers to radial or tangential surfaces, and end grain to the cross section). Edge-to-edge- and end-to-end-grain joints are widely used in glued, laminated beams and other structural products. The use of adhesives in other types of building construction is expanding. The greatest variety of joints occurs in furniture construction, where not only edge-to-edge- and end-to-end-grain joints are used, but also end-to-side-grain joints. Most of the following discussion will deal with furniture joints; for applications of glued joints in building construction and laminated beams, see *Design with Wood* and *Glued Laminated Timber*.

1. Edge-Grain Joints

Edge-to-edge-grain joints are used for furniture parts such as solid wood seats and tops, and lumber cores for panels. The material is laid up in large, revolving glue-clamp carriers called reels, which hold the glued stock under pressure while the adhesive dries. High-frequency electronic glue presses are also used. Pressure is used to secure proper alignment of the parts and to ensure intimate contact between the wood and the glue. Animal glues, cold- and hot-setting urea formaldehyde, and poly(vinyl acetate) adhesives are commonly used to bond the wood together (see *Adhesives and Adhesion*). Boards are carefully machined to mate smoothly along their entire lengths without gaps, to prevent uneven glue lines and nonuniform distribution of pressure (see *Machining Processes*). Machined edges are preferred to sanded. Machining is carried out just prior to gluing, to remove surface resins and surface oxidation products. Pressures of 0.35–2 kPa are used. Wide boards are placed towards the outsides of the panels to help distribute pressures uniformly throughout the entire assembly. Temperature of the glue, the wood and the surroundings are controlled to produce joints with thin but adequate glue lines which retain sufficient adhesive to prevent starved joints. Assembly times are held as short as possible to reduce subsequent cure times. Hold-downs are used to prevent panels from bowing while being pressed. Pneumatic torque wrenches lessen the time required to tighten clamps, and ensure that clamping forces are applied uniformly. The wood is conditioned to a uniform moisture content prior to gluing to prevent subsequent problems with unequal shrinking and swelling (see *Shrinking and Swelling*). Glued panels are allowed to set until the swollen glue lines have dried sufficiently to allow machining.

2. End-Grain Joints

Furniture joints which require the connection of end-to-end-grain or end-to-side-grain surfaces are termed assembly joints. Included are dowel, mortise-and-tenon, finger, corner-block and gusset-plate joints (see Fig. 1). An illustration of finger joints is given in the article *Glued Laminated Timber*. Several of these assembly joints are also used in light frame building construction.

2.1 Dowel Joints

Owing to their simplicity, dowel joints are the most common connection employed in furniture construction. Withdrawal strength of a dowel from the side grain of a wooden part is directly proportional to the diameter of the dowel and to the shear strengths parallel to the grain of the dowel and of the materials joined together; withdrawal strength is slightly less than proportional to depth of embedment of the dowel. For highest strengths, both the dowel and the walls of the hole are coated with glue; a close fit between dowel and hole is essential. The in-plane bending strength of joints constructed with two symmetrically spaced dowels is directly related to the withdrawal strength of the dowels and to the width of the spacing between them. The out-of-plane bending strength of dowel joints is much less than in-plane strength; thickness of the end-grain member is the dominant strength parameter (Eckelman 1979). Dowel joints also carry shear loads in furniture. Strength is limited largely by the splitting strength of the wood. Typical joints constructed with two dowels 1 cm in diameter carry loads in excess of 2.5 kN. Methods of calculating the strengths of most of the dowel joints described have been developed (Eckelman 1978, 1988); therefore, these joints can be designed to meet specified in-service requirements.

2.2 Mortise-and-Tenon Joints

Mortise-and-tenon joints have largely been displaced by dowel joints in furniture because of the relative ease

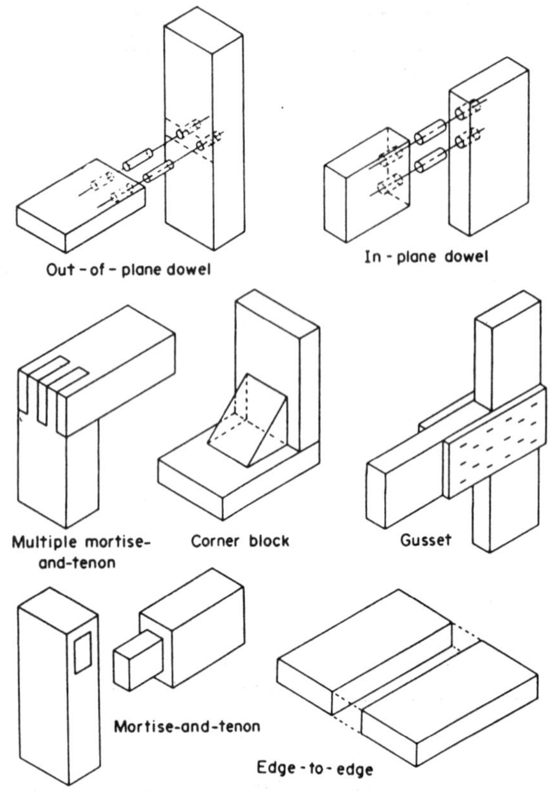

Figure 1
Common types of assembly joints

of construction of the latter. Adhesives are used in nearly all mortise-and-tenon furniture joints. Well-made joints develop somewhat more strength than comparable dowel joints (Sparkes 1968). Bending strength is related to the thickness, width and length of the tenon, and the closeness of fit of the tenon into the mortise (Eckelman 1988). Bending strength of the tenon itself provides the ultimate constraint on strength.

2.3 Corner Blocks

Corner blocks are used in furniture construction to reinforce points of high shear and bending stress. Ordinarily, corner blocks are cut with each of the two glue faces at angles of 45° to the grain; as a result, both faces form bonds intermediate to side grain and end grain. Blocks may also be cut with one face parallel to the grain to allow a side-grain-to-side-grain bond on one face. Properly fitted glue blocks can effectively resist bending loads which tend to close a joint, but may be less effective in resisting bending loads which tend to open it (Eckelman 1978, 1988).

2.4 Finger Joints

End-grain finger joints are used both to increase the longest length of lumber available and to salvage material by creating long members from short scraps (Blomquist et al. 1983). Strongest joints occur when the fingers have a small slope and are sharply pointed. Corner finger joints are more difficult to cut and clamp than end joints, although joints with slight angles are readily fabricated since they are essentially end-grain joints. Such joints are regularly used in items such as the back posts of chairs. Techniques for cutting geometrically complex high-strength finger joints have been developed (Richards 1962), but are little used commercially.

2.5 Multiple Mortise-and-Tenon Joints

Multiple mortise-and-tenon joints consist of a number of tenons with parallel sides, which are cut into the ends of two mating members. A common example of this construction is the box joint. They are also used to form visible decorative joints in fine furniture. Well-made multiple mortise-and-tenon corner joints may develop up to 50% of the strength of the wood (Richards 1962). Dovetail joints are basically similar in design, although they are not usually thought of as multiple mortise-and-tenon joints. Primary use of dovetail joints is in fine-furniture construction, where they are used for drawer construction. Close fit of the parts and use of adequate glue are necessary for high strength.

2.6 Staple Glued Plywood Gusset Joints

Staple glued plywood gussets are used in hidden areas in furniture to reinforce heavily stressed joints (Eckelman 1978, 1988). Such joints were once widely used in the construction of roof trusses, but have largely been supplanted by metal-tooth connector plates (see *Design with Wood*; *Joints with Mechanical Fastenings*). Strength of such joints is dependent on the size of gusset used and the rolling shear strength of the plywood.

Bibliography

Blomquist R F, Christiansen A W, Gillespie R H, Myers G E 1983 *Adhesive Bonding of Wood and Other Structural Materials.* Pennsylvania State University, University Park, Pennsylvania

Eckelman C A 1978 *Strength Design of Furniture.* Tim Tech, West Lafayette, Indiana

Eckelman C A 1979 Out of plane strength and stiffness of dowel joints. *For. Prod. J.* 29: 32–38

Eckelman C A 1988 *Product Engineering and Strength Design Manual for Furniture.* Tim Tech, West Lafayette, Indiana

Feirer J 1983 *Cabinetmaking and Millwork*, revised edn. C A Bennett, Peoria, Illinois

Gillespie R H, Countryman D, Blomquist R F 1978 *Adhesives in Building Construction*, US Department of Agriculture Handbook 516. US Government Printing Office, Washington, DC

Pound J 1973 *Radio Frequency Heating in the Timber Industry*, 2nd edn. Clowes, London

Richards D B 1962 High strength corner joints for wood. *For. Prod. J.* 12: 413–18
Selbo M L 1975 *Adhesive Bonding of Wood*, US Department of Agriculture Bulletin No. 1512. US Government Printing Office, Washington, DC
Slaats M A 1979 Glue bond quality in wood. *Int. J. Furniture Res.* 1: 24–26
Snider R 1975 *Gluing and Furniture Design.* Franklin Chemical Industries, Columbus, Ohio
Sparkes A J 1968 *The Strength of Mortise and Tenon Joints*, Report 33. Furniture Industry Research Association, Stevenage, UK
US Forest Products Laboratory 1981 *Proc. 1980 Symp. Wood Adhesives—Research, Application, and Needs.* US Forest Products Laboratory, Madison, Wisconsin
US Forest Products Laboratory 1987 *Wood Handbook*, US Department of Agriculture Handbook 72. US Government Printing Office, Washington, DC

C. A. Eckelman
[Purdue University, West Lafayette, Indiana, USA]

Glued Laminated Timber

Glued laminated timber used for structural purposes is an engineered stress-rated product comprising assemblies of suitably selected and prepared wood laminations securely bonded together with adhesives. The grain of all laminations is approximately parallel longitudinally. The individual laminations are of lumber thicknesses. Laminations may be composed of pieces end-joined to form any length, of pieces placed or glued edge-to-edge to make wider components, or of pieces bent to curved form during gluing.

Glued laminated timber (often referred to by the generic term glulam) was first used as early as 1893 in Europe, where laminated arches (probably glued with casein adhesives) were erected for an auditorium in Switzerland. Improvements in casein adhesives during World War I created further interest in the use of glued laminated timber structural members for aircraft and, later, as framing members for buildings. The development of durable synthetic-resin adhesives during World War II permitted the use of glued laminated timber members in bridges, truck beds and marine construction, where a high degree of resistance to severe service conditions is required. Today, synthetic-resin adhesives—primarily of the phenol–resorcinol and melamine types—are the principal adhesives used for structural laminating (see *Adhesives and Adhesion*).

Typically, glued laminated structural timbers are rectangular in cross section. They may be either straight or curved between supports. Straight beams can be designed and manufactured with horizontal laminations (load applied perpendicular to wide faces of laminations) or vertical laminations (load applied parallel to wide faces of laminations), as shown in Fig. 1. Horizontally laminated beams are the most widely used. Curved members are horizontally laminated to permit bending of the laminations to a curved form during gluing.

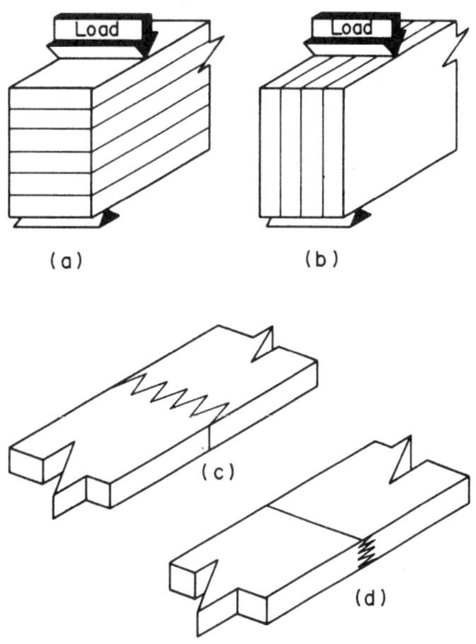

Figure 1
Glued laminated timber forms: (a) beam, horizontally laminated; (b) beam, vertically laminated; (c) end joint, vertical finger joint; (d) end joint, horizontal finger joint

The advantages of glued laminated timber construction include the provision of the ability to manufacture larger structural elements from smaller commercial sizes of lumber, giving better utilization of available lumber resources; the ability to achieve architectural effects through the use of curved shapes; and the ability to design structural elements varying in cross section between supports, and in accordance with strength requirements. Advantages also include minimization of checking or other seasoning defects associated with large sawn-wood members, because the laminations can be dried before being glued, thus permitting designs to be based on the strength of seasoned wood. In addition, lower grade lumber can be used for less highly stressed laminations without adversely affecting the structural integrity of the member.

Factors determining the assigned strength values of structural lumber will also affect glued laminated timber, since glulam is made up of individual pieces of structurally graded lumber (see *Lumber: Behavior Under Load*; *Lumber: Types and Grades*). The mere gluing together of pieces of wood does not, of itself, improve strength properties in laminating material of lumber thicknesses over those of comparable sawn timbers. Three of the benefits inherent in the laminating process do, however, form the bases for assigning higher glued laminated timber design values:

(a) drying or seasoning of the laminations; (b) positioning or placement of the laminations in the member; and (c) distribution of growth characteristics.

As wood seasons or dries under controlled conditions, it increases in strength. It is difficult to thoroughly dry pieces of wood with least dimensions greater than about 10 cm within a reasonable time, but laminated timbers are fabricated of individual laminations less than 5 cm thick, dried to a moisture content of 16% or less; therefore, regardless of the size of the finished laminated timber, it is dry, and design values can be based upon the strength of dry wood.

The bending strength of horizontally laminated timbers depends on the positioning of the various grades of lumber used as laminations. High grade laminations are placed in the outer portions of the member where high strength is effectively used, and lower grade laminations are placed in the inner portion, where low strength will not greatly affect the overall strength of the member. By selective placement of the laminations, knots are scattered and improved strength can be obtained. Studies have indicated that knots are unlikely to occur one above another in several adjacent laminations.

A comparison of the allowable design values for a typical laminated-timber grade for bending members, and a No. 1 grade of sawn timber of the size classification "beams and stringers," both made from Douglas fir, is shown in Table 1.

Principal steps in the manufacture of glulam members are as follows.

(a) Selection and preparation of laminations. Lumber used for structural laminating is commonly of the same softwood species used for other construction purposes. In the USA, for example, the softwood species most commonly used are Douglas-fir, Southern pine and hem-fir (a marketing combination of western hemlock and true firs). The lumber is graded for knot size and location, rate of growth, density, slope of grain and other characteristics that affect its use as a structural material. The laminations are surfaced prior to gluing, to assure that the wide faces can be brought into close contact for adhesive bonding.

(b) End joining of laminations. Lumber in the long lengths necessary for large laminated timbers is not normally available; therefore, the fabrication of structural end joints is an important part of the laminating process. The most common type of end joint used by the laminating industry consists of a series of intermeshed tapered "fingers" cut into the ends of laminations. These end joints, known as finger joints, can be well bonded and develop the high strengths needed for structural laminating (Fig. 1).

(c) Spreading of adhesives. Adhesives are spread on the laminations by means of rollers through which the laminations pass, or, more commonly, by an extruder which spreads beads of adhesive along the length of the lamination as it passes beneath.

(d) Clamping. After the laminations have been spread with adhesives, pressure is applied in forms that have been preset to the shape required in the finished glulam product. Pressure is applied, usually by means of a series of evenly spaced clamps, to bring the laminations into close contact, to pull the member to its final shape, to force out excess adhesive and to hold the laminations together during the adhesive curing period. After this curing period, which may range from 8 to 24 h, the glulam members are removed from the forms and are surfaced by large planing equipment to the specified width dimension.

(e) Finishing. In the finishing area of a laminating plant the completed glulam members are cut to the proper length, appropriate taper cuts are made, holes and daps for connectors are formed, end sealers are applied and wrapping material is applied as specified.

(f) Quality control. Throughout the laminating process, samples of production are regularly checked to be sure that the products are in conformance with established laminating standards. Quality control is an essential part of the laminating process, since the strength of the glulam members is dependent upon the quality of the glue joints.

Table 1
Comparison of design values, Douglas-fir

Design values	Bending (MPa)	Compression perpendicular to grain (MPa)	Shear (MPa)	Modulus of elasticity (MPa)
Glulam, typical grade	16.55	4.48	1.14	12410
Sawn timber, No. 1 "beams and stringers"	9.31	4.31	0.59	11030

Special equipment, plant facilities and manufacturing skills are needed to assure that this high quality is maintained.

The advent of techniques for glue laminating of wood provided a practical means for manufacturing wood structural members not limited by the size and shape of a tree. The use of glued laminated timber permits the construction of timber buildings with long clear spans and a variety of shapes.

See also: Glued Joints

Bibliography

American Institute of Timber Construction 1983 *Structural Glued Laminated Timber*, ANSI/AITC 190.1. American Institute of Timber Construction, Englewood, Colorado

American Institute of Timber Construction 1987 *Standard Specifications for Structural Glued Laminated Timber of Softwood Species*, AITC 117–87. American Institute of Timber Construction, Englewood, Colorado

American Society for Testing and Materials 1983 Standard method for establishing stresses for structural glued laminated timber (glulam) manufactured from visually graded lumber, ASTM D3737–83. 1983 *Annual Book of ASTM Standards*, Vol. 4: 09, *Wood*. American Society for Testing and Materials, Philadelphia, Pennsylvania, pp. 645–63

Chugg W A 1964 *Glulam: The Theory and Practice of the Manufacture of Glued Laminated Timber Structures*. Benn, London

National Forest Products Association 1986 *National Design Specification for Wood Construction*. National Forest Products Association, Washington, DC

R. P. Wibbens
[Littleton, Colorado, USA]

H

Hardboard and Insulation Board

Hardboard and insulation board are rigid or semirigid sheet materials with thicknesses of less than 1 in. (~25 mm). Summarily known as fiberboards they are a category of the large class of wood composition-boards, made by various methods of manufacture from wood elements ranging from veneers to fibers (Fig. 1). The term fiber refers to any wood or other lignocellulosic element possessing the size and shape of the wood cell, the biological building block of all woody material.

In the terminology of the American Hardboard Association the boundary between insulation board and hardboard is a board density of 480 kg m^{-3} (30 lb ft^{-3}). In the trade, however, a distinction is made between medium-density fiberboard, often referred to as MDF, and higher-density fiberboard, called hardboard in the narrower sense.

In the manufacturing process solid wood is first reduced to fiberlike elements (pulping) which are subsequently recombined in sheet form with or without the addition of binders.

Fiberboard processes utilize the lower-quality end of the available range of raw wood material. Owing to the small size of the elements, these processes achieve a higher degree of homogenization and randomization of the anisotropic raw material than any other wood utilization process, with the exception of paper.

The most important fiberboard markets are for interior and exterior cladding, and industrial applications in the furniture and automotive industries.

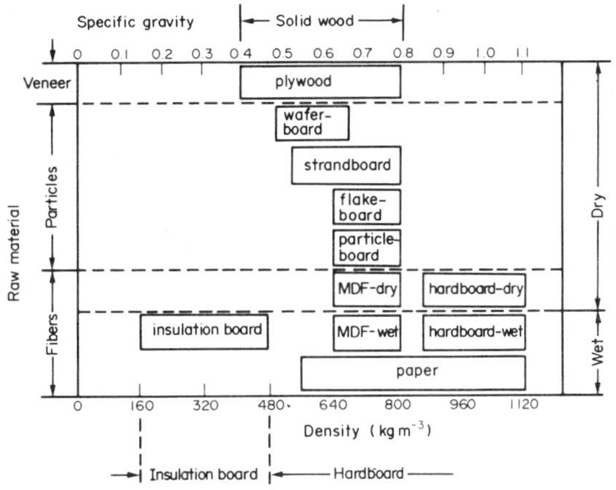

Figure 1
Classification of wood composition-boards (after Suchsland and Woodson 1986)

1. *Bonds in Fiberboard*

Although the tensile strength of the wood fiber is very high, it is generally not utilized to its fullest potential in the structural configuration of a paper sheet or a fiberboard. This is due to the total bonding area between fibers being incapable of transmitting the shear stresses required to stress the fiber to its limit, either because of the limited size of the total shear area or because of the quality of the bond. In low- and medium-density fiberboard, stress failures occur predominantly in the bond. At high board densities failures do occur in the fiber, reflecting both more intimate contact between fibers and possible modification of fiber characteristics under the severe conditions in the hot press.

The establishment of maximum quality fiber bonds is, therefore, one of the predominant goals of process design. Three different types of bonding occur in fiberboard, either individually or simultaneously, depending on the process (Back 1986).

1.1 Hydrogen Bonding

Hydrogen bonds depend on extremely short-range attractive forces associated with surface molecules, and develop between positively charged hydrogen atoms and any negatively charged atoms such as oxygen. This type of bonding occurs primarily in wet fiberboard processes, where the surface tension of the evaporating water pulls the fibers together within the range of molecular forces.

1.2 Lignin Bonding

Lignin bonds are attributed to the thermal softening of the lignin, a major wood component concentrated in the outer region of the cell wall, and a fusing together of lignin-rich surfaces under the conditions existing in the hot press (Spalt 1977). It has also been proposed that such bonds are formed by the condensation of phenolic substances derived from lignin. Lignin bonds are limited to wet-process fiberboards densified in hot presses, although they may occur in dry-process hardboard as a result of severe pulping conditions (Suchsland et al. 1983, 1987). Like the hydrogen bond, the lignin bond is unavailable to any of the other wood composition-board products.

1.3 Adhesive Bonding

Adhesive bonds formed by added binders are essential in the manufacture of dry-process fiberboard and all other wood composite products, and are used to enhance board quality in wet processes. The most common binders are phenol–formaldehyde (dry-process hardboard) and urea–formaldehyde resins (dry-process medium-density fiberboard). Other adhesives include drying oils and thermoplastic resins (Table 1).

2. *Fiberboard Processing*

Figure 2 gives a simplified outline of the entire range of fiberboard processes. Primary defiberizing of the raw material occurs by one of three mechanical pulping processes: (a) the Masonite process, subjecting the wood chips to high steam pressure (7 MPa), which upon sudden release explodes them into pulp; (b) the pressurized refiner, separating fibers between rotating grinding disks while under elevated steam pressure; and (c) the atmospheric refiner, grinding pretreated (steamed or water cooked) or untreated chips at atmospheric pressure. Chemical pulping is not used in fiberboard manufacture.

The fibers are either diluted with water and the mat formed on fourdrinier machines (wet process), or dried and formed by air-felting machines (dry process). Binders and sizing (chemicals to reduce water absorption) are added to liquid furnish and precipitated on fibers by pH reduction (wet process), or are distributed on fibers by spraying or by mechanical action (dry process).

Insulation board is made by drying the wet-formed mat. It relies almost entirely on hydrogen bonding. All other fiberboards are hot pressed. The hot press densifies the mat, provides the condition for bond development (plasticizing of lignin, curing of adhesives) and reduces the water content of the wet mat. Water and steam removal in the hot press from the wet and moist (semidry) mats is facilitated by inserting a screen between the bottom mat surface and press platen. The screen pattern is embossed on the back of the finished board (S1S—smooth one side). Dry-formed mats are pressed without screens (S2S—smooth two sides); S2S boards can also be made from dried wet-formed mats. The design of press cycles (pressure–, temperature–time relationship) influences density distribution over the board cross section and thereby affects strength properties. Thicker boards (dry medium-density boards) are particularly sensitive to such effects (Suchsland and Woodson 1976).

Table 1
Binder systems used in fiberboard manufacture (after Suchsland and Woodson 1986)

Fiberboard	Bonding (listed in increasing order of amount of binder used): Primary	Secondary
Wet process		
insulation board	hydrogen	starch, asphalt
S1S masonite	lignin	—
S1S	lignin	phenolic thermoplastics
S2S	lignin	thermoplastics, drying oils
MDF	lignin	thermoplastics, drying oils
Dry process		
S2S	phenolic	—
MDF	urea	—

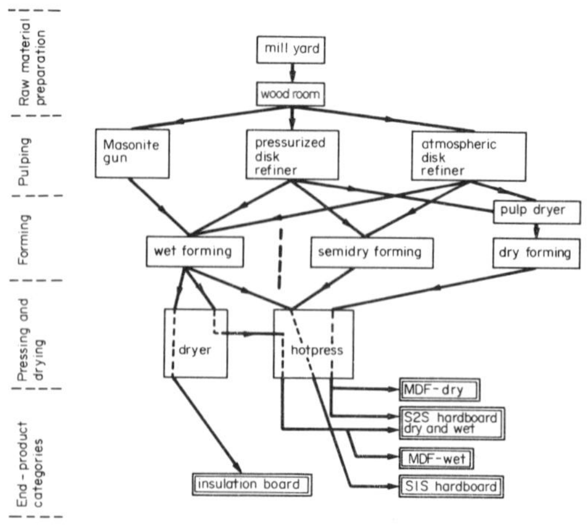

Figure 2
Simplified flowchart of fiberboard processes (after Suchsland and Woodson 1986)

Heat treatment (baking at 260 °C) of pressed boards with or without prior application of drying oils (tempering) increases bending strength and reduces water absorption and swelling. Bending strength improvement is believed to be due to fiber bond enhancement at high temperatures. Reduction of water absorption follows conversion of hemicelluloses into less hygroscopic furfural polymers (Stamm 1964). The surface application of tempering oils and subsequent heat treatment increases surface hardness substantially, a property of great importance in the manufacture of premium quality factory-finished wall panelling. Excessive heat treatment causes pyrolytic breakdown of the cell wall. Insulation board and thick medium-density fiberboard are neither heat treated nor tempered.

3. *Fiberboard Properties*

Most strength properties of fiberboard are strongly affected by raw material and manufacturing variables such as fiber characteristics, board density, binder content and heat treatment, and are, therefore, controllable within considerable margins. Real fiberboard properties are compromises between market demands and economic limitations, and are often adjusted to the requirements of particular applications. Table 2 lists minimum and maximum hardboard properties prescribed by the commercial standard.

Table 2
Classification of hardboard by thickness and physical properties[a]

Class	Nominal thickness (in.)	Water resistance (maximum average per panel): Water absorption based on weight (%)	Water resistance (maximum average per panel): Thickness swelling (%)	Modulus of rupture (minimum average per panel) (MPa)	Tensile strength (minimum average per panel): Parallel to surface (MPa)	Tensile strength (minimum average per panel): Perpendicular to surface (MPa)
Tempered	1/10					
	1/8	25	20			
	3/16			41.3	20.7	0.9
	1/4	20	15			
	5/16	15	10			
	3/8	10	9			
Standard	1/12	40	30			
	1/10					
	1/8	35	25			
	3/16			31.0	15.2	0.6
	1/4	25	20			
	5/16	20	15			
	3/8	15	10			
Service-tempered	1/8	35	30			
	3/16	30	30	35	13.8	0.5
	1/4	30	25			
	3/8	20	15			
Service	1/8	45	35			
	3/16	40	35			
	1/4	40	30			
	3/8	35	25			
	7/16	35	25			
	1/2	30	20			
	5/8	25	20	20.7	10.3	0.3
	11/16	25	20			
	3/4					
	13/16					
	7/8	20	15			
	1					
	1–1/8					
Industrialite	1/4	50	30			
	3/8	40	25			
	7/16	40	25			
	1/2	35	25			
	5/8	30	20			
	11/16	30	20	13.8	6.9	0.2
	3/4					
	13/16					
	7/18	25	20			
	1					
	1–1/8					

[a] Source: US Department of Commerce (1980a)

Properties related to the hygroscopicity of the wood cell wall are much more difficult to control. All fiberboards, by absorbing and giving off water, maintain a moisture content in equilibrium with the surrounding air (Fig. 3). Important consequences of moisture content fluctuations are dimensional changes (thickness swelling and linear expansion), of which the linear expansion is critical in fiberboard

Table 3
Property requirements for medium-density fiberboard for interior use[a]

Nominal thickness (in.)	Modulus of rupture (MPa)	Modulus of elasticity (MPa)	Internal bond (tensile strength perpendicular to surface) (MPa)	Linear expansion (%)	Screwholding Face (kg)	Screwholding Edge (kg)
13/16 and below	20	2000	0.62	0.30[b]	148	125
7/8 and above	19	1723	0.55	0.30	136	102

[a] Source: US Department of Commerce (1980) [b] For boards having nominal thicknesses of $\frac{3}{8}$ in. or less, the linear expansion value is 0.35%

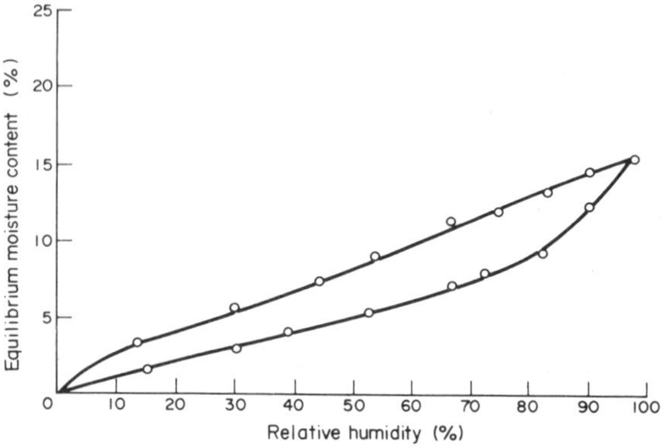

Figure 3
Adsorption and desorption isotherms between 0 and 98% RH at 20 °C for heat-treated hardboard (after McNatt 1974)

exterior siding. Also affected by moisture content changes are most strength properties, but most severely the interlaminar shear modulus. The shear modulus of $\frac{1}{4}$ in. (~6.25 mm) tempered hardboard at 80% relative humidity (RH) averaged only 70% of the modulus at 65% RH, and only slightly more than half the value at 30% RH. The modulus of elasticity in bending varied less than 20% of the value at 65% RH over the range 0–90% RH. Strength properties are similarly affected by increased temperature (Haygreen and Sauer 1969). Fiberboard, like solid wood, is viscoelastic, and the time-dependent responses to stress of fiberboard require special consideration in structural applications (McNatt 1970, Moslemi 1964, Suchsland 1965).

Thick medium-density fiberboard properties as recommended by the commercial standard are listed in Table 3. Mechanical properties are less critical for insulation board, with the exception of insulation sheathing, which requires tensile strength values up to 2.07 MPa and racking loads of at least 2360 kg. Minimum thermal conductivity values for insulation board vary from 0.0548 to 0.0692 W m^{-1} K^{-1}.

See also: Particleboard and Dry-Process Fiberboard; Paper and Paperboard

Bibliography

Back E L 1986 The bonding mechanism in hardboard manufacture—a summary. *18th IUFRO World Congr.*, Ljubljana, Sept. 7–13. Preprint

Haygreen J, Sauer D 1969 Prediction of flexural creep and stress rupture in hardboard by use of time–temperature relationship. *Wood Sci.* 1: 241–49

McNatt J D 1970 Design stresses for hardboard—Effect of rate, duration and repeated loading. *For. Prod. J.* 20: 53–60

McNatt J D 1974 Effects of equilibrium moisture content changes on hardboard properties. *For. Prod. J.* 24: 29–35

Moslemi A A 1964 Some aspects of visco-elastic behavior of hardboard. *For. Prod. J.* 14: 337–42

Spalt H A 1977 Chemical changes in wood associated with wood fiberboard manufacture. In: Goldstein I S (ed.) 1977 *Wood Technology: Chemical Aspects*, ACS Symposium Series 43. American Chemical Society, Washington, DC, pp. 193–219

Stamm A J 1964 *Wood and Cellulose Science.* Ronald, New York

Suchsland O 1965 Swelling stresses and swelling deformation in hardboard. *Mich. Agric. Exp. Stn., Q. Bull.* 47(4): 591–605

Suchsland O, Woodson G E 1976 *Properties of Medium Density Fiberboard Produced in an Oil Heated Laboratory Press*, US Department of Agriculture Forest Service Research Paper SO-116. US Government Printing Office, Washington, DC

Suchsland O, Woodson G E 1986 *Fiberboard Manufacturing Practices in the United States*, USDA Agriculture Handbook No. 640. US Government Printing Office, Washington, DC

Suchsland O, Woodson G E, McMillin C W 1983 Effects of hardboard process variables on fiberboarding. *For. Prod. J.* 33(4): 58–64

Suchsland O, Woodson G E, McMillin C W 1987 Effect of cooking conditions on fiber bonding in dry-formed binderless hardboard. *For. Prod. J.* 37(11/12): 65–69

US Department of Commerce 1973 *Cellulosic Fiber Insulating Board*, Voluntary Product Standard PS57-73. US Government Printing Office, Washington, DC

US Department of Commerce 1980a *Basic Hardboard*, American National Standard ANSI/AHA A135.4. US Government Printing Office, Washington, DC

US Department of Commerce 1980b *Medium Density Fiberboard for Interior Use*. US Government Printing Office, Washington, DC
US Department of Commerce 1982 *Prefinished Hardboard Paneling*, American National Standard ANSI/AHA A135.5. US Government Printing Office, Washington, DC
US Department of Commerce 1983 *Hardboard Siding*, American National Standard ANSI/AHA A135.6. US Government Printing Office, Washington, DC

O. Suchsland
[Michigan State University, East Lansing, Michigan, USA]

Health Hazards in Wood Processing

Wood is a very versatile material that can present risks to the health of woodworkers during processing. Woodworkers include the following occupations: sawmill workers, loggers, furniture makers and carpenters. This article examines the health risks, including known and presumed causes of the conditions linked with woodworking. Several acute and chronic reactions to wood-dust exposure have been recorded in the medical literature. Acute conditions include dermatitis (skin eruptions) and respiratory ailments such as asthma or pneumonites. Chronic disease responses include long-term respiratory conditions, impaired nasal clearance and fatal cancers. Of particular concern is the very high risk of nasal cancer among European furniture makers. Monitoring the working environment for elevated levels of wood dust is an important means of preventing some of the health hazards described here.

1. Dermatitis and Other Systemic Effects

Exposure to irritating woods can produce lesions on the skin, irritation of mucosal membranes and other clinical reactions; responses that have been described anecdotally for centuries but only recently annotated clinically. These responses are initiated by reaction to sawdust, by mechanical actions of hairs and bristles or by chemical irritation or sensitization.

1.1 Irritant Dermatitis

Skin lesions are produced by contact with the sap or powders in the bark of certain tropical woods. These trees contain irritating chemicals that cause reddening and blistering of exposed skin of loggers and timber-handlers. The irritant dermatitis may be exacerbated by high temperature, humidity and a lack of protective clothes or bathing facilities; initial lesions may be accompanied by secondary infections and other skin eruptions. Table 1 lists some of the irritating woods and the endogenous chemicals that produce skin lesions.

Table 1
Toxic substances found in woods[a]

Substance	Woods	Organs affected	Comment
Alkaloids and glycosides	Pigeon plum, greenheart, mesquite, ferro, baralocus, wenge, quatambu, Ceylon satinwood, avodire, red peroba, Knysna boxwood, opepe, aburo	Skin	Many structures and compounds are yet undetermined
Saponins	Satine, albiza, vinhaticco, black bean, mahua, makore	Skin, mucous membranes	Large quantities in satine; bassic acid from makoresaponin used to poison arrows
Phenols, catecols	Poison ivy, grevillea, ginkgo (maidenhair)	Skin	
Quinones	Lapacho, peroba, teak, angelim, partridge, tagayoson, walnut shells, ebony, mansonia, rosewood, purpleheart (possibly yellow poplar, sucupira, cocuswood, quassia, salmwood)	Skin, eyes	Mansonia and rosewood contain dalbergiones; strong sensitizers
Stilbenes	Iroko, Peru balsam	Skin, respiratory tract	
Terpenes	Pines, latex from *Euphorbia* spp.	Skin	Liverworts growing on bark account for woodcutters eczema
Furocoumarins	Satinwoods, Bergamot orange, latex from figs	Skin	Furocoumarins are sensitive to light

[a] Adapted from Woods and Calnan (1976) and Hausen (1981)

1.2 Atopic or Sensitizing Dermatitis

Atopic or sensitizing dermatitis arises when the body produces antibodies to the extracts of heartwood, and with repeated contact, a dermal reaction (including eruptions, redness, itching and scaliness) to the extract usually affecting hands, forearms, face, eyes, neck, scalp and the groin is produced. To differentiate irritant from atopic dermatitis, a physician (often a dermatologist) determines whether the patient's blood or skin shows a characteristic reaction to an extract of the wood (Woods and Calnan 1976). Both types of dermatitis may have a latent period lasting from hours to two weeks after first (or repeated) contact. Some more common tropical woods producing either sensitizing or irritant dermatitis include African mahagony (*Khaya ivorensis*), rosewoods (*Machaerium* spp. and *Dalbergia* spp.), cocobolo (*Dalbergia retusa*), iroko (*Chlorophora excelsa*), teak (*Tectonia grandis*) and mansonia (*Mansonia altissima*). The US National Institute for Occupational Safety and Health (NIOSH) estimates that 1% of workers may develop skin sensitization to sawdust of American woods (NIOSH 1987).

Occasionally, hairs or bristles from tree bark, leaves or other parts of the tree can produce skin irritations among loggers. Another condition, called "woodcutters' eczema" or "wood poisoning," can be produced by skin contact with epiphytes, lichens or liverworts growing on tree bark or surrounding shrubs. A common cause of this lesion is the leafy liverwort, *Frullania*, whose sesquiterpene lactones have been found to be skin sensitizers. Research has also shown that furocoumarins, which become skin sensitizers when exposed to sunlight, may be common in satinwoods (*Fagara flava* and *Chloroxylon swietenia*), Bergamot orange (*Citrus aurantium* var. *bergamia*), and in the latex of fig trees (*Ficus carica*).

2. *Acute Respiratory Ailments*

Exposure to wood dust can produce two acute conditions, either allergy to the wood (or its extracts) or pneumonites caused by fungal spores in the wood. Because the symptoms are similar, patients must be tested by skin patching or by inhalation challenge tests to determine whether they react to the wood or to the spores.

2.1 Woodworkers' Asthma

The clinical definition of woodworkers' asthma is the immediate or delayed hypersensitivity reaction among workers when exposed to the wood dust from exotic or hardwood lumber. After some period without exposure, the constricted breathing returns to normal. Confirmation of the reactivity of the worker to the wood dust is usually obtained by inhalation-testing the patient with an extract of the dust or the dust itself. If the woodworker responds to the dust, a decline in the patient's forced vital capacity (FVC) or forced expiratory volume in 1 second (FEV_1) confirms reactivity to the wood. Skin patch tests for antibody reactions to specific woods often show little correlation with the suspect woods; declines in either FEV_1 or FVC are a much clearer means to establish individual reactivity. Table 2 shows the symptoms and/or declines in pulmonary function associated with specific woods. Woodworkers' reactivity to some timbers has been reported only in case studies and no information was presented regarding the smoking status of the patients. Some of the more common woods include redwood (*Sequoia sempervirens*), teak (*Tectonia grandis*), oak, mahogany and cedar (Goldsmith and Shy 1988).

Western red cedar or Canadian red cedar asthma is the allergic condition that has been studied most thoroughly by clinicians and epidemiologists. This condition has been confirmed to be caused by exposure to sawdust of western red cedar (*Thuja plicata*). The clinical picture is as follows: woodworkers without familial or personal history of asthma or allergy suffer asthma symptoms when western red cedar is machined, although they work without reacting to other timbers. Initial complaints are of nasal and eye

Table 2
Pulmonary symptoms and effects from reactant woods

Observed effects	Woods
Asthma and shortness of breath	Redwood (*Sequoia sempervirens*) Cedar of Lebanon (*Cedrus libani*) Iroko (*Chlorophora excelsa*) Abiruana (*Lucuma* spp.) Zebrawood (*Microberlinia* spp.) Tanganyika aningre (*Aningeria* spp.) Central American walnut (*Juglans olanchana*) African maple (*Triplochiton scleroxylon*) Western red cedar (*Thuja plicata*)
Decline in FEV_1 or FVC	Redwood (*Sequoia sempervirens*) Oak Mahogany Cedar Iroko (*Chlorophora excelsa*) Abiruana (*Lucuma* spp.) Ramin (*Gonystylus bancanus*) Zebrawood (*Microberlinia* spp.) Teak (*Tectonia grandis*) Okume (*Aucoumea klaineana*) Tanganyika aningre (*Aningeria* spp.) Central American walnut (*Juglans olanchana*) African maple (*Triplochiton scleroxylon*) Western red cedar (*Thuja plicata*)

irritation, later becoming nasal obstruction with cough. The symptoms increase in severity with loss of exercise tolerance, development of breathlessness, wheezing and nocturnal cough including phlegm production. These symptoms become chronic and may persist for weeks after cessation of exposure to red cedar. Western red cedar asthma affects both smokers and nonsmokers alike. Bronchial challenge tests with red cedar dust or extracts of the dust produce immediate, delayed and dual hypersensitivity reactions, including loss of up to 50% of normal FEV_1 or FVC. Dr. Moira Chan-Yeung from the University of British Columbia showed that Western red cedar asthma is the specific result of pulmonary response to plicatic acid, the largest fraction in the nonvolatile extract of the timber (Chan-Yeung et al. 1973). Plicatic acid has also been implicated in the cause of asthma from Eastern white cedar (*Thuja occidentalis*) (NIOSH 1987). Chan-Yeung's work suggests that approximately 5% of mill workers handling Western red cedar have red cedar asthma (summarized in Goldsmith and Shy 1988 and NIOSH 1987).

2.2 Pneumonites and Wood Dusts

Some timbers, sawdust and wood bark may become contaminated with mold and fungal spores and produce allergic alveolitis similar to woodworkers' asthma. The symptoms and clinical signs include malaise, sweating, chills, fever, loss of appetite, breathlessness and chest tightness. In addition to positive dermal and pulmonary reactions to the spores, there may be x-ray evidence of alveolitis, which is absent in cases of woodworkers' asthma. Some of the identified pneumonites and their associated woods are shown in Table 3.

Table 3
Pneumonites and wood dusts

Type of wood	Possible agent(s)
Maple (bark)	*Cryptostroma corticale* *Saccharomonospora virdis*
Cork (suberpsis)	*Penicillium frequentans*
Wood chips or pulp (wood trimmers' disease)	*Aspergillus fumigatus* *Thermoactinomyces vulgaris* *Alternaria*
Redwood dust	*Graphium* *Pullalaria*

3. Chronic Health Risks

There are some case reports of nonmalignant pulmonary disease among occupations having wood-dust exposures (NIOSH 1987). However, the greatest concern in the medical literature has been on premalignant and malignant lesions in the nose.

3.1 Premalignant Nasal Conditions

Because European wood-furniture workers have been known since the 1960s to be at very high risk from nasal cancer (IARC 1981), several epidemiologic studies have sought to elucidate the stages of premalignancy. Goldsmith and Shy (1988) reviewed the studies of nasal symptoms among European woodworkers and showed the following: British and Danish furniture workers had impaired nasal clearance, including greater symptoms of rhinitis, middle-ear inflammation, sinusitis and bloody noses. Tissue samples from the middle nasal turbinates (where nasal tumors arise) of furniture workers from Sweden and Norway showed that they had greater dysplasia and metaplasia than controls who worked in other occupations. Furthermore, among nasal cancer patients who were woodworkers, researchers found similar irregular tissues adjacent to the tumors. Thus, it appears that in parallel with impaired nasal clearance, there is a sequence of cellular events (requiring several decades) from tissue injury to adenocarcinoma of the nasal sinuses.

3.2 Cancer Mortality Risks

Epidemiologists showed in the mid-1960s that nasal cancers (specifically adenocarcinomas of the nasal turbinates) were strongly related to the manufacture of wood furniture in the High Wycombe area of the UK (reviewed by IARC 1981). This relationship has been confirmed in Denmark, Canada, France, the USA, Sweden, Finland, Norway, Italy, the FRG, the GDR, the Netherlands, Belgium, Austria, Switzerland and Australia. Table 4 shows the range of cancer mortality risks for woodworkers by tumor site. (If the observed risk was the same as the expected risk, the value would equal 100.) None of these studies was able to adjust for differences in cigarette smoking habits or other confounding factors. It is possible that the elevated cancer risks for lymphoma, Hodgkin's disease and leukemia may be related to interactive exposure of wood dust and either solvents or paints applied to wood as preservatives or possibly to contact with herbicides having potential carcinogenic properties. The International Agency for Research on Cancer evaluated the scientific evidence and concluded the following (IARC 1981):

> There is *sufficient evidence* that nasal adenocarcinomas have been caused by employment in the furniture-making industry. The excess risk occurs mainly among those exposed to wood dust. Although adenocarcinomas predominate, an increased risk of other nasal cancers among furniture workers is also suggested . . . . No evaluation of the risk of lung cancer is possible . . . . The epidemiologic data are not sufficient to make a definitive assessment of the carcinogenic risks of employment as a carpenter or joiner. A number of studies, however raise the possibility of an increased risk of Hodgkin's disease . . . . The evidence suggesting increased risk of lung, bladder, and

Table 4
Cancer mortality risks for woodworkers

Cancer	Sawmills, logging	Furniture makers	Carpenters
Stomach	190–98	107–75	128–74
Nasal sinuses	886–76	1318	124–81
Lung	138–45	130–67	120–79
Bladder	109–62	—	160–100
Lymphoma	100–73	—	113–99
Hodgkin's disease	193–67	—	162–58
Leukemia	193–67	133–100	134–38

stomach cancer comes from large population-based occupational mortality statistical studies and is inadequate to allow an evalaution of risks for these tumors.

4. Preventing Disease Risks among Woodworkers

There are two means of reducing the risk of disease in woodworking industries: good housekeeping and medical surveillance of employees. Good housekeeping means controlling all wood dusts at or below the occupational health standards prevalent in the country where the plant operates. NIOSH (1987) describes the wood dust exposure regulations in force in many countries at the time the summary of the literature was written. It is important to recognize that toxicity of exposure depends on whether the timber is hardwood or softwood, and the specific toxic reaction for each timber, especially imported or exotic varieties. Appropriate medical surveillance includes baseline medical information when a worker begins employment and periodic evaluation of health in order to determine any clinical change in status since the previous contact. Essential in medical surveillance of woodworkers is a thorough examination of the upper respiratory tract, especially the nasal sinuses, and an estimate of the cumulative exposure to wood dust.

Bibliography

Chan-Yeung M, Barton G M, MacLean L, Grzybowski S 1973 Occupational asthma and rhinitis due to Western red cedar (*Thuja plicata*). *Am. Rev. Resp. Dis.* 108: 1094–102

Goldsmith D F, Shy C M 1988 Respiratory health effects from occupational exposure to wood dusts. *Scan. J. Work Environ. Health* 14: 1–15

Hausen B M 1981 *Woods Injurious to Human Health.* De Gruyter, Hawthorne, New York

IARC 1981 *Monographs on the Evaluation of the Carcinogenic Risk of Chemicals to Humans*, Vol 25: *Wood, Leather and some Associated Industries.* World Health Organization International Agency for Research on Cancer, Lyon

NIOSH 1987 *Health Effects of Exposure to Wood Dust: A Summary of the Literature.* U.S. Department of Health and Human Services, National Institute for Occupational Safety and Health

Woods B, Calnan C D 1976 Toxic woods. *Br. J. Dermatol.* 94 (suppl): 1–97

D. F. Goldsmith
[University of California, Davis, California, USA]

History of Timber Use

The tree and its wood have played a prominent role throughout history. The *International Book of Wood* points out that "man has no older or deeper debt" than that which he owes to trees and their wood (Bramwell 1976). "A culture is no better than its woods" writes W. H. Auden in *Woods.* The Bible records that Noah built the ark from gopherwood, the first ark of the covenant was made from acacia wood, and the cedars of Lebanon framed Solomon's temple. The best known of the early cultures developed along the Tigris–Euphrates, Nile, Indus and Yellow Rivers where forests were quickly depleted. Imports were soon heavily depended upon to supply people's needs for wood.

1. The Early Years

Wood has been one of the most important raw materials from early Paleolithic times, both for building and for the manufacture of tools, weapons and furniture. Wood was worked early in human existence because no elaborate tools were needed. At that time, however, the quality of wood products depended more on the characteristics of the wood than on the tools available for woodworking. The availability of copper tools by about 5000 BC made possible the higher degree of craftsmanship evident in some few surviving relics of the time. This high degree of craftsmanship led to the use of wood in carefully worked form for coffers and chests for the storage of precious possessions by about 2600 BC. Closely related to this was the first coopering to make barrels of various types by about 2800 BC.

2. *The Last Millenium*

From the tenth to the eighteenth centuries in Europe, wood was the material primarily used for buildings, tools, machines, mills, carts, buckets, shoes, furniture and beer barrels, to name a few of the myriad articles of wood of the time. The first printing press was made of wood and such presses continued to be made of wood for a hundred years. Most of the essential machines and inventions to establish the machine age were developed in wood during this period. Wood played a dominant role in all industrial operations and much of the art and culture of the times. In Europe, wood use reached a high plateau around the sixteenth century. About that time, however, the availability of timber began to diminish owing to the many demands for both fuel and materials and the expansion of agriculture.

North America's wood use reached a plateau during the middle to late nineteenth century, 150–200 years after that phase had peaked in Europe. The seemingly inexhaustible forests of colonial days were exploited, along with other natural resources, to feed a rapidly growing economy. Railroads, telegraph lines, steel mills and other industries were consuming wood at a dramatically increasing rate. During the latter half of the nineteenth century, the volume of sawnwood produced annually increased from 10 million to 85 million cubic meters, a level that has been maintained up to the present. Traditional uses of wood—for fuel, shipbuilding, construction of large buildings and bridges—were taken over by petroleum, coal, iron and steel, stone and brick. However, new uses for plywood, paper, poles, sleepers and chemicals, as well as the continuing demand for lumber, maintained a high level of timber production.

3. *Wood for Transportation*

Wood has played a key role in the transportation of people and their possessions, both as fuel and as a material, for thousands of years. Sledges made of wood were used for transport in northern Europe from 7000 BC. These were used for heavy loads such as stones and archaeologists believe that the massive stones of Stonehenge must have been moved on wood sledges placed on rollers. From this, the cart or wagon was created by putting the sledge on wheels. There are pictures of wheels that date from 3500 BC and actual wheeled vehicles found in tombs from 3000–2000 BC. In medieval times the spoked wheel became a great achievement of the joiner's art and in classical Greece both the spoked wooden wheel and the three-part solid wooden wheel were in common use. Solid wheels of planks were used on farm carts. Spoked wheels with 10–14 spokes were found in Roman forts in northern England. Rods of wood inserted into grooves and turning between the hub and the axle formed the first roller bearings.

In the nineteenth century in North America, railroads used wood for fuel, as well as for tracks, sleepers, cars, bridges, trestles, tunnel linings, sheds and stations. In the cities public transportation was mostly of wood, including horsecars, electric trolleys, cablecars, carriages and buggies. Roads made of planks laid across parallel rows of timbers embedded in the earth had come from Russia to London in the 1820s and spread to the USA during the period 1850–1857. When they were in good condition, these were the best roads in the country and more than 3000 km of such road were constructed in the mid-1800s. However, their demise soon came about as the result of excessive cost of maintenance.

Wood bridges employing advanced design concepts, laminated structural timbers and trusses were common in the nineteenth century. Covered bridges were common in North America and parts of Europe by the late nineteenth century, with the wood covering designed to protect the wood framework of the bridge itself. These were gradually replaced as the technology of wood preservation provided more economical means of wood protection and as iron, steel and concrete became the common materials for most bridge construction.

One of the first uses of wood for water transport was probably a raft or hollowed-out log. The earliest wooden ships copied the hull form of the reed boats that had been made by the Egyptians in about 4000 BC. Larger ships were built in Egypt using cedar imported from Lebanon, sometimes on a grand scale. The barge built for Queen Hatshepsut in 1500 BC to transport granite obelisks from Aswan to Thebes had a displacement of some 6800 t and required 30 oar-powered tugs to tow it. Theophrastus, a pupil of Aristotle, recorded that the shipbuilding woods in ancient Greece were silver fir, fir and cedar—silver fir for lightness, fir for decay resistance, and cedar in Syria and Phoenicia because of lack of fir. The Phoenicians dominated trade in the Mediterranean for a thousand years up to the time of Christ with their biremes, large galleys with two banks of oars. By 1000 AD, Vikings from Scandinavia were travelling mostly in wooden ships that were at least 25 m long. The wooden ship evolved somewhat in design up to the late nineteenth century. By this time, paddlewheel steamboats were cruising the major rivers of the USA and consuming large quantities of wood fuel in the process. American clipper ships were the fastest seagoing ships of the late nineteenth century, their sleek shape and large expanse of sail taking them to 20 miles (32 km) per hour. However, by the 1880s iron steamships dominated the fleets of most naval powers, with wooden ships passing even in North America as the economics became unfavorable.

The discovery in 1982 of a 15 m ship sunk at some time around 1400 BC at Ulu Burun, off the coast of Turkey, with valuable cargo from around the Mediterranean, added substantially to our knowledge of the shipping, ships and wood use of the time. While most of the wood of the ship had deteriorated, enough remained of fragments, along with distribution of remains of the cargo on the ocean floor, to tell much about the size and nature of the ship. Included in the cargo were wooden tablets that served as a base for the wax used in writing and logs of wood known in Egypt as "hbny" and generally considered to be ebony. However, the latter was identified as African blackwood (*Dalbergia melanoxylon*) rather than true ebony (*Diospyros* sp.).

4. Wood for Weapons and Tools

Wood was also a key material for building the war devices of the ancient world. The battering ram, the scaling ladder, the tortoise and the siege tower are examples. Another example is the catapult, used to attack enemies from a safe distance. The properties of wood made it particularly suitable for such uses. High strength-to-weight ratio was a valuable and desired characteristic of wood then, just as it is now. Instruments of war such as those mentioned were essential to the expansion of Greek and Roman civilizations and of the science and technology that developed under the tutelage and guidance of the great thinkers and teachers of the time.

One of the oldest weapons, and the oldest surviving wood relic, is the pointed end of a spear made of yew from the Lower Paleolithic era found waterlogged in an English bog. Spear handles of wood were common in the Upper Paleolithic. The bow was invented during that period and there is evidence of use of the lever concept in tools and weapons as well as of hafting stones in wood to make axes and adzes. In the Mesolithic and Neolithic eras there is much evidence of new techniques for working wood, better axes and adzes, and chisels. Woodworking improved greatly with the advent of efficient carpenter tools during the Neolithic era, about 2000 BC, and still more during the Bronze Age. The lathe spread from the Mediterranean to northern Europe in about 1500 BC, but was not in common use until iron cutting tools were available, by about 700 BC. The plow was in common use at this time, beginning with devices made from a single piece of wood and evolving to the use of various woods according to their characteristics.

5. Wood in Construction

Wood has been the most versatile and useful building material in its many forms and adaptations. Furthermore, until the relatively recent development of metallic and plastic structural materials, it was the only material from which complete structural frameworks could be fabricated readily. The type and durability of structures built at various times and places has depended on the type and quality of timber available and the conditions of use, as well as on the culture and way of life of the people concerned. Early pole structures were built in Paleolithic times from small trees growing along the rivers. The techniques used have been carried down over the ages and are very similar to those used by nomadic peoples today. However, the most significant developments in the use of wood as a building material have taken place where the culture and living conditions favored the erection of permanent structures.

In forested zones, where timber was plentiful, solid walls of tree trunks or heavy planks were built. Walls of timber houses in Neolithic Europe were frequently made of split trunks set vertically in or on the ground or a bottom sill plate, as indicated by examples excavated in moor settlements in Germany. One of these also shows a beam with mortise holes. This method of wall construction continued through the Iron Age up to Norman times. A Viking fortress built in Denmark in about 1000 AD had this type of construction, as did stave churches built during that century. Later construction used similar principles, but employed vertical squared timbers or sawn planks. This "palisade" type of house construction was brought to North America by early French settlers along the Mississippi River.

Another common style of heavy timber construction, "log cabin" style with wall timbers placed horizontally rather than vertically, can be traced back nearly three thousand years. It has been most frequently used in the northern, central and mountainous areas of Europe and North America where there have been plentiful supplies of relatively large, straight trees. In this type of construction, round, squared or sawn timbers are laid horizontally and interlocked at the corners of the building. Some large farmhouses built in Central Europe about 900 BC were apparently of this form. Vitruvius, a Roman architect writing at the time of Christ, referred to houses of this type built in a district of Asia Minor bordering the Black sea. Records of the Roman period of a few centuries before and after the birth of Christ indicated considerable use of wood for roof construction, interior storage areas and lofts, and other parts of buildings.

Gothic construction in wood was introduced during the eleventh to fourteenth centuries in Europe. In this construction the roof timbers form an integral part of the frame and the principal framing members were typically large and heavy. This type of construction persisted until the seventeenth century in Europe, but variations of it were used in the USA even into the nineteenth century. Popularity of Gothic construction peaked during the last part of that century.

Wood construction has had an interesting evolution in North America because of the relatively abundant timber resource and the scattered development of much of the country. Native Indians in the forested areas of the East and Northwest built homes and community houses from indigenous woods, the style depending on the culture and nature of wood available. In the East, structures were commonly made of poles covered with bark of birch or elm and palisades of vertical logs were used for protection. Along the northern Pacific coast, the Indians built houses from planks of split cedar or redwood and even had gabled roofs and decorative carvings. Even in some of the sparsely forested areas of the Great Plains and the western mountains, Indians built frames of timber and covered them with earth to make strong permanent dwellings.

Architecture of the early colonists evolved from that of their homelands, adapted to the climatic and cultural conditions of the times and materials availability. The log cabin was introduced into North America by Scandinavian immigrants in about 1638 and was adopted in the eighteenth century by Scottish–Irish immigrants. In the far Northwest, explorers and settlers from Russia moved south from Alaska and built houses, forts and churches of log-cabin-type construction.

The dwellings built in New England by immigrants from England during the seventeenth and eighteenth centuries followed the pattern of the English timber frame house with wattle and daub or brick between the framing members. The classical "saltbox" type of house evolved from adaptations to that pattern to provide more protection against the severe weather—covering with oak or pine clapboards, improvement of the timber frame, and exclusion of the brick. Houses in the Middle Atlantic states and the South were more likely to have exteriors of brick or stone, but elaborately decorated interior woodwork was not uncommon. The ready availability of building materials, especially wood, led to the detached or freestanding home becoming the standard dwelling in that part of the world.

Public buildings of the eighteenth century in North America followed similar trends. Churches were frequently made of wood, evolving from the simple to ornate. Balconies were supported by wood pillars, often elaborately carved. A wooden canopy was frequently placed above the carved wood pulpit as a sounding board. Public buildings in the Georgian style frequently combined wood and brick in ways that took advantage of good features of both materials. An outstanding example is Independence Hall in Philadelphia which combined with its basic brick construction a wood superstructure for the tower clock, a wood balustrade, cedar shingles and elaborate interior woodwork.

Wood remained the principal construction material in North America well into the nineteenth century and remains so for housing today, as it does in some parts of Scandinavia and other regions where timber supplies are plentiful and the tradition of wood construction remains strong. A critical element in this continued use was the development of "balloon frame" construction in the 1830s in Chicago. This was developed to increase flexibility in building construction and to overcome the problem of the lack of trained artisans needed to make the intricate connections necessary for the heavy timber and half-timber then in use. Rather than the heavy beams, posts and diagonals connected by precise and intricate joinery, the balloon frame concept employs relatively light studs held together by panels and nailed joints. Lumber and wood or other panel material can supply the necessary components inexpensively and flexibly. Following the disâstrous Chicago fire of 1871, what had been done in wood was re-created in iron. Problems of fire and the need for high-rise buildings have contributed to the demise of wood-frame construction in the centers of large cities, but it is still the major form of house construction in most of the country.

6. *Wood for Furniture*

Wooden furniture appears to have been developed first in Egypt, beginning in the early Dynastic times (c. 3000 BC). It was rare until copper tools became available, though considerable refinement in woodworking had been made possible by the development of polished stone tools. Reeds and rushes had earlier been used for furniture and early wood furniture designs copied the light style of that material. Shortages of furniture wood in that part of the world led to early developments in economical wood use. Planks cut from the same log were laid side by side and carefully fitted together. Defects were cut out and replaced with patches or plugs. Short pieces were joined with scarf joints to make longer ones. A toilet casket of Amenhetep II (1447–1420 BC) also shows the use of precious wood as a face veneer over common wood. Mortising, tenoning and dovetailing to make strong attractive furniture joints appeared by 2600 BC. Parts of furniture were originally lashed together with linen cords or rawhide thongs. Later copper bands were used, especially on large coffins. Bronze pins or nails entered the picture in about 1440 BC.

Carvers of wood, who had evolved from carpenters or joiners, were found frequently in Europe during the middle ages. By the early thirteenth century, painting and gilding of wood had evolved into a separate craft. The joiner gradually became a complete furniture maker, while turners concentrated on turned furniture, which had been developed during Roman times and moved into northern Europe. By the sixteenth century, however, the two crafts came back together as turners worked with joiners to decorate higher-quality chairs and bed frames.

At that time furniture- and cabinet-making reached a very high level in several parts of Europe, with new designs and styles developed and executed by outstanding designers and furniture makers. As the Dark Ages passed into the Middle Ages with the Renaissance, new styles of furniture developed in Italy, England and France. Furnishings used in colonial America were at first quite simple and usually homemade. However, the rich array of woods available for furniture and other uses in the New World, together with highly skilled craftsmen, many of whom had apprenticed in Europe, soon led to high-quality manufacture in that region in styles that copied those in Europe as well as some original styles. By the latter part of the nineteenth century, the earlier small furniture shops in the USA had been largely superseded by large furniture factories that could take advantage of the new planers, mortisers, carving machines, saws, veneer lathes and veneer slicers that were becoming available.

7. *Plywood and Veneer*

The material we now know as plywood can be traced back to ancient times, even though the term "plywood" did not appear in an English language dictionary until 1927. The tomb of King Tutankhamun from 1325 BC, discovered in 1922, revealed some of the world's oldest examples of plywood manufacture. However, gluing of veneers to make plywood was an old art even at that time. Historians generally place the art of veneering at about 3000 BC. The purpose of decorative veneering was then, as it is largely now, to extend as far as possible the utilization of the valuable, attractive woods. Such woods had to be imported into the centers of artistic wood products manufacture in Egypt and other countries and were undoubtedly expensive, and supplies were uncertain. A tablet from Amenhotep III to King Harundaradu of Arzawa, found at Tell-el-Amarna, states "one hundred pieces of ebony I have dispatched." Ornamental woods found in Egypt in that time were limited to palms, sycamore, tamarisk and acacia. More exotic woods, such as ebony, walnut, rosewood and teak, were imported from India.

The ancient Greeks also used veneer for ornamentation, but not to the same extent as the Egyptians. The highest degree of craftsmanship and artistry was first reserved for the furnishings of temples and public buildings, but later the furnishings used by private citizens also became more luxurious.

The Romans were fond of using figured veneers on their furniture, particularly on their tables. The most expensive Roman tables were those made with the veneer of citrus trees grown in Africa. While the Romans desired comfort and utility, first consideration was given to style and quality of workmanship, with the decorative value of fine veneer being highly prized.

As compared with decorative plywood, manufactured primarily from hardwoods, softwood plywood is of relatively recent origin. It began in the early 1900s in the USA and the industry is still concentrated primarily in that country. It was developed as an alternative, and in some ways an improvement, to lumber by gluing together thin layers of knife-cut wood (cut usually on a rotary lathe) with the grain of alternate layers at right angles. Early development of the industry was in response to the need for wood panels for doors, but the industry has now expanded to produce a wide array of building panels. An exhibit at the Lewis and Clark Exposition in Portland, Oregon, in 1905, provided the interest and awareness on which was based the growth of the industry over the next decades.

The increased demands for structural wood during World War I provided the first basis for rapid growth of the softwood plywood industry. The new design and application potential of this panel product led to its being adopted for many structural applications. Exterior applications were limited, however, until the availability of synthetic resin adhesives in 1935. Those superior adhesives, coupled with improvements in manufacturing and quality control, set the stage for a further rapid expansion of the industry as a means of overcoming the declining size and quality of the timber resource. This led to further development in response to the needs of World War II and to subsequent additional expansion of the industry. Softwood plywood was made in the northwestern USA originally from Douglas fir, but in 1911 a plant was established in California to make it from ponderosa pine. Plywood made from southern pine entered the market in 1963. Several other softwoods are now also used in plywood manufacture.

Production of plywood panels has increased substantially during the past few decades in most parts of the world other than Europe. From 1964 to 1985, world production doubled from 22 to 44 million cubic meters per year. Currently, something over 40% of that is produced in the USA.

8. *Wood-Based Composites*

The development of wood-based composite materials, mostly within this century, has had a significant effect on wood use and opened new opportunities for creative products. The capability to make engineered structural products and a variety of forms and combinations with resins and other materials and the opportunity to use wood residues from other types of production provide incentives for application of this concept.

Wet-process fiberboard was developed during the latter part of the nineteenth century and commercially produced as insulating board early in the twentieth. It was based on papermaking technology. The board ranges in specific gravity from 0.16 to 0.40 and is widely used for sheathing, interior panelling, roof insulation and siding.

Wet-process hardboard was the next type of composite board to enter the picture. William H. Mason discovered the process in 1924, leading to the process and board type marketed as masonite. This is a hot-pressed board with a specific gravity of about 1.00.

Platen-pressed particleboard evolved from a series of developments in the late nineteenth and early twentieth centuries using shavings, sawdust or thin wood particles. The concept was further refined through the early part of the twentieth century, leading to commercial production after World War II. Featured in development since that time has been the concept of board made from flakes or strands of wood and suitable for structural use as well as for decorative use. This type of board, called flakeboard, waferboard or oriented strandboard has grown into competition with softwood plywood during the period since 1960. It has typically been manufactured from either softwood or medium-density hardwood, such as aspen (*Populus* sp.).

This is a field that is still evolving rapidly. Medium-density fiberboard came along in the 1960s and is used primarily as core stock in the furniture industry. Mineral-bonded building products made of excelsior (wood wool) and various mineral products, commonly cement, are suitable for a wide variety of structural products. Products molded from wood particles are relatively common in Europe and may become so in other parts of the world if the economics are favorable. World production of particleboard has increased spectacularly in recent decades, from 7.6 million cubic meters in 1964 to 43.5 million cubic meters—essentially equal to production of plywood—in 1984. Expansion of the particleboard industry has been strongest in Europe, which accounts for half of the world's production. Fiberboard production has increased relatively little over the same period, amounting to 16 million cubic meters.

9. *Wood for Paper*

Up to the middle of the nineteenth century, paper had been made primarily from rags. Their increasing cost and decreasing availability, together with a rapidly increasing demand for paper, had led to studies of other possible raw materials. The naturally fibrous structure of wood and its ready availability made wood a prime candidate and by the middle of the century three commercially feasible processes for making paper from wood had been developed. Hugh Burgess in the UK and Morris Kean in the USA developed the soda process, which could yield pulp suitable to mix with that from rags or straw in papermaking, and established a paper company in Philadelphia to use the process. Heinrich Voelter of Germany perfected the groundwood process, which became the principal method for producing woodpulp. In this process which is still the primary method of producing pulp for newsprint, the yield is high but the strength of the paper is low. The sulfite process was perfected in Sweden and became the basis for development in the USA by Benjamin Tilghman, who had previously developed a way to grind rags mechanically as an improvement in industrial papermaking.

Paper from woodpulp greatly reduced the cost of books and papers and encouraged literacy. It also led to manufacture of paperboard, wallpaper and paper collars and bonnets. Wood is the source of pulp for most of the paper made and used in the world today. Creative new approaches to pulping and papermaking during recent years are leading to increased efficiency in use of the raw material, paper products with new use capabilities, including structural applications, and reduced environmental impacts.

See also: Resources of Timber Worldwide

Bibliography

Bass G F 1987 Oldest known shipwreck reveals wonders of the Bronze Age. *Natnl. Geogr.* 172(6): 693–733
Bramwell M (ed.) 1976 *The International Book of Wood.* Simon and Schuster, New York
Daumas M 1969–79 *A History of Technology & Invention,* Vols. 1–3. Crown, New York
Davey N 1963 *A History of Building Materials.* Phoenix House, London, pp. 32–48
Derry T K, Williams T I 1961 *A Short History of Technology.* Oxford University Press, Oxford
Hindle B (ed.) 1975 *America's Wooden Age: Aspects of Its Early Technology.* Sleepy Hollow Restorations. Tarrytown, New York
Mumford L 1963 *Technics and Civilization.* Brace and World, New York
Peters T F 1987 The rise of the skyscraper fron the ashes of Chicago. *Am. Heritage Invent. Technol.* 3(2): 14–23
Singer C, Holmyard E J, Hall A R, Williams T I (eds.) 1954–59 *History of Technology,* Vols. 1–5. Clarendon Press, Oxford
Youngquist W G, Fleischer H O 1977 *Wood in American Life: 1776–2076.* Forest Products Research Society, Madison, Wisconsin
Youngs R L 1982 Every age, the age of wood. *Interdisciplin. Sci. Rev.* 7(3): 211–19

R. L. Youngs
[Virginia Polytechnic Institute and State University, Blacksburg, Virginia, USA]

Hygroscopicity and Water Sorption

In common with other lignocellulosic materials, wood is hygroscopic; that is, it can adsorb or desorb water in response to changes in the relative vapor pressure of the atmosphere surrounding it. This affinity of wood for water is due primarily to hydroxyl groups which are accessible within the lignocellulosic cell wall of wood.

1. *Definitions*

Hygroscopicity of wood is expressed quantitatively in terms of the wood moisture content m, which is usually defined on a dry weight basis (kg of water per kg of dry wood) at equilibrium with a given relative vapor pressure h or percentage relative humidity H, where $H=100h$. The curve relating m to h is defined as the sorption isotherm, and the equilibrium value of m for a given h is designated as the equilibrium moisture content.

Moisture content based on wet weight is related to m by

$$m_w = m/(1+m) \quad (1)$$

Moisture content is frequently expressed on a percentage basis, designated as M, where $M = 100m$. It may also be expressed in terms of concentration c, defined as $c = \rho_0 m$, where ρ_0 is the wood density based on dry mass and volume at m.

Water in green wood (i.e., fresh from the living tree) occurs in three forms: "free" or capillary water contained in the cell cavities, "bound" or hygroscopic water contained in the cell walls, and water vapor contained in the air spaces in those cell cavities not completely filled with liquid water. When green wood is subjected to normal atmospheric conditions, it loses all of the water in the cell cavities as it dries to a moisture content at equilibrium with the water vapor in the atmosphere.

The moisture content at which all of the cell cavity or free water has been lost but the cell walls are still saturated with water has been designated by Tiemann as the fiber-saturation point m_f or M_f. This is a critical moisture content because below M_f changes occur in most of the important physical properties of wood which vary with moisture content. These include changes in dimensions, mechanical properties, electrical properties and treatability with preservatives. This article is concerned primarily with wood containing hygroscopic or bound water only (except for the small amount of water vapor in the cell cavities), since this is the moisture condition in which wood is generally used.

The fiber saturation point M_f is of the order of 25–30% of the dry wood weight for most woods of the USA, but may be considerably lower for woods with high extractive content such as many tropical woods. M_f is not a precise point since capillary condensation may occur in fine interstices in the cell wall at high relative vapor pressure close to unity. Furthermore, M_f is lower for adsorption than for desorption and tends to decrease with increasing temperature at a rate of approximately 1% per 10 °C temperature rise above 0 °C. Below 0 °C the reverse phenomenon occurs, that is, M_f decreases with decreasing temperature.

2. *Measurement of Wood Moisture Content*

Wood moisture content has been measured by more than a dozen different methods. Only those more commonly used are mentioned here; these include the oven-drying, distillation and electrical measurement methods.

In the conventional oven-drying method, the weight loss from a moist sample is determined gravimetrically after it has been dried in an oven at 103 ± 2 °C to constant weight. The weight loss, presumed to be equal to the moisture loss, is divided by the oven-dry weight to obtain m, or M on a percentage basis. There are several errors which may be associated with this method, such as the loss of volatile extractives which cannot be distinguished from moisture loss and the presence of residual water vapor in the drying oven, which may be reduced by vacuum drying.

The distillation method is a modification of the oven-drying method wherein the moist wood is heated in a water-immiscible liquid which is a solvent for the volatile extractives. The evaporated water is condensed in a calibrated trap and measured volumetrically, thus eliminating errors due to extractives.

Electrical moisture meters are often used to measure the moisture content of wood because they give immediate results and are nondestructive to the wood. Two principal types are resistance meters, which are generally dc operated, and dielectric or power-loss meters, which use ac, often at radio or microwave frequencies. The resistance type is most effective from $M = 6$ to M_f. The dielectric meters are effective at all moisture contents but are sensitive to wood density variations. Both types require correction for wood temperature and are sensitive to moisture gradients. They also require calibration for wood species, although this is less important for the resistance meters than for the dielectric or power-absorption types, since the latter are affected by wood density.

3. *Factors Affecting Moisture-Sorption Isotherms*

If green wood is allowed to equilibrate at each of several successively decreasing relative vapor pressures, to zero, all at constant temperature, the sorption isotherm obtained is defined as the initial desorption isotherm. Subsequent equilibration at successively higher values of h, to unity, gives the adsorption isotherm, which is always lower than the desorption isotherm. Repeating the desorption process produces the secondary desorption isotherm which lies below the initial desorption at higher values of h, but tends to converge toward it as h decreases (see Fig. 1).

Further repetition of the sorption cycle results in replication of the adsorption and secondary desorption isotherms. The difference between the reproducible adsorption and desorption isotherms is termed sorption hysteresis, usually expressed in terms of the ratio M_a/M_d, where M_a is the adsorption moisture-content value and M_d the repetitive desorption value for a given value of h. The M_a/M_d ratio usually ranges between 0.75 and 0.85 (Fig. 2).

In general, moisture-sorption isotherms are measured at room temperature. An increase in the temperature affects the hygroscopicity of wood in two ways. The first or immediate effect is to reduce the equilibrium moisture content at any constant relative vapor pressure as is shown in Fig. 3. This is a more or less reversible effect. The second or long-term effect of

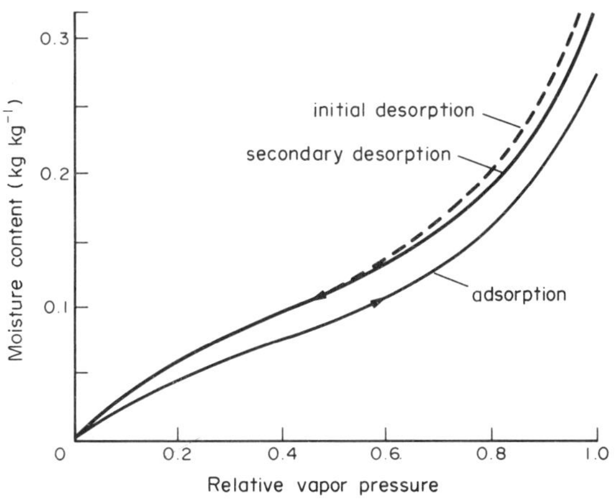

Figure 1
Typical initial desorption, adsorption and secondary desorption moisture isotherms for wood at 25 °C

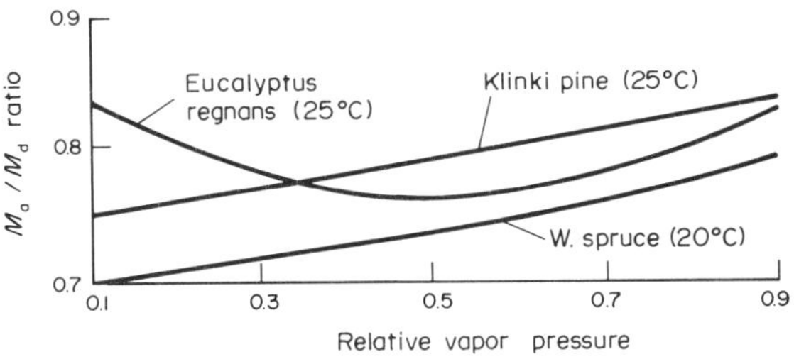

Figure 2
Adsorption–desorption ratios (M_a/M_d) for three woods at 20–25°C, based on data summarized by Stamm (1964)

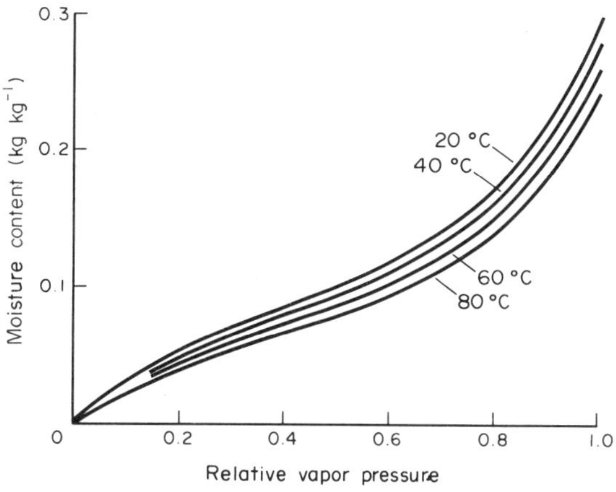

Figure 3
Effect of temperature on typical moisture isotherms for wood

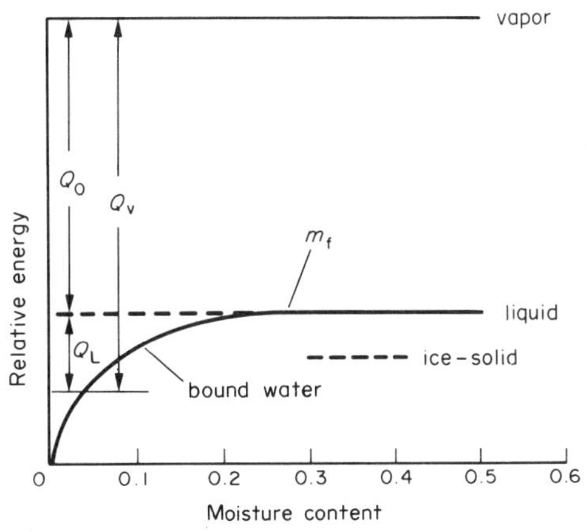

Figure 4
Relative energy levels of water vapor, liquid water and bound water in wood as functions of wood moisture content

exposing wood to high temperatures, particularly above 100 °C, is a permanent reduction in hygroscopicity when the wood is brought back to room temperature. Unfortunately, the mechanical properties of wood are also reduced by long-term exposure to high temperatures (Stamm 1964).

4. Thermodynamics of Moisture Sorption

The three forms of water found in wood differ in their energy levels, as is shown schematically in Fig. 4. Water vapor has the highest level, followed by liquid water in the cell cavities with essentially the same energy as ordinary liquid water. Bound water in the cell walls of wood is in the lowest energy state, the level decreasing with decreasing moisture content below m_f. At moisture contents above m_f the energy level is essentially equal to that of liquid water.

The energy level difference Q_0 between water vapor and liquid water is the heat of vaporization of liquid water. The difference Q_v is the heat of vaporization of bound water, equivalent to Q_0+Q_L, where Q_L is the differential heat of sorption of liquid water by wood. This is defined as the energy (kJ) released when 1 kg of water is taken up by wood of sufficiently large mass that *m* remains essentially constant. Two different methods are used to measure Q_L for wood, the isosteric or thermodynamic method and the calorimetric method.

The isosteric method for measuring Q_L depends on the fact that a linear relationship results when the sorption isotherms shown in Fig. 3 are replotted as ln *h* vs the reciprocal of the absolute temperature *T*, the resulting curves being isosteres of constant moisture content. The slope of one of these linear isosteres is related to Q_L by

$$Q_L = -(R/0.018)(\partial \ln h/\partial T^{-1})_m \qquad (2)$$

where R is the molar gas constant (kJ mol^{-1} K^{-1}) and 0.018 is the mass (kg) of water per mole. The slopes decrease with increasing m and the derived value of Q_L is a maximum for dry wood, decreasing exponentially as m increases (Fig. 5). An approximate empirical equation which applies up to m_f is

$$Q_L = \exp(a - bm) \quad (3)$$

where a and b are empirical constants, approximately equal to 7.07 and 15, respectively, when Q_L is in kJ kg^{-1}.

In the calorimetric method, Q_L is not measured directly because of the experimental difficulty involved in distributing a small increment of water uniformly throughout a large quantity of wood. Rather, it is derived from measurement of the heat of wetting W, which is defined as the heat (kJ) released when 1 kg (dry weight) of finely divided wood particles, conditioned to a uniform moisture content, is wetted in an excess of water to a moisture content much greater than m_f. The relationship of Q_L and W is given by

$$Q_L = -dW/dm \quad (4)$$

If Q_L is known, W can be evaluated as

$$W = \int_m^\infty Q_L \, dm \quad (5)$$

or, using Eqn. (3),

$$W = Q_L/b = (1/b) \exp(a - bm) \quad (6)$$

It is less difficult to measure W experimentally by calorimetric methods than to measure Q_L by the isosteric method. This is because of the difficulty in measuring sorption isotherms over a range of temperatures as is required to measure Q_L.

The total heat of wetting $\bar{W}$ is the value W when $m=0$. Therefore, $\bar{W} = (1/b) \exp(a)$. It gives a measure of the overall hygroscopicity of the wood and appears to be proportional to the fiber saturation point m_f.

The difference $(\bar{W} - W)$ between the total heat of wetting and that measured at m greater than zero is known as the integral heat of sorption. It is the complement of W with respect to variation with m (Fig. 5).

The free-energy change ΔG (kJ kg^{-1}) associated with moisture sorption can be calculated from

$$\Delta G = (RT/0.018)(\ln h) \quad (7)$$

Since h is less than unity, ΔG is negative, the absolute magnitude increasing with decreasing h (Fig. 5). The corresponding decrease in entropy ΔS associated with sorption is given by

$$\Delta S = (Q_L - \Delta G)/T \quad (8)$$

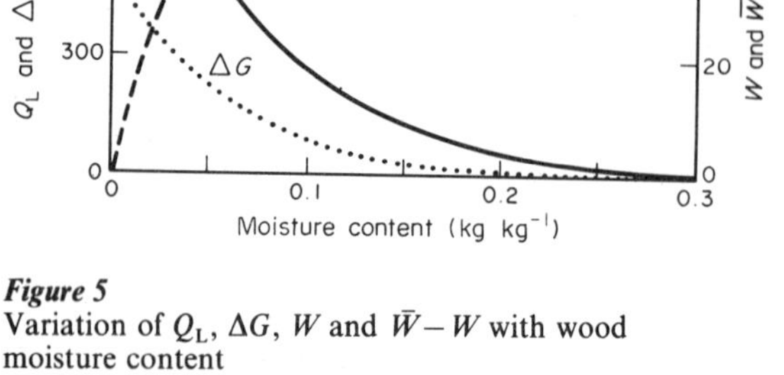

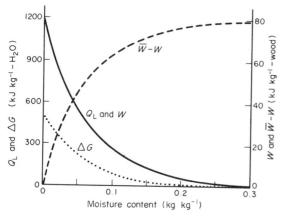

Figure 5
Variation of Q_L, ΔG, W and $\bar{W} - W$ with wood moisture content

5. Theories of Moisture Sorption

The several theories which have been proposed to explain the mechanism of water sorption by wood and hygroscopic polymers fall into two general categories. In one of these, the water molecules are assumed to be condensed in one or more layers on sorption sites or internal surfaces within the wood cell wall. In the second category, the polymer–water system is treated as a solution in which some of the water molecules form hydrates with sorption sites within the cell wall and the remaining water molecules form a solution.

The surface-sorption theory most often applied to wood is the Brunauer–Emmett–Teller (1938) or BET theory. This has recently been modified by Dent (1977) into a theory which better fits the sorption isotherms of wood as well as those of other natural hygroscopic materials such as cotton and wool, for which it was originally developed. The Dent theory will, therefore, be discussed here rather than the better-known BET model.

The Dent theory, in common with the BET theory, postulates that water is sorbed in two forms. The first is in the form of "primary" water molecules strongly attached to specific or primary sorption sites such as hydroxyl groups in accessible portions of the cell wall. The second form is as "secondary" water molecules attached to sites already occupied by primary water molecules or other secondary molecules. These are held by much weaker forces than those attached directly to primary sorption sites. Three fundamental constants determine the sorption isotherm according to Dent. Two of these are equilibrium constants: k_1 for the equilibrium between the primary water and liquid water, and k_2 for the equilibrium between the secondary water and liquid water. The third constant, designated here as m_0, is the moisture content corresponding to complete occupation of all the primary sorption sites with primary water (i.e., one molecule on each available site).

The Dent model predicts a sigmoid sorption isotherm of the form

$$m=h/(A+Bh-Ch^2) \quad (9)$$

where A, B and C are constants determined empirically from the experimental sorption isotherm, plotted as h/m vs h. Figure 6 shows a typical sorption isotherm. The constants are related to the three fundamental constants k_1, k_2 and m_0 as follows:

$$\left.\begin{aligned} A&=1/(k_1 m_0)\\ B&=(k_1-2k_2)/(k_1 m_0)\\ C&=(k_1-k_2)/(k_1 m_0)\end{aligned}\right\} \quad (10)$$

The moisture content m_1 corresponding to the primary water content at any value of h can be calculated from

$$m_1=m_0 k_1 h/(1-k_2 h+k_1 h) \quad (11)$$

Likewise, m_2 is calculated from

$$m_2=m_1 k_2 h/(1-k_2 h) \quad (12)$$

The sum of m_1 and m_2 gives the total moisture content m. Figure 6 shows m_1 and m_2 plotted against h, as well as the total isotherm of m vs h.

The free-energy changes ($kJ\,kg^{-1}$) associated with the sorption of primary and secondary water are given by

$$\left.\begin{aligned} \Delta G_1&=-(RT/0.018)\ln k_1\\ \Delta G_2&=-(RT/0.018)\ln k_2\end{aligned}\right\} \quad (13)$$

The values of k_1 and k_2 for wood at room temperature are approximately 10 and 0.75, respectively, and the corresponding values of ΔG_1 and ΔG_2 are -319 and $+40\,kJ\,kg^{-1}$, respectively. The total free-energy change associated with sorption at any moisture content m can be calculated from

$$\Delta G=\Delta G_1\ (\partial m_1/\partial m)+\Delta G_2(\partial m_2/\partial m) \quad (14)$$

where

$$\partial m_1/\partial m=(1-k_2 h)^2/[1+k_2 h^2(k_1-k_2)] \quad (15)$$

and

$$\partial m_2/\partial m=1-(\partial m_1/\partial m) \quad (16)$$

The solution theory most often applied to the moisture-sorption isotherm for wood is the Hailwood–Horrobin (1946) theory. Although this is a solution theory in contrast to the Dent theory, it has certain common features with the latter. It predicts the identical sorption isotherm (Eqn. (9)) as the Dent theory with identical values for the constants A, B and C. Furthermore, two of the three fundamental constants are identical with k_2 and m_0 of the Dent theory although the third constant is somewhat larger than k_1. It also predicts two forms of water: water of hydration (m_h), analogous to the primary water (m_1) of the Dent model, and water of solution (m_s), similar to the secondary water (m_2) of the Dent model. However, the magnitudes of m_h and m_s are slightly different from m_1 and m_2 at all values of m, although the general shapes of the corresponding curves are similar. Figure 6 includes curves of m_h and m_s, together with m_1 and m_2 predicted by the Dent model for the same isotherm.

Several other sorption theories have been applied to wood and similar natural polymers. Van den Berg and Bruin (1981), for example, compiled a list of 77 isotherm equations which have been applied to fibrous biological materials, including wood. Although they were primarily interested in materials used for food, such as cereals and meats, it seems probable that some of these would also apply to wood and its components.

See also: Drying Processes; Fluid Transport; Shrinking and Swelling

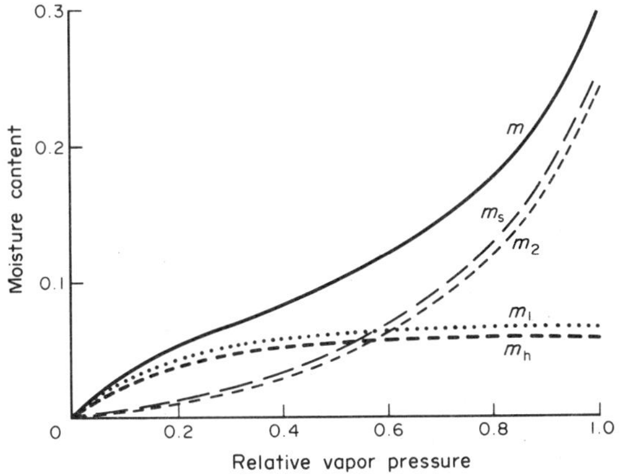

Figure 6
Typical sorption isotherms for wood (at 25 °C) and for the components m_1, m_2, m_h and m_s

Bibliography

Brunauer S, Emmett P H, Teller E 1938 Adsorption of gases in multimolecular layers. *J. Am. Chem. Soc.* 60: 309–19

Dent R W 1977 A multilayer theory for gas sorption. Part I: Sorption of a single gas. *Textile Res. J.* 47: 145–52

Hailwood A J, Horrobin S 1946 Absorption of water by polymers: Analysis in terms of a single model. *Trans. Faraday Soc.* 42B: 84–92

Skaar C 1988 *Wood–Water Relations.* Springer, Berlin

Stamm A J 1964 *Wood and Cellulose Science.* Ronald Press, New York

Van den Berg C, Bruin S 1981 Water activity and estimation in food systems: Theoretical aspects. In: Rockland L B, Stewart G F (eds.) 1981 *Water Activity: Influences on Food Quality.* Academic Press, New York, pp. 1–61

C. Skaar
[Virginia Polytechnic and State University, Blacksburg, Virginia, USA]

I

Industries Based on Wood in the USA: Economic Structure in a Worldwide Context

The US wood-based industry grows and processes timber, a renewable forest resource. The effectiveness of the industry is largely a function of how it is organized to undertake these activities. Important considerations are the character of major groups and subindustries that form the wood-based industry, the number of establishments that are available to manufacture timber into usable products, and the degree to which assets, value added and employees are concentrated with few companies. Identification of the role and importance of specific corporations in the industry also enhances comprehension of the industry's complex structural character. Finally, since timber is such an important product to industrial and final consumer groups, the extent of land devoted to timber production and the volume of such material produced become major considerations to a sound understanding of the role of the wood-based industry as a processor of material.

1. US Wood-Based Industry in General

The wood-based industry is a major sector of the US economy and a processor of nearly 453 million cubic meters of industrial and consumer material, namely wood fiber. In 1982, the industry's activities added 55.1 billion US dollars in value to shipments that totalled 131.0 billion US dollars. Such figures represent nearly 7% of the value added and shipment values generated by all manufacturers in 1982. In order to process wood fiber as a material, new capital expenditures by the industry approached 6.7 billion US dollars in 1982, or 1.3 times the amount expended in 1977 (Table 1). The 45 209 establishments engaged in the manufacture of products from wood fiber represented approximately 13% of all US manufacturing establishments, a decline of over 5000 since 1977. Employment in the wood-based industry was just under 1.4 million in 1982, a decline of 12.5% from the 1977 level.

2. Structural Characteristics of the Industry

2.1 Major Groups and Industries

The US wood-based industry is composed of three major groups and nearly 40 industries which make up the groups (Office of Management and Budget 1972). The major groups and selected industries within them are listed below.

(a) Lumber and wood products: (i) logging camps and contractors; (ii) sawmills and planing mills; (iii) millwork, plywood and structural wood members; (iv) wood containers; (v) wood buildings and mobile homes; and (vi) miscellaneous wood products.

(b) Paper and allied products: (i) pulpmills; (ii) paper mills (except for building paper); (iii) paperboard mills; (iv) converted paper products; (v) paperboard containers; and (vi) building paper and board mills.

(c) Wood furniture and fixtures: (i) household furniture; (ii) wood office furniture; and (iii) wood partitions and fixtures.

2.2 Establishments

An establishment is an economic unit, generally at a single physical location where business is conducted or where services or industrial operations are performed. The number of establishments which process wood fiber is a major structural character of the industry. The major group, lumber and wood products, accounted for more establishments (32 984) in 1982 than existed in the remainder of the wood-based industry (Table 1). Between 1977 and 1982 the group experienced a 12% decline in establishments. Within the group, the logging camps and contractors industry is most common (i.e., 35% of the establishments). The fastest growing segment of the group is the pallets and skids industry which reached 1649 establishments in 1982—an increase of 366 over 1977 levels and an increase of 1193 over 1972 levels.

Establishments in the paper and allied products group increased by 1% annually between 1954 and 1977 (Table 1). In 1982, 6381 manufacturing establishments operated in the paper and allied products field. The largest number of establishments in the group exist in the corrugated and solid fiber box industry (i.e., 1492 in 1982). The latter industry has been steadily adding establishments since 1950 when they numbered 805.

The third major group (wood furniture and fixtures) accounted for 13% of the establishments in the wood-based industry in 1982, namely 5844 (Table 1). Since 1954, the group's establishments have increased by a factor of 2.2. The wood household furniture industry accounted for the largest number of establishments (2607) in the group in 1982.

2.3 Concentration

Concentration is a key indicator of economic structure and an important factor in judging materials processing efficiency. Concentration is typically measured

Table 1
Trends in general economic characteristics of the US wood-based industry, by selected years 1954–1982

Items and year	Industry: Lumber and wood products	Paper and allied products	Wood furniture and fixtures
Establishments			
1982	32984	6381	5844
1977	37302	6545	6475
1972	33[illegible]48	6038	5464
1967	36795	5890	3292
1963	36150	5713	3517
1958	37789	5271	4727
1954	41484	5004	2677
Value added by manufacture (million dollars)			
1982	15376.6	33375.7	6333.9
1977	16222.8	22170.9	4825.2
1972	10309.4	13064.1	3443.1
1967	4973.4	9756.3	1043.7
1963	4020.6	7395.7	782.9
1958	3176.6	5707.5	1134.2
1954	3241.6	4630.2	946.0
New capital expenditures (million dollars)			
1982	1342.4	5098.3	255.1
1977	1562.6	3295.0	182.0
1972	930.9	1335.3	179.9
1967	426.0	1585.3	30.9
1963	394.7	708.5	22.0
1958	277.0	618.5	38.1
1954	217.4	532.7	24.7
Employees in operating establishments (thousands)			
1982	576.4	605.6	252.2
1977	692.4	628.7	286.0
1972	691.0	633.4	263.2
1967	554.0	638.9	108.7
1963	563.1	588.0	98.6
1958	585.4	551.3	192.6
1954	645.9	527.7	191.9

Source: Bureau of the Census 1986a

by "concentration ratios" (i.e., percentage of an industry's shipment values, total assets or total employment which is accounted for by the industry's largest firms). With some exceptions, the wood-based industry is not highly concentrated. Consider the value of industry shipments (Table 2). The 1982 four-company concentration ratios for industries in the lumber and wood products group ranged from 5% in the wood pallets and skids industry to 43% in the particleboard industry. Nearly half the industries in this group have four-company ratios of less than 20%. Industries in the wood furniture and fixture group are also "unconcentrated." The group's major industry, wood household furniture, increased to 16% at the four-company level in 1982, after experiencing no change between 1972 and 1977 (i.e., 14%). The paper and allied products group is characterized by substantially higher concentrations ratios. At the four-company level, the ratios range from 15% in the set-up paperboard box industry to 93% in the molded pulp goods industry. The pulpmill industry exhibits substantial concentration at the 20-company level (99%). This is in part reflective of the large capital investments required to achieve economies of scale necessary to process pulp material.

Assets can be another focus of concentration. In the wood-based industry, asset concentration is substantial. In the lumber and wood products group, just 12

Table 2 Share of shipment values accounted for by largest companies, by selected US wood-based companies

Industry and year	Percent of industry shipment values accounted for by: 4 largest companies	8 largest companies	20 largest companies	50 largest companies	Total shipment values (million US$)
Logging camps and logging contractors					
1982	30	37	47	52	8274.0
1977	29	36	45	50	6230.1
1972	18	25	35	41	2529.5
1967	14	22	30	35	1476.3
1963	11	19	25	31	1154.7
1958	13	19	25	32	833.8
Sawmills and planing mills					
1982	17	23	34	46	10065.2
1977	17	23	36	49	10866.7
1972	18	23	33	45	6420.8
1967	11	15	22	31	3506.4
1963	11	14	20	29	3156.3
Millwork					
1982	15	20	30	44	4248.3
1977	14	20	31	46	3928.1
1972	10	15	25	41	2426.8
Hardwood veneer and plywood					
1982	25	39	57	74	1304.4
1977	27	40	60	78	1250.4
1972	34	43	61	77	911.8
Softwood veneer and plywood					
1982	41	56	74	92	3221.5
1977	36	53	72	91	3804.8
1972	37	50	67	87	2011.5
Particleboard					
1982	43	69	94	100	547.4
1977	48	70	95	100	452.9
1972	48	69	92	100	295.4
Pulpmills					
1982	45	70	99	100	3110.4
1977	48	76	99	100	2091.1
1972	59	83	99	100	709.9
1967	45	70	97	100	730.5
1963	48	72	97	100	609.1
1958	46	68	94	100	452.0
Paper mills (except building paper)					
1982	22	40	71	93	20994.6
1977	23	42	70	92	12613.3
1972	24	40	66	88	6395.2
1967	26	43	65	86	4844.0
1963	26	42	63	85	3824.9
Paperboard mills					
1982	28	45	75	96	9531.1
1977	27	42	70	95	7124.3
1972	29	44	69	93	4153.5
1967	27	42	67	92	2907.0
1963	27	43	67	90	2315.2
Corrugated and solid fiber boxes					
1982	19	33	58	73	10558.1
1977	19	33	58	75	7351.1
1972	18	32	58	76	4308.8
1967	18	33	61	78	2959.6
1963	20	36	63	79	2166.1
Wood household furniture					
1982	16	23	37	54	5056.6
1977	14	23	39	56	4148.8
1972	14	22	40	60	2870.0

Source: Bureau of the Census 1986b

corporations (and in the paper and allied products group, just 23 corporations) accounted for approximately two-thirds of the assets and taxable income in their respective groups in the early 1980s. Over three-quarters of the corporations in the lumber and wood products group had total assets less than US$500 000 and accounted for only 5.6% of the group's total. At the other extreme, only 0.1% of the corporations had assets at or above US$250 million, yet they (12 in number) accounted for 61.6% of the group's total assets. In the paper and allied products group, the picture was much the same in 1981–1982. Over 90% of the group's corporations had under US$5.0 million in assets and accounted for only 7.3% of group-wide assets. In contrast, 0.7% (12 corporations) of the group's corporations accounted for 68.1% of total assets in the group.

Concentration by employee size class is yet another measure of structure. It provides insight to the economic scale at which firms must operate to process materials efficiently. In the lumber and wood products group, 57% of all establishments in 1977 had fewer than 5 employees and accounted for only 4% of employment, value added, and materials purchased by the group. Large firms with 1000 or more employees accounted for less than 0.1% of the group's establishments and less than 4% of group-wide value added, materials purchased and employees in the same year. In the middle of these extremes were the majority of the companies in the group (i.e., 60% of the group's employees were in medium-sized establish-

Table 3
Twenty leading lumber producing companies, 1985

Company	Lumber production (10^3 m^3)
Weyerhaeuser	6156.7
Louisiana–Pacific	4712.5
Champion International	4335.9
Georgia–Pacific	3973.9
Boise Cascade	1892.6
International Paper	1597.6
Sierra Pacific	1477.8
Pope and Talbot	1448.0
Simpson Timber	1176.6
Roseburg Forest Products	1033.6
Publishers Paper	932.1
Union Camp	885.2
DAW Forest Products	856.1
Plum Creek Timber	849.8
Crown Zellerbach	833.2
Willamette Industries	807.0
WTD Industries	776.8
Bohemia	765.0
Southwest Forest Industries	759.2
Federal Paper Board	757.5

Source: *Forest Industries* 1986

Table 4
Twenty sales-leading US paper and allied product companies, 1984

Company	Sales (million US$)
Champion International	4267
Kimberly-Clarke	3870
International Paper	3706
Scott Paper	2970
Mead Corporation	2728
Crown Zellerbach	2604
Boise Cascade	2466
James River	2425
Georgia–Pacific	2134
Weyerhaeuser	2080
Great Northern Nekoosa	1814
Hammermill Paper	1746
Container Corp. of America	1552
Westvaco	1545
Union Camp	1422
Fort Howard Paper	1363
Stone Container Corporation	1175
Temple-Inland	941
Packaging Corp. of America	815
Bowater	795

Source: Vance Publishing Corporation 1987

Table 5
Seventeen leading Canadian paper and allied product companies, 1983 (note: company sales figures include wood products other than paper and allied products)

Company	Company total sales (million Canadian dollars)	Pulp capacity rank
MacMillan Bloedel	2044	2
Domtar	1820	5
Abitibi–Price	1660	1
Consolidated–Bathurst	1393	4
Canfor	995	11
British Columbia Forest Products	900	7
Crown Forest Industries	727	9
Great Lakes Forest Products	495	6
Fraser	436	17
Eddy Paper Company	430	21
Donohue	357	14
Westar Timber	339	13
Ontario Paper Company	338	8
Northwood Pulp & Timber	291	16
Eurocan Pulp & Paper	181	19
James Maclaren Industries	165	22
Cascades	62	23

Source: Miller Freeman Publications 1985

ments with 20–249 employees). A similar situation existed in the paper and allied product group. The very small and very large firms accounted for relatively few employees group-wide and accounted for a small percentage of the group's key economic variables just discussed. In 1977, over three-quarters of the group's employees occurred in just over one-quarter of the establishments, with 100–2499 employees.

3. *Major Corporations in the US Industry*

The wood-based industry is composed of public and private companies. In the aggregate, little is known about the latter. Of the 9831 public companies that filed annual reports with the Securities and Exchange Commission in 1980, 193 were in the lumber and wood products or paper and allied products groups. The 1985 leading 20 US lumber producers accounted for approximately 22% of 67.85 billion board feet of lumber produced in 1985 (Table 3). Paper and allied product companies in 1984 were led in sales by Champion International (Table 4). The twentieth ranked company (Bowater) had paper and allied product sales that were only 19% of the industry's sales leader. Wood-based companies are frequently involved in the processing of wood fiber for both paper and lumber products. In 1978, paper sales made up almost three-quarters of the wood-based sales of the largest companies in the industry.

From a nationwide perspective, the lumber and paper sectors of the wood-based industry contributed almost 2% of the US national income in 1978. The pulp and paper sector contributed US$17.1 billion, only 5% more than the lumber sector's US$16.2

Table 6
Large wood-based firms operating outside the USA and Canada, 1983

Company	Country	Sales (10^6 US$)	Assets (10^6 US$)	Net income (10^6 US$)	Employees	Product specialty
Flick Group	FRG	3894780	2791406	114116	42560	Paper and wood products; chemicals
Reed International	UK	3023112	1735266	34760	57800	Publishing, printing, paper, pulp
Bowater	UK	2459820	1909448	(83677)	30500	Paper and wood products
Oji Paper	Japan	1954376	1885255	27427	10031	Paper, pulp and wood products
Beghin–Say	France	1392986	1265351	16497	8098	Paper products, food products
Jujo Paper	Japan	1561087	1585728	140	7211	Paper, pulp and wood products
Daishowa Paper	Japan	1328270	1602296	7321	5147	Paper, pulp and wood products
Svenska Cellulosa	Sweden	1273101	1509238	32714	15250	Paper and wood products
Honshu Paper	Japan	1117792	1616205	200	5364	Paper, pulp and wood products
Swedish Match	Sweden	1101582	804245	20593	18350	Wood products, chemicals, building materials
Rauma–Repola	Finland	1058342	1389370	8278	18099	Paper, pulp; wood and metal products
Sanyo–Kokusaka Pulp	Japan	1031021	1265921	4460	4833	Paper, pulp and wood products
Enso–Sutzeit	Finland	1027232	1775538	(14336)	13444	Paper and wood products
PWA	FRG	1023890	633899	5038	10407	Paper and wood products
Jefferson Smurfit Group	Ireland	968643	794407	24996	12119	Pulp, paper and wood products
Rengo	Japan	867755	505377	10908	4300	Paper and wood products
Royal Packaging Industries VanLeer	Netherlands	863782	652538	17405	13997	Paper, pulp; wood and metal products
Kymmene–Stromberg	Finland	834319	1170309	(6853)	16087	Paper, fiber, wood products
DRG	UK	823730	443578	18191	15878	Paper and wood products
Stora Kopparbergs Bergslags	Sweden	761619	1054844	26210	9396	Paper and wood products

Source: *Fortune* 1984

billion. National income in 1978 was 81% of gross national product.

4. Major Worldwide Corporations

4.1 Canada

The wood-based industry is an integral part of the economic fabric of Canada. The following 10 firms lead the Canadian lumber industry in terms of 1985 lumber production (million cubic meters):

> Canadian Forest Products (3.1); British Columbia Forest Products (2.8); West Fraser Mills (2.4); MacMillan Bloedel (2.4); Weldwood of Canada (2.1); Westar Timber (1.7); Northwood Pulp and Timber (1.4); Whonnock Industries (1.2); Balfour Forest Products (1.2); and Doman Industries (1.2).

Such firms accounted for approximately one-third of 1985 Canadian lumber production. The leading Canadian producer of paper and allied products in 1983 in terms of sales was MacMillan Bloedel (Table 5). In terms of pulp producing capacity, however, Abitibi–Price was most noticeable.

4.2 Worldwide

Manufacture and sale of wood products is a significant economic activity of some of the world's leading corporations. In 1983, the Flick Group (FRG) and Reed International (UK) were sales leaders among wood-based firms headquartered outside the USA and Canada. Of 20 firms identified (Table 6), all are large, fully integrated operations involving the manufacture of a wide variety of products from wood fiber.

5. Industrial Control of Timber as a Raw Material

5.1 Industry-Wide

Timber as a raw material is an essential ingredient to the manufacture of products produced by the wood-based industry. In 1977, industrial ownership of commercial timberland in the USA accounted for 14% of the nation's total or 2.784×10^5 km^2. Commercial timberland is land capable of producing at least 141.5 m^3 of industrial wood per square kilometer and not reserved for uses that are incompatible with timber production. Regionally, the South has the largest concentration of industrially owned timberland. The 1.469×10^5 km^2 of industrial timberland in that region is 1.54×10^4 km^2 more than industry owns in all other regions combined. Nationally there has been a slight upward trend in industrial timberland ownership. The 1977 level is 15.5% larger than that occurring in 1952. Regional ownership trends—upward or downward—have been slight and are not expected to change dramatically in the years ahead. This attests to the growing difficulties involved in locating large tracts of timberland and securing financial resources for their purchase.

Table 7
Hardwood and softwood growing stock on forest industry land by region; 1952, 1962, 1970, 1976, 1990, 2010 and 2030

	Growing stock (10^6 m^3)						
					Projections		
Region	1952	1962	1970	1976	1990	2010	2030
Softwood							
North	0.18	0.22	0.32	0.36	0.45	0.57	0.64
South	0.46	0.59	0.61	0.66	0.75	0.91	1.04
Pacific Coast	1.36	1.16	1.02	0.94	0.72	0.58	0.50
Rocky Mountains	0.19	0.18	0.17	0.14	0.11	0.10	0.09
Total softwood	2.19	2.15	2.12	2.11	2.04	2.16	2.26
Hardwood							
North	0.20	0.24	0.29	0.32	0.41	0.49	0.50
South	0.31	0.37	0.40	0.46	0.59	0.68	0.63
Pacific Coast	0.06	0.09	0.11	0.11	0.14	0.14	0.16
Rocky Mountains	0.003	0.003	0.003	—[a]	0.003	0.003	0.006
Total hardwood	0.57	0.70	0.81	0.90	1.15	1.33	1.29
Total industry	2.76	2.85	2.93	3.01	3.19	3.49	3.55
Total US stock	17.09	18.34	19.27	20.10	22.21	24.66	25.19

Source: Forest Service 1980b [a] Less than 10^6 m^3

Commercial timberland in the US contains over 2×10^{10} m^3 of growing stock—raw wood-fiber material. Forest industry accounts for 3×10^9 m^3 or 15% of the national total. In 1976 it was regionally distributed as follows: 23% North, 37% South, 35% Pacific Coast, and 5% Rocky Mountains. Growing stock volumes on industrial timberland are expected to increase 18% by 2030 (Table 7). In absolute terms, hardwood volumes will make the largest contributions to this increase (3.96×10^8 m^3).

Approximately 19% of the 6.23×10^8 m^3 of annual timber growth occurred on industrially owned forest land. Softwood growth—especially in the South—exceeded hardwood growth by 8.5×10^6 m^3. Regionally, growth in Southern industrial forest land is expected to approach 5.48×10^7 m^3 annually by 2030, an increase of 26% over 1976. By 2030, 58% of industrial net annual timber growth will originate in the South. Timber removal from growing stock and growth on industrial timberland are substantial. Removals are primarily softwoods (86%) which originate in the South and the Pacific Coast states. Nationally, 29% of timber removed from forest land originates from that owned by forest industry, i.e., 1.16×10^8 m^3. Considering industrial timberlands nationwide, timber removal in 1976 was equally balanced with net annual growth (Table 8). Species and regional disparities are, however, quite dramatic. Hardwood growth on industry lands was twice annual growth in the same year, while softwood growth replaced only 80% of volumes removed. Regionally, the softwood growth-removal ratios are lowest for the Pacific Coast and Rocky Mountain regions (i.e., 0.5 and 0.6). Softwood growth in the North region was twice 1976 removal from industrially owned land.

Table 8
Ratio of net annual growth to removals for hardwoods and softwoods on forest industry land by region

Region	1952	1962	1970	1976
Softwood				
North	1.6	2.4	2.4	2.0
South	1.3	2.3	1.4	1.1
Pacific Coast	0.3	0.4	0.4	0.5
Rocky Mountains	0.8	0.7	0.6	0.6
Total softwood	0.7	1.0	0.8	0.8
Hardwood				
North	1.9	2.7	1.9	1.8
South	1.0	0.8	1.2	2.0
Pacific Coast	4.1	4.0	3.0	3.5
Rocky Mountains	—[a]	—[a]	—[a]	0.1
Total hardwood	1.3	1.3	1.5	2.0
Total industry	0.8	1.1	0.9	1.0
Total USA[b]	1.2	1.4	1.4	1.5

Source: Forest Service 1980b [a] < 0.1 [b] Projected ratios for total USA are: 1990, 1.3; 2010, 1.1; and 2030, 1.0

Nationwide, annual timber growth per hectare of timberland averaged 3.15 m^3 in 1976. Industrial

Table 9
Average net annual growth and potential growth (m^3 ha^{-1}) in the USA by owner and region, 1976

	Ownership				
Region	All owners	National forest	Other public	Forest industry	Farmer and other private
North					
Current	2.45	3.01	2.52	3.08	2.31
Potential	4.62	4.41	4.13	5.18	4.62
South					
Current	3.99	3.99	3.78	4.20	3.92
Potential	5.39	4.97	4.97	5.81	5.39
Rocky Mountains					
Current	2.03	2.10	1.75	3.50	1.75
Potential	4.20	4.48	3.85	5.18	3.57
Pacific Coast					
Current	3.43	2.10	3.71	5.60	4.34
Potential	6.79	6.37	6.16	8.33	6.93
Total					
Current	3.15	2.45	2.94	4.13	3.15
Potential	5.18	5.18	4.76	6.09	5.04

Source: Forest Service 1980b

timberland exceeded this national average by 1 m^3 (Table 9). Regionally, industrial timberland growth rates range from a low of 3.08 m^3 per hectare in the North to 5.6 m^3 per hectare in the South. In all regions, growth rates of industrial timberland exceeded that occurring in all other ownerships. Growth rates for all ownerships and regions are below potentials characterized by fully stocked natural stands. Nationally, 61% of potential growth was being realized in 1976—industrial lands were growing at nearly 69% of their potential. Southern industrial forests were being managed so as to capture 72% of their potential.

5.2 Corporate Patterns

Wood-based companies may control access to timber by a variety of means, including fee ownership, timber purchase agreements, public timber sales contracts and indirect influence over other private timberland. Determining the nature and extent of timberland

Table 10
The forty largest industrial forest land ownerships in the USA in 1979

	Area owned (km^2)					
Company	Northwest[a]	Intermountain[a]	South	Central	New England	Total
International Paper	1655	417	19850	0	6851	28773
Weyerhaeuser	9069	2428	12472	0	0	23970
Georgia–Pacific	3213	0	9907	1340	2254	16714
St Regis Paper	619	1255	6022	1930	3039	12865
Champion International	1129	3517	5767	1756	0	12169
Great Northern Nekoosa	0	0	1194	1044	8737	10975
Boise Cascade	1153	4650	2064	1323	1493	10684
Scott Paper	979	0	2974	0	3484	7438
Crown Zellerbach	2845	737	3456	0	0	7038
Union Camp	0	0	6969	0	0	6969
Time, Inc.	0	0	6192	0	0	6192
Burlington Northern	506	5532	0	0	0	6038
Continental Group	0	0	5957	0	0	5957
Diamond International	0	1655	0	631	3594	5880
Mead	0	0	2946	2456	0	5403
Potlatch	0	2096	2210	1000	0	5305
Westvaco	0	0	2590	2351	0	4941
St Joe Paper	0	0	4452	0	0	4452
ITT (Rayonier)	1416	0	2922	0	0	4338
Bowater	0	0	2631	1619	0	4249
Owens-Illinois	0	0	2958	1093	0	4051
Procter and Gamble	0	0	3913	0	0	3913
Louisiana-Pacific	639	2072	559	194	0	3464
Mobil (Container Corporation)	0	0	3233	0	0	3233
Santa Fe Industries (Kirby)	0	0	2647	0	0	2647
Gulf and Western	0	0	0	0	2448	2448
Johns-Manville (Olinkraft)	0	0	2363	0	0	2363
Kimberley-Clark	0	0	2270	0	0	2270
Willamette	874	0	1028	316	0	2212
American Can	0	0	947	1077	0	2023
Masonite	433	0	1421	158	0	2011
Southwest Forest Industries	0	49	1720	85	0	1854
Southern Pacific	0	1821	0	0	0	1821
Tenneco (Packaging Corporation of America)	0	0	1206	490	0	1696
Longview Fibre	1651	0	0	0	0	1651
Federal Paperboard	0	0	1473	0	0	1473
Chesapeake Corporation of Virginia	0	0	1433	0	0	1433
Hammermill Paper	0	0	708	724	0	1433
Southern Natural Resources	0	0	1327	0	0	1327
Cleveland Cliffs	0	0	0	1255	0	1255

Source: Ellefson and Stone 1984 [a] Northwest region is limited to Washington and Oregon (west of the Cascades) and California Redwood Region. Intermountain region includes eastern Washington and Oregon and remaining portion of California

control for specific companies is very difficult. In 1979, the total area of forest land owned by the 40 largest industrial timberland owners was estimated to be over $2.347 \times 10^5\ km^2$ or 84% of the nation's commercial timberland (Table 10). The top 90 companies in that year accounted for approximately 91% of the timberland held by industry. Since 1979, there have been numerous shifts in ownership of industrial timberland among companies.

Concentrations of fee-owned timberland can be significant. In 1979, the four largest landowning companies accounted for nearly three out of every ten square kilometers of timberland owned by wood-based firms, while the largest 40 landowning companies retained nearly 85% of such land. Considering all timberland ownership in the USA, the largest wood-based firms accounted for a relatively small portion in 1979—the largest 40 owners owning only 12% of the nation's commercial timberland.

Although land is the base for the wood-based industry's raw material (timber), it is not a particularly good reflector of industrial control over wood fiber. Timber inventory on the land is much more relevant. For example, even though International Paper Company owned the largest acreage of timberland in 1979, the volume and value of timber owned by other companies was substantially greater. The US$9.5 billion worth of Weyerhaeuser timber in 1979 was nearly equal to the combined timber values of International Paper, Georgia–Pacific, and Crown Zellerbach, yet the latter three had nearly twice the fee-owned land of Weyerhaeuser. Weyerhaeuser's highly valuable old-growth timber explained this discrepancy.

In 1979, the 40 largest corporate timber holdings had a combined volume of nearly $2.3 \times 10^9\ m^3$ or 75% of industry-wide growing stock. The largest fee owner of timber was Weyerhaeuser—$3.26 \times 10^8\ m^3$ meters valued at US$9.5 billion. Relatively few Northwest regional holdings exceeded $2.8 \times 10^7\ m^3$, while 13 of the top 40 had Southern timber inventories that ranged from 37 to $156 \times 10^6\ m^3$.

Of the $3 \times 10^9\ m^3$ of industrial fee-owned timber in the USA, nearly 30% was accounted for by the four largest timber inventory holders in 1979—Weyerhaeuser, International Paper, Georgia–Pacific, and Crown Zellerbach. The eight largest accounted for 42%, while 75% of the industrially owned timber volumes could be attributed to the 40 largest companies. Of the timber inventory on all ownerships nationwide in 1979, only 11% was owned in fee by the largest 40 timber-owning companies.

See also: Future Availability; Resources of Timber Worldwide; Trade, Prices and Consumption of Timber in the USA

Bibliography

Bilek E M, Ellefson P V 1987 *Organizational Arrangements Used by U.S. Wood-Based Companies Involved in Direct Foreign Investment: An Evaluation*, Minnesota Agricultural Experiment Station Bulletin 526–1987 (AD-SD-3207). University of Minnesota, St Paul, Minnesota

Bureau of the Census 1986a *1982 Census of manufacturers (General Summary—industry, product class, and geographic statistics)*. US Department of Commerce, Washington, DC

Bureau of the Census 1986b *1982 Census of Manufacturers (Concentration Ratios in Manufacturing)*. US Department of Commerce, Washington, DC

Celphane T P, Carroll J 1980 *Timber Ownership, Valuation, and Consumption Analysis for 87 Forest Products, Paper, and Diversified Companies.* Morgan Stanley, New York

Clawson M C 1975 *Forest for Whom and for What.* Johns Hopkins University Press, Baltimore, Maryland

DeBraal J P, Majchrowicz T A 1981 *Foreign Ownership of US Agricultural Land (1979–1980)*, Agricultural Information Bulletin 448. Economics and Statistics Service, US Department of Agriculture, Washington, DC

Ellefson P V, Chopp M C 1978 *Systematic Analysis of the Economic Structure of the Wood-based Industry*, Staff paper No. 3. College of Forestry, University of Minnesota, St Paul, Minnesota

Ellefson P V, Stone R N 1984 *U.S. Wood-based Industry: Industrial Organization and Performance.* Praeger, New York

Enk G A 1975 A description and analysis of strategic and land-use decision making by large corporations in the forest products industry. *Diss. Abstr. Int. A* 36: 2997

Forbes 1980, 125 (10): 145–154. Appraised value: the stuff of dreams

Forest Industries 1986, 113(7): 14–21. Lumber outlook improves

Forest Service 1980a *An Assessment of the Forest and Rangeland Situations in the United States.* US Department of Agriculture, Washington, DC

Forest Service 1980b *An Analysis of the Timber Supply Situation in the United States, 1952–2030.* US Department of Agriculture, Washington, DC

Fortune 1984 (August). Directory of world-wide corporations

Irland L C 1976 Foreign ownership and control of U.S. timberland and forest industry. In: *Foreign Investment in US Real Estate.* Economic Research Service, US Department of Agriculture, Washington, DC

LeMaster DC 1977 *Mergers Among the Largest Forest Products Firms, 1950–1970*, College of Agriculture Research Center Bulletin 854. Washington State University, Pullman, Washington

McKeever D B, Meyer G W 1984 *The Softwood Plywood Industry in the United States, 1965–82.* Forest Products Laboratory, Forest Service, US Department of Agriculture, Madison, Wisconsin

Mead W J 1966 *Competition and Oligopsony in the Douglas-fir Lumber Industry.* University of California Press, Berkeley, California

Meyer P 1979 Land rush: a survey of America's land—who controls it, how much is left? *Harpers* 258(1544): 45–60

Miller Freeman Publications 1985 *North American Pulp and Paper Fact Book.* Miller Freeman, San Francisco, California

Office of Management and Budget 1972 *Standard Industrial Classification Manual.* US Government Printing Office, Washington, DC

O'Laughlin J, Ellefson P V 1982 *New Diversified Entrants among U.S. Wood-Based Companies: A Study of Economic*

Structure and Corporate Strategy, Minnesota Agricultural Experiment Station Bulletin 541. University of Minnesota, St Paul, Minnesota

O'Laughlin J, Ellefson P V 1981 U.S. wood-based industry structure: top 40 companies. *For. Prod. J.* 31(10): 55–62

O'Laughlin J, Ellefson P V 1982 Strategies for corporate timberland ownership and management. *J. For.* 80(12): 784–88

Phelps R B 1980 *Timber in the U.S. Economy, 1963, 1967, and 1972*, US Forest Service General Technical Report WO-21. US Government Printing Office, Washington, DC

Scherer F M 1980 *Industrial Market Structure and Economic Performance*. Houghton Mifflin, Burlin, Massachusetts

Vance Publishing Corporation 1987 *Lockwood's Directory of the Paper and Allied Trades*. Vance, New York

P. V. Ellefson

[University of Minnesota, St Paul, Minnesota, USA]

J

Joints with Mechanical Fastenings

The strength and stability of any structure depend heavily on the fastenings that hold its parts together. A prime advantage of wood as a structural material is the ease with which wood structural parts can be joined together with a wide variety of fastenings—nails, spikes, screws, bolts, lag screws, staples, and metal connectors of different types. For utmost rigidity, strength and service, each type of fastening requires careful design. This article is concerned with the standard mechanical fasteners and some of the factors that need consideration in the design of joints using them.

1. Nails

Nails are the most common mechanical fastening used in wood construction. There are many types, sizes, and forms of nails. These include the "common" nail design and annularly grooved, helically grooved, cement-coated and galvanized designs, many of which have special areas of use.

Resistance of nailed connections can be developed by direct withdrawal (load parallel to nail axis) from side grain of a wood member, or laterally (load perpendicular to nail axis), when driven into side or end grain of a wood member. Withdrawal resistance of nails from end grain is unreliable and usually such loading is avoided.

The maximum load to withdraw a bright, common wire nail driven into the side grain of seasoned wood or unseasoned wood that will remain wet in use, as shown by test (US Forest Products Laboratory 1987), is given by the empirical formula

$$p_w = 54.1\, G^{5/2} DL \qquad (1)$$

where p_w is the maximum load in newtons; L is the depth of penetration of the nail in the member holding the nail point in millimeters; G is the specific gravity of the wood based on oven-dry weight and volume at 12% moisture content; and D is the diameter of the nail in millimeters. For design purposes a safety factor of about 6 is applied to Eqn. (1).

Loads at a joint slip of 0.38 mm for bright, common wire nails in lateral resistance driven into side grain of seasoned wood, are expressed (US Forest Products Laboratory 1987) by

$$p_l = KD^{3/2} \qquad (2)$$

where p_l is the lateral load in newtons at 0.38 mm slip and K is a constant. Values of K are listed in Table 1 for a range of softwoods and hardwoods. For design purposes a safety factor of about 1.6 is applied to Eqn. (2).

Table 1
Coefficients for computing loads for fasteners in seasoned (15% moisture content) wood

Specific gravity range	Lateral load coefficient K		
	Nails	Screws	Lag screws
Hardwoods			
0.33–0.47	1400	3360	3820
0.48–0.56	2000	4640	4280
0.57–0.74	2720	6400	4950
Softwoods			
0.29–0.42	1440	3360	3380
0.43–0.47	1800	4320	3820
0.48–0.52	2200	5280	4280

Factors that affect the load carried by nails, other than those expressed in Eqns. (1) and (2), include moisture content, nail shank, nail point, and nail-head pull through. Most of these affect withdrawal resistance much more than lateral resistance.

2. Screws

Both wood screws and tapping screws are used in wood construction and are available in a wide range of materials and head types.

The maximum withdrawal load p_q in newtons for wood screws inserted in the side grain of seasoned wood may be expressed as

$$p_q = 108.2 G^2 DL \qquad (3)$$

where D is the shank diameter of the screw in millimeters and L is the length of penetration of the threaded part of the screw in millimeters. This formula is applicable in softwoods when screw lead holes have a diameter of about 70% of the root diameter of the threads, and in hardwoods of about 90%. For design purposes a safety factor of about 6 is applied to Eqn. (3); it accounts for variability and duration of load effects.

The withdrawal resistance of tapping screws is generally about 10% higher than for wood screws of comparable diameter and length of threaded portion.

The proportional limit load p_q in newtons in lateral resistance for wood screws in the side grain of seasoned wood is given by

$$p_p = KD^2 \qquad (4)$$

where D is the diameter of the screw shank in millimeters and K is a coefficient based on specific gravity for hardwoods and softwoods (Table 1). Slip values associated with the limiting loads will be from 0.18 to 0.25 mm. For design purposes a safety factor of about 1.6 is applied to Eqn. (4).

3. *Lag Screws*

Maximum withdrawal load p_r in newtons for lag screws from seasoned wood is given by

$$p_r = 125.4G^{3/2}D^{3/4}L \tag{5}$$

Lag screws require lead holes which vary from about 40% of the root diameter for very lightweight woods to about 85% for dense woods. For design purposes a safety factor of about 5 is applied to Eqn. (5).

Lateral loads for lag screws depend upon whether the load is applied parallel or perpendicular to the grain. For parallel loading, the proportional limit load p_t in newtons is given by

$$p_t = KD^2 \tag{6}$$

This equation applies when the attached member is more than about 3.5 times the shank diameter and depths of penetration in the main member are from 7 to 11 times the shank diameter. The depth of penetration depends on species density. For design purposes a safety factor of about 2.25 is applied to Eqn. (6).

When the lag screw is inserted in side grain and loaded perpendicular to the grain, the lateral resistance formula is multiplied by a factor of 1.00 for a 4.8 mm ($\frac{3}{16}$ in.) diameter lag screw, down to a factor of 0.50 for a 25.4 mm (1 in.) diameter lag screw.

General requirements for moisture content, location, spacing, and so on, in the design of joints using lag screws should be checked in the various design manuals in the Bibliography.

4. *Bolts*

Lateral loads for bolts are related to the compressive strength parallel or perpendicular to the grain of the wood, depending on the direction of the applied load.

The load a bolted joint carries depends on the amount of bending in the bolt. For small ratios of length L, in the center member of a three-member joint, to bolt diameter D (i.e., the L/D ratio), the bearing stress under the bolt is fairly uniform. For L/D ratios greater than 2 for parallel to grain loading and 4 for perpendicular to grain loading, the bolt bends and there is an increasing reduction in proportional limit load with increasing L/D ratio.

Many details such as spacing, edge and end distances, member thickness, member species, and moisture content need consideration in the design of a bolted joint. Design manuals should be checked for these requirements.

5. *Timber Connectors*

Timber connectors are made in a variety of designs and sizes. The most popular are split rings (Fig. 1) and shear plates (Fig. 2). These types of connectors are

Figure 1
Joint with split-ring connector showing connector, precut groove, bolt washer and nut

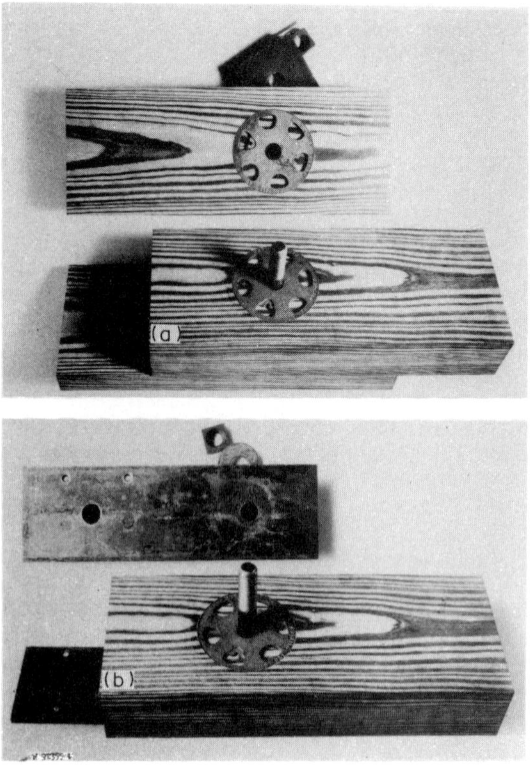

Figure 2
Joints with shear-plate connectors: (a) with wood side plates, (b) with steel side plates

located at the contacting surfaces between joint members and help to distribute the high concentration of stress present in these areas with bolted joints. Their primary use is in heavy timber construction where higher loads are experienced.

The same sort of variables that affect the load capacity of timber connector joints affect the capacity of bolted joints. These include wood species and moisture content, thickness of members and the spacing and location of connectors. Design manuals should be consulted when designing connector joints.

6. *Sheet Metal Connectors*

Many specialty-type connectors are made for use in wood construction. These include such devices as framing anchors, joist and beam anchors, and rafter anchors. These types of devices generally require nails to attach them to the wood and are thus subject to the same variables that affect nailed joints.

The most prevalent sheet-metal connectors are truss plates. These are sheet-metal plates with punched teeth or barbs which are pressed into the wood to form the joint. These connectors are used primarily in the construction of roof and floor trusses. Joints made with these connectors have similar characteristics to laterally loaded nailed joints and are generally affected by the same variables. The manufacturers of these plates have been responsible for establishing their design values and methods of use following industry-established standards.

See also: Building with Wood; Design with Wood; Glued Joints

Bibliography

American Institute of Timber Construction 1985 *Timber Construction Manual*, 3rd edn. Wiley, New York

Gurfinkel G 1981 *Wood Engineering*, 2nd edn. Kendall/Hunt, Dubuque, Iowa, pp. 121–78

National Forest Products Association 1986 *National Design Specification for Wood Construction.* National Forest Products Association, Washington, DC

US Forest Products Laboratory 1987 *Wood Handbook: Wood as an Engineering Material*, USDA Handbook No. 72. US Government Printing Office, Washington, DC, Chap. 7

T. L. Wilkinson
[US Forest Products Laboratory, Madison, Wisconsin, USA]

L

Laminated Veneer Lumber

Wood is a biological material with the capability to optimize itself for survival in the form of a tapered cylindrical stem with branches. Consequently, defects (e.g., knots, slope of grain) occur naturally when logs are sawn into lumber. Clear (defect-free) wood is composed, by nature, of well-oriented cells with a wall structure of helically wound cellulose microfibrils (see *Ultrastructure*). It has, therefore, the highest strength-to-weight ratio in tension among common structural materials. In order to utilize this excellent property of wood at a higher yield and in a greater range of dimensions, the elimination or dispersion of defects by glue jointing or glue lamination is necessary (see *Glued Laminated Timber*; *Glued Joints*).

Laminated veneer lumber (LVL), also known as parallel-laminated veneer (PLV), is one of the most suitable products for this purpose and can be processed with higher yield, and less time and labor, than glued laminated timber or plywood. LVL produced with a continuous press has been approved as an engineered material with reliable strength and stiffness.

1. History of Development

Development of LVL started with the production of high-strength wood aircraft members in the 1940s. Veneers were sometimes impregnated with phenolic resin to meet other requirements such as stability, hardness and screw-holding strength. In the following decades, the machinability and uniformity of mechanical properties of LVL have been appreciated by the furniture industry and used in the production of curved furniture parts.

As the supply of high-quality sawlogs diminished, the importance of LVL increased in many countries due to its higher yield potential, the introduction of automatic production methods, and its adaptability to engineering end-use design. Since the beginning of the 1970s, LVL products have been used as structural members (e.g., tension chords of trusses and outer laminations of glued laminated beams) in place of lumber components because of their reliability in strength.

2. Processing

LVL is processed in a manner similar to plywood (see *Plywood*) but contains only parallel laminations. Rotary-cut veneers are mostly used for LVL laminations. The yield of rotary veneer has much influence on the yield of LVL. Increases in veneer yield have become necessary as the gluability and diameter of available logs diminishes. Newly developed lathes with peripheral driving devices make it possible to peel veneer down to 50 mm core diameter. It has been reported that LVL yield was at least 47% more than sawn lumber yield when green veneer yield was more than 65% of bolt volume. These percentages will increase with the use of new types of lathes.

Veneer is sometimes treated with chemicals such as preservatives and fire retardants before drying. Mixing these chemicals with the adhesive is also effective when thin veneers are used.

Normally, veneer is dried to around 5% moisture content in either conventional hot-air circulating or jet dryers. It is advantageous for LVL production that veneer can be dried with less energy and in less time than an equal volume of laminations for glued laminated timber. Press dryers are also useful in this process and it was found that the drying time required is proportional to the 1.4th power of the veneer thickness. Thinner veneers, therefore, can be dried more economically than thicker veneers. However, production of LVL with thinner veneers requires more adhesive. The balance of these factors as well as the improved properties of LVL achieved with thinner veneers should be considered in the choice of veneer thickness.

Phenolic resin or other adhesives of the same quality are used in producing LVL for structural purposes, while urea resin adhesive is generally used for nonstructural purposes (see *Adhesives and Adhesion*).

Butt joints of veneer ends are significant defects in LVL. The joints, therefore, must be staggered and distributed as evenly as possible. The acceptability of butt joints in LVL depends on the number of laminations. Scarf joints of veneer ends are also employed in some cases. Though LVL with scarf-jointed veneers shows better strength and appearance, the process is not as simple as with butt joints. To obtain high tensile strength with simple process, crushed lap joints (in which lap joints are staggered and high processing pressure is applied to crush overlapping veneer ends) have been used on commercial LVL products (see *Glued Joints*).

Since LVL production started as an extension or variation of the plywood process, most LVLs are hot-pressed in a multiplaten press. In this case, LVL contains no veneer end-joints but it is limited in length to less than 2.5 m. These short LVL members are then scarf- or finger-jointed into longer pieces.

Modern plants for the production of structural LVL tend to employ a continuous system of veneer assembly and hot pressing in which joints are well distributed. This makes the LVL process more labor saving than either the plywood or glued laminated timber processes. A caterpillar press with electrical heating units is employed in the production of LVL (Micro-Lam) flanges with high tensile strength for commercial structural trusses. Roller belt type continuous presses are rather common, one of which is employed in a proposed on high-efficiency production of LVL from thick veneers (Press-Lam) by utilizing residual heat of veneer drying. Other types of continuous presses employ heating devices with radiofrequency heating units or hot platens behind the belt. With these continuous presses, high yield and high efficiency production of endless LVL is feasible.

3. *Mechanical Properties*

Figure 1 shows how variations of strength are reduced in LVL compared to solid lumber from the same logs. The reliability in strength increases with increases in the number of laminations. The improvement in strength is higher when the quality of logs is lower. The same is true with stiffness of LVL and solid lumber.

The allowable design stress f_a can be estimated by the following approximate form:

$$f_a = (f_m - n\sigma)/2.1$$

where f_m is the average strength, σ is the standard deviation, n is a coefficient depending on the shape of the strength distribution, and the denominator 2.1 includes the effect of long-term loading, a safety factor, etc. In general, the 5% exclusion limit f_5 (the stress value below which probability of failure occurrence is 5%) is taken to be $(f_m - n\sigma)$. If the distribution is Gaussian, n is 1.645. A number of experiments with LVL showed that n should be taken as 2.0, considering safety factors in use (Koch 1973, Bodig et al. 1980).

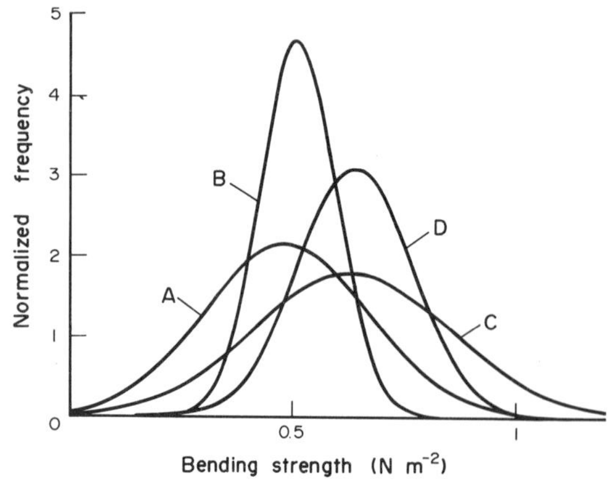

Figure 1
Comparison of bending strength between solid lumber and LVL processed from the same logs. A: solid sawn, 2″ × 4″ Siberian larch; B: 8-ply LVL, same log as A; C: solid sawn, 2″ × 4″, southern pine; D: 6-ply LVL, same log as C

As can be seen in Fig. 1, the estimated allowable stress of LVL will be far greater than that of solid lumber. LVL commercially manufactured from 2.5–3.0 mm thick Douglas-fir veneers (Micro-Lam) with 13 or more plies and crushed lap joints has been officially approved as a material having an allowable stress of 1509 N cm^{-2} in tension and 1921 N cm^{-2} in bending.

Butt joints are serious defects in LVL. However, their effect decreases with increases in the number of plies to just a few percent in 9-ply products and to being negligibly small for 12 or more plies. As derived from fracture mechanics theory, the tensile strength of LVL is inversely proportional to the square root of the lamination thickness. A lateral distance of 16 times the lamination thickness between adjacent butt joints prevents stress interaction (Laufenberg 1983).

The tensile strength of LVL produced by end-jointing shorter pieces is, in general, more than 90% of the strength of unjointed LVL for scarf joints with a 1:9 slope, more than 70% for finger joints of 27.5 mm length, and more than 60% for mini-finger joints of 10 mm length.

Low strength perpendicular to the long axis (i.e., cleavage and shear in the plane parallel to the fibers) is a drawback of LVL. This relates to the lathe checks of veneer created during peeling. However, the average strength increases and the variation decreases with increasing number of plies, and thus the allowable stress perpendicular to grain can be estimated to be the same value as for solid lumber.

4. *Grading and Quality Control*

One of the most promising characteristics of LVL is the capability to ensure its structural performance by grading veneer according to quality and selecting the optimum combination or placement of the veneers. Visual grading of veneers which has been widely used in the plywood industry is not totally satisfactory for screening veneer used in stress-rated LVL. Mechanical methods of testing modulus of elasticity of veneer in tension are used in research, but commercial application is not feasible.

The stress-wave timing (SWT) technique is one of the most promising nondestructive methods of estimating properties of each veneer piece before it is laminated into LVL in the industrial process (Jung 1982). An ultrasonic timer with 50 kHz piezoelectric transducers, for instance, is used to determine the time required for a stress wave to travel the length of each veneer. The dynamic modulus of elasticity is computed as the product of gross mass density of veneer and square of velocity. By this method, the modulus of elasticity of LVL can be predicted with reasonable accuracy in bending applications from the measured stiffness of the veneers composing the member. However, the correlation between stiffness and ultimate strength of LVL is not always good. A grading system for commercial products employs the SWT method to categorize veneers into four stiffness classes.

5. Applications

The furniture and fittings industries prefer LVL because of its ease of processing, uniform properties, good adaptability to curved parts with small radius, stability and better edge appearance. Arch door frames, staircases, chairs, beds, cabinets, counter tables, wardrobes, etc., are popular applications of this material.

Uniform mechanical properties of LVL, and especially its high allowable stress in tension compared to its weight, make this material useful as the flanges of I-beams combined with plywood webs or steel-pipe lattice webs. Such I-beams are widely used as joists in house construction. Open-web trusses combining LVL-steel pipe are used as roof trusses and second-floor joists in large buildings such as factories and warehouses. These open-web trusses are also utilized for arch roofs of very large structures such as a 120 m-span football stadium and exhibition halls. The fields of application are expanding to stringers and decks of highway bridges, stringers of trailer cars, crossties, scaffold planks and to other engineered members in wood construction (Kunesh 1978).

Bibliography

Bodig J, Jayne B A 1982 *Mechanics of Wood and Wood Composites.* Van Nostrand Reinhold, New York

Jung J 1982 Properties of parallel-laminated veneer from stress-wave-tested veneers. *For. Prod. J.* 32(7): 30–35

Koch P 1973 Structural lumber laminated from 1/4-inch rotary-peeled southern pine veneer. *For. Prod. J.* 23(7): 17–25

Koch P 1985 *Utilization of Hardwoods Growing on Southern Pine Sites,* Vol. 3: Agriculture Handbook No. 605. US Government Printing Office, Washington, DC

Kunesh R H 1978 Micro-Lam: Structural laminated veneer lumber. *For. Prod. J.* 28(7): 41–44

Laufenberg T L 1983 Parallel-laminated veneer: processing and performance research review. *For. Prod. J.* 33(9): 21–28

H. Sasaki
[Kyoto University, Kyoto, Japan]

Lignin-Based Polymers

Lignin is nature's second most abundant polymer, behind cellulose, and its annual biosynthesis rate has been estimated to be approximately 2×10^{10} t (see *Chemical Composition*). It is an amorphous polyphenolic component common to vascular plants for which it serves as a reinforcing agent. Its function as a natural adhesive for cellulosic fibers has attracted the attention of the polymer industry, especially of manufacturers of adhesives for the wood composites market (see *Adhesives and Adhesion*).

This article reviews sources and actual and potential uses of lignin in water-soluble and structural polymer systems.

1. Lignin Sources for Polymers

Scientific and technological interest in lignin has focused primarily on the role lignin plays in relation to commercial pulpmaking and papermaking processes. Isolated lignin is best known as the cell wall component that is removed by dissolution from wood during pulping (Sarkanen and Ludwig 1971, Glasser and Kelley 1987). The interest in lignin as a chemical raw material concerns those lignin fractions that can be isolated from "spent pulping liquors," i.e., from solutions in the pulping process that contain dissolved lignin. There are, in principle, two types of commercial pulping processes, and a third which is currently undergoing pilot plant testing (see *Paper and Paperboard*).

1.1 Lignin Sulfonates

Lignin from the traditional (acid) sulfite pulping process is generated as a chemically modified, sulfonated lignin derivative, which is water-soluble by virtue of having sulfonate equivalent weights in the region of 400–600 (or sulfur contents of 6–8%) (Glennie 1971, Rydholm 1965). Solubility in water is achieved by chemical modification with a hydrophilic functional group that is about 10 times as polar and hydrophilic as a hydroxyl group. Chemical (hydrolytic) degradation plays an apparently less dramatic role in lignin dissolution, since the molecular weights of lignin sulfonates remain extraordinarily high. The lesser contribution of molecular fractionation to solubility in water as compared to chemical modification is also reflected in low phenolic hydroxyl contents (i.e., little hydrolytic depolymerization). Lignin sulfonates behave as inorganic, ionic polymers that are virtually insoluble in organic solvents. Certain lignin sulfonate fractions are, however, soluble in mixtures of polar organic solvents with water (e.g., water-saturated butanol). No glass transition T_g or melting T_m temperatures have been reported for lignin sulfonates, and thermal decomposition appears to occur before softening. Lignin sulfonates are generated in mixture with water-soluble carbohydrate degradation products, from which they cannot easily be separated. Industrially, fermentation of reducing C-6 (hexose) sugars and yeasting of C-5 (pentose) sugars are the most widely used "purification" methods (McGovern and Casey 1980, Forss 1968). Less-common treatments involve ultrafiltration, reverse osmosis, electrodialysis and ion exchange separations (Adhihetty 1983, Woerner and McCarthy 1986, Dubey et al. 1965).

Lignin sulfonates can be upgraded by ion exchange, where sodium and ammonium forms are more desirable than calcium and magnesium. Quality specifications for lignin sulfonates concern ionic form, purity (as determined by methoxy and reducing sugar content), molecular weight characteristics and degree of sulfonation.

1.2 Kraft Lignin

The more modern kraft pulping process dissolves lignin from wood with aqueous mixtures of NaOH and Na_2S (Bryce 1980). This treatment involves both derivatization and molecular degradation (depolymerization), where the latter predominates. This typically results in lower molecular weights, narrower molecular weight distributions, and lignins with higher phenolic hydroxy content. Chemical depolymerization involves principally alkyl aryl ether hydrolysis. Uptake of reduced sulfur generates thiol (mercaptan) functionality, and this is the reason why kraft lignin is also sometimes called "thio lignin." Sulfur content reaches 2–3% (Marton 1971). Isolation of kraft lignin from spent kraft pulp liquor involves acidification with carbon dioxide and/or sulfuric acid followed by filtration of the acid-insoluble lignin. This treatment generates a 99.5% pure, completely organic, phenolic polymer with number average molecular weight of ~2500 amu; with a glass transition temperature of ~160 °C; with an ash content of <0.5%; and with limited solubility in organic solvents (aqueous acetone, pyridine, DMF, etc.). Quality specifications for kraft lignin concern sulfur, phenolic hydroxy and methoxy contents, molecular weight, and purity.

1.3 Novel Bioconversion Lignins

The separation of polysaccharides and lignin from woody plant tissues can also be achieved by treatments with such organic solvents as aqueous methanol, ethanol, butanol or phenol (Glasser and Kelley 1987, Lora and Aziz 1985), and with high-pressure steam followed by extraction with alkali or solvent (Chum et al. 1984, Wallis and Wearne 1985). Both process options are sometimes pursued in conjunction with acidic or alkaline solvent mixtures. Lignins derived from these processes are termed "organosolv" and "steam explosion" lignin. Both options generate lignin free of sulfur-containing functionality; with molecular weights of around 800–1000 amu; and with vastly superior solubility in organic solvents as compared to conventional pulp lignin. These, mostly hardwood-derived, bioconversion lignins have high phenolic hydroxy functionality, often high purity, and they are said to be highly reactive. The latter must probably be attributed to their greater solubility in organic solvents, their lower overall molecular weights, and their lower glass transition temperatures as compared to those of other, pulping-derived lignins. These lignins may in the future become available as uncontaminated, organic polymers (or oligomers) with good solubility and thermoplasticity. Quality specifications involve methoxy content, phenolic hydroxy content, molecular weight, molecular weight distribution, glass transition temperature and solubility in organic solvents.

Table 1 summarizes chemical and molecular characteristics of lignins from different sources.

2. *Water-Soluble Polymers*

Lignin sulfonates have achieved prominence among industrial surfactants, and they are the highest volume class of cationic surfactant known to industry. In the late 1970s, sales amounted to approximately 5×10^5 t per year (Chum et al. 1985). While lignin sulfonates from sulfite pulping make up the vast majority of water-soluble lignin derivatives, other types are in use and have been explored as well. Among them are sulfomethylated kraft lignins (de Groote et al. 1987). These have the advantage of better controlled surface active properties by virtue of a lower degree of sulfonation, and greater purity owing to the removal of contaminants during the isolation of parent kraft lignin. There is an approximately 4 to 1 price advantage in favor of the sulfite-pulping-derived raw material. Other water-soluble lignin-based polymers include those with greater concentrations of organic ionic, especially carboxylic, functionalities as a consequence of carboxylation or oxidation (Glasser and Kelley 1987, Lin 1983). The introduction of carboxyl groups has been based on reactions with dicarboxylic anhydrides such as maleic and succinic anhydride, and with chloroacetic acid. Oxidation reactions have been based on the reaction of lignin in aqueous alkali with molecular oxygen under pressure, or with hydrogen peroxide or ozone (Lin 1983). Undesirable secondary reactions often result in repolymerization of lignin leading to higher molecular weights.

A novel route to water-soluble lignin derivatives is based on graft copolymerization of lignin sulfonates with 2-propenamide using anhydrous $CaCl_2$ or cerium (Ce^{4+}) ions for initiation (Meister and Patil 1985). This reaction is illustrated in Fig. 1. Other initiation systems have been examined as well.

3. *Structural Polymers*

For lignin to contribute to the performance of a structural material, lignin must be soluble in a common solvent with other material components, or in the melt of a thermoplastic polymer, or it must be present as finely dispersed particles in a continuous rubber-like matrix. This solubility, miscibility or dispersibility is often the limiting factor in lignin's performance in structural materials. Three general types of structural polymers are distinguished, and these are thermoplastic, thermosetting and filled systems.

Table 1
Characteristics of lignins from different sources

	Lignin sulfonates	Kraft lignin	Novel bioconversion lignin
Purity (%)	55–90	95–>99	90–>95
Sulfur (%)	5–9	2–9	0
Methoxy (%)	9–11	12–14	18–22
Phenolic OH (%)	2–4	4–6	4–6
$\bar{M} \times 10^3$ (amu)	>10	2–4	0.8–1.2
$\bar{M}_w/M_n$	~10	~10	~3
$\bar{T}_g$ (°C)		155–180	90–140
Solvent solubility	water	aq. acetone, dioxane, DMF, DMSO	aq. ethanol, acetone, chloroform, dioxane, DMF, DMSO

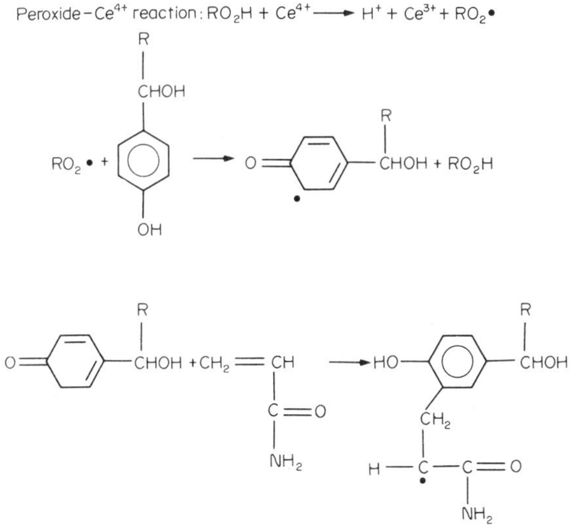

Figure 1
Schematic graft copolymerization of lignin with 2-propenamide (after Sarkanen and Ludwig 1971. © Wiley, New York. Reproduced with permission)

3.1 Thermoplastic Polymer Systems (Polyblends)

Thermoplastic polymer systems involve mixtures of linear polymers which undergo melt flow at elevated temperature, with lignin or a lignin derivative in a uniform, continuous phase (Walsh et al. 1985). Upon cooling, component demixing usually occurs, and multiphase materials, polymer blends or polyblends, are formed in which the two components are separated to a greater or lesser extent. Polyblends can be produced by solution casting, in which both polymer components are dissolved in a common solvent and the solvent is removed from the mixture by (slow) evaporation; or by injection molding of a homogeneous melt mixture. The two polymer components contribute synergistically to each others' properties if some type of polymer–polymer interaction is achieved resulting in "compatibility." Compatibility and miscibility are governed by chemical and molecular factors, such as solubility parameter and molecular weight (Walsh et al. 1985). Thus, both chemical modification and fractionation according to molecular weight may be employed to improve superior blend behavior.

Polyblends of lignin and lignin derivatives have been studied in conjunction with polyethylene, with ethylene-vinyl acetate copolymer, with poly(methyl methacrylate), with poly(vinyl alcohol), with hydroxypropyl cellulose, and with several other (thermoplastic) cellulose derivatives (Ciemniecki and Glasser 1988, Rials and Glasser 1988). Results, in general, have revealed that lignin may contribute stiffness to linear, thermoplastic polymers; and that it may serve as either plasticizer or antiplasticizer, depending on degree of interaction, molecular weight parameters, T_g of the individual components, etc. In general, the degree of interaction was found to be related to the presence of polar functionality in the thermoplastic cocomponent. Thus, whereas polyethylene resulted in complete phase separation with no sign of polymer–polymer interaction and with poor physical properties, vinyl acetate groups, carbonyl groups, and alcohol groups all produced improved interaction (Glasser et al. 1988). Cellulose derivatives displayed a propensity for forming organized morphological features resembling liquid crystal mesophases (Rials and Glasser 1988).

3.2 Thermosetting Systems

The molecular weight and functionality features of lignin enhance its potential as backbone component of network polymers. Among options for cross-linking lignins are phenol-formaldehyde, urethanes, epoxides and acrylates (Glasser 1988). Although not of industrial significance at present, all systems have been described in the recent literature.

The polyphenolic nature of lignin lends itself to association with phenol formaldehyde resins (Nimz 1983). Isolated lignin can be added to commercial phenol formaldehyde resin systems, either as powder, as alkaline solution, or as metholylated or phenolated lignin derivative. All options have been examined, and the contribution by lignin was found to depend on method of formulation. Phenol replacement levels of up to 60% have been reported (Muller et al. 1984), but most common substitution levels which produce acceptable adhesive performance range from 25% to 30%. Phenol formaldehyde type resins with higher substitution levels require water soluble lignin derivatives, mostly lignin sulfonates, which are cross-linkable with either hydrogen peroxide or sulfuric acid (Nimz 1983). These resins cannot be used interchangeably with commercial phenol formaldehyde resin formulations.

Lignin-containing polyurethane products have been described on the basis of lignin fractions soluble in low molecular weight, aliphatic glycols which are then jointly cross-linked with diisocyanates (Yoshida et al. 1987); and based on chemically modified lignin derivatives soluble in cross-linking agent or in a common solvent with the cross-linker (Saraf and Glasser 1984). In wood composites, emulsion systems involving lignin derivatives and polymeric diisocyanates or diamines have also been examined (Newman and Glasser 1985). For many applications, miscibility of lignin or lignin derivative with the cross-linking agent during gelation has been identified as the critical parameter (Glasser 1988).

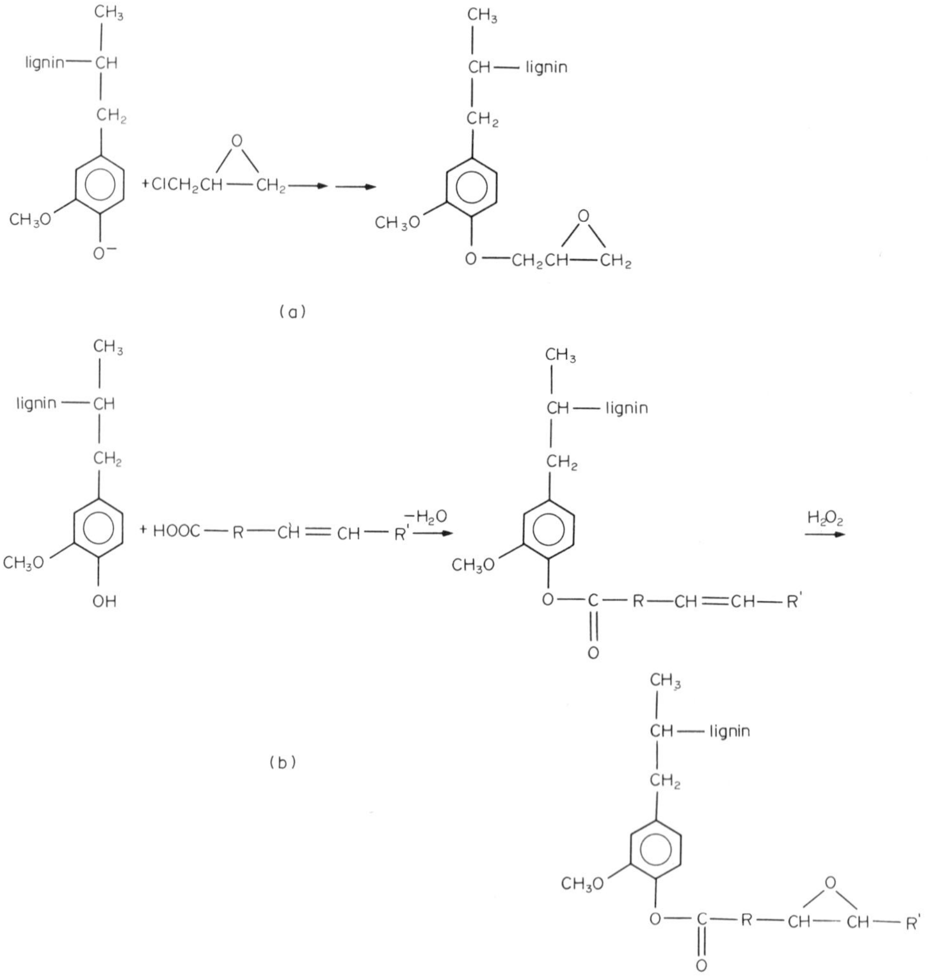

Figure 2
Lignin epoxidation: (a) with epichlorohydrin; (b) by peroxide reaction of an unsaturated ester group (after Glasser and Kelley 1987. © Wiley, New York. Reproduced with permission)

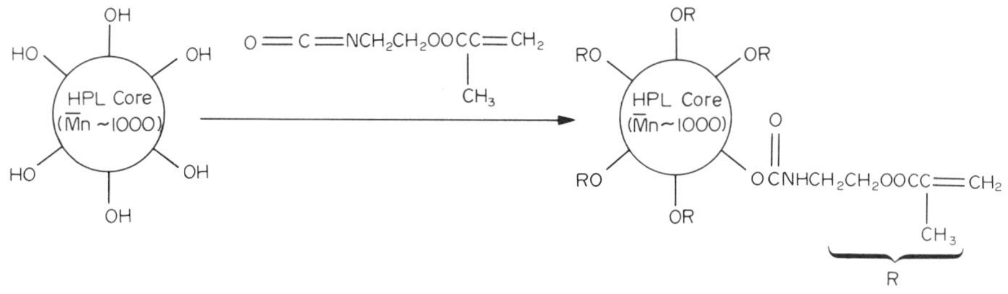

Figure 3
Schematic lignin acrylation reaction

Efforts to convert lignin into epoxy resin systems have concentrated on kraft lignin, phenolated kraft lignin and nonphenolic hydroxy alkyl lignin derivatives by reaction with epichlorohydrin (Tai et al. 1967); and kraft lignin containing unsaturated functionality by reaction with hydrogen peroxide (Holsopple et al. 1981). This is illustrated schematically in Fig. 2. In all cases polymeric, multifunctional epoxy derivatives are formed which can be cross-linked with conventional diamine or anhydride cross-linkers. Solubility and miscibility with a cross-linking agent become again critical performance parameters.

Acrylated lignin derivatives have been prepared from kraft lignin by reaction with acrylic acid chloride and anhydride (Naveau 1975); and by reaction of solid and liquid hydroxy alkyl lignin derivatives with isocyanate-functional methyl methacrylate derivative (Glasser et al. 1988) (Fig. 3). Lignin derivatives with acrylate functionality can be copolymerized with conventional vinyl monomers following normal initiation, such as with peroxides (Glasser et al. 1988). Reactivity ratios have been determined for certain cases, and azeotropic compositions have been established. Results suggest a preference for the formation of alternating copolymers in which copolymerization is favored over homopolymerization of any component.

Isolated lignin usually is a stiff (high modulus of elasticity), glassy polymer which contributes rigidity to polymer systems. Stiff materials typically have low impact resistance, and this has successfully been addressed in segmented materials by the use of toughness-building components (Saraf et al. 1985a,b). This approach, which has been adopted for several thermosetting polymer systems involving lignin, was made possible by the commercial availability of hydroxide, amine, and vinyl terminated polyether and liquid rubber segments. Phase separation was found to be prevalent in most applications, and the desired effect of rubber toughening was easily achieved.

3.3 Filled Systems

Kraft lignin has been found to display surface active properties *vis-a-vis* vulcanized rubber that permit the production of particulate-filled reinforced styrene–butadiene copolymer competitive with carbon-filled tire products (Sirianni and Puddington 1972, Dimitri 1976). Careful control over isolation, post-treatment and co-precipitation methods was found to be required for lignin to achieve the desired dispersibility and reinforcing characteristic.

See also: Cellulose: Chemistry and Technology; Cellulose: Nature and Applications; Chemicals and Liquid Fuels from Wood

Bibliography

Adhihetty T L D 1983 Utilization of spent sulfite liquor. *Cell. Chem. Technol.* 17: 395–99

Bryce J R G 1980 Alkaline pulping. In: Casey J P (ed.) 1980 *Pulp and Paper-Chemistry and Chemical Technology*, 3rd edn. Wiley–Interscience, New York, pp. 377–492

Chum H L, Parker S K, Feinberg D A, Wright J D, Rice P A, Sinclair S A, Glasser W G 1985 *The Economic Contribution of Lignins to Ethanol Production from Biomass*, SERI/TR-231-2488. Solar Energy Research Institute, Golden, Colorado

Chum H L, Ratcliff M, Schroeder H A, Sopher D W 1984 Electrochemistry of biomass-derived materials. I. Characterization, fractionation, and reductive electrolysis of ethanol-extracted explosively-depressurized aspen lignin. *Wood Chem. Technol.* 4(4): 505–32

Ciemniecki S L, Glasser W G 1988 Multiphase materials with lignin. I. Blends of hydroxypropyl lignin with poly (methyl methacrylate). *Polymer* 29: 1021–29

de Groote R A M C, Neumann M G, Lechat J R, Curvelo A A S, Alaburda J 1987 The sulfomethylation of lignin. *Tappi J.* 70(3): 139–140

Dimitri M S 1976 Lignin reinforced polymers. US Patent 3,991,022

Dubey G A, McElhinney T R, Wiley A J 1965 Electrodialysis—Unit operation for recovery of values from spent sulfite lignor. *Tappi* 48(2): 95–98

Fross K 1967 Spent sulfite liquor—An industrial raw material. *Ind. Chem. Belge* 32 (Spec. No.): 405–10

Glasser W G 1988 Crosslinking options for lignins. In: Hemingway R W, Conner A H (eds.) 1988 *Adhesives from Renewable Resources*, ACS Symposium Series No. 385. American Chemical Society, Washington, DC

Glasser W G, Kelley S S 1987 Lignin. In: *Encyclopedia of Polymer Science and Engineering*, Vol. 8, 2nd edn. Wiley, New York, pp. 795–852

Glasser W G, Knudsen J S, Chang C -S, 1988a Multiphase materials with lignin. 3. Polyblends with ethylene-vinyl acetate copolymers. *J. Wood Chem. Tech.* 8(2), 221–34
Glasser W G, Nieh W, Kelley S S, de Oliveira W 1988b Method of producing prepolymers from hydroxyalkyl lignin derivatives. US Patent Appl. 183,213
Glennie D W 1971 Reactions in sulfite pulping. In: Sarkanen K V, Ludwig C H (eds.) 1971 *Lignins: Occurrence, Formation, Structure and Reactions.* Wiley, New York, pp. 695–768
Holsopple D B, Kurple W W, Kurple W M, Kurple K R 1981 Epoxide-lignin resins. US Patent 4,265,809
Lin S Y 1983 Lignin utilization: potential and challenge. *Progr. Biomass Convers.* 4: 31–78
Lora J H, Aziz S 1985 Organosolv pulping—A versatile approach to wood refining. *Tappi J.* 68(8): 94–97
McGovern J N 1980 Silvichemicals. In: Casey J P (ed.) 1980 *Pulp and Paper—Chemistry and Chemical Technology*, 3rd edn. Wiley–Interscience, New York, pp. 492–503
Marton J 1971 Reactions in alkaline pulping. In: Sarkanen K V, Ludwig C H (eds.) 1971 *Lignins: Occurrence, Formation, Structure and Reactions.* Wiley, New York, pp. 639–94
Meister J J, Patil D R 1985 Solvent effects and initiation mechanisms for graft-polymerization on pine lignin. *Macromolecules* 18: 1559–1564
Muller P C, Kelley S S, Glasser W G 1984 Engineering plastics from lignin. 9. Phenolic resin characterization and performance. *J. Adhes.* 17(3): 185–206
Naveau H D 1975 Methacrylic derivatives of lignin. *Cell. Chem. Technol.* 9: 71–77
Newman W H, Glasser W G 1985 Engineering plastics from lignin. 12. Synthesis and performance of lignin adhesives with isocyanates and melamine. *Holzforschung* 39(6): 345–53
Nimz H H 1983 Lignin-based wood adhesives. In: Pizzi A (ed.) 1983 *Wood Adhesives—Chemistry and Applications.* Dekker, New York, pp. 248–88
Rials T G, Glasser W G 1988 Multiphase materials with lignin. 4. Blends of hydroxypropyl cellulose with lignin. *J. Appl. Polym. Sci.* (in press)
Rydholm S A 1965 *Pulping Processes.* Wiley, New York
Saraf W P, Glasser W G 1984 Engineering plastics from lignin. 3. Structure property relationships in solution cast polyurethane films. *J. Appl. Polym. Sci.* 29(5): 1831–41
Saraf V P, Glasser W G, Wilkes G L 1985 Engineering plastics from lignin. 7. Structure property relationships of poly(butadiene glycol)-containing polyurethane networks. *J. Appl. Polym. Sci.* 30: 3809–23
Saraf W P, Glasser W G, Wilkes G L, McGrath J E 1985 Engineering plastics from lignin. 6. Structure property relationships of peg-containing polyurethane networks. *J. Appl. Polym. Sci.* 30: 2207–24
Sarkanen K V, Ludwig C H (eds.) 1971 *Lignins: Occurrence, Formation, Structure and Reactions.* Wiley, New York
Sirianni A F, Barker C M, Barker G R, Puddington I E 1972 Lignin reinforcement of rubber. *Rubber World* 166(1): 40–41, 44–45
Tai S, Nagata M, Nakano J, Migita M, 1967 Lignin. 51. Utilization of lignin. 4. Epoxidation of thiolignin. *Mokuzai Gakkaishi* 13: (3) 102–7
Wallis A F A, Wearne R H 1985 Fractionation of the polymeric components of hardwoods by autohydrolysis–explosion–extraction. *Appita* 38(6): 432–37
Walsh D J, Higgins J S, Maconnachie A (eds.) 1985 *Polymer Blends and Mixtures.* Nijhoff, The Hague
Whittaker R H, Likens G E 1975 The biosphere and man. In: Lieth H, Whittaker R H (eds.) 1975 *Primary Productivity of the Biosphere*, Ecological Studies 14. Springer, Berlin, pp. 305–28
Woerner D L, McCarthy J L 1986 The effect of manipulatable variables on fractionation by ultrafiltration. *AIChE Symp. Ser.* 82: 77–86
Yoshida, H, Morck R, Kringstad K P, Hatakeyama H 1987 Kraft lignin in polyurethanes. 1. Mechanical properties of polyurethanes from a kraft lignin polyether triol polymeric MDI system. *J. Appl. Polym. Sci.* 34: 1187–98

W. G. Glasser
[Virginia Polytechnic Institute and State University, Blacksburg, Virginia, USA]

Lumber: Behavior Under Load

Traditional methods for characterizing and predicting the behavior of lumber under loads have been based on extrapolation of the behavior of small clear wood test specimens by utilizing a series of correction factors. In the USA, techniques for the evaluation of the correction factors were developed by the American Society for Testing and Materials (1986a, b, c). In other countries, both small clear wood specimens and full-size lumber tests are currently used as the basis for characterizing lumber behavior.

Recently the concept of extrapolating lumber behavior from test data on small clear wood specimens has been questioned and in North America several programs for testing full-size lumber have evolved (Bodig 1977, Madsen 1977, Madsen and Barrett 1976). Such full-scale test programs, along with recent developments in improving assessment of the variability of strength properties of wood members, have contributed significantly to the available methodology for evaluation of lumber behavior under load.

1. Characterization of Lumber as a Structural Material

To understand the behavior of lumber as a structural material, it is important to keep in mind the basic cellular structure, the spatial arrangement of wood substances in the cell wall, and the biological origin of the material (see *Ultrastructure*; *Macroscopic Anatomy*). Various conditions associated with geographical origin, soil type and precipitation, for example, all contribute to variations in basic material properties related to structural use.

The term lumber denotes the commercial product sawn from logs into dimensions such as 2 × 4s (50 × 100 mm) and 1 × 10s (25 × 250 mm) and so on. Historically, lumber has been a primary product of much of the timber harvested. It is used in construction of all types, particularly in light-frame construction of houses and commercial buildings of limited height. The major constituent of lumber is clear wood, but it also contains numerous natural and manufacturing characteristics often termed "defects." Lumber is classified by species and according to some system of sorting, termed "grading" (see *Lumber: Types and Grades*). The effects of defects, grading and additional factors make the characterization of lumber a complex problem. A list of some of the more important influencing factors in addition to those of species and clear wood properties is presented in Table 1. Examples of the large differences in properties of lumber used in structural design and of clear wood are given in Table 2 for two selected commercially important species.

2. *Experimental Studies of Lumber Behavior*

Much current research is aimed at theoretical and experimental developments which involve the direct assessment of lumber behavior under loads as opposed to present indirect assessment using results of small clear specimen tests. Extensive programs involving "in-grade" testing and evaluation of lumber behavior have been developed in recent years in North America. These programs have been aimed at obtaining characteristic values of the strengths and stiffnesses of the various grades. Both laboratory and field (at lumber mill sites) investigations have been and are currently being conducted. Although several different testing techniques have been involved, a number of general patterns have emerged.

Table 1
Factors which influence lumber behavior in addition to basic clear wood properties

Factor	General effect
Presence of growth characteristics (e.g., knots, cross grain, checks)	reduction of strength and to a lesser degree stiffness
Presence of manufacturing defects (e.g., drying splits, wane)	reduction of strength and to a lesser degree stiffness
Size effects	reduction in strength with volume and/or depth of member increase
Grading	variable strength and stiffness effects depending on grading method
Duration of load	effects of load duration on small clear wood members are not reproduced in lumber, especially for lower grades
Treatment and environmental effects (e.g., moisture, decay, weathering)	varying strength and stiffness reductions depending on type of treatment and severity of environment

Table 2
Comparison of selected clear wood and lumber design properties[a]

Species	Modulus of rupture (MPa)	Modulus of elasticity (GPa)	Compression[b] (MPa)
Douglas-fir (coast)			
clear wood[c]	67.5	11.9	36.5
no. 2 lumber[d]	10	11.7	7
Southern pine (loblolly)			
clear wood[c]	67	10.8	35
no. 2 lumber[d]	9.5	11.0	6.5

[a] At 19% moisture content [b] Parallel to grain [c] Data from Forest Products Laboratory (1987) [d] Data from National Forest Products Association (1986)

Typical bending test failures of two lumber members are shown in Fig. 1. Note how failure is associated with knots and cross grain. Results of these tests tend to show significant differences between the "select" (upper) and the lower grades, but do not substantiate large differences among the lower grades. A trend line illustrating this effect is shown in Fig. 2. Results from in-grade testing of lumber have raised additional questions concerning the applicability of current methods of grading. The use of a "strength ratio"—the ratio of the strength of the member reduced in cross section by the largest knot (permitted for the grade) to that of the member without the knot—has traditionally been the basis of lumber strength values. Tests of lumber, particularly of larger sizes, have shown a very large scatter in actual "strength ratios" as opposed to the trends predicted.

Size and volume effects have been noted in many tests of lumber and other wood products. In lumber, the effects of size have been evidenced in the results of in-grade testing programs. A trend line for this effect is shown in Fig. 3. Effects of size depend on grading practices, volume of the lumber member, species and other variables which have not been fully quantified. Reductions of as much as 30% have been noted in comparing 2 × 12s with 2 × 4s (50 × 300 mm with 50 × 100 mm), for example.

Duration of load has traditionally been considered in assigning lumber strength values (US Forest Products laboratory 1987). The quantification of this effect was based on clear wood tests which indicated strength reductions of as much as 60% for ten year load duration, as opposed to short-term test results of a few minutes' duration. Tests of lumber have not produced similar results, particularly for the lower grades. It appears that the strengths of these lower-grade members are controlled by knots and cross grain; such members thus behave less critically under

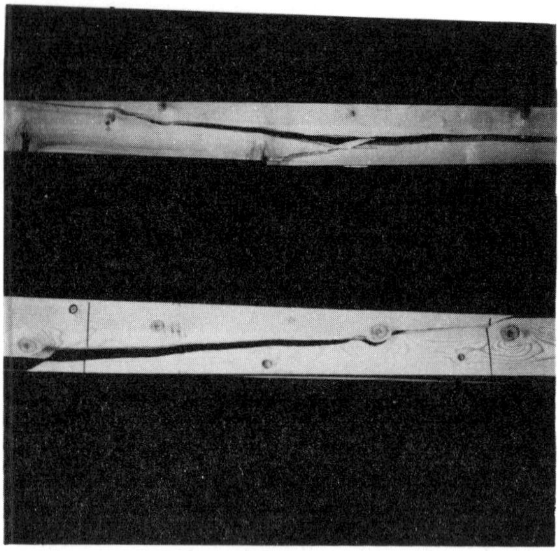

Figure 1
Typical bending-test failures of softwood lumber members

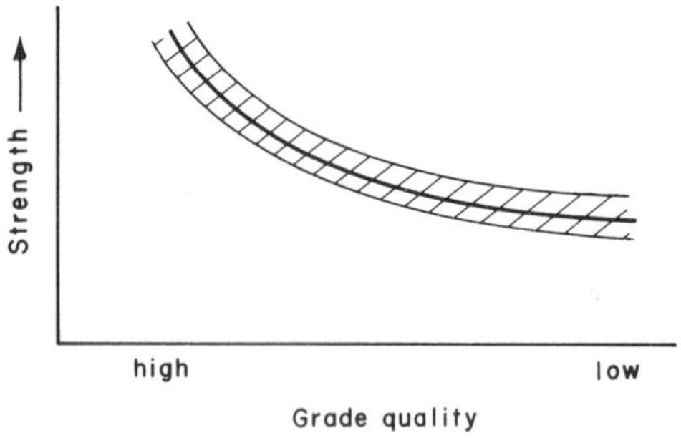

Figure 2
Relation of strength to grade quality for visually graded lumber

long-term loads than the clear wood samples. Research is continuing on duration-of-load effects for full-size lumber members.

Effects of moisture content on the strength behavior of lumber members can be considerable (up to 16% assigned in traditional grading). Although trends for clear wood have been known for some time (US Forest Products Laboratory 1987), characterization of lumber behavior is not fully established. Test results for lumber indicate that the stronger pieces generally follow established trends for clear wood while the lower-strength pieces do not show substantial reductions in strength with increased moisture content.

Another inherent concern regarding use of lumber for structural applications relates to its variability of strength and stiffness. An example of the variability of bending strength of 2 × 8 (50 × 200 mm) lumber is

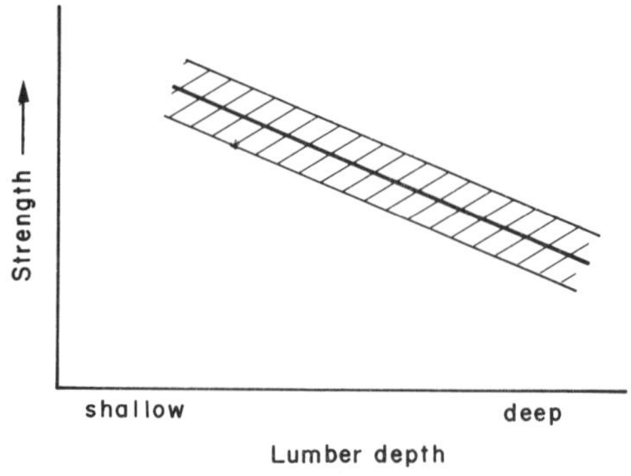

Figure 3
Relation of bending strength to beam depth for lumber

shown in Fig. 4. Design stresses have traditionally been based on the lower 5% exclusion limit (i.e., only 5% of lumber pieces are of lesser strength), while stiffness properties have been taken to be the mean values. Characterization of variability of lumber using statistical methods is becoming increasingly important with the advent of reliability-based design methods. A statistical distribution has been fitted to the histogram shown in Fig. 4, which also illustrates the 5% exclusion limit for this case. Further research to evaluate the variability of lumber properties is continuing and the trend is to give greater recognition to this variability in design.

Although much information is still being collected on "in-grade" behavior of lumber, on variations in load (particularly for light-frame structures) and on strength of wood structural systems, a unified approach is emerging. An exciting new era in which designers will have the tools with which to predict more accurately the true performance of wood structures has arrived with the advent of reliability analysis methods. Further efforts to characterize lumber and wood structural system behavior are needed to bring the use of these methods to full fruition.

Recent efforts to improve the characterization of lumber behavior have expanded the scope of mathematical modelling and nondestructive evaluation. The automated measurement of localized slope-of-grain of an individual piece of lumber has enabled quantification of the stress distribution around knots and the resulting effect on material behavior (McDonald and Bendtsen 1985). Quantification of the stress distribution around knots has been assessed using finite element/fracture mechanics techniques that enabled the determination of both failure mechanism and ultimate load in a knot-containing wood member under tension (Cramer and Goodman 1985). Such advances in modelling behavior under load is likely to lead to nondestructive evaluation procedures which can provide a direct, reliable assessment of the strength of individual pieces of lumber. More accurate

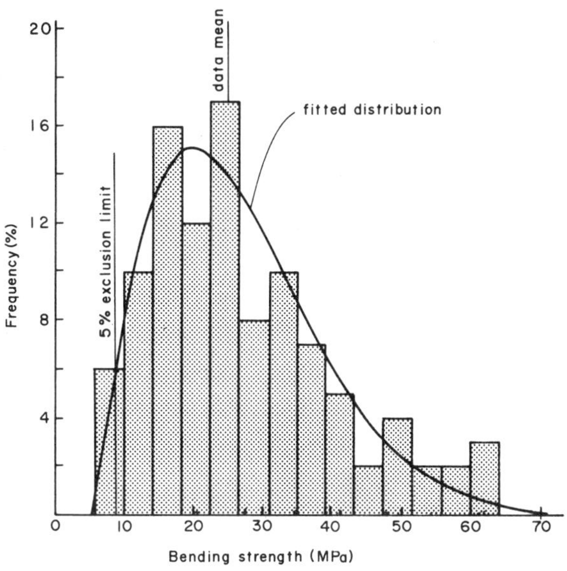

Figure 4
Frequency distribution for bending strength of 2 × 8 (50 × 200 mm) Douglas-fir larch lumber

means of assigning values of lumber strength and stiffness become increasingly important as the wood industry moves toward reliability-based design procedures.

3. *Traditional Methods for Assigning Values of Lumber Strength and Stiffness*

Behavior of lumber under loads has traditionally been predicted by assigning "allowable stresses" and mean values of stiffness (modulus of elasticity). The basis for the assignment of strength properties has been the application of a system of sorting called stress grading, either through a process of visual grading or more recently by mechanical procedures. Each stress grade is characterized by a set of sorting (grading) criteria, a grade name and a set of assigned strength and stiffness properties.

In the USA the principles of visual stress grading of lumber are embodied in ASTM D 245 (American Society for Testing and Materials 1986b). Starting with small clear wood specimen test results, strength values are modified by a number of factors accounting for variability: duration of load, k_t; safety, k_s; special conditions, k_p; special grading, k_g; moisture, k_m; and strength-reducing defects, k_d. Variability is incorporated by using the lower 5% exclusion limit, l_5, as the basis of computation. The reduction factor for defects, k_d, is termed the "strength ratio" as previously defined. Determination of the strength ratios is based mostly on experimental results, with some extrapolation for the particular characteristics permitted for the grade. Thus, in most cases, the allowable stress F is derived by the relationship

$$F = l_5 k_t k_s k_p k_d k_g k_m \quad (1)$$

For compression perpendicular to the grain, the allowable stress is based on the mean proportional limit stress. Similarly, the mean modulus of elasticity (including a shear deformation component of approximately 10%) is used. The various correction factors are provided in ASTM D245; different values are assigned for softwoods and hardwoods.

In describing a stress grade, lumber is classified according to its use, a description of permissible growth characteristics, and allowable properties for the grade related to its strength ratio. The following cases have been tabulated in ASTM D245 for determining strength ratios.

(a) slope of grain for bending, and tension and compression parallel to grain;

(b) size and location of knots for bending and compression members;

(c) size of shakes, checks and splits for horizontal shear in bending members.

The tabulated strength ratios are obtained as described below for bending and compression members.

(a) Bending members
  (i) Knots: strength ratio is derived as the ratio of the moment-carrying capacity of the member reduced in cross section by the largest knot to that without defect.
  (ii) Shakes, checks and splits: strength ratio is derived similarly to knots, except that the effect is confined to horizontal shear only; the critical cross section of the member with defect is reduced in area by subtracting the area of the defect in question from the gross section in deriving the strength ratio.

(b) Compression members
  (i) Knots: strength ratio for compression parallel to the grain is derived similarly to that derived for bending members.
  (ii) All defects: for compression prependicular to the grain, the strength ratio is assumed to be 100%; that is, no effect of defects is considered.

Allowable properties and unit stresses for design for the stress grade are obtained following the established procedures, once the limiting characteristics have been determined for the stress grade involved. The derived allowable unit stresses (design stresses) are tabulated in codes such as the National Design Specification for Wood Construction (National Forest Products Association 1986) for the USA, or similar codes for other countries. Additional correction factors for the use of

Table 3
Example design values for visually graded structural lumber[a] (adapted from National Forest Products Association 1986)

Commercial grade	Size classification	Extreme fiber in bending (MPa)		Tension parallel to grain (MPa)	Horizontal shear (kPa)	Compression perpendicular to grain (MPa)	Compression parallel to grain (MPa)	Modulus of elasticity (GPa)
		Single-member uses	Repetitive-member uses					
Dense select structural	50-100 mm wide and thick	17	19.5	9.5	655	3.15	13	13
Select structural		14.5	16.5	8.5	655	2.65	11	12.5
Dense no. 1		14	16.5	8.5	655	3.15	10	13
No. 1		12	14	7	655	2.65	8.5	12.5
Dense no. 2		11.5	13.5	7	655	3.15	8	11.5
No. 2		10	11.5	6	655	2.65	7	11.5
No. 3		5.5	6.5	3.25	655	2.65	4	10.5
Appearance		12	14	7	655	2.65	10.5	12.5
Stud		5.5	6.5	3.25	655	2.65	4	10.5
Construction	50–100 mm thick, 100 mm wide	7	8.5	4.25	655	2.65	8	10.5
Standard		4	4.5	2.5	655	2.65	6.5	10.5
Utility		2	2.25	1.25	655	2.65	4	10.5

[a] Douglas fir-larch (surfaced dry or surfaced green, used at 19% max. m.c.)

lumber in conditions other than those assumed in the derivation of allowable unit stress values are also provided. Example values for visually graded lumber are listed in Table 3. Data on lumber and clear wood properties may also be found in numerous other publications (American Institute of Timber Construction 1985, Kollmann and Côté 1968, Forest Products Laboratory 1987, Bodig and Goodman 1973, Bodig and Jayne 1982).

As an alternative to visual stress grading, machine stress rating has also been employed. In this method, a nondestructive evaluation of lumber stiffness is made, most often by deflecting or vibrating the piece, usually in a flatwise mode. This is then coupled with a visual override to check for edge and end defects and the strength of the accepted pieces is set by inference, using a previously established correlation between strength and stiffness.

In mechanical grading, several of the factors listed for visually graded lumber (Eqn. (1)) are automatically incorporated into the evaluation procedure, since the actual material is tested. Thus, in this case the list of corrections is reduced to

$$F = l_5 k_t k_s k_p \tag{2}$$

Example values for machine stress-rated lumber are provided in Table 4.

4. Load and Resistance Factor Design for Engineered Wood Construction

Concepts for structural design have undergone radical changes in philosophy, especially in recent years. Major research efforts have been mounted by professional engineering and academic communities to develop procedures to apply reliability-based design (RBD) to structures. Reliability-based design recognizes that, in reality, both the material resistance and the loads placed on a structure are variable quantities. The probability of failure of a structure is a function of load and resistance distributions. The area of overlap

Table 4
Design values for machine stress-rated structural lumber[a] (adapted from National Forest Products Association 1986)

Grade designation	Extreme fiber in bending (MPa): Single-member uses	Extreme fiber in bending (MPa): Repetitive-member uses	Tension parallel to grain (MPa)	Compression parallel to grain (MPa)	Modulus of elasticity (GPa)
900f-1.0E	6	7	2.5	5	7
1200f-1.2E	8.5	9.5	4	6.5	8.5
1350f-1.3E	9.5	10.5	5	7.5	9
1450f-1.3E	10	11.5	5.5	8	9
1500f-1.3E	10.5	12	6	8.5	9
1500f-1.4E	10.5	12	6	8.5	9.5
1650f-1.4E	11.5	13	7	9	9.5
1650f-1.5E	11.5	13	7	9	10.5
1800f-1.6E	12.5	14	8	10	11
1950f-1.5E	13.5	15.5	9.5	10.5	10.5
1950f-1.7E	13.5	15.5	9.5	10.5	11.5
2100f-1.8E	14.5	16.5	11	11.5	12.5
2250f-1.6E	15.5	18	12	12.5	11
2250f-1.9E	15.5	18	12	12.5	13
2400f-1.7E	16.5	19	13.5	13.5	11.5
2400f-1.0E	16.5	19	13.5	13.5	14
2550f-2.1E	17.5	20.5	14	14	14.5
2700f-2.2E	18.5	21.5	15	15	15
2850f-2.3E	19.5	23	16	16	16
3000f-2.4E	20.5	24	16.5	16.5	16.5
3150f-2.5E	21.5	25	17	17	17
3300f-2.6E	23	26	18.5	18.5	18

[a] Machine rated lumber 50 mm thick or less; all widths

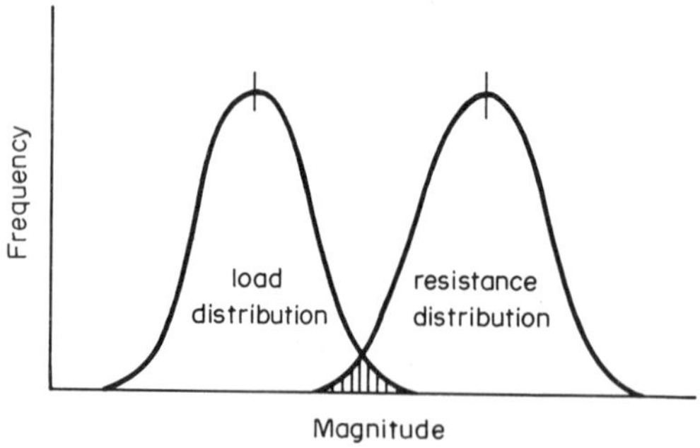

Figure 5
Load effects–resistance relationships used in reliability-based design

of the two distributions shown in Fig. 5 relates to the probability of failure. Hence, fundamental to the RBD process is the definition of material resistance distribution.

Reliability-based design of steel structures was developed through an extensive industry-sponsored effort and through the American Institute of Steel Construction (AISC 1986). In addition, the concrete building design code (ACI 1983) gives the appearance of being up-to-date since it has a load and resistance factor design (LRFD) format and uses an ultimate strength approach. The wood industry is currently evaluating design procedures which represent this modern approach to solving structural problems.

Numerous RBD research and development efforts have been conducted, including the completion of a prestandardization document under the auspices of the American Society of Civil Engineers Committee on Wood (ASCE 1987). The availability of this publication by early 1988 makes it a valuable resource for use as background for an RBD-based LRFD code for engineered wood construction.

The current direction of RBD code development and promotion has reached the level of unstoppable forward inertia, both in the USA and elsewhere. Nearly every major country already has or is establishing an RBD code. European (Crubile et al. 1985) and Canadian (CSA 1984) LRFD-type codes are already in existence for engineered wood products, although they are not RBD-based in every case. In the USA a designer is faced with using an RBD-based LRFD procedure for steel, a pseudo-RBD LRFD procedure for concrete, and an allowable stress, non-LRFD procedure for wood products. The completion of the in-grade testing program will provide material resistance data for the development of an LRFD design procedure for the wood industry.

See also: Deformation Under Load; Nondestructive Evaluation of Wood and Wood Products; Strength; Structure, Stiffness and Strength

Bibliography

American Concrete Institute Committee 318 1983 *Building Code Requirements for Reinforced Concrete*, ACI 318–83. ACI, Detroit, Michigan

American Institute of Steel Construction 1986 *Load and Resistance Factor Design Specification for Structural Steel Buildings.* AISC, Chicago, Illinois

American Institute of Timber Construction 1985 *Timber Construction Manual; A Manual for Architects, Engineers, Contractors, Laminators, and Fabricators concerned with Engineered Timber Buildings and other Structures*, 3rd edn. Wiley–Interscience, New York

American Society for Testing and Materials 1986a *Standard Methods of Testing Small Clear Specimens of Timber*, ASTM Designation D143-83. American Society for Testing and Materials, Philadelphia, Pennsylvania

American Society for Testing and Materials 1986b *Standard Methods for Establishing Structural Grades and Related Allowable Properties for Visually Graded Lumber*, ASTM Designation D245-81. American Society for Testing and Materials, Philadelphia, Pennsylvania

American Society for Testing and Materials 1986c *Standard Methods for Establishing Clear Wood Strength Values*, ASTM Designation D2555-81. American Society for Testing and Materials, Philadelphia, Pennsylvania

American Society of Civil Engineers 1987 *Reliability-Based Design of Engineered Wood Structures*, Task Committee of Committee on Wood. ASCE, New York

American Society of Civil Engineers Structural Division 1975 *Wood Structures: A Design Guide and Commentary.* American Society of Civil Engineers, New York

Bodig J 1977 *Bending Properties of Douglas-Fir, Larch and Hem-Fir Dimension Lumber*, Special Report No. 6888. Department of Forest and Wood Sciences, Colorado State University, Fort Collins, Colorado

Bodig J, Goodman J R 1973 Prediction of elastic parameters for wood. *Wood Sci.* 5: 249–64

Bodig J, Jayne B A 1982 *Mechanics of Wood and Wood Composites.* Van-Nostrand Reinhold Company, New York

Canadian Standards Association 1984 *Engineering Design in Wood* (Limit States Design), ISSN 0317-5669

Cramer S M, Goodman, J R 1985 Predicting tensile strength of lumber. In: *5th Nondestructive Testing of Wood Symp.* Washington State University, Pullman, Washington

Crubile P, Ehlbeck J, Brunighoff H, Larsen H J, and Sunley J 1985 *Common Unified Rules for Timber Structures*, EUROCODE 5, Report prepared for the European Communities

Gurfinkel G 1981 *Wood Engineering*, 2nd edn. Kendall–Hunt, Dubuque, Iowa

Hoyle R J Jr 1978 *Wood Technology in the Design of Structures*, 4th edn. Mountain-Press, Missoula, Montana

Kollmann F E P, Côté W A Jr 1968 *Principles of Wood Science and Technology*, Vol. I. Springer, Berlin

McDonald K A, Bendtsen B A 1985 Localized slope of grain—Its importance and measurement, In: *5th Nondestructive Testing of Wood Symp.* Washington State University, Pullman, Washington

Madsen B 1977 *In-Grade Testing: Problem Analysis*, Structural Research Series; Report No. 18. Department of Civil Engineering, University of British Columbia, Vancouver

Madsen B, Barrett J D 1976 *Time-Strength Relationship for Lumber*, Structural Research Series; Report No. 13. University of British Columbia, Vancouver

National Forest Products Association 1986 *National Design Specification for Wood Construction*. National Forest Products Association, Washington, DC
US Forest Products Laboratory 1987 *Wood Handbook: Wood as an Engineering Material*, Agriculture Handbook 72. US Government Printing Office, Washington, DC
Wood L W 1951 *Relation of Strength of Wood to Duration of Load*, US Forest Products Laboratory Report No. R 1916. US Forest Products Laboratory, Madison, Wisconsin

R. W. Anthony
[Engineering Data Management, Fort Collins, Colorado, USA]

J. R. Goodman
[Colorado State University, Fort Collins, Colorado, USA]

Lumber: Types and Grades

The custom of sorting lumber into grades has a long history of progressive evolution, beginning with formal rules published in Sweden in 1764 and in the USA in the 1830s. However, it is quite possible that the practice of lumber grading predates these periods since it is hard to imagine a material as old as timber being used without some effort to appraise its structural character from the earliest times. The rules developed in Sweden and in the state of Maine have been continuously revised and enlarged in scope to keep pace with the changing economy and the demand for material for more and more exacting use and reliability. Lumber manufacturers have played the principal role in grade development in the USA, but government has often been instrumental in motivating rule development, since it is a frequent large consumer of wood products for military and civil projects. Industries which use large amounts of wood have sometimes prepared grade specifications which have ultimately become incorporated into manufacturers' grades.

Since wood is a commodity, grades must be standardized if comparable products are to be produced by the many mills which are characteristic of the lumber business. This is essential to free trade across broad regions. The standards are not international, but tend to be centered in such trading areas as the countries of the European Economic Community and North America. The North American systems employ the English system of units of measurement, whereas the other countries of the world nearly all use the metric system. Plans to convert English to metric units have been developed and could be put into use whenever consumer demand prevails. This article is mainly concerned with the North American systems.

Lumber is usually graded by visual inspection for conformance to written descriptions. This is almost universal in hardwood lumber grading and is basic to most softwood lumber grading. Since about 1960 the use of nondestructive evaluation, to measure the stiffness of each piece of lumber and thereby provide an added measure of quality, has gained widespread acceptance in many parts of the world (see *Nondestructive Evaluation of Wood and Wood Products*). This has added to the reliability and uniformity of certain structural wood products. The derivation of properties for the visual grading methods is discussed elsewhere (see *Lumber: Behavior Under Load*).

1. Nonstructural Lumber

There is a large market for many lumber products where structural performance is not the major requirement. Most hardwood and many softwood lumber uses depend on other features. Users who plan to cut lumber into smaller pieces for other products are interested in the size and quantity of relatively clear pieces obtainable from the boards purchased. The yield of clear cuttings is therefore the first consideration. Other users who may not reduce the lumber to smaller pieces may have a very particular end use in mind which is not structural, flooring being a good example.

1.1 Hardwood Lumber Grades

Hardwood lumber grades fall into three general categories: factory grades; hardwood dimension grades; and finished market products.

Factory grades, also called "cutting" grades, are produced in thicknesses up to 150 mm and sorted according to the percentage of pieces of a given quality which the lumber will yield. The pieces are produced in length increments of 300 mm, with the minimum piece width depending on grade level. The sawmill may provide a variety of widths in filling an order, as long as the individual pieces meet the grade requirements. Shipments may also be of random lengths, provided that the percentage of clear cuttings of the required size can be obtained from the boards. Standard hardwood cutting grades are listed in Table 1, based on rules of the US National Hardwood Lumber Association.

These grades can be obtained either seasoned or unseasoned, at the customer's option. Many buyers prefer to do their own drying to ensure the dryness of the material prior to surfacing and cutting to final size. Product requirements also determine the moisture-content specifications which the buyer wishes to control. Usually these grades are purchased with sawn rather than dressed (regularized) surfaces. Furniture lumber is a good example of such requirements.

Hardwood dimension grades are lumber that has already been cut from factory grades and is nearly ready to use in the sizes furnished by the supplier. It is produced in three general classes: solid dimension flat

Table 1
Standard hardwood cutting grades (adapted from Forest Products Laboratory 1987)

Grade[a] and lengths allowed (m)	Widths allowed (mm)	Surface measure of pieces (m^2)	Amount of each piece that must work into clear-face cuttings (%)	Number of maximum cuttings allowed	Minimum size of cuttings required (mm)
Firsts: 2.4–4.9	⩾150	0.37–0.84	91.67	1	100 × 1520 or 75 × 2130
		0.93–1.30	91.67	2	
		⩾1.39	91.67	3	
Seconds: 2.4–4.9	⩾150	0.37, 0.46	83.33	1	100 × 1520 or 75 × 2130
		0.56, 0.65	83.33	1	
		0.56, 0.65	91.67	2	
		0.74–1.02	83.33	2	
		0.74–1.02	91.67	3	
		1.11–1.39	83.5	3	
		1.11–1.39	91.67	4	
		⩾1.69	83.33	4	
Selects: 1.8–4.9	⩾100	0.19, 0.28	91.67[b]	1	100 × 1520 or 75 × 2130
		⩾0.37			
No. 1 Common: 1.2–4.9	⩾75	0.09	100	0	100 × 610 or 75 × 915
		0.19	75	1	
		0.28, 0.37	66.67	1	
		0.28, 0.37	75	2	
		0.46–0.65	66.67	2	
		0.46–0.65	75	3	
		0.74–0.93	66.67	3	
		1.02–1.21	66.67	4	
		⩾1.30	66.67	5	
No. 2 Common: 1.2–4.9	⩾75	0.09	66.67	1	75 × 610
		0.19, 0.28	50	1	
		0.19, 0.28	66.67	2	
		0.37, 0.46	50	2	
		0.37, 0.46	66.67	3	
		0.56, 0.65	50	3	
		0.56, 0.65	66.67	4	
		0.74, 0.84	50	4	
		0.93, 1.02	50	5	
		1.11, 1.21	50	6	
		⩾1.30	50	7	
No. 3A Common: 1.2–4.9	⩾75	⩾0.09	33.33	unlimited	75 × 610
No. 3B Common: 1.2–4.9	⩾75	⩾0.09	25	unlimited	40 × 610

[a] Firsts and Seconds are combined as one grade (FAS); all grades allow maximum percentages below certain lengths, these tolerances distinguishing Nos. 1, 2 and 3 Common [b] Same as for Seconds

stock; kiln-dried dimension flat stock; and solid dimension "squares". It may have sawn or smoothly surfaced faces, at the buyer's option.

There are five grades of flat stock: Clear, Clear-one-face, Paint, Core and Sound. The two clear grades must be clear on the faces, ends and edges. Paint grade admits defects that can be covered by opaque paint. Core grade permits tight, solid knots, small worm holes and small seasoning checks and may have plugged or patched portions. The objective is a solid material with voids that can be bridged by an overlay of veneer or composition material. Sound grade permits larger defects and surface skips in dressed pieces.

Dimension "squares" may be up to twice as wide as they are thick. There are four grades: Clear, Select, Paint and Sound. Select grade must have two clear faces which are adjacent, with the other face clear for one-third the length from either end and small, sound defects in the remaining two-thirds. The other grade descriptions are as for flat stock.

Hardwood dimension can be made from smaller pieces, end and/or edge glued to make larger pieces. In some cases dimension is furnished as completely finished parts, ready to assemble.

Finished market products of hardwood are graded to fit the end use. Typical of such products are hardwood flooring, railroad-car flooring, architectural moldings, tool handles, railway ties (sleepers) stair treads, construction boards and box and pallet materials. There are individual grade rules for such products. Softwood finished market products include all of the above plus such items as scaffold plank, ladder materials, tank and pipe materials, house siding (cladding) and trim.

1.2 Softwood Lumber

Nonstructural softwood lumber includes a large variety of grades for purposes similar to those for nonstructural hardwoods, but directed at a different group of end-use products. The general types are Appearance grades and Factory or Shop grades.

Appearance grades are manufactured in two groups: Appearance Framing and Selects, and Finish and Boards. Appearance Framing is dimension lumber (50–100 mm nominal thickness). It is available in one grade which is basically No. 1 Structural Light Framing or Structural Joist and Plank, with more limiting appearance features. Sizes are 50–100 mm nominal thickness and 50–350 mm nominal width. It is stress-rated to the same values as the No. 1 grades mentioned above, so is really both a nonstructural and a structural grade. It is used where structural members which will be exposed must be of high appearance quality.

Selects are nominally 25–100 mm thick and 50 mm or more wide. The grades are B and Better (also called 1 and 2 Clear), C-Select and D-Select. They are used for interior walls, architectural woodwork, trim, moldings and cabinets.

Finish grades are the same size as Selects or may be special patterns for such uses as ceilings, door casings, base boards, siding (cladding), flooring, panelling, lath, stepping (stair treads) and other styles. The basic grades for rectangular stock are: Superior, Prime and E-finish. Superior and Prime can be obtained as flat grain (FG), vertical grain (VG) or mixed grain (MG). Some grading agencies use other names such as C and Better, D and E.

Boards are lower in quality than Selects or Finish and are available in the same sizes. The grades in descending order of quality are: No. 1 Common, No. 2 Common, No. 3 Common, No. 4 Common and No. 5 Common. Some agencies use the grade names Select Merchantable, Construction, Standard, Utility and Economy, in order of decreasing quality. The grades of different agencies are not exact counterparts of one another but show a rough equivalence.

Factory and Shop grades are cutting grades. Shop grades 25 mm thick are: Select, No. 1, No. 2 and No. 3. Many thicknesses and widths are manufactured, largely oriented to the needs of the millwork industry, doors, window frames and sash, and all the various components of these products. The grade names are too numerous to summarize.

Industrial Clears are very high-quality cutting grades, suited to cabinet and door manufacturing. The grades are B and Better Industrial, C-Industrial and D-Industrial. They are Clear-one-face with flat, vertical or mixed grain options to describe annual-ring orientation. They are not included in all agency rules, however.

Nonstructural softwood lumber grades are not standardized to the extent of hardwood or structural softwood grades. The practices reflect the needs of the consumers served by the different regional agencies and are affected by the species which their member mills produce and their concepts of how best to serve their customers.

2. Structural Lumber

Structural lumber is the most uniformly and highly standardized type of wood product because it is used almost exclusively for buildings and other structures. The grade descriptions in the USA and Canada are virtually identical. Structural lumber grades have complete sets of allowable design properties for strength and stiffness. The physical description of each grade is the same for all species of wood. This means that wood of low-density species has lower structural properties than more-dense species of the same grade. Exceptions to this are the machine stress-rated grades, which have the same bending, tension and compression-parallel-to-grain properties and the same elastic modulus for each grade, independent of species. In consequence, the lower-density woods of a given grade of this kind of lumber tend to have smaller defects and possibly a slightly better appearance. Furthermore, the higher grades of machine stress-rated lumber are unavailable in the lower-density species. Machine stress-rated lumber is used primarily for structural products such as trusses, glued laminated timber, I and box beams, ladders, scaffolds, electric-utility cross arms and cooling-tower columns. The machine stress rating of lumber is discussed elsewhere (see *Nondestructive Evaluation of Wood and Wood Products*). A table of allowable stresses for machine stress-rated lumber can be found in the article *Lumber: Behavior under Load*.

Visually stress-graded lumber accounts for most of the structural softwood lumber in common use. There are three general types:

(a) boards—lumber less than 50 mm in nominal thickness;

(b) dimension—lumber from 50 to 100 mm in nominal thickness;

(c) timbers—lumber 125 mm or more in nominal thickness (least dimension).

The nominal thickness is not the real thickness but is a size used in determining the quality of lumber for ordering and pricing purposes. It is a rough approximation of the unseasoned and rough-sawn dimension of the piece prior to drying and dressing (regularizing). The actual dimensions are listed in Table 2 for North American lumber. This table shows that the unseasoned (green) sizes for boards and dimension are slightly larger than the seasoned sizes. This arises from the fact that in any particular environment wood will reach the same final moisture content without regard to its moisture content at the time it is manufactured to the standard size. To permit sawmills without seasoning facilities to manufacture and sell lumber in competition with mills which have those facilities, the standards provide for two types of lumber:

(a) S-dry—lumber manufactured after being seasoned to an average of 15% and a maximum of 19% moisture content;

(b) S-green—lumber manufactured unseasoned or only partially seasoned, at a moisture content exceeding 19%.

To ensure that the consumer will receive the same full measure of wood, the S-green lumber, which has not experienced shrinkage at the time it is manufactured to size, must be sized to a larger dimension. Therefore, either S-dry or S-green lumber will be of adequate size to justify the published allowable strength properties when at final equilibrium with the environment.

Table 2 also reveals that timbers (as opposed to boards and dimension) have no dry sizes. This is because timbers season very slowly—too slowly to permit prompt delivery of an order of dry timbers. Timbers are therefore standardized only in the green condition.

Table 2
American Standard lumber sizes for stress-graded and non-stress-graded lumber for construction

	Thickness (mm)			Face width (mm)		
		Minimum dressed			Minimum dressed	
	Nominal	Dry	Green	Nominal	Dry	Green
Boards	25	19	20	51	38	40
	32	25	26	76	64	65
	38	32	33	102	89	90
				127	114	117
				152	140	143
				178	165	168
				203	184	191
				229	210	216
				254	235	241
				279	260	267
				305	286	292
				356	387	343
				406	387	394
Dimension	51	38	40	51	38	40
	64	51	52	76	64	65
	76	64	65	102	89	90
	89	76	78	127	114	117
	102	89	90	152	140	143
	114	102	103	203	184	191
				254	235	241
				305	286	292
				356	337	343
				406	387	394
Timbers	⩾127		13 less than nominal	⩾127		13 less than nominal

Structural lumber is usually sold dressed to size. It can be obtained rough-sawn but, except for timbers, this is rare in North America. In Europe, however, rough-sawn lumber is more common. It is carefully sawn to uniform width and thickness, whereas in North America rough-sawn lumber may not be so regular. The standard size of rough-sawn lumber is 3 mm larger than unseasoned dressed lumber, to permit future dressing to size.

The grades of visually graded structural lumber are listed in Table 3. There are five basic categories in this table, and the grades are listed in descending order of strength in each category.

Light Framing and Studs are 50 × 50 to 100 × 100 mm in nominal size. The name grades listed in Table 3 are peculiar to this category. These grades are lower in strength than their counterparts in Structural Light Framing: Select Structural, No. 1, No. 2, and No. 3. Light Framing and Studs are sometimes called Yard Lumber. They are commonly used in home building on spans designated in building regulations. They are a widely used, cheap grade of utilitarian wood, but are not often used where structural requirements are critical.

Structural Light Framing grades are also 50 × 50 to 100 × 100 mm in nominal size. The dense grades in Table 3 apply only to Douglas fir, western larch and southern pine, and are not commonly available.

Structural Joists and Plank are like Structural Light Framing but wider. Their allowable properties are similar but not always identical. Both categories are graded so that they may be used as bending members placed either on edge (joist) or flat (plank).

All the above grades are "dimension" lumber. Boards are available in the same grade and width categories but are less than 50 mm in nominal thickness. All these grades can be cut to shorter lengths without affecting the grading, but they must be used at the same cross-sectional dimensions that they possessed when graded: they cannot be reduced in width without destroying their grading.

Beams and Stringers, and Posts and Timbers, are timbers. They are only available unseasoned. They cannot be reduced in length or in cross section without affecting the validity of their grading. They are also produced in grades Nos. 3 and 4 but have no assigned structural properties in these grades. Beams and Stringers are intended mainly for use as bending members placed on edge, and Posts and Timbers as compression members or columns.

Structural grading rules in North America are published for nearly 60 species. They are often grouped when the species are structurally similar. The principal species for designers are southern pine, Douglas-fir–larch, Hem-fir (western hemlock and true fir) and spruce–pine–fir (a species group of white spruce, various pines and true firs according to the National Lumber Grades Authority of Canada).

The North American structural rules are also used to grade some hardwoods, which are locally available

Table 3
Visual stress grades of lumber in the USA and Canada

Light Framing and Studs	*Beams and Stringers*
50–100 mm thick; 50–100 mm wide	⩾125 mm thick; width = thickness + 50 mm or more
Construction	Dense Select Structural
Standard	Select Structural
Utility	Dense No. 1
Stud	No. 1 Dense SR[a]
Structural Light Framing	No. 1
50–100 mm thick; 50–100 mm wide	No. 1 SR[a]
Dense Select Structural	No. 2 Dense SR[a]
Select Structural	No. 2 SR[a]
Dense No. 1	
No. 1/Appearance	*Posts and Timbers*
Dense No. 2	⩾125 mm thick; width = thickness + 50 mm or less
No. 2	Same grade names as Beams and
No. 3	Stringers, but not the same
Stud	property values
Structural Joists and Planks	*Special Structural Grades*
50–100 mm thick; ⩾125 mm wide	50–100 mm thick; any width
Same grade names as	Dense Structural 86[a]
Structural Light Framing	Dense Structural 72[a]
but not the same property	Dense Structural 65[a]
values	

[a] Southern pine only

Table 4
Grade stresses and moduli of elasticity for softwood strength classes: for the dry exposure condition

Strength class	Bending parallel to grain ($N\,mm^{-2}$)	Tension parallel to grain ($N\,mm^{-2}$)	Compression parallel to grain ($N\,mm^{-2}$)	Compression perpendicular to grain[a] ($N\,mm^{-2}$)	Shear parallel to grain ($N\,mm^{-2}$)	Modulus of elasticity ($N\,mm^{-2}$)		Approximate density[b] ($kg\,m^{-3}$)
						Mean	Minimum	
SC1	2.8	2.2	3.5	2.1/1.2	0.46	6800	4500	540
SC2	4.1	2.5	5.3	2.1/1.6	0.66	8000	5000	540
SC3	5.3	3.2	6.8	2.2/1.7	0.67	8800	5800	540
SC4	7.5	4.5	7.9	2.4/1.9	0.71	9900	6600	590
SC5	10.0	6.0	8.7	2.8/2.4	1.00	10700	7100	590/760

[a] When the specification specifically prohibits wane at bearing areas, the higher values of compression perpendicular to the grain stress may be used, otherwise the lower values apply [b] Since many species may contribute to any of the strength classes, the values of density given in this table may be considered only crude approximations. When a more accurate value is required, it may be necessary to identify individual species and use the values given in Appendix A of BS 5268: Part 2

and valuable. These are aspen, black cottonwood and yellow poplar. Hardwoods are often suitable for structural use. Only price and higher value for other purposes keep hardwoods from more common structural use.

Structural lumber is a medium-price wood material. It is never used clear except for occasional pieces. No. 2 is the commonly used grade and No. 1 the next most common grade specified.

In some countries, notably in tropical Central America, the most common building materials are hardwoods. Several countries have grading systems for structural hardwoods. Hardwoods are extensively graded for structural use in Australia.

In the UK the types of grade follow a pattern similar to those in North America, although there appears to be a more limited number of grades and variety of species. The UK is a wood-importing country, which probably accounts for the smaller number of grades utilized; moreover, the designation of grades which are so much alike that they are difficult to distinguish is avoided. The British rules include four Joinery grades: Classes 1, 1S, 2 and 3. Structural grades in use in the UK are:

(a) visually graded—SS (special structural), GS (general structural);

(b) machine grades—M75 (75% strength ratio), M50 (50% strength ratio), MSS and MGS (machine-graded SS and GS).

Softwood grading practice in the UK provides five standardized strength classes, SC1 through SC5, with dry use (18% maximum moisture content) allowable stresses and elastic moduli listed in Table 4. UK structural grades are designed to achieve particular ratios of class strength to clear, straight-grained wood strength. This is the case for both visual (SS and GS) and machine (MSS, MGS, M75, and M50) grades. Allowable properties are species dependent for both visual and machine grades. Table 5 keys grade names to allowable property classes. MSS and MGS are equivalent to SS and GS, respectively. M75 and M50 are not exact equivalents of other grades, as a perusal of Table 5 will show.

Table 6 shows the UK strength classes applicable to certain North American species and grades utilized in

Table 5
Strength classes: BS 4978 stress grade/species combinations grouped under the BS 5268 strength classes

Standard name	Strength class[a]				
	SC1	SC2	SC3	SC4	SC5
Imported					
Parana pine			GS	SS	
Pitch pine (Caribbean)			GS		SS
Redwood			GS/M50	SS	M75
Whitewood			GS/M50	SS	M75
Western red cedar	GS	SS			
Douglas-fir–larch (N. America)			GS	SS	
Hem-fir (Canada)			GS/M50	SS	M75
Hem-fir (USA)			GS	SS	
Spruce–pine–fir (Canada)			GS/M50	SS/M75	
Western whitewoods (USA)	GS		SS		
Southern pine (USA)			GS	SS	
British grown					
Douglas fir		GS	M50/SS		M75
Larch			GS	SS	
Scots pine			GS/M50	SS	M75
Corsican pine		GS	M50	SS	M75
European spruce	GS	M50/SS	M75		
Sitka spruce	GS	M50/SS	M75		

[a] When specifying a strength class, certain species may need to be listed as exceptions if joint strength or durability are critical. See BS 5268 Parts 2 and 5 for the relevant information.

Table 6
Strength classes for Hem-fir (from Canada and USA) and Spruce–pine–fir (from Canada)

Grade	Section size (mm)	Strength class SC1	SC2	SC3	SC4	SC5
Structural light framing	38 × 89, 38 × 38 and 38 × 63	No. 3[a]		No. 1, No. 2	Sel	
	63 × 63	No. 3		Sel		
	63 × 89, 89 × 89			Sel		
Light framing	38 × 89	Std, Stud, Util	Const			
	38 × 38	Stud	Const			
	38 × 63, 63 × 63	Const, Std, Stud				
	63 × 89, 89 × 89	Std, Util	Const			
	38 × 114, 38 × 140	Stud				
Joist and plank	not less than 38 × 114 mm	No. 3		No. 1, No. 2	Sel	

[a] No. 3 grade should not be used for tension members Key: Sel—Select structural; Std—Standard; Const—Construction; Util—Utility

the UK. Higher strength classes (SC6 through SC9) are defined in BS 5268: Part 2, and are used in the case of denser hardwood species, but not for softwoods.

Sweden, Finland, Poland and the USSR have stress-grading rules and rules for nonstructural lumber, but they are not as widely publicized.

See also: Building with Wood

Bibliography

Brown N C, Bethel J S 1958 *Lumber*, 2nd edn. Wiley, New York

Building Research Establishment 1984 *Specifying Structural Timber*, Digest 287. BRE, Princes Risborough Laboratory, Aylesbury

British Standards Institution 1973 *Timber Grades for Structural Use*, BS 4978. BSI, London

British Standards Institution 1978 *Specification for Dimensions of Softwood*, Part 1: 1978. *Sizes of Sawn and Planed Timber*, BS 4471. BSI, London

British Standards Institution 1984 *Permissible Stress Design, Materials and Workmanship*, Code of Practice for the Structural Use of Timber, BS 5286. BSI, London

Forest Products Laboratory 1987 *Wood Handbook: Wood as an Engineering Material.* US Government Printing Office, Washington, DC

Hoyle R J Jr 1982 Lumber: Grade, sizes, species. In: Dietz A G H, Schaeffer E L, Gromala D S (eds.) 1982 *Wood as a Structural Material*, Clark C Heritage Memorial Series on Wood, Vol. II. Pennsylvania State University, University Park, Pennsylvania, pp. 145–52

Hoyle R J Jr, Woeste F E 1988 *Wood Technology in the Design of Structures*, 5th edn. Iowa State University Press, Ames, Iowa

Ozelton E C, Baird J A 1976 *Timber Designers' Manual.* Crosby Lockwood Staples, London

R. J. Hoyle Jr.
[Lewiston, Idaho, USA]

M

Machining Processes

Wood machining is the process of manufacturing wood products such as lumber, veneer and furniture parts. The objective of wood machining is to produce a desired shape and dimension with requisite accuracy and surface quality in the most economical way. Machining processes in the manufacture of wood products may be classified as follows: sawing; peeling and slicing; planing, molding and shaping; turning and boring; sanding; and nontraditional machining processes such as cutting with laser beams, high-velocity liquid jets and vibrating cutters.

Optimization of wood machining processes involves attempts to reduce losses of machined material and wear of cutting tools; improve the accuracy of dimensions and surface quality; increase production output and reduce cost; and improve worker safety. A major development has been the increased application of microprocessor and computer control systems (Maier 1987).

1. Sawing Technology

Sawing is the most important and most frequent cutting process. Sawing machines are classified according to the basic machine design, that is, sash gang saws (reciprocating, multiple blade frame saws), circular saws, band saws and chain saws. Saws are designated ripsaws if they are designed to cut along the grain, as bucking or trim saws if they are designed to cut across the grain, or as combination saws if designed to cut along and across the grain, as well as at a certain angle to the grain (e.g., miter saws). Sawing machines are further classified according to their use. For example, a bucking saw is used for cutting logs to length, a head rig or head saw for primary log breakdown, a resaw for resawing cants into boards, an edger for edging boards, a trimmer for cutting boards to length, and a scroll saw for general-purpose sawing in furniture plants.

Saw blades are made from cold-rolled, hardened and tempered steel. A high carbon content, nickel-alloyed saw steel is used in most cases (e.g., Uddelholm Steel UHB 15N20: 0.75% C and 2.0% Ni). Other saw steel alloys may contain manganese, chromium and vanadium. Circular saws typically range from 1.0 to 5.0 mm in thickness and from 100 to 1800 mm in diameter. The thickness of band-saw and gang-saw blades may range from 0.40 to 2.1 mm.

Band-saw width ranges from 6 to 50 mm for the narrow band saws used in furniture manufacturing, and from 60 to 360 mm for saws used in lumber manufacturing. Band-saw thickness and width depend on the saw wheel diameter and width. As a rule, the saw blade thickness should not exceed 0.1% of the wheel diameter, and the band-saw width should not be greater than wheel width plus gullet depth and an additional 5 mm. The typical gang saws are 2.00 mm in thickness and approximately 175 mm in width.

Saws vary considerably with regard to tooth and gullet design. The primary design considerations include tooth strength and gullet loading capacity, the function of the gullet being sawdust removal. Other important factors are tooth wear and noise generation. The typical band-saw tooth geometry is described by specifying rake and clearance angle as depicted in Fig. 1. If the saw tooth has a face and/or top bevel, those angles should be also specified. The optimum tooth geometry, as determined from the measurement of power requirements, depends mainly upon cutting direction, wood density and moisture content. Tooth geometry may vary considerably: the rake angle for crosscut circular saws ranges from $+10°$ to $-30°$. In the case of circular ripsaws and band saws, the rake angle will vary from 10° for high-density hardwoods to 30° for softwood species. The clearance angle may range from 8° for dense hardwoods to approximately 10° for softwoods.

The side clearance or set, which is required to reduce friction between the saw blade and generated surface, is usually provided by setting, that is, either by deflecting alternate teeth (spring-setting), or by spreading the cutting edge (swage-setting), as in band saws (Fig. 1). The side clearance for wide band saws may range from 0.30 to 0.35 mm for hardwoods and from 0.50 to 0.60 mm for softwoods. Certain specialty circular saws such as smooth-trimmer or miter saws are tapered (hollow ground) to provide side clearance. Inserted-tooth saws, carbide- or stellite-tipped saws and chain-saw teeth are designed so that a desired side clearance

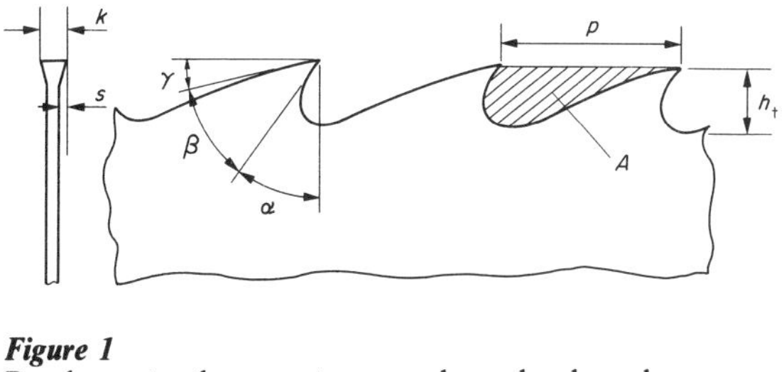

Figure 1
Band-saw tooth geometry: α = rake or hook angle, p = pitch or tooth spacing, β = sharpness angle, h_t = depth of gullet or tooth height, γ = clearance angle, A = gullet area, k = kerf width (theoretical), s = side clearance or set

can be provided, and consequently the setting is not required. The purpose of tipping saw teeth with hard alloys is to increase their wear resistance, which prolongs the useful life of the blade. Methods of improving wear resistance of saw teeth have been described in detail by Kirbach (1979).

The actual size of the gullet is determined by the pitch or tooth spacing and the size of the gullet controls the maximum cutting height and feed space. For a typical wide band saw the gullet area is approximately equal to the product of the pitch and the depth divided by 1.75 (Koch 1964, Williston 1978). Each single tooth will remove a certain volume of wood given by the feed per tooth and the cutting height. This volume should correspond to the chosen gullet capacity (Kollmann and Côté 1968). The feed per tooth t is given by $t=p(F/C)$ where p is the pitch (mm), F is the rate of feed ($m\,min^{-1}$), and C is the cutting speed ($m\,min^{-1}$). The average blade velocity C is about $2500\,m\,min^{-1}$.

The use of thin circular saws (thickness of cut or kerf 3 mm or less) has proved to be very beneficial to industry in the reduction of kerf losses, as long as saw stability is maintained through blade design and on-line vibration control. One of the principal manifestations of circular-saw instability is standing-wave resonance (Mote and Szymani 1977) and the rotation speed at which a standing wave is formed is called the critical speed. All in-plane or membrane stresses (i.e., stresses due to temperature gradients, rotation, cutting forces and tensioning or prestressing) shift the saw natural frequencies and alter its critical speed accordingly. Computer codes are available for estimating the critical speed and the critical speed margin (the difference between critical and operating speeds) based on design and operation variables. The saw operating speed should be at least 15% below critical speed. The sawing accuracy improves with the increase of the critical speed margin (Szymani and Mote 1977). In the case of band saws, currently available computer codes can analyze band-saw vibrations and be used to predict natural frequency and evaluate band-saw design relative to band vibration and stability (Ulsoy et al. 1978, Szymani 1986).

The effective stiffness and stability of saws can be increased by introducing radial slots, by prestressing or tensioning, by using guiding systems, by on-line cooling near the cutting edge, and by heating near the center (i.e., thermal tensioning) (Szymani and Mote 1977, Mote et al. 1982, Szymani 1986). Radial slots in circular saws reduce compression hoop stresses at the saw periphery due to temperature gradients, introduce asymmetry into the saw-blade design and consequently reduce transverse vibration and reduce noise. The application of various guiding systems in conjunction with the use of floating-collar or splined-arbor saws, which can float on the arbor, is a common and particularly effective method used for stability control of thin-kerf circular saws. Conventional guide systems are either contacting or noncontacting displacement-limiting devices normally placed near the cutting edge. In the case of band saws, in addition to prestressing and the use of saw guiding systems, the type of straining mechanism for providing axial tension and its response will significantly affect saw stability and consequently sawing accuracy. The saw blade must operate under maximum applied tension force, consistent with the endurance strength of the saw material, in order to maximize stiffness and critical edge-buckling load. Regardless of the operating conditions, the stress level in the saw blade must be kept constant. This can be achieved, for example, by on-line thermal tensioning (introduction of a thermal stress state beneficial to saw stability) and/or feed rate control. The on-line control of circular and band-saw stability basically consists of either modifying the forces exciting the blade or altering the effective saw-blade stiffness and damping to reduce vibration.

2. *Rotary Cutting and Slicing*

Rotary cutting (peeling) and slicing of wood are used in the manufacture of veneer. At least 95% of veneer is produced by peeling for which a veneer lathe is used, and about 3–4% by slicing, for which a horizontal or vertical slicer is used. The remaining 1–2% is produced by sawing. The primary components of any lathe or slicer are the knife and pressure bar (Baldwin 1975, Lutz 1978). They are similar in both machines and perform the same function. The cross section of a lathe presented in Fig. 2 illustrates the position of the knife and the pressure bar. The most common knife thickness for a lathe is 16 mm, and for the face veneer slicer, 19 mm. The European horizontal slicer may use a knife 15 mm in thickness. The knife's Rockwell hardness on the C scale may vary from 56 to 60.

While the knife severs the veneer from the bolt or flitch, the pressure bar compresses the wood and thus

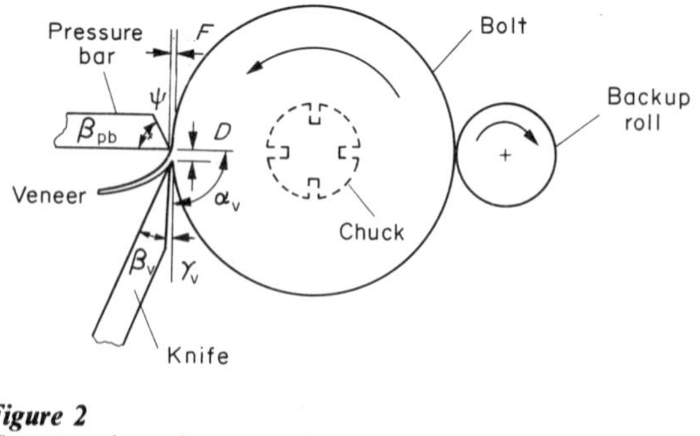

Figure 2
Cross section of a veneer lathe with fixed pressure bar: α_v = knife angle, β_v = knife bevel angle, γ_v = clearance angle, D = lead or vertical gap, β_{pb} = pressure-bar bevel angle, ψ = pressure-bar compression angle, F = gap (horizontal gap)

reduces splitting of the wood ahead of the knife. The amount of compression depends on wood density and will vary from species to species. For Douglas fir, it is about 15% of the veneer thickness; for western white spruce about 8%. In both lathe and slicer, the pressure bar is important in controlling roughness, depth of checks, and thickness of the veneer. The slicer has a fixed nosebar while the lathe may have either a fixed nosebar or a rotating roller bar. The flitch on a slicer is backed by the flitch table while support for a veneer bolt may be provided by a powered backup roll. For maximum yield of rotary peeled veneer it is essential that bolts are chucked in the geometric center. Opto-electronic scanning and computerized centering systems are currently available with modern lathe chargers which allow determination of chucking centers for best yield of each individual block. Other developments in veneer manufacturing equipment include high-speed veneer lathes having spindle speeds of 500 rpm and over, digital carriage drive which eliminates the mechanical clutch assembly, hydraulic powered backup rolls and dual hydraulic spindles.

The production of high-quality veneer requires proper pretreatment of the wood prior to cutting. This is done by heating green wood in water or steam or by using steam-heated knives. Heating wood above 50 °C makes it more plastic and reduces veneer checking during peeling or slicing. The recommended peeling temperature ranges from 50 to 90 °C and will vary with the average specific gravity of the species (Lutz 1978). The temperatures required for proper cutting of softwoods are generally higher than those required for hardwoods of comparable density.

3. *Planing, Molding, Shaping and Routing*

Planing refers to the peripheral milling of wood. Its purpose is to smooth one or more surfaces of the workpiece and at the same time bring the workpiece to some predetermined dimension. The machinery used for planing operations includes: (a) surfacers designed to smooth one or two sides of the workpiece and reduce it to a predetermined thickness, and (b) planers and matchers defined as double surfacers which are further equipped with two opposed profile side heads that can simultaneously machine the two edges to the desired pattern or profile.

The molding operation aims to machine lumber into forms of various cross-sectional shapes, for example, picture-frame moldings. Both planing and molding machines employ rotating cutterheads. By definition, a molder differs from a planer in that the molder sideheads are staggered instead of directly opposed. The typical operating speed for multiknife cutterheads ranges from 3600 to 6000 rpm. The number of spindles may range from one to ten. Molding machines can be equipped with variable feed rate typically ranging from 6 to 60 m min^{-1}.

Shaping involves machining an edge profile or edge pattern on the side and/or periphery of a workpiece. The basic types of shapers include single and double spindle shapers, double head automatic shaper and center profiler. The shaper spindle speeds range from 7200 to 10 000 rpm.

Routing is similar to the shaping operation. While a shaper always shapes the periphery, a router is used to make a variety of cuts such as mortises, irregularly shaped holes and three-dimensional plunge cuts using computer numerical control (CNC). Most router spindle speeds are 10 000 or 20 000 rpm, depending on the diameter of the cutter. When machining abrasive, composite wood products, there is a trend to use polycrystalline diamond cutting tools in routing and shaping operations (Ettelt 1987).

In all three operations, it is of prime importance to adjust the operating conditions and knife geometry so that machining defects are maintained at a satisfactorily low level (Koch 1964, Willard 1980). The most commonly encountered defects are torn or fuzzy grain, raised and loosened grain, and chip marks. These defects are caused by improper cutting angles, chip thickness which is too large, dull knives, low density species and often the presence of reaction wood. In the case where the torn-grain defect is highly probable, the most important variable is the number of marks per centimeter or inch (reciprocal of the feed per cutter). The marks per centimeter should be between three and five for rough planing operations and between five and six for finishing cuts. The clearance angle should in all cases exceed a value of 10°. The optimum cutting angle (angle between the knife face or knife bevel and a radius of the cutter head) lies between 20° and 30° for most planing situations; however, in the cases of interlocked or wavy grain, it may be necessary to reduce the cutting angle to 15° or even 10°.

4. *Turning and Boring*

Turning of wood is a machining process for generating cylindrical forms by removing wood, usually with a single-point cutting tool. The turning machines include single- and multiple-spindle lathes. The tools used for turning on the lathe perform operations primarily directed to machining the outer surfaces of the workpiece. From practical experience and experimental investigations (Kollmann and Côté 1968) lathe clearance angles between 12° and 18° offer optimum cutting conditions. In practice, a lip or wedge angle between 20° and 30° is recommended for softwoods, the wedge angle corresponding to the sharpness angle in the case of saw teeth and to the knife bevel angle in case of a veneer knife. For hardwoods, wedge angles between 50° and 60° are recommended. The quality of surfaces of most turned wood articles is of the utmost importance, for example, for tool handles. The roughness perpendicular to the grain increases with the feed speed. The specific pressure of the tool on the turned surface also has a remarkable influence on the roughness.

Most machines which will perform turning operations can also perform boring operations, although machines are available which will perform boring, drilling, and other related operations. Boring machines can have many configurations, ranging from the simple vertical, single-spindle boring machine to complex transfer machines involving multiple vertical, horizontal, and angular spindles. There are many specialized boring-bit designs in use. The common bit types include (a) double-spur, double-lip solid-center bit on which the spurs cut ahead of the lips; (b) double-spur, double-twist bit on which the spurs cut after the lips; and (c) twist drill. The first is a fast-boring general-purpose bit, the second bit is particularly suited for boring to extreme depth. The twist drill is frequently used on machine boring equipment for drilling in end grain and for boring dowel holes (Koch 1964, Willard 1980). The quality of finish produced by a twist drill may be inferior to that produced by a bit equipped with spurs.

5. *Sanding Technology*

Sanding is the abrasive machining of wood surfaces to obtain smooth surface quality. The abrasive tool consists of a backing material to which abrasive grains are bonded by an adhesive coat. The abrasive or sanding tool is specified by the sanding and backing materials. Sanding materials vary according to type, size and form of grain. Typical abrasive materials for woodworking applications are garnet, aluminum oxide and silicon carbide. Garnet is the most commonly used because of its low cost and acceptable working qualities. It is used with all types of machines for sanding softwoods. Aluminum oxide abrasives are used extensively for sanding hardwood, particleboard and hardboard. Silicon carbide abrasives are used for sanding and polishing between coating operations and for sanding softwoods where the removal of raised fibers is a problem. The size of the abrasive particles is specified by the mesh number (i.e., the approximate number of openings per linear inch in the screen through which particles will pass); mesh numbers range from about 600 to 12.

Backing materials vary according to the strength, flexibility and required spacing of the sanding tool and are made of paper, cotton or polyester cloth, or cloth–paper combinations. Bonding materials are generally animal glues, urea resins, or phenolic resins. The choice of these materials depends upon the required flexibility of the tool and the work rate required of the tool. Animal-glue bonds are the most flexible, whereas resin bonds are harder, more moisture and heat resistant and have superior grain retention.

Sanding machines include multiple-drum sanders, wide-belt sanders, automatic-stroke sanders and contact wheel disk sanders. The drum sander is probably the oldest of all the woodworking machines using coated abrasives, and it is used in solid wood furniture manufacturing.

The drum sanding machine is used following the planer or veneer press. Multiple-drum sanders are of the endless-bed or roll-feed type and have from two to six drums. The abrasive is usually a heavy paper-backed aluminum oxide product. In very heavy sanding operations, a fiber-backed abrasive is recommended. A sequence of 60, 80 and 100 mesh abrasive is frequently used on a three-drum endless-bed sander.

Wide-belt sanders use an abrasive belt at least 30 cm wide and are commonly used in board plants (plywood, particleboard, hardboard). Silicon carbide is normally used as the abrasive. They have higher production rates and greater accuracy than multiple-drum sanders.

Heavy-duty high-speed wide-belt sanders are called abrasive planers when used for dimensioning and surfacing. Abrasive planers are used for dimensioning of plywood, particleboard for furniture, as well as for dimensioning of accurately sawn, kiln-dried lumber. In comparison with the knife planer, the abrasive planer has in general higher production rates, a lower noise level, and virtually no machining defects. New developments in wide-belt sanding include the use of antistatic belts and sanding with aerostatically (air cushion) supported belts. It is critical when using sanders and abrasive planers to have an adequate dust removal system.

Surface finish during the sanding process is for the most part independent of pressure and cutting speed. The optimum belt speed as determined by the specific quantities of abrasion is about $30\,\mathrm{m\,s^{-1}}$ for particle size 60 and slightly less than $30\,\mathrm{m\,s^{-1}}$ for particle size 120 (Pahlitzsch 1970).

Automatic-stroke sanders use a narrow abrasive belt and a reciprocating shoe which creates contact between the abrasive and the workpiece. This sanding machine is commonly used in furniture plants for final sanding operations and touch-up sanding.

The contact-wheel sander also uses a narrow abrasive belt. Contact wheels normally range from 150 mm to 350 mm in diameter. A typical application is the sanding head on an edge banding machine where the edging tape is given a finish after application to the board. Cloth belts are usually preferred because of their durability.

The disk sander consists of a revolving back plate to which a coated abrasive disk of paper or cloth is attached by an adhesive. It usually incorporates tilting action for angle or miter sanding. The major disadvantage of this method is a pattern of circular scratches which have to be removed by other means before finishing.

6. Nontraditional Machining Processes

Various cutting techniques have been investigated for possible use in the wood industry in an effort to reduce, or eliminate, kerf losses. Those with potential applications include the laser beam, the high-energy liquid jet and, to a lesser degree, vibrating cutters (Szymani and Dickinson 1975).

A wide variety of material can be cut using a continuous carbon dioxide laser. The laser beam produces a very narrow kerf, in most cases under 1 mm. The major disadvantages of cutting wood and wood-based panels with the laser are capital cost and maintenance and low feed rate, resulting in high cost per unit of lineal cut, and the charring of the generated surface. Major advantages of a laser beam include the ability to cut intricate patterns, high cutting accuracy and the possibility of numerical control. Therefore, the operations most economically justified are laser engraving and the automatic preparation of wooden die blocks for the folding-carton industry.

The potential application of the liquid jet as a cutting tool depends on the availability of high-pressure pumping equipment capable of generating a high-velocity continuous jet. For the generation of a high-energy continuous flow, a pressure level of about 700 MPa is required. The nozzles range from 50 to 375 μm in diameter and are made from commercial sapphire orifice jewels. The liquid jet, like the laser, approaches the ideal single-point cutting tool, which can follow highly complicated patterns. It eliminates crushing or deformation of the material and generation of dust. The cutting speeds compare favorably with most other methods when cutting soft materials. However, when cutting harder materials such as wood and wood-based products, the cutting speed depends on the depth of cut, and the practical application of the high-velocity liquid jet becomes limited. The greatest potential use of liquid-jet cutting is in the paper and paperboard industry where it has been quite successful in cutting laminated paperboard into upholstery frames.

Cutting with a vibrating cutter (blade) has been investigated since 1944 in an effort to eliminate kerf losses. It has been found that cutting perpendicular to the grain is feasible whereas cutting along the grain poses difficulties.

A vibrating inclined blade produces superior end-grain surfaces, presumably by the kinematic reduction of rake angle. Because of the limited knowledge of kerfless cutting by means of vibrating cutters, there is speculation among researchers and engineers as to its feasibility.

Two other cutting techniques which can eliminate sawdust are currently under investigation. One, a longitudinal wood slicer used for manufacturing lumber from cants, has been developed in Canada. The other, a knife-like wood slitter resembling the circular saw, is being developed in the USA.

The wood slicer consists of a stationery knife blade and a set of wheels held at high pressure which force the cant against the blade and exert compressive forces on the longitudinal portion of the wood through which the knife blade is passing. Recent pilot plant trials have revealed that the slicer can produce lumber up to 15 cm wide at 90 m min^{-1} without generating sawdust.

The investigations on the application of the wood slitter are less advanced. One potential application is in lumber edging where the slitter–edger could replace existing edgers.

See also: Forming and Bending

Bibliography

Baldwin R F 1975 *Plywood Manufacturing Practices.* Miller Freeman, San Francisco, California

Ettelt B 1987 *Sawing, Milling, Planing, Boring: Wood Machining and Cutting Tools* (in German). DRW-Verlag, Stuttgart

Johnes C W 1987 *Cutterheads and Knives for Machining Wood.* Jones, Seattle

Kirbach E D 1979 *Methods of Improving Wear Resistance and Maintenance of Saw Teeth.* Technical Report No. 3. Western Forest Products Laboratory, Vancouver

Koch P 1964 *Wood Machining Processes.* Ronald Press, New York

Kollmann F F P, Côté W A Jr 1968 *Principles of Wood Science and Technology*, Vol. 1: *Solid Wood.* Springer, New York, pp. 475–554

Lutz J F 1978 *Wood Veneer: Log Selection Cutting and Drying*, Technical Bulletin No. 1577. US Department of Agriculture Forest Products Laboratory, Madison, Wisconsin

Maier G 1987 *Woodworking Machines: Requirements, Concepts, Machine Elements, Construction* (in German). DRW-Verlag, Stuttgart

Mote C D Jr, Szymani R 1977 Principal developments in thin circular saw vibration and control research, Part 1: Vibration of circular saws. *Holz Roh- Werkst.* 35: 189–96

Mote C D Jr, Schajer G S, Wu W Z 1982 Band saw and circular saw vibration and stability. *Shock Vibr. Dig.* 14(2): 19–25

Pahlitzsch G 1970 The international state of research in the field of wood sanding (in German). *Holz Roh- Werkst.* 28: 329–43

Szymani R 1986 Status report on the technology of saws. *For. Prod. J.* 36(4): 15–19

Szymani R, Dickinson F E 1975 Recent developments in wood machining processes: Novel cutting techniques. *Wood Sci. Technol.* 9: 113–28

Szymani R, Mote C D Jr 1977 Principal developments in thin circular saw vibration and control research, Part 2: Reduction and control of saw vibration. *Holz Roh- Werkst.* 35: 219–25

Ulsoy A G, Mote C D Jr, Szymani R 1978 Principal developments in band saw vibration and stability research. *Holz Roh- Werkst.* 36: 273–80

Willard R 1980 *Production Woodworking Equipment*, 4th edn. North Carolina State University, Raleigh, North Carolina

Williston E M 1978 *Saws: Design, Selection, Operation, Maintenance*. Miller Freeman, San Francisco, California

R. Szymani
[Wood Machining Institute, Berkeley, California, USA]

Macroscopic Anatomy

This article covers the gross characteristics of wood after first examining the origins of the material. It then addresses the special terminology that relates to wood and its properties, the macroscopic structure (that is, visible with the unaided eye), anatomical features, cell types and cell distribution patterns, sapwood–heartwood differences, and features that distinguish softwoods (coniferous woods) from hardwoods (angiosperms or broadleaved species).

Among the major materials available to the engineer today, wood is unique because it is a biologically produced renewable material. Since it is produced in living trees, wood cannot be made to exact specifications as is the case for steel and other materials. It is also variable in its structure and properties, even within a single species, and although efforts to control the ultimate properties of wood through genetic selection have been successful, the specific properties desired for a particular application are best provided by choice of species and/or density.

1. Characteristics

Wood is a cellular composite. The orientation of its cellular components and the organization of its tissue systems contribute greatly to its behavior in service. For example, the tendency to split lengthwise (along the grain rather than across) is due to the vertical orientation of the majority of wood cells which are individually elongated elements. They may be likened to tubes with tapered and usually closed ends, with a cell cavity referred to as the lumen. In softwoods these cells are longitudinal tracheids less than 5 mm in length (Fig. 1a), whereas in hardwoods they are vessel elements (Fig. 1b, d, e), fiber tracheids, vasicentric tracheids and libriform fibers (Fig. 1c), all comparatively shorter than the coniferous tracheids.

The surface of a softwood, exposed when a tree stem is cut perpendicular to its long axis, is shown in Fig. 2; the organization of a generally cylindrical stem is evident. The concentric bands are annual growth rings or growth increments which, for most woods grown in temperate zones, indicate the amount of wood produced in a growing season. In tropical regions, such increments are not always clear and may represent wet and dry seasons. Note that the wood or xylem of the tree stem is protected by an outer layer of bark or phloem. The bark thickness varies greatly depending upon species and the age of the tree.

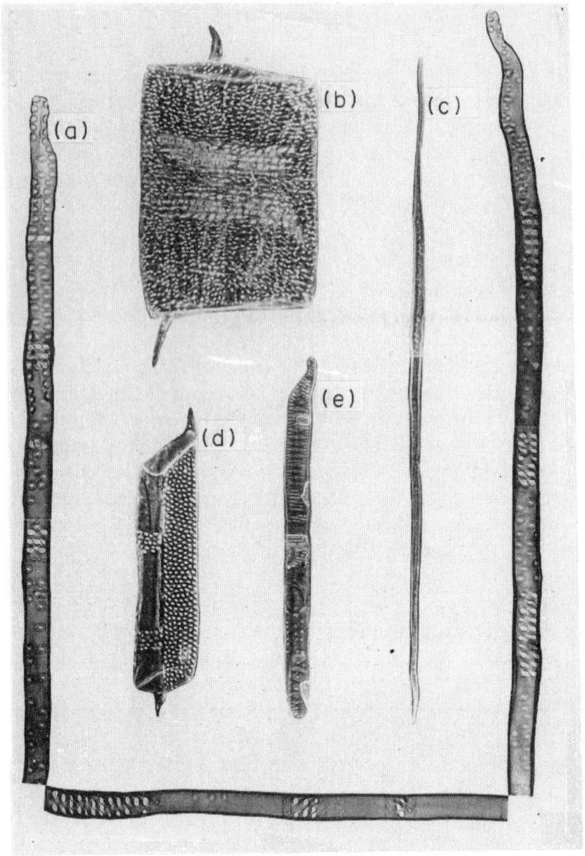

Figure 1
A composite of common wood cell types found in softwoods and hardwoods: (a) a typical coniferous tracheid which is several times longer than any hardwood element; (b), (d) and (e) are vessel elements; and (c) is a libriform fiber. Note the pitting on the cell walls (after Core et al. 1979)

At the interface between the wood and bark is a thin layer of growing cells called the cambium (Fig. 3). This initiating zone produces both bark and wood cells as the tree grows in diameter and circumference, beginning at the central pith. In effect most of the cells in a tree are dead and contain only water, except in the cambial zone and in portions of the ray tissue.

Unlike most materials widely used in engineering, wood is anisotropic, as might already be surmised from the first illustrations. Indeed, the structure is quite complex and can be described as having three aspects: the cross section X, the radial section R (from pith to bark) and a tangential section of surface T (cut tangent to the bark or cambium) (Fig. 4). These designations are necessary in describing wood properties because the material behaves differently in each direction.

Figure 2
Transverse surface of a softwood log (Douglas fir)

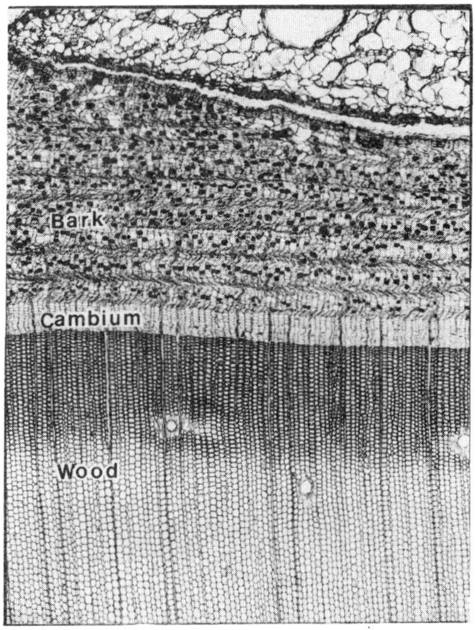

Figure 3
Photomicrograph of a small portion of a growing eastern white pine. Cell division in the cambial zone provides cells for both bark and wood

2. *Sapwood and Heartwood*

In many wood species color differences are apparent between growth increments near the center of the tree stem and those nearer the bark. This is due to a

Figure 4
Wedge-shaped segment of hardwood: cross-sectional (X), radial (R) and tangential (T) surfaces

darkening of the wood as it changes from sapwood to heartwood. Good examples of this are apparent in black walnut and black locust among the hardwoods, and redwood and Douglas fir in the softwood category. In these species, the heartwood becomes dark whereas the sapwood remains light-colored or almost white. Figure 5 illustrates this contrast very clearly.

The true boundary between sapwood and heartwood is the point in a ray where the parenchyma cells have ceased functioning. This may or may not coincide exactly with the color boundary described above. Also, conversion from sapwood to heartwood is not accompanied by a color change in all species. Though not a universal feature, heartwood is often less susceptible to deterioration by fungi or other biological agents because of the toxic nature of the organic substances deposited in the heartwood cell walls.

3. *Texture and Grain*

Many wood species are readily identified, even in lumber form, without benefit of hand lens or microscope. Distinctive color, odor or other gross characteristics make this possible. If the wood is surfaced (planed) or finished (coated), then such features as

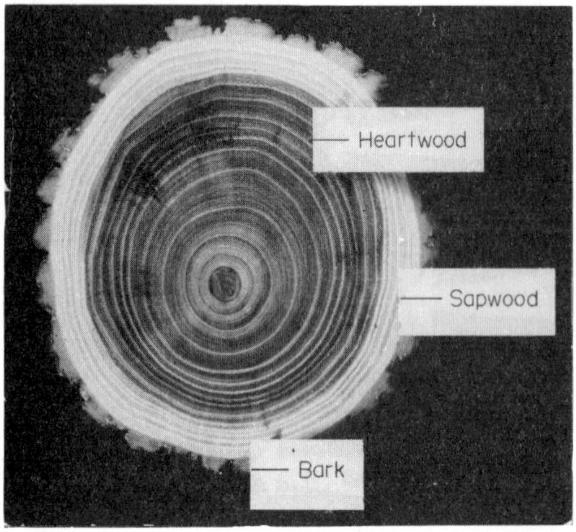

Figure 5
Transverse surface of a hardwood stem with dark-colored heartwood and light-colored sapwood (after Core et al. 1979)

texture and grain become more obvious and are helpful gross characteristics. Texture refers to the size and distribution of cells; fine-textured wood is composed of cells having small dimensions and small lumens. Grain is often confused with texture; however, it refers to the orientation of the elements with respect to the surface of the wood. For example, cross grain, birdseye, curly or spiral grain all refer to the deviation of the cells, whereas straight grained refers to their orientation parallel to the edges of a board. Grain can contribute to the creation of figure, that is, the patterns or markings that can be seen on wood surfaces as a result of either structural or color variations.

4. Anatomy

The cellular makeup of wood cannot be appreciated unless some magnification is used. Even a hand lens is useful to see relative cell sizes and shapes, as well as cell distribution patterns (Fig. 6). The transverse surface is usually smoothed with a sharp knife so that the detail can be observed.

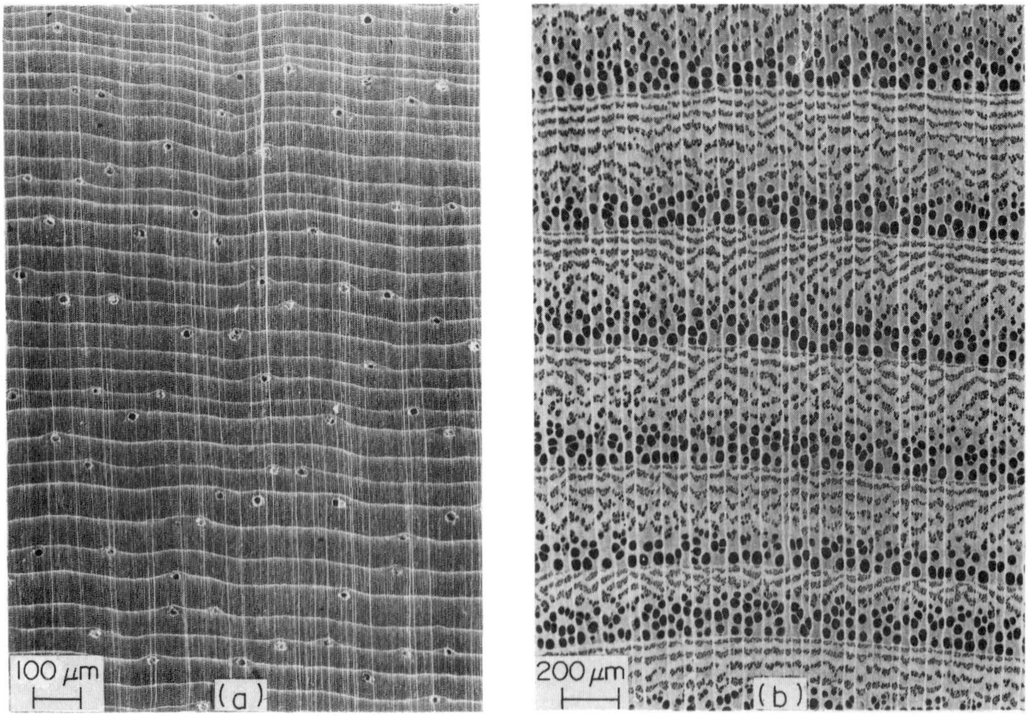

Figure 6
Cross-sectional views of typical softwood and hardwood at hand-lens magnification: (a) sugar pine—growth ring shows gradual transition and large resin canals; and (b) hackberry—ring porous with earlywood pores in 2–5 rows and latewood pores in wavy, elm-like pattern

For greater detail it is necessary to cut thin sections of water-softened wood for mounting on a glass microscope slide. Generally, three sections are prepared, one for each of the three views or aspects of wood: cross, radial and tangential sections. Each section can be examined individually with the light microscope, as structural evidence is critical for definitive identification of a wood species. To recreate the three-dimensional structure of wood requires imagination and experience.

Recently, however, it has become possible to view all three aspects of a small wood cube in a magnified three-dimensional field with the aid of a scanning electron microscope. The advantageous depth-of-field this technique provides has revolutionized microscopy of wood. Figure 7 shows relatively low-magnification views, which afford the investigator high resolution of structural detail, for a coniferous wood and for the wood of a broad-leaved species. Figure 7a, for a typical softwood, is characterized by longitudinal tracheids aligned in radial rows as viewed on the cross section. The tangential surface is interrupted by rays seen in sectional view. The rays also appear on the radial surface where the horizontal bands of parenchyma cells can be seen as an aggregate of brick-shaped cells. In Fig. 7b, the highly ordered structure of softwood is lacking and the characteristic element is the vessel (or pore as viewed on the transverse surface). This is a large-diameter cell specialized for vertical conduction. In a longitudinal view, visible on both the tangential and radial surfaces, one can see how a vessel consists of individual short cells joined end-to-end in tube-like fashion. Sieve plates of various types can be found at the junction between vessel elements.

Surrounding the pores, or vessels, are the various types of hardwood elements. The libriform fibers are narrow-lumened, thick-walled cells, probably nonconductive but specialized for support. Between these two extremes of vessel element and libriform fiber are the vasicentric tracheid, the fiber tracheid and the vessel element produced late in the growing season and therefore smaller in diameter than the earlywood vessels (see Fig. 1).

5. *Wood Rays*

Although both hardwoods and softwoods consist largely of longitudinally oriented elongated cells, they both contain significant numbers of elements which are oriented transversely or perpendicular to the long axis of the tree. These cells are ray parenchyma and other associated specialized cell types which are organized into radial bands called rays (see Fig. 7b). In softwoods the wood rays rarely constitute more than 10% of the wood volume, but in certain hardwoods the ray volume may exceed 25%. Ray parenchyma cells are usually considered to be storage cells, but

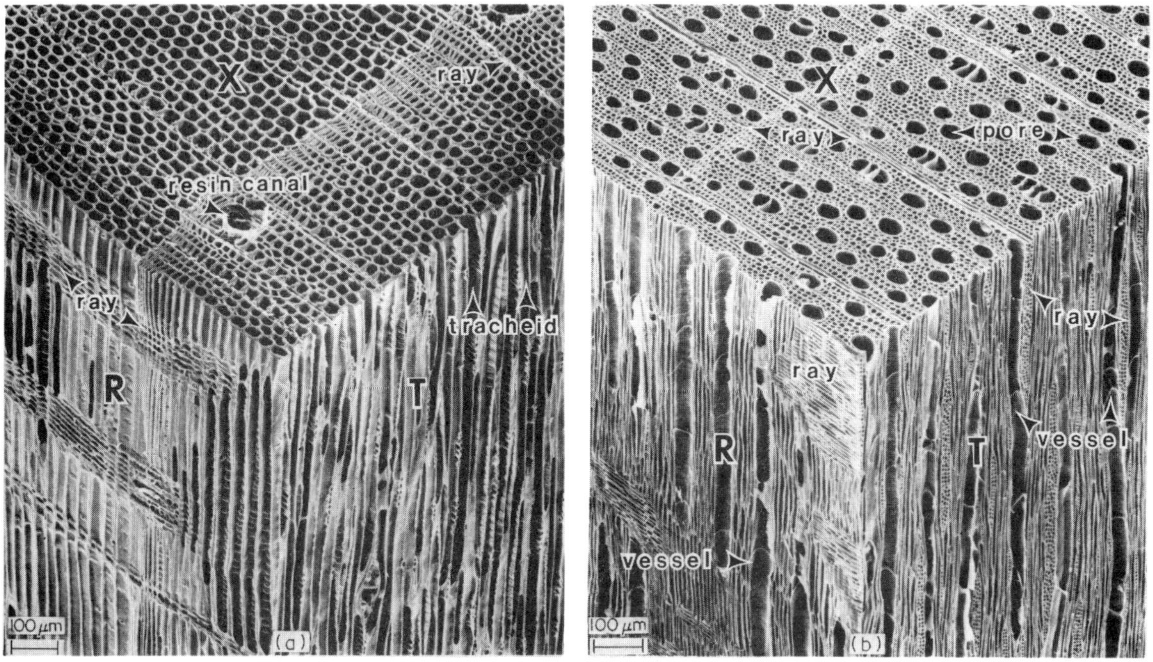

Figure 7
Scanning electron micrographs showing cross-sectional (X), radial (R) and tangential (T) surfaces of small cubes of wood: (a) a typical softwood (eastern white pine) and (b) a typical hardwood (red maple)

other cells associated with the ray parenchyma, particularly the ray tracheids found in some of the softwoods, may be conductive elements.

Depending on the way a log is cut, rays can contribute to figure on the surface of a wood panel or board. Rays are also important contributors to wood behavior. They are considered to restrain radial shrinkage and swelling since dimensional changes in this direction average only about one-half of those in the tangential direction. Rays may also be considered as contributors to planes of weakness in wood because end checks (splits) on seasoned lumber will usually follow or parallel the rays.

6. *Earlywood and Latewood*

As a biologically produced plant material, wood production reflects seasonal influences. As previously mentioned, tropical species react to wet and dry seasons. In temperate climates some species in both the softwood and hardwood categories vary the nature of their cell production as the growing season progresses. With the flush of growth in the spring, large diameter vessels are produced. Then, in subsequent weeks, vessel diameter may be reduced.

In those instances, such as in the oaks where the change is abrupt from earlywood to latewood, the term ring porous wood is applied. Where vessel size remains approximately the same throughout the growth increment, the wood is classified as diffuse porous. In some cases there is a pattern that is neither ring nor diffuse porous and the wood may be called semi-ring porous or semi-diffuse porous.

For coniferous species, where vessels are lacking, tracheid diameter and cell wall thickness vary from early-season to late-season production. In some species the change is very gradual, leading naturally to the term gradual transition. In others, there is a sudden change from thin-walled, large-lumened tracheids to narrow-lumened, thicker-walled cells. This is an abrupt transition softwood. Figures 8 and 9 show these features in examples from both softwood and hardwood categories.

7. *Cell Distribution Patterns*

Coniferous wood is relatively simpler in organization than the hardwoods. There is only one major type of longitudinal element, the tracheid, aligned in radial rows. Nevertheless, there are certain differentiating features among the softwoods: cell diameter varies, which influences texture; the transition from earlywood to latewood may be gradual or abrupt; the latewood zone may be wide or narrow compared with the earlywood zone; longitudinal parenchyma cells may be visible and show a characteristic distribution; and resin canals may occur.

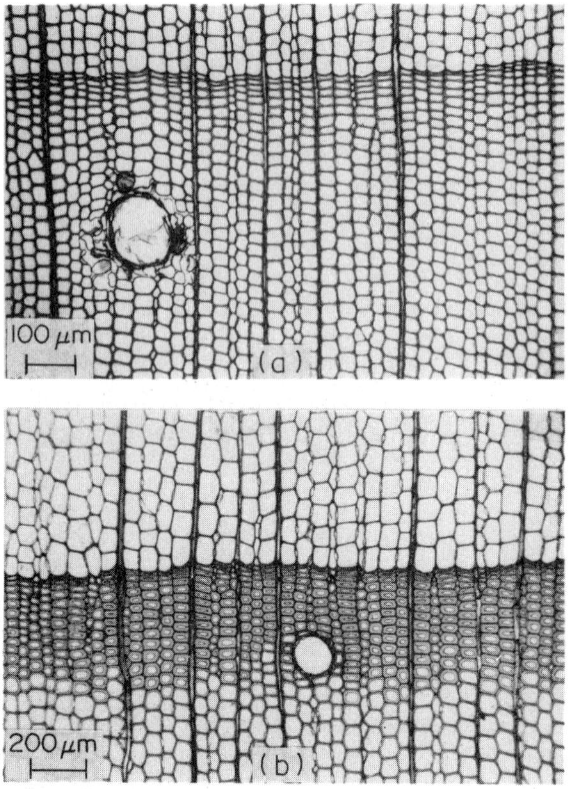

Figure 8
Cross sections of two softwoods showing earlywood–latewood transition: (a) gradual transition of tracheid sizes in western white pine and (b) abrupt transition of tracheid size and cell wall thickness in western larch (after Core et al. 1979)

When present, the parenchyma cells will show as dark filled cells in micrographs and may be found in random distribution, or as terminal cells (i.e., near the end of a growth ring). Resin canals are present, for example, in the pines and spruces, but lacking in cedars, true firs and redwood. When present they may be large or small, solitary or in groups or rows. They may be found in the outer part of the ring exclusively or distributed sporadically within the growth increment. The presence of normal longitudinal resin canals indicates that fusiform rays (rays incorporating a horizontal resin canal) may be observed on the tangential surface, or occasionally on the radial surface.

Among the hardwoods there are many distinctive patterns of cell distribution that are characteristic for a genus or a species. The ring-porous, diffuse-porous and semi-ring-porous classifications (Fig. 9) refer only to general vessel or pore distribution. There are also very specific pore patterns such as the wavy bands of summerwood pore groupings in elm or the dendritic groupings in oak.

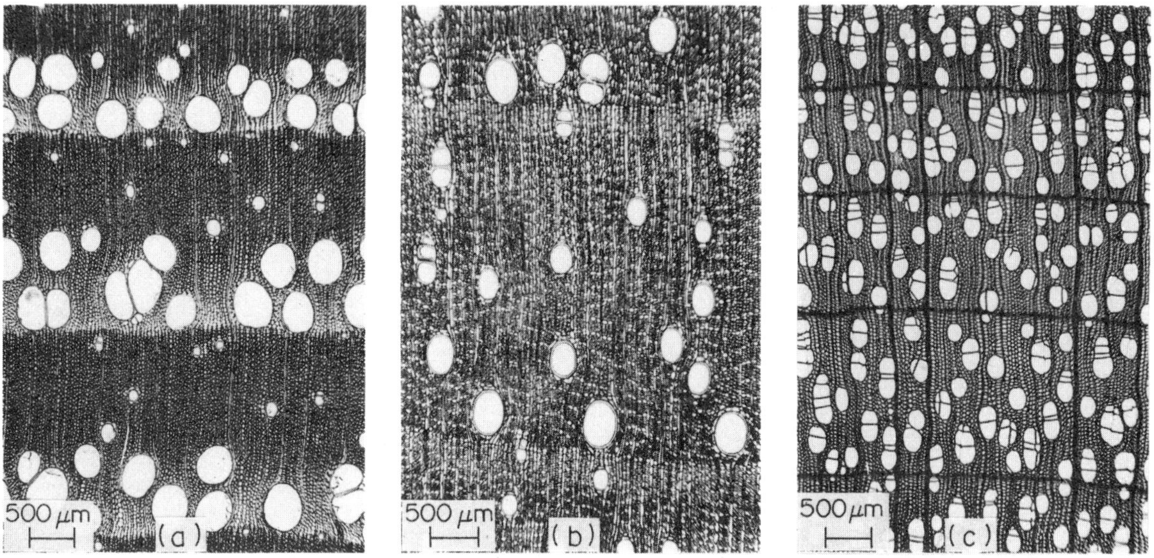

Figure 9
Cross sections of three hardwoods showing pore distribution patterns: (a) ring-porous hardwood (black ash), (b) semi-ring-porous hardwood (persimmon) and (c) diffuse-porous hardwood (yellow birch) (after Core et al. 1979)

The contribution of longitudinal parenchyma to the distinctive patterns of anatomy in hardwoods is extensive. These cells may appear in horizontal rows, in groups surrounding vessels, or marginally. Frequently, these patterns are unique, making the identification very simple for the experienced investigator.

Ray tissue can also be quite different in hardwoods than in softwoods. The rays in softwoods are generally narrow, uniseriate or biseriate (one or two cells wide), except when a resin canal forms a part of the ray, the so-called fusiform ray. In hardwoods some species always contain some multiseriate rays whereas others have uniseriate rays exclusively. There are also aggregate or oak-type rays that are readily seen even without magnification.

The composition of coniferous rays was previously described as being of parenchyma cells, with ray tracheids present in some cases. In hardwoods the parenchyma cells may be procumbent or upright. If only one type is present, it is a homocellular ray, whereas if both appear in a single ray, it is called a heterocellular ray. In some species the ray cells may have inclusions such as crystals, or specialized cells such as oil cells may be incorporated into the ray structure.

8. *Irregularities, Abnormalities and Defects*

Up to this point, the impression may have been given that wood is a uniform though anisotropic aggregation of elongated cellular elements. Unfortunately, this is not the case, as it is often difficult to find a straight-grained piece of wood that is free of all irregularities, abnormalities and defects.

Irregularities in wood may result from natural growth patterns. One of the most common of such interruptions is the knot. A knot is merely a tree branch that is included in the wood of the stem. The cambium of the trunk is continuous with the cambium of the branch so that wood produced in the region where a branch emerges is an intergrown or tight knot while the branch is living. Once the branch dies, the continuity is broken and any growth of wood around the branch produces an encased or loose knot. The size, nature and location of knots are important considerations when lumber is graded, especially for structural applications.

Irregularities of grain may be natural or caused by the sawmill operator. Curly, wavy or interlocked grain all result from fiber alignment on a relatively small scale. However, spiral grain is the result of helical orientation of cells around the tree stem along its entire length. Diagonal grain can be due to poor cutting practice when there is considerable taper in a log and it is cut parallel to the pith rather than to the bark. Other grain irregularities may also be traced to log form.

Abnormalities can also be of natural origin such as in reaction wood. This abnormal wood is formed in leaning trees. In conifers it is called compression wood because it appears on the underside of a tree stem, while in hardwoods it is termed tension wood, being found on the upper side of the leaning tree. Both forms of reaction wood have properties which detract from its normal use. For example, compression wood is very brash and can break suddenly under load. Tension wood warps, twists and bends severely during seasoning. Consequently, it is important to be able to recognize these abnormalities in wood before it is utilized in a critical application. Structural variations of reaction wood are discussed in the article *Ultrastructure*, while modifications in chemical composition are described in the article *Chemical Composition.*

Defects in wood can occur as a result of insect attack in the living tree or after the tree has been felled and processed. Wood rot or decay is caused by fungal attack which leaves the wood in weakened or deteriorated condition. Stain and mold may also appear because of attack by staining fungi (see *Decay During Use*).

See also: Constituents of Wood: Physical Nature and Structural Function; Deterioration by Insects and Other Animals During Use

Bibliography

Core H A, Côté W A, Day A C 1979 *Wood Structure and Identification*, 2nd edn. Syracuse University Press, Syracuse, New York

Jane F W 1970 *The Structure of Wood*, 2nd edn. Black, London

Kollmann F F P, Côté W A 1968 *Principles of Wood Science and Technology*, Vol. 1, *Solid Wood*. Springer, Berlin

Panshin A J, de Zeeuw C 1980 *Textbook of Wood Technology: Structure, Identification, Properties, and Uses of the Commerical Woods of the United States and Canada*, 4th edn. McGraw-Hill, New York

W. A. Côté
[State University of New York, Syracuse, New York, USA]

Mineral-Bonded Wood Composites

Mineral-bonded wood composites are molded panels or boards which contain about 10–70 wt% of wood particles or fibers and about 30–90 wt% of mineral binder. The properties of the composites are significantly influenced by the matrix and also by the amount and nature of the woody material and the density of the composites. Mineral binders used are portland cement, magnesia cement (Sorel), gypsum and mixed mineral binders which contain at least two components of the aforementioned binders. The composites are classified according to their density and their binder (Tables 1, 2 and 3).

1. Manufacturing Methods

The various stages in the manufacturing process are: preparation of the woody materials (Kollmann 1955, Maloney 1977); mixing these materials with the binder components, water and additives; mat forming; pressing; hardening or curing; demolding; further curing; drying; and sizing.

Three processes—the wet, the casting, and the semidry method—are used to manufacture the composites. The classification is based on the amount of water used for preparing the furnish, and also on the method to compact it to a coherent material.

In the wet process the thin slurry made of wood fibers, the binder, the aggregates, additives and water are fed to a sieve which is dewatered using vacuum chambers. The process can be continuous or discontinuous. In the continuous process a thin web is formed, e.g., using a Hatschek machine, which is wound on a cylinder until the desired thickness is reached. The web is then split and transferred to a steel caul. The green boards are compacted by pressure before the binder is allowed to harden. Since springback forces of the refined wood fibers are negligible after pressing, the boards are hardened without further compression. This method is used to manufacture flat and corrugated boards and molded products.

In the casting process, all the water needed to prepare the furnish is applied to the mixture. The furnish is a thick and fluid mixture and is compacted by gravitation or vibration or by a slight pressure, before it is allowed to harden. According to Joklik (1983) a mixture of wood chips, water containing a foaming agent, cement and sand are cast into wall elements using wooden frames as casting molds. The total water/binder ratio in this process is about 0.8. It is desirable to keep the water/binder ratio as low as possible in the manufacture of mineral-bonded wood composites, because excess water will leave the binder after hardening and curing, creating micropores which reduce the mechanical strength of the material. The extra water also prolongs the curing time needed to reach the desired strength. The total amount of water added exceeds the theoretical amount for the hydration (hardening) process, as water is also needed as homogenizing agent. Portland cement requires about 25%, magnesia cement 20–30% and gypsum hemihydrate 17% of water. The influence of the water/binder ratio is demonstrated by the following example: hardened portland cement of water/cement ratios of 0.4 and 0.8 resulted in a density of 2.0 and 1.6 $g\,cm^{-3}$, respectively.

In the semidry process all the water is applied to the wood furnish. However, the water becomes available to the binder after pressing (at 1–3 MPa) the furnish to the desired thickness, and forming a coherent material. The total water/binder ratio is about 0.4, and the woody material acts as a water reservoir. The furnish is spreadable, and can be formed into a mat, applying forming devices commonly used in manufacturing of particleboards. After pressing to the desired thickness, the pressure can be released or sustained until the boards gain enough strength and can be demolded without damage. Mineral-bonded fiberboards made

Table 1
Some characteristics and material requirements of medium-density mineral-bonded wood composites

Characteristic	Medium-density composites	
Structure of material	coarse-pored	porous
Wood particle/fiber	wood wool	flakes, shavings, slivers, sawdust
Particle dimensions		
length (cm)	50	0.4–7
width (mm)	3–6	2–15
thickness (mm)	0.2–0.5	up to 6
Binders	portland cement, magnesium oxysulfate	portland cement
Products	boards	boards, molding
Density (g cm^{-3})	0.36–0.57	0.45–0.80
Bending strength (MPa)	0.4–1.7	0.4–2.0
Elastic modulus (MPa)	300	
Compression strength parallel to surface (MPa)	0.4–1.2	0.5–2.5
Thickness swelling after 24 h in water (%)	(3.5)	1.2–1.8
Length swelling by humidity, RH 35–90% (%)	(0.7)	
Thermal conductivity ($W m^{-1} K^{-1}$)	0.09–0.14	0.1–0.20
Sound insulation by DIN 52210 (dB)	31[a]	31[a]
Resistance to wood-destroying fungi and insects	resistant	resistant
Fire test by DIN4102[b]	B1	B1
Material requirement per m^3		
wood (kg)	120–180	200–300
binder (kg)	180–250	200–300
additives (kg)	6–8	5–10
water (l)	200–290	150–300
Trade names	wood-wool boards	Arbolit (USSR) Velox (Austria) Durisol (Switzerland)

[a] 5 cm thick boards and 0.4 cm plaster on each side [b] A2 incombustible; B1 low flammability

according to the semidry method are hardened without further pressure, whereas particleboards have to be clamped. In a newly developed extrusion process, wood particles, fly ash, cement and water-containing waterglass are formed into hollow core panels. Due to the low content of wood particles and also the addition of sand, such panels can be piled after the extrusion process (Anon 1987e, f). The hardening time of most of the inorganic binders is long, compared with those of synthetic resins commonly used to manufacture particleboards and medium-density fiberboards. Portland cement requires about 8–24 h, magnesia cement 4–6 h and gypsum hemihydrate 15–30 minutes at ambient temperatures. The final strengths are attained after curing for 28 days. New developments aim to reduce the hardening time of mineral-bonded composites to minutes and enable the manufacture of composites, akin to that of particleboards and medium density fiberboards (Anon 1987a, b, c).

Another feature of mineral-bonded composites is the high ratio (1–10) of binder to woody material, whereas wood–polymer composites have a ratio of 0.08–0.12. Although the prices of inorganic binders per kilogram are rather low, due to the high amounts needed, the total cost of such binders per cubic meter of end product can be high.

1.1 Portland-Cement-Bonded Composites

Portland cement is a hydraulic binder and the bonding agent in hydrated cement is a calcium silicate. Cement is standardized in many countries. The classification is based mainly on the minimum compression strength (MPa) of specified specimens made of cement, sand and water. Generally three classes of cements are distinguished, e.g., according to German Industrial Standards PZ 35, PZ 45 and PZ 55. Wood–cement composites made of the same wood and with cement of different strength classes exhibit the same trends in

Table 2
Some characteristics and material requirements of medium-high density mineral-bonded wood composites

Characteristic	Medium-high-density composites				
Structure of material	compact	compact	compact	compact	compact
Wood particle/fiber	flakes, shavings strands	flakes, shavings	wood pulp (recycled pulp) and inorganic fibers	flakes	wood pulp (recycled pulp)
Particle dimensions		same as resin-bonded particleboard		same as resin-bonded particleboard	
length (cm)	1–3				
width (mm)	1–3				
thickness (mm)	nearly 0.3; fine particles for surface				
Binders	portland cement	magnesium oxysulfate, magnesium oxychloride	magnesium oxycarbonate	gypsum hemihydrate	gypsum hemihydrate
Products	boards, moldings	boards	boards	boards	boards
Density (g cm^{-3})	1.0–1.35	0.9–1.25	0.7–1.1	1.0–1.2	1.0
Bending strength (MPa)	6–15	7–14	8–10	6–9	4–7
Elastic modulus (MPa)	3000–5000	3000	3000	3000–5000	4000–5000
Compression strength (MPa) parallel to surface	10–15	10			
Thickness swelling after 24 h in water (%)	1.2–1.8	4–8		(destroyed)	(destroyed)
Length swelling by humidity, RH 35–90% (%)	0.12–0.25 . . . 0.4	0.25	0.10	0.05–0.07	0.04
Thermal conductivity (W m^{-1} K^{-1})	0.19–0.26	0.2–0.25	0.18	0.2	0.29
Sound insulation by DIN 52210 (dB)	32 (12 mm)	30 (16 mm)	29	41 (2 × 10 mm)	31 (10 mm)
Resistance to wood-destroying fungi and insects	resistant	resistant	resistant	—	—
Fire test by DIN 4102[b]	B1 A2 (spec. boards)	B1 A2 (spec.boards)	A2	B1 A2 (spec.boards)	A2
Material requirement per m^3					
wood (kg)	280	340	100	200	190
binder (kg)	800	500 MgO	500	800	870
additives (kg)	45	50 $MgSO_4$ or $MgCl_2$		5	500
water (l)	300	200		200	500
Trade names	Century board (Japan) Duripanel (Switzerland) Golden Board (Japan)	Ilves mineral (Finland)	Golden Siding Board (Japan)	Sasmox (Finland) Arborex (Norway)	Fermacel (Germany)

[a] 5 cm thick boards and 0.4 cm plaster on each side [b] A2 incombustible; B1 low flammability

strength development as shown by concrete. The hydration of portland cement is retarded by about 0.5% of wood starch or 0.25% of free monosaccharides or oligosaccharides or phenolic compounds (based on the weight of cement), making demolding within a reasonable length of time impracticable. Wood with a high alkali-soluble content (e.g., a high hemicellulose content) or wood deteriorated by fungi or excess of light or heat, may have the same effect. In general softwoods are more suitable than hardwoods. Methods to improve the suitability of woods are: natural seasoning (2–3 months); addition of accelerators; extraction of wood particles with water; addition of diffusion-retarding compounds; drying of the

Table 3
Some characteristics and material requirements of high-density mineral-bonded wood composites

Characteristic	High-density composites	
Structure of material	compact	compact
Wood particle/fiber	cellulose fiber	flakes, shavings 0.2–2 cm
Binders	cement and quartz	cement, fly ash
Products	boards, moldings	hollow core panels[a]
Density ($g\,cm^{-3}$)	1.35	1.4
Bending strength (MPa)	21	3–6
Elastic modulus (MPa)	6000	3400
Compression strength parallel to surface (MPa)		10–15
Thickness swelling after 24 h in water (%)	0.25	0.2
Length swelling by humidity, RH 35–90% (%)	0.15	0.2
Thermal conductivity ($W\,m^{-1}\,K^{-1}$)	0.35	0.40
Sound insulation by DIN 62210 (dB)		high
Resistance to wood-destroying fungi and insects	resistant	resistant
Fire test by DIN4102[b]	A2	A2
Material requirement per m^3		
wood (kg)	100	100
binder (kg)	1000	390 cement 300 fly ash 430 sand 10 waterglass 320 water
Trade names	e.g., Internit (Germany)	Fibrecon (Singapore) Ecopanel (Finland)

[a] extrusion process [b] A2 incombustible

wood particles; use of rapid-hardening cement; and addition of fumed silica or rice hull ash (Simatupang et al. 1987). Accelerators used in the casting and semidry processes are: calcium or magnesium chloride; a mixture of aluminum sulfate and calcium hydroxide or waterglass; waterglass; calcium formate or acetate; sodium or potassium silicofluoride; and sodium fluoride. In the wet process only waterglass is effective (Kitani 1985). The amount of accelerator used is in the range of 0.5–3% (based on the weight of cement). Curing of cement can be performed at ambient temperature or at higher temperature and higher pressure, e.g., hydrothermal treatment of cellulose cement fiberboards (Coutts and Michel 1983).

1.2 Magnesia-Cement-Bonded Composites

The bonding agent is the basic salt of magnesium chloride, sulfate, nitrate or phosphate with the general formula

$$x\mathrm{Mg(OH)_2}\,.\,y\mathrm{MgAR}\,.\,z\mathrm{H_2O}$$

where x, y and z are integers and AR denotes an acid radical. The various basic salts have different properties. Magnesia cement or Sorel cement composed of magnesium salts of hydrochloric, sulfuric or carbonic acids are mostly used. The cement is hard and tough, but is not weather-resistant. It is prepared by mixing calcined magnesium carbonate or magnesium oxide with a concentrated solution (12–25%) of magnesium chloride or magnesium sulfate (Mörtl and Leopold 1978). Due to the rapid hardening of magnesium oxysulfate at higher temperatures, even in the presence of higher concentrations of wood extractives, boards bonded with such cements can be manufactured continuously, e.g., Heraklithboards (Kollmann 1955, Maloney 1977). Magnesia-bonded particleboards are manufactured in Finland in a slightly modified particleboard plant, using the process described by Simatupang (1985).

Magnesium oxycarbonate is prepared by heating magnesium carbonate trihydrate. The latter is obtained by treating a magnesium hydroxide slurry with flue gases (free of SO_2) containing carbon dioxide (Kirk and Othmer 1978). Magnesium-oxycarbonate-bonded fiberboards are manufactured in Japan using a wet process (Anon 1987).

1.3 Gypsum-Bonded Composites

The binder used is gypsum hemihydrate ($CaSO_4 . \frac{1}{2}H_2O$) which is obtained by the calcination of naturally occurring gypsum, waste gypsum from chemical processes, e.g., phosphogypsum or gypsum from desulfurizing of flue gases (Wirsching 1978, 1984). There exist two phases (α and β) of gypsum hemihydrate. The β-hemihydrate, plaster of Paris, is mostly used as binder for composites. The binder components are rehydrated gypsum dihydrate crystals ($CaSO_4 . 2H_2O$) which crystallize from a supersaturated solution. Wood extractives generally inhibit the formation of perfect crystals, and cause a reduction of the shear strength between wood and hardened gypsum (Seddig and Simatupang 1988). The gypsum bond is not water resistant. Due to the short hardening time, plaster of Paris is used for the continuous manufacture of cardboards (Wirsching 1978) and fiberboards (Williams 1987). Gypsum-bonded particleboards according to the Kossatz process are currently manufactured in two plants in Finland and Norway (Hübner 1985, Wilke 1988).

New developments aim to extend this semidry process to gypsum fiberboards using recycled paper as reinforcing material (Anon 1988).

2. Utilization

The medium-density composites have been used extensively for insulation purposes and as fire-resistant materials. However, due to the availability of newer building materials, production has decreased worldwide. Recently the concrete casing method using boards or precast wood–cement moldings has found increasing application.

Cement-bonded particleboards, made according to Elmendorf (1966), are used as wall elements and siding materials in prefabricated houses. The properties and durability of such materials have been reviewed (Dinwoodie and Paxton 1983, Simatupang 1987). It is feasible to construct such buildings entirely from wood–cement panels by gluing the panels into folded shapes that form load-bearing structural elements (Anon 1978). Worldwide 27 cement particleboard plants exist. Of these, 12 are in the USSR and 5 in Japan. The rest are distributed in 9 countries.

The replacement of asbestos fibers in cement-bonded fiberboards with less harmful materials is making great progress. In Australia, New Zealand (Sharman and Vautier 1986) and Finland, cellulose fibers are used for such purposes. In Germany such fiberboards are used indoors, whereas outdoors a combination of synthetic polymer and cellulose fibers is utilized. Alkali-resistant glass fibers are also used as asbestos substitutes. Cellulose fibers have to be added, because the synthetic fibers are too coarse to retain the cement particles from the slurry during manufacturing. The nature of this natural fiber also has an influence on the properties of such boards (Simatupang and Lange 1987).

Magnesia particleboards, mostly overlaid with veneer, are used as low-flammability ceilings and walls in theaters and hotels. Gypsum fiberboards and particleboards have the same utilization as gypsum cardboards, with which they compete.

See also: Drying Processes; Particleboard and Dry-Process Fiberboard; Wood–Polymer Composites

Bibliography

Anon 1978 The Bison folding system (BFS). *Build. Prefab.* 3: 18–19

Anon 1987a *Neue, Unweltfreundliche Technologien zur Herstellung Zementgebundener Holzspanplatten*, Bison Information, Forschung und Entwickelung. Bisonsystem, Springe, FRG

Anon 1987b *Rauma-Repola and Falco Cement Flake Board Technology.* Rauma-Repola Engineering Works, Finland

Anon 1987c *The Ecocem Process for Fast and Economic Production of Cement Bonded Particleboards.* RAUTE Wood Processing Machinery, Raute Oy, Lahti, Finland

Anon 1987d *Golden Siding Board.* Nippon Hardboard, Nagoya, Japan

Anon 1987e *Natural Fibre Concrete's Mechanical and Physical Properties.* ACOTEC (Advanced Construction Technology). Salpakangas, Finland

Anon 1987f *ECOPANEL.* Raute Oy, Lahti, Finland

Anon 1988 *Information about Gypsum Fibre Board System Würtex.* Würtex Maschinebau, Uhingen, FRG

Coutts R S P, Michel A J 1983 Wood pulp fiber-cement composites. *J. Appl. Polym. Sci. Appl. Polym. Symp.* 37: 829–44

Dinwoodie J M, Paxton B H 1983 *Wood Cement Particleboards—A Technical Assessment*, Building Research Establishment Information Paper IP 4/83. Building Research Establishment, Princes Risborough, UK

Elemendorf A 1966 Method of making a non-porous board composed of strands of wood and portland cement. US Patent No. 3, 271, 492

Hübner J E 1985 The industrial production of gypsum boards with reinforcing wood flakes. *TIZ-Fachberichte* 109(129): 908–16

Kirk R E, Othmer D F 1978 *Encyclopedia of Chemical Technology*, Vol. 14. Wiley, New York, pp. 619–34

Kitani Y 1985 Inhibition of pulp cement board hardening by pulp and its relieving methods. *Mokuzai Kogyo* 40(5): 19–23

Kollmann F F P 1955 *Technologie des Holzes und der Holzwerkstoffe*, Vol. 2. Springer, Berlin, pp. 468–88

Maloney T M 1977 *Modern Particleboard and Dry-Process Fiberboard Manufacturing.* Miller Freeman, San Francisco, California

Mörtl G, Leopold F 1978 Caustic magnesia for the building industry. In: Cooper B M (ed.) 1978 *Proc. 3rd Industrial Minerals Int. Congr.* Metal Bulletin, London

Seddig N, Simatupang M H 1988 A shear method for testing the suitability of various wood species for gypsum-bonded wood composites and SEM-microscopy of the shear areas. *Holz Roh- Werkst.* 46: 9–13

Sharman W R, Vautier B P 1986 Accelerated durability testing of autoclaved wood-fiber-reinforced cement sheet composites. *Durability Build. Mater.* 3: 255–75

Simatupang M H 1985 Caustic magnesia as a binder for particleboards. *Proc. Symp. Forest Products Research International Achievements*, Vol. 6. SCIR Centre, Pretoria, pp. 1–19

Simatupang M H 1987 Manufacturing process and durability of cement-bonded wood composites. *Proc. 4th Int. Conf. on Durability of Building Materials and Building Components*, Vol. 1. Pergamon, Oxford, pp. 128–35

Simatupang M H, Lange L 1987 Lignocellulosic and plastic fibres for manufacturing of fibre cement boards. *Int. J. Cement Compos. Lightw. Concr.* 9(2): 109–11

Simatupang M H, Lange H, Neubauer A 1987 Influence of seasoning of poplar, birch, oak and larch and the addition of condensed silica fume on the bending strength of cement-bonded particleboards. *Holz Roh- Werkstoff.* 45: 131–36

Wilke D 1988 Die Gipsplatte kommt vorannunmehr auch in Norwegen. *Holz-Zentralbl.* 114(33): 502–3

Williams W 1987 Gypsum fiberboard proves success for Germany's Fels-Werke. *Wood Based Panels N. Am.* 7(2): 26–29

Wirsching F 1978 Gypsum. In: *Ullmanns Encyklopädie der Technischen Chemie*, Vol. 12 Verlag Chemie, Weinheim

Wirsching F 1984 Drying and agglomeration of flue gas gypsum. In: Kuntze R A (ed.) 1984 *The Chemistry and Technology of Gypsum*, ASTM STP 861. American Society for Testing and Materials, Philadelphia, Pennsylvania, pp. 160–73

M. H. Simatupang
[Institut für Holzchemie und Chemische Technologie des Holzes, Hamburg, FRG]

N

Nondestructive Evaluation of Wood and Wood Products

Nondestructive evaluation of wood and wood products is defined as any method by which information can be obtained without altering the end-use potential of the members evaluated. Many methods fall within this broad definition, such as conventional determination of density and moisture content. More sophisticated methods include defect detection by (a) optical scanning to locate knots and surface flaws, (b) ultrasonic sensing to determine elastic properties and to investigate wood internally, (c) microwave testing to determine moisture content, and (d) radiation testing to detect internal variations and/or insect damage.

This article discusses basic properties of wood and wood products and how these properties distinguish NDE methods for wood products from those used with other materials. It also discusses several NDE methods currently in use or being extensively researched for use with wood products.

1. Characteristics of Wood and Wood Products

Solid sawn structural members of wood consist of a nonhomogeneous material with anisotropic properties and a wide degree of variability. The features of wood that bring about its variable nature and its anisotropic properties can be explained by envisioning the cross section of a tree. Starting at the center (or the pith), the tree grows outward, adding a growth ring each year. This growth ring is usually composed of two types of tissue: springwood, formed in the period of fast growth, and summerwood, formed in the period of slow growth.

These two types of tissue have different densities and properties, and are one source of variability in the tree. The ratio of springwood to summerwood is dependent on species and environment, but commonly varies from one growth ring to the next. Further variability is due to the presence of knots, the bases of the tree's branches within the stem. Knots are a serious source of weakness in structural lumber.

The basic structural element of wood, the fiber, is generally aligned longitudinally with the tree stem and determines the direction of its grain. In a material of this construction, we may expect the longitudinal compressive and tensile strength to be high, the transverse values to be lower, and shear between bundles of fibers to be a plane of relative weakness. Tree growth and product-cutting practice, as for lumber, do not always produce grain parallel to the long dimension. The result, known as cross grain, produces a plane of weakness in the product.

Wood-based composite materials are manufactured by bonding wood elements (fibers, flakes, particles or veneers) together with an adhesive under elevated temperatures and pressures. These products have lower variability and lower strength and stiffness than wood parallel to grain.

2. Nondestructive Evaluation Methods

2.1 Visual Grading of Dimension Lumber

During the early years of lumber stress-grading development, efficiency in the cutting, grading and distribution of lumber was not as critical as it is today, and therefore grading rules were highly conservative. These rules, based on limited sampling and on visual inspection, have been subject to review periodically ever since, in step with the increasing competition and changes in mill practice, yet basic visual-grading concepts remain largely intact (Galligan and Pellerin 1964).

Visual grading is a method in which the average strength value of clear wood of each species is used as a basis for assessing the strength of the lumber. To arrive at design-stress values, the clear-wood values are first adjusted downward by factors which recognize the variability of clear wood, the effects of long-term loading in use, and a "safety" or accidental overload factor. The adjusted strength values then are multiplied by a further reduction factor, termed the strength ratio, which estimates the effect that a certain size of knot or other defect will have on strength. This procedure is outlined in ASTM Standard D 245-81 (1987).

Several problems are inherent in the visual-grading technique. Some of these are technical problems based on the fundamentals of the method; others are practical problems present in the mill. Basically, visual grading is a method that tends to ensure safety at the expense of efficient grading by undervaluing all pieces of a species through the use of a "variability factor" not based on the actual sample being visually tested. The strength properties of clear wood have a wide degree of variation, and this must be taken into consideration in the method since no test of the clear-wood properties is made on the production line. Modulus of elasticity values, for example, can vary by a factor greater than two, and strength values have approximately the same variation. Further, the application of a strength ratio to the clear-wood strength value assumes that the visual grader can assess at a glance which of the competing flaws in the lumber is controlling and that this decision is sufficient to assign proper grade to that particular piece of lumber. The variability and inaccuracy resulting from these procedures must be compensated for by the methods used in arriving at design-stress values.

This discussion of visual-grading methods has been meant to be critical purely from the technical point of view of nondestructive testing, that is, from the position that an adequate nondestructive test must be accurate with regard to each piece.

2.2 Bending Modulus of Elasticity Measurement

Traditionally, the only widespread nondestructive evaluation of lumber other than visual grading has been measurement of moisture content through the use of electrical resistance and capacitance methods. With these exceptions, nondestructive evaluation did not make its influence felt on the lumber industry until the advent of machine graders and the method called machine-stress rating (MSR) in 1962. These machines measure the modulus of elasticity E of individual lumber specimens that are passed through them endwise. They accomplish this by measuring the bending deflection resulting from a known load or by measuring the load required to accomplish a given amount of deflection. Special features of some machines include sensitivity to low-point elasticity, that is, the ability to detect sections of the lumber which may have modulus of elasticity much lower than the average of the piece.

Design stresses are determined from the machine E values by a very useful regression between E and modulus of rupture R. In the USA and Canada, visual-grading restrictions are placed on the machine results to achieve even greater accuracy. The method is the result of years of research, and study continues in an effort to further improve the working correlation.

The machine used almost exclusively in the USA and Canada is the Continuous Lumber Tester, commonly referred to as the CLT-1, whereas in the European countries the Plessy Computermatic is the machine most commonly used.

A brief description of the CLT-1 is presented for a better understanding of MSR. The lumber to be evaluated passes through a series of rollers which cause the lumber to be deflected or bent, first downward by a fixed amount and then upward by nominally the same fixed amount. The forces required to achieve this bending are measured and then averaged together to provide a force measurement that is compensated for kink and bow in the piece. Before combining measurements in this way, the force measurement from the downward-bending section is delayed to correspond with the force measurement from the upward-bending section for any given location along the piece being evaluated.

The compensated force is measured continuously as the lumber passes through the machine. The lowest value of E (low-point E) and the average value (average E) are used to determine the machine grade for each piece of lumber. To meet the requirements for a particular MSR grade, the average E value must be above the average E threshold for that grade and simultaneously the low-point E must be above the low-point E threshold for the grade. The CLT identifies each piece of lumber with a spray mark according to the machine grade for the piece.

The MSR grade stamp is applied manually by a visual grader who first determines if the piece meets the visual requirements for the grade identified by the machine mark. The visual grader can agree with the machine grade or can downgrade the piece but in no case is allowed to improve above the grade the machine grade. In addition, lumber produced by this machine/man combination must, on a sampling basis, meet off-line quality control tests for stiffness and strength.

The largest unknown in the analysis for an individual mill is the yield or production estimate for each MSR grade. These yield estimates can be expected to vary between mills, and within a mill, with changes in the log supply. A comprehensive report by Galligan et al. (1977) discusses in considerable detail the procedures for estimating yield.

Measurement of the flexural rigidity (*EI* product) of composite panel products has received considerable attention recently as a potential NDE method. A machine similar to that used for MSR lumber production is used. As with MSR lumber grading, an estimate of a panel's modulus of rupture and maximum moment-carrying capacity are derived from established regression equations.

2.3 Transverse Vibration

This method is based on transverse oscillations that permit measurement of two fundamental properties of materials, namely energy storage and energy dissipation. These fundamental properties were hypothesized by Jayne (1959) to be related to the same mechanisms that control the mechanical properties, i.e., modulus of elasticity and modulus of rupture, respectively.

The outward manifestation of energy storage of material is its natural frequency in vibration and, through well-established equations relating natural frequency to modulus of elasticity, a direct calculation of this modulus can be made. For a beam, the relationship is expressed for transverse vibration as follows:

$$E_v = \frac{f_n^2 W L^3}{k^2 I g} \tag{1}$$

where E_v is the dynamic modulus of elasticity, f_n the natural frequency, W the weight, L the span between supports, k a constant dependent on mode of support, I the moment of inertia and g the acceleration due to gravity.

In contrast, the mathematical relationship involving energy dissipation and modulus of rupture has not been well established. However, Jayne's hypothesis of causal relationship indicated the possibility of developing a system by which energy dissipation could be used to predict modulus of rupture. Thus, the problem was first to develop a means of measuring energy dissipation, and second, to develop correlative information sufficient to prove the existence of a useful relationship between energy dissipation and strength.

The method developed for this purpose involved the use of free transverse vibration. In this case, energy dissipation is measured as the rate of decay, or logarithmic decrement δ, of the amplitude of the vibration of the vibrating body and is expressed in the form:

$$\delta = \ln(A_0/A_n)/n \qquad (2)$$

where A_0 and A_n are the amplitudes of two oscillations n cycles apart.

2.4 Stress-Wave Propagation

Like transverse vibration, the stress-wave propagation method is a dynamic analysis, but it differs in that the direction of the excitation is longitudinal and the frequency of excitation is about 1000 times higher than for transverse vibration. Despite these basic differences, both methods yield very similar values of modulus of elasticity for construction lumber.

The propagation of longitudinal stress waves in solids is influenced in a complex manner by both the mechanical and physical properties of the medium. To describe the propagation for practical use, the complex expressions commonly are simplified to elementary one-dimensional wave propagation theory as applied to an isotropic, homogeneous material. If a specimen has lateral dimensions which are small compared with the wavelength of the propagating wave, this simplified theory yields the following equation relating the modulus of elasticity E with the velocity of propagation c and density ρ:

$$E = c^2\rho \qquad (3)$$

When wave-velocity measurements were made on wood products, Eqn. (3) yielded values of E which agreed closely with the corresponding static moduli.

Stress-wave NDE methods have also been intensively investigated for use with reconstituted wood-based materials and are currently used for mechanical property evaluations in veneer grading programs for use in laminated veneer lumber manufacture as well as for defect (delamination) detection. Strong relationships between stress-wave parameters, such as wave speed and attenuation, and the mechanical properties of wood-based composites (internal bond, and tensile and flexural stiffness and strength), have been shown to exist (Ross and Pellerin 1988, Ross and Vogt 1985).

The vibration methods described above have been commercially utilized to evaluate wood products such as lumber, particleboard and veneer. The methods have also been successfully used to evaluate the properties of poles, logs, plywood, laminated beams, furniture components and paper. In addition, the stress-wave method has been used to detect the presence of decay in large structural members of buildings.

Results from these methods and others are reported in the *Proceedings of the 4th, 5th, and 6th Nondestructive Testing of Wood Symposia* (Pellerin and Galligan, 1978, Ross and Vogt 1985, Beall 1987). The need for further research still exists, however, on methods that can examine anisotropic, heterogeneous lumber to estimate its basic design values.

The advent of the machine graders and presence of new dynamic methods is a beginning. High efficiency will probably result from a marriage of the best features of the visual system and the piece-by-piece thoroughness of the automated nondestructive test.

2.5 Acoustic Emission/Acousto-Ultrasonics

The use of acoustic emission (AE) technology has also received considerable attention (see *Acoustic Emission and Acousto-Ultrasonic Characteristics*). Beall (1987) has shown strong correlative relationships between several AE parameters and the internal bond strength of particleboard, medium-density fiberboard and oriented strandboard products. Most promising parameters include acoustic emission events to failure and the attenuation of pulsed ultrasonic waves injected into the material.

See also: Lumber: Behavior Under Load; Lumber: Types and Grades

Bibliography

American Society for Testing and Materials 1987 Establishing structural grades and related allowable properties for visually graded lumber, ASTM Standard D 245-81 In: *1987 Annual Book of ASTM Standards.* ASTM, Philadelphia, Pennsylvania

Beall F C 1987 Acoustic emission and acousto-ultrasonics for application to wood products. *Proc. 6th Nondestructive Testing of Wood Symp.* Washington State University, Pullman, Washington

Beaton J, White W B, Berry F H 1972 Radiography of truss and wood products. *Mater. Eval.* 30(10): 14A–17A

Galligan W L 1965 *Proc. 2nd Symp. on the Nondestructive Testing of Wood.* Engineering Extension Service, Washington State University, Pullman, Washington

Galligan W L, Pellerin R F 1964 Nondestructive testing of structural lumber. *Mater. Eval.* 22(4): 15–20

Galligan W L, Snodgrass D V, Crow G W 1977 Machine stress rating: Practical concerns for lumber producers, FPL 7. USDA Forest Products Laboratory, Madison, Wisconsin

Jayne B A 1959 The vibrational properties of wood as indices of quality. *For. Prod. J.* 9(11): 413–16

Konarski B, Wazny J 1974 Use of ultrasonic waves in testing wood attacked by fungi. *RILEM 2nd Int. Symp. New Devel. Nondestruct. Test. Non-Metall. Mater.* 1: 11–18

Lakatosh B K 1966 *Defectoscopy of Wood.* Lesnaya Promyshlennost, Moscow (Translated from Russian and published for the USDA Forest Service, contract NSF-C466)

NASA SP-5113 1973 *Nondestructive Testing—A Survey*. US Government Printing Office, Washington, DC
Nearn W T, Bassett K 1968 X-ray determination and use of surface-to-surface density profile in fiberboard. *For. Prod. J.* 18(1): 73–74
Pellerin R F, Galligan W L 1978 *Proc. 4th Nondestructive Testing of Wood Symp.* Engineering Extension Service, Washington State University, Pullman, Washington
Ross R J, Vogt J J 1985 Nondestructive evaluation of wood based particle and fiber composites with longitudinal stress waves. *Proc. 5th Nondestructive Testing of Wood Symp.* Washington University, Pullman, Washington
Ross R J, Pellerin R F 1988 NDE of wood composites with longitudinal stress waves. *For. Prod. J.* 38(5): 39–45

R. F. Pellerin
[Pullman, Washington, USA]

R. J. Ross
[Trus Joist Corporation, Boise, Idaho, USA]

P

Paper and Paperboard

Paper and paperboard production is an environmentally sound, integrated approach to converting natural resources (wood, water, energy, chemicals) into a myriad of useful packaging, communication and specialty products. This article relates the raw materials and processing steps in the manufacture of paper; discusses major paper properties; summarizes applications, production levels and geographic consumption; and indicates environmental, energy and recycling considerations.

1. Raw Materials

Paper is an interlocking network of cellulose fibers enhanced with filler materials and binders. Cellulose fibers, which are the dominant component in paper or paperboard products, have a strong influence on the practical range of physical and mechanical properties of the product. The provision of fiber represents the highest raw material cost for a typical paper mill.

The average "whole" fiber in a paper product is hollow (10–50% void) and ribbon-like with dimensions of 1 mm × 25 μm × 5 μm (length × width × thickness) and weighs 2×10^{-7} g (Dodson 1976). The native cellulose polymer is composed of anhydroglucose monomer units (molecular weight 162) with an overall original degree of polymerization of 2500–8000. Fiber-source and papermaking-process variables combine to produce a wide range of fiber lengths in a paper product. The major effect of wood species (deciduous and coniferous) upon fiber dimensions is shown in Fig. 1. A typical length for hardwood fiber is 0.7 mm and for softwood 2.5 mm.

In contrast to man-made fibers, which are uniform in composition, the wood fiber structure contains four distinct concentric layers (Fig. 2). Each layer has its own orientation of small fibrous elements, fibrils and chemical makeup (see *Ultrastructure*). The primary wall, which is the first to be removed in either chemical pulping or mechanical treatment, is very thin with random fibrils enmeshed in lignin and intracellular adhesives. There are three sublayers in the adjoining secondary wall. The first or outer layer (S1) is a moderately thick layer composed of fibrils in laminae, wound at precise large angles relative to the fiber axis and crossing one another for balanced structural strength. The S1 and the primary layers are rich in lignin. The secondary wall middle layer (S2) is very thick, lamellar and has a high cellulose content. Fibrils in S2 are wound at a single small angle (about 5–20°) relative to the fiber axis. The S2 layer provides most of the characteristic mechanical properties (especially

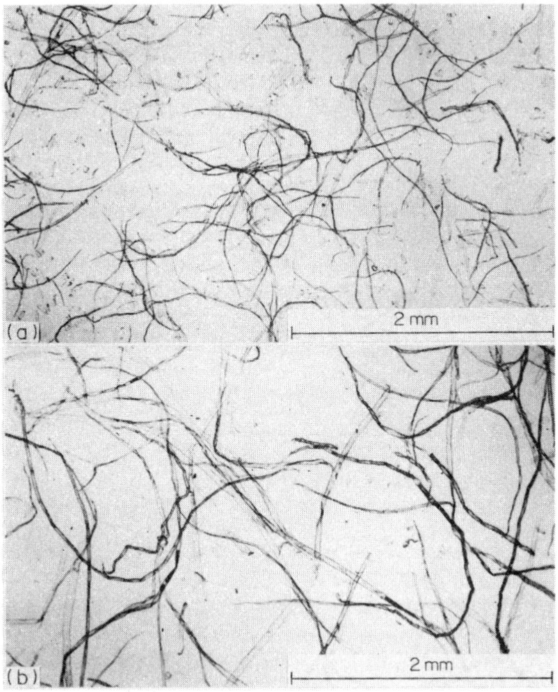

Figure 1
(a) Hardwood (deciduous) fiber for papermaking, and
(b) softwood (coniferous) fiber for papermaking

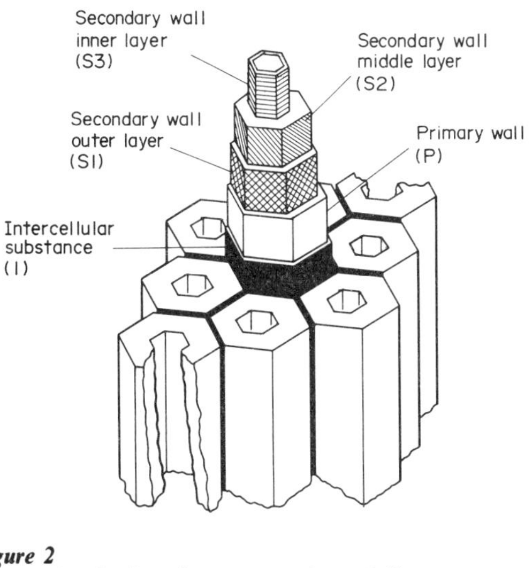

Figure 2
Schematic of cell wall structure of wood fiber

tensile strength) associated with wood fibers and paper. The innermost layer of the secondary wall (S3) resembles the S1 layer in thickness but has fibrils only wound in a single large angle relative to the axis.

Fiber sources are chosen to obtain the desired blend of strength and physical properties for the product. In North America, long, thin softwood fibers are included for tear resistance, while short, stubby hardwood fibers enhance bulk and surface smoothness. In developing countries, where labor costs are relatively low, native vegetable fibers such as bamboo and annual crops (sugar cane bagasse and straw) may be the fiber of choice. These fibers can have exceptional fiber length and strength, but their large-scale use in paper awaits economic production techniques for processing and dispersing them (Clark 1969, McGovern 1982). In industrialized countries, very long man-made polymer fibers such as polypropylene or glass may be blended with cellulose to confer tear resistance to wallpapers and envelopes.

Fiber substitution for traditional hardwoods and softwoods has become increasingly common in certain industrialized regions such as Europe. Eucalyptus pulp has captured more than 40% of the European market for paper-grade hardwood (Gallep 1987). It is exported worldwide mainly from Portugal, Spain and Brazil at levels of 0.5–1 Mt per year from each country. The short, thin eucalyptus fibers offer softness and printability benefits in tissue and paper grades (McGrath 1987).

An increasing source of paper fiber is secondary or recycled post-consumer fiber. Major sources of recycled fiber are corrugated cartons, newspapers and shredded business papers. Recycled fiber must be freed of extraneous matter such as inks, coatings and polymeric films. Generally, papers made from recycled fibers have low tensile strength but have increased opacity and better formation than papers made from virgin fibers, because the repeated processing shortens and mechanically disrupts the fibers (Altieri and Wendell 1967).

Waste-paper recycle has risen from a 1968 level of 20% to the 25–30% level in the 1980s, and is projected to approach 35% by the year 2000. Major applications for recycled fiber include newspapers, corrugated paperboard and tissue (Franklin 1986).

A wide range of organic and inorganic additives are used to modify the strength, aesthetics and physical characteristics of paper. Except for dry or wet strength agents, additives generally reduce overall mechanical strength of paper by interfering with hydrogen bonding.

Three major additive categories are fillers, internal size and surface size. Both inorganic (clay, calcium carbonate and titanium dioxide) and organic (urea formaldehyde) fillers are added to paper in order to improve optical, surface and bulk characteristics. Rosin or synthetic materials (alkyl succinic anhydride and alkyl ketene dimer) are added as internal sizing agents to reduce the sensitivity of the fiber web to moisture during printing. Surface sizing with a binder such as starch enhances both surface strength for printing and bulk strength properties for conversion and end-use requirements.

To enhance appearance and printability of paperboard, a very thin layer of pigment and binder is applied. Fine-particle-size clay, titanium dioxide and calcium carbonate pigments are major inorganic coating ingredients. Polystyrene coating pigment enchances gloss. Primary coating binders are starch, styrene butadiene and poly(vinyl acetate).

Large quantities of water and fuel are required to produce a tonne of paper. Water is essential in transporting and treating pulp slurries, and in developing strength in paper products (Campbell effect). Chemical recovery and paper-drying processes are the major consumers of energy.

2. *Production of Pulp*

2.1 Pulping

Pulping of wood can be done mechanically with or without steam and/or chemicals to obtain fibers in high yield (80–95+%). High-yield fibers have considerable lignin and hemicellulose associated with the cellulose, which causes the fibers to be stiff, to bond relatively poorly and to produce a bulky fiber web. Papers from mechanical pulps have low tensile and bursting strengths.

The pressurized groundwood (PGW) process, developed by Finnish papermakers, has found application in lightweight coated (LWC) and super calendered (SC) grades, while Swedish-developed processes for chemithermomechanical pulp (CTMP) and chemimechanical pulp (CMP) have proven attractive for making fluff pulp, tissue and bleached paperboard (Breck and Styan 1985).

The "kraft" or sulfate process, illustrated in Fig. 3, is the dominant commercial pulping method (40–70% yield). The pulp has higher cellulose content than mechanical pulp and provides better hydrogen bonding among fibers. The high-pH kraft process typically uses sodium hydroxide (caustic) and sodium sulfide to delignify wood chips under high pressure and temperature in a digester. Unbleached products using kraft pulp include sacks, bags, corrugated board, and saturating paper for laminates. Bleached kraft products included fine papers and printed folding cartons.

The third pulping method, the sulfite process, uses sulfur dioxide combined as a salt (sodium, calcium, magnesium or ammonium). Normal operating ranges during pulping are 120–145 °C and pH 1–9. Sulfite pulps range from greaseproof with relatively high lignin (4%) and high hemicellulose (14%), to dissolving pulp with high alpha-cellulose, low lignin (1%) and low hemicellulose (6%) for cellulose derivatives.

Modern woodyards stress greater pulp uniformity, beginning with screening wood chips by thickness rather than by length which typically reduces digester rejects by a factor of 1.5–3.0 (Christie 1987). A growing trend is pulping in the presence of anthraquinone to increase pulping rate and interfiber bonding potential (Tay et al. 1985).

Nonpaper pulps (Gross 1982) account for about 5% of all the pulp produced and appear chiefly as: (a) chemical cellulose or dissolving pulp, (b) fluff pulp for diapers and other high-bulk products, and (c) speciality pulps used in products ranging from battery separators to plastic moldings.

Chemical coproducts from the papermaking process include speciality chemicals used in pollution control, agriculture, dyestuffs, cement, rubber, protective coatings and plastics. A wide range of activated carbon, lignin and tall (pine) oil products derived from wood are commercially available.

2.2 Bleaching

The common bleaching method (Fig. 3) involves removal of lignin and hemicellulose from naturally white cellulose in kraft and sulfite pulps. The process can involve combinations of chlorine, chloride dioxide, hypochlorite, and less frequently oxygen, in a series of two to six stages. An alkaline wash or extraction stage between bleaching stages removes chlorinated by-products. Excess bleaching can degrade pulp strength.

Whitening of high-yield (chemimechanical) pulps employs hydrogen peroxide or sodium hydrosulfite which bleach lignin. These agents do not significantly weaken the fiber, but their "bleaching" effect can be reversed by ultraviolet light (newspaper yellowing in sunlight).

The use of chlorine compounds has been reduced in new plants because kraft and sulfite pulp mills have shown increasing interest in first-stage treatment by oxygen as a means of reducing effluent loads and

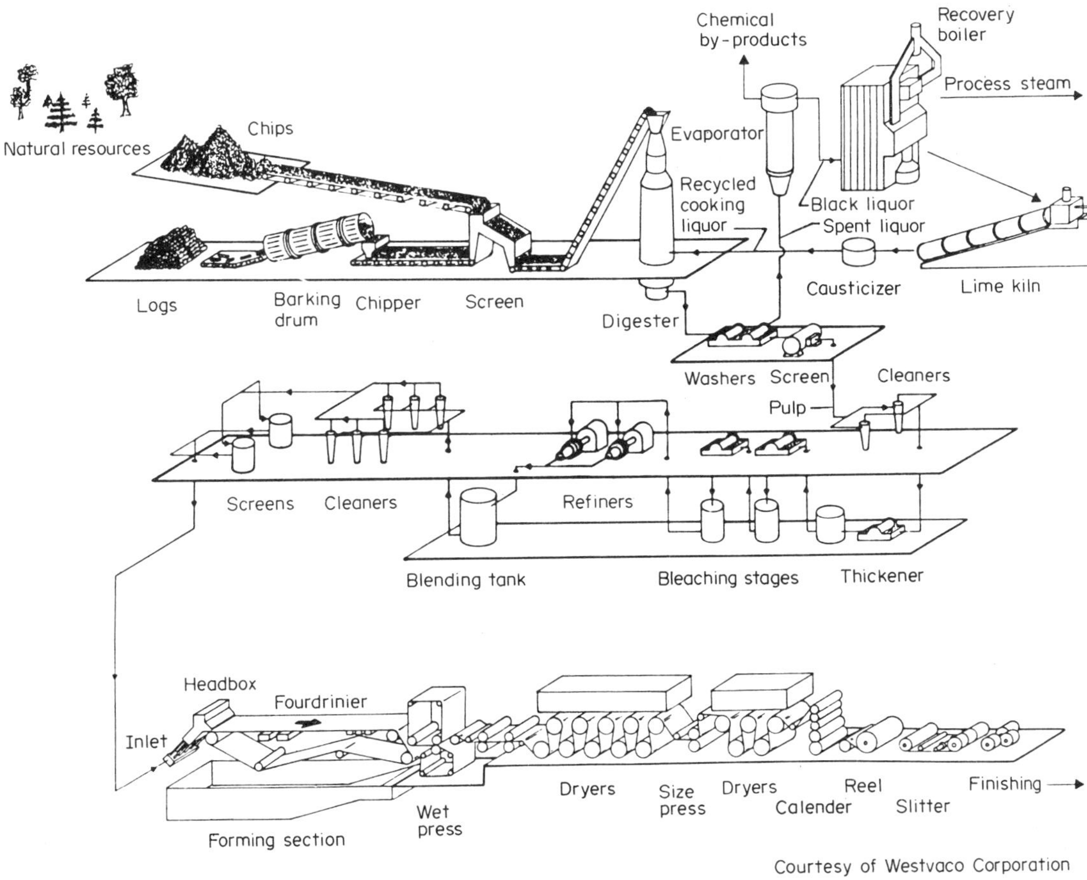

Figure 3
The papermaking process (after Wardrop and Bland 1959)

operating costs (Ducey 1986). Expansion in production of CTMP/CMP utilizes more sodium hydrosulfite and hydrogen peroxide stages to bleach, rather than remove, lignin for these high-yield pulps (Carmichael 1986).

3. Production of Paper and Paperboard

Cellulose fiber is the major ingredient in paper products, but water is the predominant material used in its manufacture. Water is essential in developing the characteristic strength of the fiber network through hydrogen bonding. The furnish for paper is supplied at a low consistency (0.1 to 1.0%) to equipment which forms, consolidates and dries the web at up to 1700 m min^{-1}.

3.1 Stock Preparation

Prior to forming the wet fiber web, the furnish for the process is subjected to mechanical refining which can fibrillate, cut and/or modify the fibers to increase fiber/fiber contact and thereby enhance strength and hydrogen bonding in the product. Recycled paper is redispersed prior to refining. The dispersion and refining operations occur at 3–12% consistency.

Following refining, the furnish is cleaned and screened to remove foreign materials which detract from quality and runnability. Filler materials added to the low consistency furnish for smoothness and optical properties include clay, titanium dioxide and calcium carbonate. Internal additives for styling and strength (both wet and dry) can be added during stock preparation. Alkaline papermaking has not advanced in North America as rapidly as projected (Wuerl 1986).

3.2 Forming

There are three categories of equipment for forming a wet paper web: fourdrinier, cylinder and twinwire. The most common, the fourdrinier system, is described below and illustrated in Fig. 3.

Formation and consolidation of the fiber network are major areas of technical development in the paper industry. During forming and initial consolidation on a fourdrinier machine, 95% of the water is removed to yield a self-supporting web which still contains 4 kg of water per kg of solid material. Fiber orientation and filler and fines distribution can be adjusted through operating conditions during forming.

The first steps in forming a fiber network involve: uniform delivery of furnish to headbox, adequately dispersing the furnish, and extruding the dilute slurry from the "slice". The stock jet impinges onto a fast-moving continuous fabric to begin dewatering. Fiber orientation in the plane of the web is affected by the relative speeds of the stock jet and the forming fabric. Direction (in twin-wire systems) and rate of water removal are used to control filler and fiber fines distribution through the web thickness. Drainage elements which enhance water removal can also promote turbulence and improve stock uniformity.

Relatively recent technical developments allow high-speed formation of the web in simultaneous or sequential multilayer structures. Formation achieved between two conventional wires can minimize web surface differences typical of paper made on a fourdrinier machine, and thereby produce a more balanced product. These developments originated in key papermaking countries including Finland, Germany, Sweden and the USA, and to a lesser extent in Australia, Canada and Japan (Wahlström 1981).

Initial growth in multilayer paper and board production primarily involved tissue and linerboard grades. It is spreading to the mid-range of basis weights to include printing and writing grades. Cited as reasons for interest in multi-ply products are flexibility in furnish utilization, for example, high-grade chemical pulps in surface layers with mechanical or recycled pulp in core layers, as well as flexibility in product design (O'Brian 1987).

3.3 Pressing

Drainage on a fourdrinier increases web consistency to approximately 20%. A web at this moisture level can be lifted from the supporting fabric and transported to the press section for additional water removal. The brief residence in the press nips doubles the web solids level to 40%. Vacuum, press loading, temperature and felts are parameters used to enhance water removal. A multiple-nip press section is shown in Fig. 3. Laboratory studies of extended nip pressing indicate solids levels as high as 55% are possible where high density papers are acceptable. This level of dewatering could greatly enhance production capacity (Wahlström 1981). In the press section, the wet web can experience a significant draw or machine-direction stress which accentuates fiber alignment and promotes greater strength in this direction at the expense of cross-machine direction strength.

The achievement of extended or wide-nip wet pressing resulted from the combined efforts of research institutes and equipment and materials suppliers. In addition to the overall press design, advances in materials for press rolls and new fabric designs were implemented. Claimed benefits from wide-nip pressing include higher machine speeds by increasing dewatering, as well as potential for improved machine-direction and cross-machine direction paper strength properties (Helm 1987). Feltmakers are improving the uniformity of pressure distribution and dewatering capability in the press nip by redesigning the fabric (Coan 1984).

A relatively recent development is press-drying, a pilot process for drying a web while it is restrained in the thickness direction. A key to effectiveness of the process is the use of unbleached kraft. A hardwood furnish can be made to behave like the more expensive softwood furnish in the liner or medium of corrugated boxes, giving: (a) reduced compression creep, (b) higher burst strength, and (c) greater tensile energy absorption (Setterholm 1979).

3.4 Drying

Excessive moisture in the web after pressing is thermally removed. This approach is expensive compared to the preceding mechanical procedures. Available drying methods depend on web porosity and basis weight. Most paper and board grades are dried on steam-heated cylinders. The two sides of the web alternately contact a series of dryers as the moisture level is reduced to approximately 5%. Felts promote web contact with drums and increase drying rate. Figure 3 shows a multicylinder dryer section.

Lightweight tissues and towels can be pressed and dried on a single "Yankee" dryer drum from which the web can be released by a doctor blade. Doctoring also provides creping which increases sheet extensibility, bulk and absorbency. Lightweight paper can also be processed in a "through-dryer" or air float dryer.

3.5 Finishing

Many alternatives are available to enhance base stock surface. It is generally calendered to improve smoothness. Aqueous, solvent or extruded polymeric coatings can be applied to meet end-use requirements. In fine papers and folding cartons, the coating and finishing steps smooth the relatively rough-surfaced base stock to meet high print-quality requirements.

Calcium carbonate is the major coating pigment in European paperboard applications because of its availability, low cost and reported benefits of improved smoothness from higher solids content (Wintgen 1987).

Coating and calendering operations can either be performed "on" or "off-machine." Extruding a polymeric film over the paper, or metallizing the paper surface (Carter and Beardow 1983) are done off the paper machine to accommodate differences in speeds and reel size, and special handling requirements.

3.6 Converting

Mechanical compaction is used for extensible paper applications such as sacks for agricultural and chemical products which require stretch and energy absorption. "Clupak" extensible paper was developed in 1958 from an invention by Sanford Cluett, who also discovered the "Sanforized" process for textiles.

Paper and board for packaging is converted by such steps as printing, cutting, folding and glueing to yield folded cartons or corrugated boxes. Fiber drums are formed in wrapped, glued multiple layers, and fastened with lids. Other formed products are pressed—for example, egg cartons and special casings. Saturating kraft sheets are impregnated with resin before laminating.

4. Properties of Paper and Paperboard

Paper and board properties derive from the raw materials properties and papermaking processes. This section presents: (a) theories on paper mechanical properties, (b) mechanical properties of paper relative to other composite materials, (c) mechanical requirements for paper cartons and boxes, (d) online measurement of paper and board properties, and (e) optical, surface, aesthetic and specialty properties.

4.1 Theoretical Models for Paper Mechanical Properties

Paper is generally considered to be a viscoelastic, anisotropic nonlinear composite material. Theoretical models describing mechanical properties of paper include those based on: (a) molecular bonds between fibers, (b) fiber networks, and (c) an orthotropic continuum.

Nissan applied hydrogen-bonding theory to explain changes in Young's modulus for paper in the presence of different solvents, and the effects on paper properties of chemical substitution in cellulose fibers. As noted by Nissan (Mark and Murakami 1983), there is at present no model covering the intermediate scale of microfibril interactions which are believed to be important in bond development and disruption during paper drying and calendering.

A number of models for paper are based on the fiber network scale interactions. Examples include models by Kallmes and Perkins. Other models in this area were presented by Cox (1952), Page and Seth (1980) and Van den Akker (1970).

Fiber network models consider polymeric and physical properties of fibers, bonding areas, geometrical network structure, and fracture mechanics. Network models omit consideration of network nonhomogeneity, nonplanar fiber orientation, and deformation of the web in the thickness direction. Despite the limitations, these models can predict certain mechanical behavior, for example edgewise compressive strength (Perkins and McEvoy 1981).

Another level of mechanical model considers paper an orthotropic plate or continuum. Because Baum used ultrasonic waves with an appropriate wavelength to paper thickness relationship, he could describe paper in a plate form (see Baum et al. 1981). The ultrasonically determined mechanical property values are higher than the corresponding values determined destructively at lower strain rates, but the two approaches provide similar mechanical descriptions. Ultrasonically measured fundamental properties have been correlated with traditional paper tests such as mullen (burst) and edgewise compression strength (Fleischman et al. 1982).

4.2 Fundamental Mechanical Measurements

The theory of composites (Cox 1952) predicts properties such as Young's modulus for paper from fiber characteristics:

$$E_p = 1/3\ E_f\{[1 - W/(L\,RBA)][E_f/(2G_f)]^{1/2}\}$$

where E_p is the sheet Young's modulus, E_f is the fiber modulus in the axial direction, W is the fiber width, L is the fiber length, RBA is the relative bonded area of sheet, and G_f is the fiber shear modulus of deformation. For well-bonded paper sheets (highly refined and pressed well-delignified pulp) of straight fibers, the in-plane elastic modulus of paper is one-third of the elastic modulus of the component fibers (Page and Seth 1980).

Paper properties related to fracture have been quantified (Dodson 1976). Levels of fracture energy per unit new area are of the order:

(a) zero span tensile: 5×10^7 J cm^{-2} (from area under load–extension curve)

(b) tearing: 0.5 J cm^{-2}

(c) splitting: 5×10^{-3} J cm^{-2}

Paper zero-span tensile strength is directly dependent upon individual fiber strength. A change in zero-span tensile with paper orientation indicates a change in fiber orientation. Fiber-to-fiber bonds and frictional forces contribute to tear. The relatively low value for splitting reflects only fiber-to-fiber bonding since fiber orientation in the thickness direction is low and frictional forces are minimal. The energy consumed in propagating a fracture across a paper strip is believed to be of the order of 10^5 J cm^{-1}, with a typical Griffith creeping crack velocity (in a sample of tracing paper) estimated at 6×10^{-4} cm s^{-1} (Dodson 1976).

Edgewise compressive strength of paperboard, which is 30–40% of tensile strength, is important in corrugated boxes and folding cartons. For process conditions producing low fiber bonding (high-yield fiber, low refining and wet pressing), edgewise compressive strength is proportional to bonding, but as the degree of bonding is increased, fiber compressive strength becomes controlling (Seth et al. 1979). These workers found the intrinsic compressive strength of laboratory-made paper to be 3600–7200 N m^{-1}, depending directly on the level of wet pressing over the range 10^2–10^4 kPa.

4.3 Comparison of Paper with Other Composites

Table 1 reveals that paper fibers and papers as composites have quite respectable mechanical properties in comparison to other natural materials (cotton, wood), and when used in an acceptable environment are in some ways superior to man-made "structural materials" such as plastics and reinforced plastic. Like other natural fibers, paper fibers are moderately low in density. They excel in stiffness—on a weight basis (specific stiffness) they are comparable to nylon, or in some cases glass fiber. Paper fiber strength declines when wet, as does the strength of nylon and glass, but its specific dry strength exceeds that of polyethylene,

Table 1
Comparative properties of fibers and composites[a] (values are upper bounds)

Material	Density (kg m^{-3})	Initial stiffness (Gpa)	Specific stiffness (km)	Strength: Dry (Mpa)	Strength: Wet (Mpa)	Specific Strength (km)	Breaking strain (%)
Fibers							
Cotton	1360	11	825	860	1070	64	10
Wool	1175	4	347	190	180	17	45
Paper							
unbleached pine kraft	1500	20	1360	1570	1030	107	35
spruce kraft, 30° fibril	1450	30	2110	1000	660	70	
spruce kraft, 5° fibril	1450	70	4926	1800	1200	127	
High-density polyethylene	860	10	1186	640	640	76	20
Nylon	1030	11	1090	420	360	42	19
Glass	2260	70	3160	3000	600	135	4.5
Carbon (graphite)	1720	500	29600	3200	3200	190	1.5
Steel	7150	210	3000	4000	4000	57	10
Composites							
Oak wood (axial/radial)	550	10/2	1850/370	90		17	
Glass-reinforced polyethylene	1630	30	1880	600	300	38	
Paper							
newsprint (MD/CD)[b]	600			22/11	3.2/1.6	3,7/1.8	1
kraft linerboard (MD/CD)[b]	721	8.7/4.1	1230/580	51/26		7.2/3.7	1.1/1.4

[a] Data provided by Dr Gary Baum, Institute of Paper Chemistry [b] MD, machine direction; CD, cross direction

nylon and steel fibers. In the composite called paper, the low density is retained and the specific stiffness is comparable to glass-reinforced polyethylene. Like some other composite materials, the specific strength of paper is lower than for individual fibers. On a specific (dry) strength basis, paper has approximately one-tenth to one-third the strength of wood, and one-twentieth to one-sixth that of glass-reinforced polyethylene.

Cellulose fibers and paper, when utilized in a manner consistent with their sensitivity to moisture, are therefore remarkably good structural materials for use in low-density applications where stiffness rather than strength is important.

4.4 Overall Carton and Box Mechanical Properties

Numerous "quality control" tests have been devised to predict paper and paperboard performance when it is subsequently formed into folding cartons, or when used in corrugated boxes. In the folding carton "block compression" test, paperboard strips are fitted vertically into slots in top and bottom metal plates and compressed to determine crush resistance. A direct correlation between the cross-machine direction block compression test on paperboard and the top-to-bottom carton compression performance has been shown (Cope 1961).

Koning (1975, 1978) developed a theoretical model and subsequently confirmed its validity for predicting the compressive properties of linerboard as related to the compression strength of corrugated containers. The load at which the container would fail was found to be related to in-plane moduli of elasticity and dimensions of liner and medium.

4.5 On-Line Determination of Properties

An exciting developing area in paper technology is the application of sensing and computing systems to determine on-line the physical uniformity of paper (using lasers), moisture content (radionuclide beams) and elastic moduli (ultrasonic velocities). The potential for rapid feedback or even feedforward control is very appealing.

Ultrasonically-determined elastic constants of paper, such as extensional stiffness (elastic modulus × caliper) and shear stiffness (shear modulus × caliper), and out-of-plane properties have been measured in the laboratory, with estimates of tensile strength (Baum and Habeger 1980, Baum et al. 1981).

The paper industry is moving to increase both its process effectiveness and product uniformity by application of statistical process control (SPC) which relies on conclusions from computed model parameters rather than from original raw data from sensors and instrumentation (Mendel 1987).

4.6 Surface Properties

Because paper and paperboard are frequently used for communicating the written word and/or colorful images, the visual aesthetic properties of paper—especially brightness, color and gloss—are often as important to end-users as the mechanical characteristics.

Paper used in household and personal products must have acceptable levels of absorbency, softness, bulk and stiffness, surface roughness and "handle" or feel. For applications where paper or paperboard is a barrier, for example, to light, sound or thermal energy, or where paper's electrical or dielectric properties are employed, specialty papers have been developed.

Among the most frequently considered nonmechanical end-use properties are optical properties. Corte (1976) notes that, strictly speaking, optical properties are related to "appearance," which is beyond the scope of physical measurement because it involves physiological and psychological factors. Nevertheless, measurement of optical properties is done using sophisticated instrumentation and optical reference materials to indicate opacity, brightness and whiteness, color, gloss and even formation, which is related to local optical contrasts in a sheet.

5. *Applications of Paper and Paperboard*

Paper has expanded into many markets beyond its initial use in writing and printing, and now competes with cloth, plastic, and insulating products. Paper fiber, combined with man-made fibers and resins, can be molded, pressed, or dry formed. Familiar applications are laminated counter tops, asphaltic roofing shingles and felts, flooring and hardboard, wallpapers, battery separators, and interior autopanelling. In these applications, paper is an inexpensive, structurally important reinforcing fiber. Whenever paper fibers are associated with wax, asphalt or other polymers, the product will be less amenable to recycling, and some mechanical properties (e.g., stiffness and tear) and aesthetics will be altered.

Important large structural paper products are: (a) corrugated paper, (b) boxes, fiberboard, spirally-wound drums, and (c) molded fiber products such as luggage cases, egg cartons, and combustible cartridge cases. These products may be printed or coated. They are designed for load-bearing under highly adverse conditions related to the material enclosed and the external environment. The ability to make these products inexpensively for shipping or transmitting such diverse materials as powdered chemicals, eggs, and refrigerators is a challenge for papermakers and converters (see *Hardboard and Insulation Board*).

Opportunities for paper and paperboard products have been changing. Paper is now often used with other materials to achieve an improved combination of properties and economy.

Plastic in the form of bags, cup stock, plates and containers is challenging paper products in markets where extended resistance to moisture or chemicals is required. In the entertainment, communication and business fields, electronic and video systems are being developed which can bypass "hard-copy" paper products. It has been projected that by 1990 about 7% of homes in the USA will rely on videotext in place of printed catalogs (Goodstein 1982).

In other markets, paper fibers are finding increasing roles. In convenience packs for microwave cooking, ovenable paperboard is challenging a market formerly dominated by glass, foil and plastic containers. Forecasts predict a 9% annual growth in food containers, and in paper containers for foods in particular (Technology Forecasts 1981). In multiwall bags for corrosive, granular chemicals, and in electrical and decorative paper-based laminates, paper plies contribute significantly to strength and/or bulk at low cost.

A trend is combining paper with man-made plastic and/or metal films, particularly for food applications. Paperboard provides a significant fraction of the stiffness and bulk for sterilized laminate used in aseptic packaging which can keep unrefrigerated juice, milk, and specialty food products fresh for up to six months (Allan 1982, Mies 1982).

Metallized paper is making inroads in the label market as well as for personal-care products, gift wrap, cigarette inner liners, food cans and dairy wrappers. The product requires an exceptionally high gloss paper, a lacquer which assures smoothness and adhesion for the metallic finish, and metal foil, generally vacuum deposited aluminum. Conventional laminate has a 2:1 paper-to-foil weight ratio, but metallized paper has over 200 parts paper to 1 part foil, considerably reducing raw material cost for the same thickness of packaging material (Carter and Beardow 1983).

6. Production and Consumption of Paper Products

World paper and paperboard production for the first time exceeded 200 Mt in 1986, giving a per capita value of 42 kg. Certain papermaking countries such as the Nordic group had an annual increase in 1986 as low as 1%, while developing areas such as Latin America and Asia reached 7–8% growth levels (Sutton et al. 1987).

Based on the reported 57.8 Mt of paper and paperboard produced in the USA in 1980, it can be computed that for a population of just over 220 million persons (US Department of Commerce 1980), the USA in 1980 used about 260 kg of paper and paperboard products for every man, woman and child in the country. Similar high consumption levels exist in other industrialized countries, and an increase in paper use levels is expected in developing market countries.

The distribution of paper and paperboard world capacity was 76% in regions of developed market economies, 14% in centrally planned economies and 10% in developing market economies. The 1987 United Nations Food and Agriculture Organization (FAO) predictions to 1995 indicate between 2.6 and 2.9% growth in both demand and production of paper and board worldwide (FAO 1987). World trade in paper and board will continue to represent 1.5% of the world's total merchandise export (Brusslan 1987).

Table 2 classifies the world pulp and paperboard capacities in five-year increments from 1976 through the projections in 1991. The disparity between the total wood pulp produced and total paper and paperboard manufactured reflects such items as nonfibrous components used in paper and boards, and the use of recycled fibers.

Chemical pulping will remain the predominant pulping method. Both chemical pulp and pulps made by various mechanical means will contribute up to a third more tonnes per year by 1991 than they did in 1976 (Food and Agriculture Organization 1987).

Newsprint is about 15% of the total paper and paperboard production, while printing and writing

Table 2
Total world pulp and paper/paperboard capacity 1976–1991

	Capacity (Mt)			
Pulp and Paper Classification	1976	1981	1986	1991
Total wood pulp, paper grades	130	142	155	166
mechanical, TMP	31	34	38	43
semichemical, chemigroundwood	11	16	10	10
chemical	88	98	107	113
Other fiber pulp	10	12	14	17
Dissolving pulp	6	6	6	6
Total paper and paperboard	180	203	226	248
newsprint	25	29	32	34
printing and writing	40	50	60	70
other paper and paperboards	115	124	134	144

Source: Food and Agriculture Organization 1987

papers will grow to about 30% by 1991. The largest production category is composed mainly of wrapping and packaging papers and boards.

The Technical Association of the Pulp and Paper Industry (TAPPI) has predicted paper and paperboard consumption for the years 1980, 1990 and 2000 (Tables 3 and 4). Predictions through the year 2000 estimate a shift from North America and Western Europe to Japan and the developing market regions. This is expected partially through market saturation in North America.

7. Energy, Recycling and Environmental Considerations in Papermaking

Three environmentally attractive characteristics of the paper industry are: (a) it uses a renewable polymer resource, i.e., wood fiber, (b) about half its energy use comes from "waste" sources such as the lignin removed during cooking of chips and bark removed from logs, and (c) its end-products are generally recyclable.

In 1981, the US pulp and paper industry consumed about 2.3×10^{18} J of which roughly 50% was generated in the paper mill. It came from cooking liquor (37.6% of total), hogged fuel (6.7%) and bark (5%). The industry has been converting from using purchased to self-generated energy sources. The self-generated level increased from 40.7 to 50.2% during the period 1972 to 1981 (Grant and Slinn 1982).

Table 3
Total paper and paperboard consumption, 1980–2000

Region	Consumption (Mt) 1980	1990	2000	%/Year
World	176.0	265.3	396.9	4.2
North America	67.9	91.6	119.3	2.9
Western Europe	42.9	63.1	89.6	3.8
Japan	18.6	31.8	49.6	5.0
Other	46.5	78.8	137.4	5.6

Source: Hagemeyer and Holt (1982)

Table 4
Regional paper and paperboard consumption, 1980–2000

Region	Consumption (% of total) 1980	1990	2000
North America	38.6	34.5	30.0
Western Europe	24.4	23.8	22.6
Japan	10.6	12.0	12.5
Other	26.4	29.7	34.9

Source: Hagemeyer and Holt (1982)

The energy per tonne of product varies. A range is from about $3.9–4.9 \times 10^{10}$ J t^{-1} for printing and writing paper (which require bleaching, sizing and, in some cases, coating) to 3.26×10^4 J t^{-1} for linerboard (which is neither bleached nor typically sized and coated) (Hersh 1981). In the paper mill, the highest single energy consumer is the drying process. Steam-heated cylinders use $6.7 \times 10^9–1.3 \times 10^{10}$ J t^{-1}. This can correspond to one-fourth to one-third of the total energy requirement for an integrated mill (Chiogioji 1979).

Economics of the fiber market require that postconsumer materials and "waste" fiber be recycled with a minimum of transportation. Recycling mills face an environmental problem not encountered in mills using virgin fiber, namely, disposition of contaminant, nonfibrous materials, including heavy metals from inks (Wrist 1982).

The greatest commercial use of recycled paper is in corrugated paper for boxes, containers, packing and low-density structural core. In general, recycled products are lower in some properties such as tensile strength.

Environmental interests in papermaking encompass all unit processes from forestry to control of stock gas and mill effluent quality. The balance between environmental and economic issues must be thoroughly evaluated. Wood harvesters optimize fiber quality, harvesting procedures, product mix (the proportion of lumber, pulp mill chips and fuel residuals) to achieve maximum yield of biomass per hectare. Alternatives for higher yield and/or quality are being evaluated also. High-yield pulping, such as thermomechanical, chemimechanical and neutral sulfite, offers a way to obtain more fiber per tonne of wood where end-use requirements allow.

The recovery cycle of the paper mill recycles the major portion of chemicals which pulp the wood while burning lignin for fuel. Efforts are underway to further improve energy recovery through better understanding of black liquor and recovery boiler operation (see Fig. 3). Reductions in water and energy use are being pursued through a more closed processes. Developments include: (a) reduction of water requirements through medium-to-high consistency pumping, cleaning and refining; (b) innovative efficient dewatering of paper webs; and (c) advances in drying.

See also: Cellulose: Chemistry and Technology; Cellulose: Nature and Applications; Chemical Composition

Bibliography

Allan D R 1982 U.S. bleached board supply/demand. *Pulp Pap.* 56(10): 164–68

Altieri A M, Wendell J W Jr 1967 *TAPPI Monograph No. 31: Deinking of Waste Paper.* Technical Association of the Paper and Pulp Industry, Atlanta, Georgia

Baum G A, Habeger C C 1980 On-line measurement of paper mechanical properties. *Tappi* 63(7): 63–66

Baum G A, Brennan D C, Habeger C C 1981 Orthotropic elastic constants of paper. *Tappi* 64(8): 97–101

Breck D H, Styan G E 1985 Explaining the increased use of mechanical pulps in high-value papers. *Tappi J.* 68(7): 40–44

Brusslan C 1987 World paper supply will meet 1995's demand. *Am. Papermaker* 50(7): 24–26

Carmichael D L 1986 Uses for high-brightness CMP expand thanks to new bleaching methods. *Pulp Pap.* 60(7): 66–70

Carter J H, Beardow T 1983 A lesson in making metallized paper. *Pap., Film Foil Converter* 57(3): 45–47

Chiogioji M H 1979 *Industrial Energy Conservation.* Dekker, New York

Christie D 1987 Chip screening for pulping uniformity. *Tappi J.* 70(4): 113–117

Clark T F 1969 Annual crop fibers and the bamboos. In: MacDonald R G, Franklin J N (eds.) 1969 *Pulp and Paper Manufacture*, 2nd edn., Vol. 2. McGraw-Hill, New York, pp. 1–74.

Coan B 1987 Paper machine clothing: Strategies for maximum performance, profit. *Pulp Pap.* 58(4): 55–68

Cope P 1961 Measuring and specifying bulge and crush resistance in cartons and carton board. *Tappi* 44(9): 633–36

Corte H 1976 Perception of the optical properties of paper. In: Bolam F (ed.) 1976 *The Fundamental Properties of Paper Related to its Uses*, Vol. 2. Technical Division of the British Paper and Board Industries Federation, London, pp. 626–61

Cox H L 1952 Elasticity and strength of paper and other fibrous materials. *Br. J. Appl. Phys.* 3(3): 72–9

Dodson C T J 1976 A survey of paper mechanics in fundamental terms. In: Bolam F (ed.) 1976 *The Fundamental Properties of Paper Related to its Uses*, Vol. 1. Technical Division of the British Paper and Board Industry Federation, London, pp. 202–26

Ducey M J 1986 Efforts in chemical pulp bleaching technology emphasize cutting costs. *Pulp Pap.* 60(7): 47–50

Fleischman E H, Baum G A, Habeger C C 1982 A study of the elastic and dielectric anisotropy of paper. *Tappi* 65(10): 115–8

Food and Agriculture Organization 1987 *Pulp and Paper Capacities, Survey 1986–1991.* Food and Argiculture Organization of the United Nations, Rome

Franklin W E 1986 Trends in recovery and utilization of waste paper in recycling mills, and other uses of waste paper, 1970–2000. *Tappi J.* 69(2): 28–31

Gallep G 1987 Eucalyptus trade signals increased competition for paper companies. *PIMA* 69(4): 4–5

Goodstein D H 1982 Electronics in the catalog arena: what will be their impact. *Am. Printer Lithogr.* 190(3): 48–49

Grant T J, Slinn R J 1982 *Patterns of Fuel and Energy Consumption in the U.S. Pulp and Paper Industry 1972–1981.* American Paper Institute, New York

Gross R M 1982 Prospects for nonpaper pulp. *Pulp Pap.* 57(9): 45

Hagemeyer R W, Holt S G 1982 A prediction of world printing and writing paper consumption. *Tappi* 65(11): 37–40

Helm D J 1987 New tandem-ENP ups speed, output of O-I corrugating medium machine. *Pulp Pap.* 61(6):128

Hersh H N 1981 *Energy and material flows in the production of pulp and paper*, ANL/CNSV-16. Argonne National Laboratory, Argonne, Illinois

Hunter D 1943 *Papermaking: The History and Technique of an Ancient Craft.* Knopf, New York

Koning J W Jr 1975 Compressive properties of linerboard as related to corrugated fiberboard containers: a theoretical model. *Tappi* 58(12): 105–08

Koning J W Jr 1978 Compressive properties of linerboard as related to corrugated fiberboard containers: theoretical model verification. *Tappi* 61(8): 69–71

MacDonald R G, Franklin J N (eds.) 1970 *Pulp and Paper Manufacture*, 2nd edn., Vol. 3. McGraw-Hill, New York

McGovern J N 1982 Fibers used in early writing papers. *Tappi* 65(12): 57–58

McGrath R 1987 U.S. companies looking for ways to jump on the eucalyptus bandwagon. *Pulp Pap.* 61(7): 92–93

Mark R E, Murakami K 1983, 1984 *Handbook of Physical and Mechanical Testing of Paper and Paperboard*, Vols. I, II. Dekker, New York

Mendel J M 1987 Statistical process control—Basis principles and techniques. *Tappi J.* 70(3): 83–87

Mies W 1982 Aseptic cartons off to a fast start in U.S. *Pulp Pap.* 56(10): 168–69

Nissan A H 1977 *Lectures on Fiber Science in Paper.* Technical Association of the Paper and Pulp Industry, Atlanta, Georgia

O'Brian H 1987 Three-layers give many possibilities. *PPI* 29(4): 73–74

Page D H, Seth R S 1980 The elastic modulus of paper II—the importance of fiber modulus, bonding and fiber length *Tappi* 63(6): 113–16

Perkins R W, McEvoy R P Jr 1981 The mechanics of the edgewise compressive strength of paper. *Tappi* 64(2): 99–102

Seth R S, Soszynski R M, Page D H 1979 Intrinsic edgewise compressive strength of paper—effect of some papermaking variables. *Tappi* 62(3): 45–46

Setterholm V C 1979 An overview of press drying. *Tappi* 62(3): 45–46

Sutton P, Pearson J, O'Brian H 1987 Paper production and consumption records broken in '86: '87 looks good. *Pulp Pap.* 61(8): 47–55

Tay C H, Fairchild R S, Imada S E 1985 A neutral-sulfite/SAQ chemimechanical pulp for newsprint. *Tappi J.* 68(8): 98–103

Technology Forecasts and Technology Surveys December 1981 *Packaged Food and Consumer Spending Projections.* PWG Publications, Beverly Hills, California, pp. 5–9

US Department of Commerce, Bureau of the Census 1980 *Statistical Abstract of the United States.* US Government Printing Office, Washington, DC

Van den Akker J A 1970 Structure and tensile characteristics of paper. *Tappi* 53(3): 388–400

Wahlström B 1981 Developments in paper technology in a global perspective. *Sven. Papperstidn.* 84(18): 32–39

Wintgen M 1987 Board coating with natural ground calcium carbonate in Europe: Technology today. *Tappi J.* 70(5): 79–83

Wrist P E 1982 The direction of [paper] production technology development in the 1980s. *Tappi* 65(11): 41–45

Wuerl P 1986 Alkaline papermaking dominates papermaking and coating chemicals scene. *Pap. Trade J.* 170(7): 41

J. W. Glomb
[Westvaco, New York, USA]

D. D. Mulligan
[Westvaco, Covington, Virginia, USA]

Particleboard and Dry-Process Fiberboard

Particleboard and dry-process fiberboard are relatively new materials of the type in which the properties of lignocellulosic biological matter are manipulated at the level of fibers or larger elements composed of many fibers. These materials were conceived in the laboratory and are closely tied to science and technology.

This article is devoted to materials made with external bonding agents, such as synthetic resins, and produced in the dry state. Fiberboard materials produced without bonding agents, using water as the processing medium, are discussed elsewhere (see *Hardboard and Insulation Board*).

Particleboard and dry-process fiberboard are classed under the generic term composition board, which also includes hardboard, insulation board, cement-bonded board and molded products. These materials are now mostly used in panel form, but it is possible to produce them in other shapes such as I-beams and corrugations. Commercial materials of comminuted wood or other lignocellulosic matter have some very desirable characteristics, such as availability in large sheets, smooth surfaces, uniformity in properties from sheet to sheet and freedom from localized defects.

Most of these materials have so far been largely excluded from primary structural applications because they have been unable to equal sawn lumber or plywood in longitudinal stiffness, dimensional stability and long-term load-carrying ability. However, the quality and availability of logs is decreasing throughout the world and it is impossible to provide large building members, as in the past, from small raw material. Fortunately, recent developments, leading to combinations of particle alignment with the use of particles deliberately manufactured with optimum geometry, can provide materials equal to or surpassing conventional building materials in structural capability and reliability. These relatively new composition-board materials can also be easily modified by treating the particles with fire retardants, preservatives and stabilizing impregnations.

1. The Material

Composition boards have many different names and definitions throughout the world and even within countries. In general, wood-base fiber and particle panel materials is the generic term applied to a group of board materials manufactured from wood or other lignocellulosic fibers or particles to which binding agents and other materials may be added during manufacture to obtain or improve certain properties. These materials are primarily made from discrete pieces which are combined with a synthetic resin or other suitable binder and bonded together under heat and pressure by a process in which the entire interparticle bond is created by the added binder. Mineral-bonded boards use inorganic cement as the binder (see *Mineral-Bonded Wood Composites*). In the USA, hardboard (made of fiber) and particleboard range in density from about 500 to over 800 $kg\,m^{-3}$ (ANSI A208.1). Medium-density fiberboard is between 500 and 800 $kg\,m^{-3}$ in density (ANSI A208.2).

One of the most significant trends in the board industry is the elimination of the previously distinct differences between particleboard and hardboard, thus making it difficult to employ clear definitions of material types. A product which is manufactured with a dry-process core and wet-process faces, which are formulated as a fibrous slurry made from old newsprint, also causes further difficulties in defining board type.

2. The General Process

Many different materials can be produced within the general field of dry-process composition boards. The major ones are made according to the platen-pressed system as opposed to the extruded system where the product is made by forcing the particles through heated dies. Extrusion, at first glance, seems to be the appropriate way to make these products; however, small capacities and inherent property defects have limited the development of this part of the industry. An exception to standard pressing is in molding. Several molded-materials plants have operated in Europe and the USA for many years and new ones have been built to produce pallets.

Figure 1 illustrates the interactions of many of the factors involved in producing lignocellulosic composition materials. By understanding and using these interactions to advantage, materials having predetermined engineering properties can be manufactured. Further research will lead to the development of better materials, either as the present combination of wood and adhesives or as combinations of these raw materials and others such as carbon and glass fibers.

Figure 2 illustrates a typical process line for producing lignocellulosic composition materials using the dry process. The raw material is first brought into the plant and reduced to the desired particle type using the proper equipment, which includes flakers for cutting flakes and wafers, and pressurized attrition mills for generating fiber for use in either particleboard or dry-process fiberboard. Figure 3 shows some of the typical particles used in these materials. The most important types used at present are flakes (which includes wafers and strands), planer shavings and fibers. Many different sizes of each are found throughout the industry.

In most cases, the furnish, which is the prepared lignocellulosic raw material, is dried to about 4% moisture content. When the commonly employed urea–formaldehyde or phenol–formaldehyde ad-

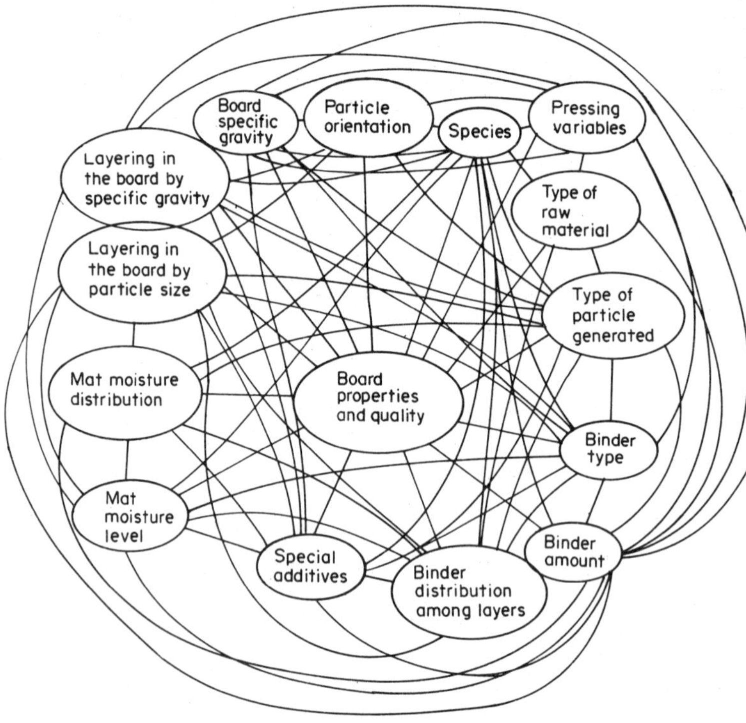

Figure 1
Some of the interactions affecting board properties and quality (after Maloney 1977)

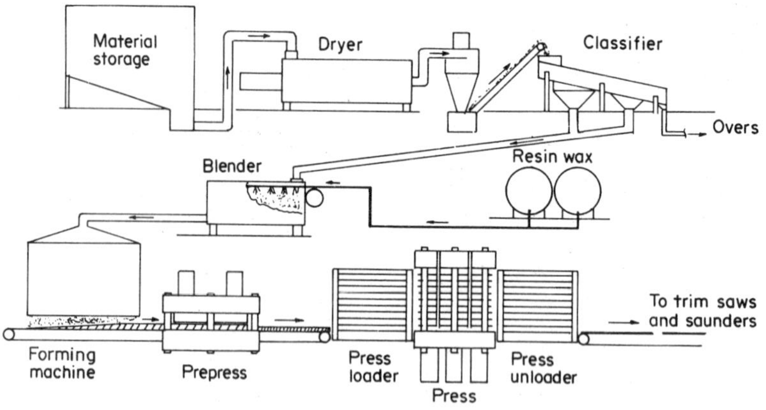

Figure 2
Flow diagram of a typical composition-board plant (after Maloney 1977)

hesives are used, some moisture is added to the furnish in the blending step as part of the binder except when powdered phenol–formaldehyde is used. Isocyanates are used to a limited degree at the present time and are being considered for expanded use in spite of their higher costs (see *Adhesives and Adhesion*).

Before blending, the furnish is usually classified into predetermined sizes as required for the particular production process. Next, the material is blended with the aforementioned resins and usually with a small amount of wax. Urea–formaldehyde resins are used in products which are manufactured for protected inter-

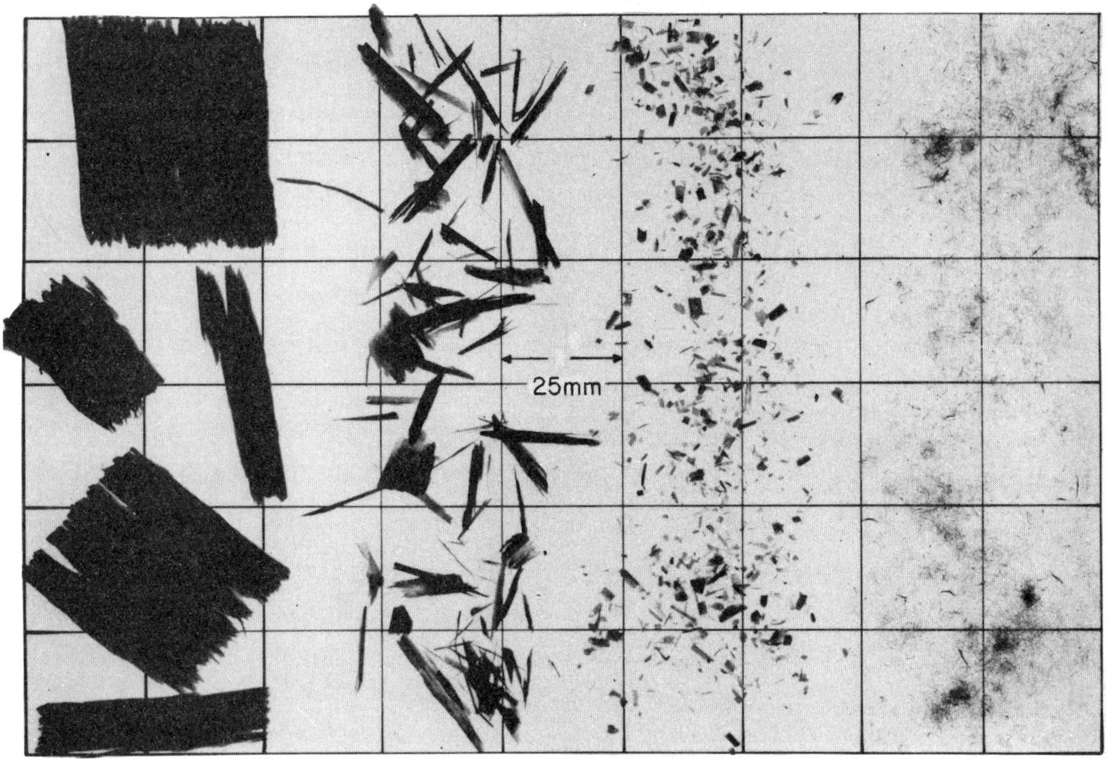

Figure 3
Particles typically used in production; left to right: wafers, ring-cut flakes, planer shavings, pressure-refined fiber (courtesy of Washington State University)

ior applications. Phenol–formaldehyde is usually used for structural applications. The resin content in the final product normally ranges from 4 to 8% for liquid and from 2 to 3% for powdered resin. The wax, usually at levels of approximately 0.25 to 1%, is used to inhibit the pickup of water by the finished product. The addition of wax may not be part of the process in all countries.

After blending, the furnish is formed into loose mats of particles. The finished mat may range in thickness from less than 25 mm to over 300 mm, depending upon the final product requirements and the type of particle being used. This mat may be prepressed to reduce its thickness before pressing.

Pressing in most plants is accomplished in a multi-opening press. A typical press may have twenty openings, each 1.5 by 5.5 m in size. Large, single-opening presses, up to 2.5 by 32 m in size, are also used. A limited number of continuous presses operate throughout the world, particularly for producing thin boards approximately 3 mm in thickness. There are new developments in continuous presses for thicker boards (3–25 mm). A number of these continuous presses are now operational and it appears that many more will be installed over the next few years. A 15 mm thick particleboard panel, for example, made with urea–formaldehyde resin can be pressed in about 3 min. Products with isocyanate binders can be pressed at about the same times and temperatures as for the urea products, whereas phenolics usually take longer to cure. Press temperatures range from approximately 160 °C for urea-bonded and isocyanate-bonded boards to 210 °C for phenolic-bonded boards.

3. Uses

Figure 4 shows some of the typical boards manufactured. The two major uses of medium-density fiberboard are in furniture and siding. The major uses of particleboard are in furniture, floor underlayment, cabinets, mobile-home decking and sheathing. Particleboard can also serve as cores of composites made with wood veneer faces. Molded products are found in doors, furniture parts, house parts and pallets. Myriad other uses are found for these materials, limited only by the user's imagination and the properties of the material.

Figure 4
Major composition-board materials manufactured at present; left to right: waferboard, flakeboard, oriented strandboard, industrial particleboard, medium-density fiberboard, hardboard (courtesy Washington State University)

4. Major Physical Properties

The strength properties of composition boards include static bending, tensile strength parallel to the surface, tensile strength perpendicular to the surface (usually called internal bond), compression strength parallel to the surface, shear strength in the plane of the board, glue-line shear, impact, interlaminar shear and edgewise shear. Important fastener holding properties are lateral nail resistance, nail withdrawal, nailhead pull-through, and direct screw withdrawal. Other properties include hardness and abrasion resistance. Properties associated with moisture include water absorption and thickness swelling, linear variation with change in moisture content, edge thickness swell, resistance to strength-property changes under severe exposure conditions, cupping, and twisting.

Not all of these properties are important for the composition boards manufactured today. Those of most importance at present are static bending (modulus of rupture and modulus of elasticity), internal bond, water absorption, thickness swell and linear variation with change in moisture content. Products going into structural applications must have good fastener holding properties. Furniture products must have good face and edge screw-holding properties. Surface smoothness is of importance when secondary finishing with paints, films and printing is part of the process. Structural products for roof and floor panels must have resistance to impact. Products exposed to the weather need to have a certain degree of durability, which is normally measured by resistance to accelerated aging.

Thermal conductivity, duration of load (static fatigue), and creep are becoming more important as these types of materials are used more in structural applications. Fire retardancy of these products is not considered a property, but it is a measure of their fire resistance or fire hazard. Today, the demand for applications where fire retardancy is important is only a minor part of the total board production. However, fire-retardant products are expected to become more important in the future.

A number of different methods are used throughout the world for evaluating the properties of composition-board materials. Table 1 presents some of the important properties of hardboard, medium-density fiberboard and particleboard. These are values for the USA but may be taken as representative of those found throughout the world. They are not necessarily design properties which have to be established separ-

Table 1
Minimum property requirements of representative composition boards, illustrating the differences found between board types

Panel description	Modulus of elasticity (MPa)	Modulus of rupture (GPa)	Tensile strength (MPa) parallel to surface	Tensile strength (MPa) perpendicular to surface
Tempered hardboard[a]	48.3		24.1	1.03
Standard hardboard[a]	34.5		17.2	0.69
Medium-density fiberboard[b]	24.1	2.09		0.69
Interior-type glue particleboard, grade 1-M-2[c]	14.8	2.24		0.41
Water-resistant-type glue particleboard, grade 2-M-2[c]	17.2	3.10		0.41
Water-resistant-type glue particleboard, grade 2-MF[c]	20.7	3.45		0.34

[a] American National Standards Institute (1982) [b] American National Standards Institute (1980) [c] American National Standards Institute (1986)

ately for the various end uses of the products concerned. Table 1 also shows the differences found between the material types shown. This illustrates how properties can be engineered into these materials by changes in particle geometry, species, resin level and type, specific gravity of the material, and all the other factors shown in Fig. 1.

Much research and development is under way in this exciting field of material science and great improvements in processing and properties are expected.

See also: Wood–Polymer Composites

Bibliography

American National Standards Institute 1980 *American National Standard for Medium Density Fiberboard*, ANSI A208.2-1980. National Particleboard Association, Silver Spring, Maryland

American National Standards Institute 1982 *American National Standard for Basic Hardboard*, ANSI/AHA A135.4-1982. American Hardboard Association, Palatine, Illinois

American National Standards Institute 1986 *American National Standard for Mat-Formed Wood Particleboard*, ANSI A208.1-1979 (R1986). National Particleboard Association, Silver Spring, Maryland

American Society for Testing and Materials 1986 *Definitions of Terms Relating to Wood-Base Fiber and Particle Panel Materials*, ASTM Specification D1554. American Society for Testing and Materials, Philadelphia, Pennsylvania

Deppe H-J, Ernst K 1977 *Taschenbuch der Spanplattentechnik*. DRW, Stuttgart

Fédération Européenne des Syndicats de Fabricants de Panneaux de Particules 1979 *Particleboard—Today and Tomorrow, Proc. 1978 FESYP Int. Particleboard Symposium*. DRW, Stuttgart

Food and Agriculture Organization of the United Nations 1976 *Proc. World Consultation on Wood-Based Panels*. Miller Freeman, Brussels

Jayne B A (ed.) 1972 *Theory and Design of Wood and Fiber Composite Materials*. Syracuse University Press, Syracuse, New York

Maloney T M (ed.) 1967–1988 *Proc. Washington State University Symposium Particleboard/Composite Materials* Nos. 1–22. Washington State University, Pullman, Washington

Maloney T M 1977 *Modern Particleboard and Dry-Process Fiberboard Manufacturing*. Miller Freeman, San Francisco, California

Moslemi A A 1974a *Particleboard*, Vol. 1, *Materials*. Southern Illinois University Press, Carbondale, Illinois

Moslemi A A 1974b Particleboard, Vol. 2, *Technology*. Southern Illinois University Press, Carbondale, Illinois

T. M. Maloney
[Washington State University, Pullman, Washington, USA]

Plywood

Plywood comprises a number of thin layers of wood called veneers which are bonded together by an adhesive. Each layer is placed so that its grain direction is at right angles to that of the adjacent layer (Fig. 1). This cross-lamination gives plywood its characteristics and makes it a versatile building material. Plywood has been used for centuries. It has been found in Egyptian tombs and was in use during the height of the Greek and Roman civilizations.

The modern plywood industry is divided into the so-called hardwood and softwood plywood industries. While the names are truly misnomers, they are in common use, the former referring to industry manufacturing decorative panelling and the latter referring to that which serves building construction and industrial uses. This article will be concerned primarily with the plywood manufactured for construction and for industrial uses in North America.

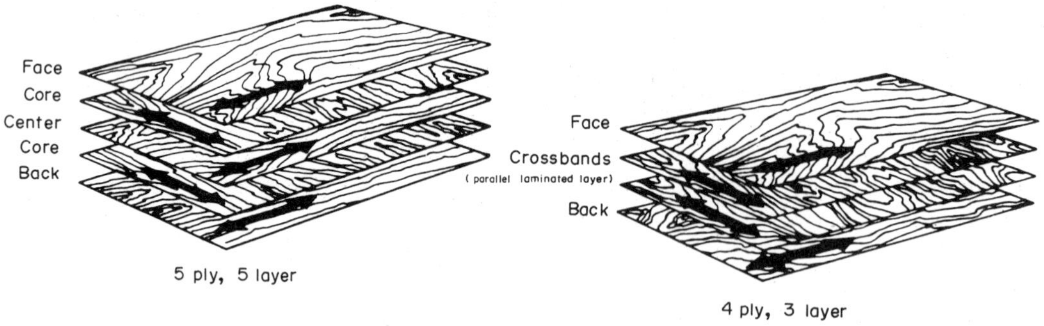

Figure 1
Ply-layer construction of plywood (courtesy of the American Plywood Association)

1. Manufacture

The fundamentals of plywood manufacture are the same for decorative and for construction grades, the primary differences being in the visual quality of the veneer faces.

1.1 Log Preparation

Logs received at the plywood plant are either stored in water or stacked in tiers in the mill yard. From there they are transported to a deck where they are debarked and cut into suitable lengths. The cut sections of logs are referred to as blocks. Debarking a log is most often accomplished by conveying it through an enclosure that contains a power-driven "ring" of scraper knives. The whirling "ring" has the flexibility to adjust to varying log diameters and scrapes away the bark.

Before transforming the blocks into veneer, many manufacturers find it helpful to soften the wood fibers by steaming or soaking the blocks in hot-water vats. The block conditioning process is usually performed after debarking. The time and temperature requirements for softening the wood fibers will vary according to wood species and desired heat penetration. Block conditioning usually results in smoother and higher-quality veneer.

1.2 Conversion into Veneer

The most common method of producing veneer is by rotary-cutting the blocks on a lathe. Rotary lathes are equipped with chucks attached to spindles and are capable of revolving the blocks against a knife which is bolted to a movable carriage. New "spindleless" lathes are being introduced which drive the block against a fixed knife with exterior power-driven rollers. Once in place, the chucks revolve the block against the knife and the process of "peeling" veneer begins. The first contact with the lathe knife produces veneer which is not full length. This is due to the fact that the blocks are never perfectly round and are usually tapered to some extent. It is therefore necessary to "round up" the block before full veneer yield is realized. Once this step is complete, the veneer comes away in a continuous sheet, in much the same manner as paper is unwound from a roll. Veneer of the highest grade and quality is usually obtained at the early cutting, knots becoming more frequent when the block diameter is reduced. A knot is a branch base embedded in the trunk of a tree (see *Macroscopic Anatomy*), and is usually considered undesirable.

All veneer has a tight and a loose side. The side which is opposite to that of the knife is the tight side because of pressure from the machinery holding it against the knife. The knife side is where the lathe checks occur and is called the loose side. The loose side is always checked to some degree and the depths of the lathe checks will depend upon the pressure exerted by the pressure bar opposite the knife. (see *Machining Processes*). Various thicknesses of veneer are peeled according to the type and grade of plywood to be produced.

Wood blocks are cut at high speed, and for this reason veneer is fed into a system of conveyors. At the opposite end of the conveyor is a machine called the clipper. Veneer coming from the lathe through the system is passed under a clipper where it is cut into prescribed widths. Defective or unusable veneer can be cut away. The clipper usually has a long knife attached to air cylinders. The knife action is a rapid up and down movement, cutting through the veneer quickly. Undried veneer must be cut oversized to allow for shrinkage and eventual trimming. Up to this point the wood has retained a great deal of its natural moisture and is referred to as green. Accordingly, this whole process before drying is called the green end of the mill.

1.3 Veneer Drying and Grading

The function of a veneer dryer is to reduce the moisture content of the stock to a predetermined percentage and to produce flat and pliable veneer. Most veneer dryers carry the veneer stock through the dryer by a series of rolls. The rolls operate in pairs, one above the other with the veneer between, in contact with each roll. The amount of moisture in the dryers is controlled by dampers and venting stacks. Moisture inside the dryer is essential to uniform drying, since this keeps the surface pores of the wood open. The air mixture is kept in constant circulation by powerful fans. The temperature and speed of travel through the dryer is controlled by the operator and is dependent upon the thickness of the veneer species and other important factors (see *Drying Processes*).

Dry veneer must be graded, then stacked according to width and grade. Veneer is visually graded by individuals who have been trained to gauge the size of defects, the number of defects and the grain characteristics of various veneer pieces. Veneer grades depend upon the standard under which they are graded, but in North America follow a letter designation A, B, C and D, where A has the fewest growth characteristics and D the most.

1.4 Veneer Joining and Repair

For some grades of plywood, and especially where large panels may be necessary, pieces are cut with a specialized machine, glue is applied and the veneer strips are edge-glued together. The machine usually utilizes a series of chains which crowd the edges of the veneer tightly together, thus creating a long continuous sheet. Other methods of joining veneer may also be employed, such as splicing, stringing and stitching. Splicing is an edge-gluing process in which only two pieces of veneer are joined in a single operation. Stringing is similar to the edge-gluing process except the crowder applies string coated with a hot melt adhesive to hold the veneers temporarily together before pressing. Stitching is a variation of stringing in which industrial-type sewing machines are used to stitch rows of string through the veneer. This system has the advantage of being suitable for either green or dry stock.

The appearance of various grades of veneer may be improved by eliminating knots, pitch pockets and other defects by replacing them with sound veneer of similar color and texture. Machines can cut a patch hole in the veneer and simultaneously replace it with a sound piece of veneer. While the size and shape varies, boat-shaped and dog-bone-shaped veneer repairs are most common.

1.5 Adhesives

The most common adhesives in the plywood industry are based on phenolic resins, blood or soybeans. Other adhesives such as urea-based resorcinol, polyvinyl and melamine are used to a lesser degree for operations such as edge gluing, panel patching and scarfing.

Phenolic resin is synthetically produced from phenol and formaldehyde. It hardens or cures under heat and must be hot pressed. When curing, phenolic resin goes through chemical changes which make it waterproof and impervious to attack by microorganisms. Phenolic resin is used in the production of interior plywood, but is capable in certain mixtures of being subjected to permanent exterior exposure. Over 90% of all softwood plywood produced in the USA is manufactured with this type of adhesive. Another class of phenolics is called extended resins, which have been extended with various substances to reduce the resin solids content. This procedure reduces the cost of the glueline and results in a gluebond suitable for interior or protected-use plywood.

Soybean glue is a protein type made from soybean meal. It is often blended with blood and used in both cold pressing and hot pressing. Blood glue is another protein type made from animal blood collected at slaughter houses. The blood is spray dried and applied to the plywood and supplied in powder form. Blood and blood–soybean blends may be cold pressed or hot pressed. Protein-based glues (blood or soybean) are not waterproof and therefore are used in the production of interior-use plywood. Protein glues are not currently in common use.

The glue is applied to the veneers by a variety of techniques including roller spreaders, spraylines, curtain coaters and a more recent development, the foam glue extruder. Each of these techniques has its advantages and disadvantages, depending upon the type of manufacturing operation under consideration (see *Adhesives and Adhesion*).

1.6 Assembly, Pressing and Panel Construction

Assembly of veneers into plywood panels takes place immediately after adhesive application. Workmanship and panel assembly must be rapid, yet careful. Speed in assembly is necessary because adhesives must be placed under pressure with the wood within certain time limits or they will dry out and become ineffective. Careful workmanship also avoids excessive gaps between veneers and veneers which lap and cause a ridge in the panel. Prior to hot pressing, many mills prepress assembled panels. This is performed in a cold press which consists of a stationary platten and one connected to hydraulic rams. The load is held under pressure for several minutes to develop consolidation of the veneers. The purpose of pre-pressing is to allow the wet adhesive to tack the veneers together, which provides easier press loading and eliminates breaking and shifting of veneers when loaded into the hot press. The hot press comprises heated plattens with spaces between them known as openings. The number of openings is the guide to press capacity, with 20 or 30 openings being the most common, although presses with as many as 50 openings are in use. When the press is loaded, hydraulic rams push the plattens together, exerting a pressure of 1.2–1.4 MPa. The temperature of the plattens is set at a certain level, usually in the range 100–165°C.

2. Standards

Construction plywood in the USA is manufactured under US Product Standard PS 1 for Construction and Industrial Plywood (US Department of Commerce 1983) or a Performance Standard such as those promulgated by the American Plywood Association. For more information on performance standards see the article *Structural-Use Panels*. The most current edition of the Product Standard for plywood includes provisions for performance rating as well. The standard for construction and industrial plywood includes over 70 species of wood which are separated into five groups based on their mechanical properties. This grouping results in fewer grades and a simpler procedure.

Certain panel grades are usually classified as sheathing and include C-D, C-C, and Structural I C-D and C-C. These panels are sold and used in the unsanded or "rough" state, and are typically used in framed construction, rough carpentry and many industrial applications. The two-letter system refers to the face and back veneer grade. A classification system provides for application as roof sheathing or subflooring. A fraction denotes the maximum allowable support spacing for rafters (in inches) and the maximum joist spacing for subfloors (in inches) under standard conditions. Thus, a 32/16 panel would be suitable over roof rafter supports at a maximum spacing of 32 in. (800 mm) or as a subfloor panel over joists spaced 16 in. (400 mm) on center. Common "span ratings" include 24/0, 24/16, 32/16, 40/20 and 48/24.

Many grades of panels must be sanded on the face and/or back to fulfill the requirements of their end use. Sanding most often takes place on plywood panels with A- or B-grade faces and some C "plugged" faces are sanded as well. Sanded panels are graded according to face veneer grade, such as A-B, A-C or B-C. Panels intended for concrete forming are most often sanded as well and are usually B-B panels. Panel durability for plywood is most often designated as Exterior or Exposure 1. Exterior plywood consists of a higher solids-content glueline and a minimum veneer grade of C in any ply of the panel. Such panels can be permanently exposed to exterior exposure without fear of deterioration of the glueline. Exposure 1 panels differ in the fact that D-grade veneers are allowed. Exposure 1 panels are most often made with an extended phenolic glueline and are suitable for protected exposure and short exterior exposure during construction delays.

3. Physical Properties

Plywood described here is assumed to be manufactured in accordance with US Product Standard PS 1, Construction and Industrial Plywood (US Department of Commerce 1983). The physical property data were collected over a period of years (O'Halloran 1975).

3.1 Effects of Moisture Content

Many of the physical properties of plywood are affected by the amount of moisture present in the wood. Wood is a hygroscopic material which nearly always contains a certain amount of water (see *Hygroscopicity and Water Sorption*). When plywood is exposed to a constant relative humidity, it will eventually reach an equilibrium moisture content (EMC). The EMC of plywood is highly dependent on relative humidity, but is essentially independent of temperature between 0°C and 85°C. Examples of values for plywood at 25°C include 6% EMC at 40% relative humidity (RH), 10% EMC at 70% RH and 28% EMC at 100% RH.

Plywood exhibits greater dimensional stability than most other wood-based building products. Shrinkage of solid wood along the grain with changes in moisture content is about 2.5–5% of that across the grain (see *Shrinking and Swelling*). The tendency of individual veneers to shrink or swell crosswise is restricted by the relative longitudinal stability of the adjacent plies, aided also by the much greater stiffness of wood parallel as compared to perpendicular to grain. The average coefficient of hygroscopic expansion or contraction in length and width for plywood panels with about the same amount of wood in parallel and perpendicular plies is about 0.002 mm mm^{-1} for each 10% change in RH. The total change from the dry state to the fiber saturation point averages about 0.2%. Thickness swelling is independent of panel size and thickness of veneers. The average coefficient of hygroscopic expansion in thickness is about 0.003 mm mm^{-1} for each 1% change in moisture content below the fiber saturation point.

The dimensional stability of panels exposed to liquid water also varies. Tests were conducted with panels exposed to wetting on one side as would be typical of a rain-delayed construction site (Bengelsdorf 1981). Such exposure to continuous wetting on one side of the panel for 14 days resulted in about 0.13% expansion across the face grain direction and 0.07% along the panel. The worst-case situation is reflected by testing from oven-dry to soaking in water under vacuum and pressure conditions. Results for a set of plywood similar to those tested on one side showed approximately 0.3% expansion across the panel face grain and 0.15% expansion along the panel. These can be considered the theoretical maximum that any panel could experience.

3.2 Thermal Properties

Heat has a number of important effects on plywood. Temperature affects the equilibrium moisture content and the rate of absorption and desorption of water. Heat below 90°C has limited long-term effect on the mechanical properties of wood. Very high temperature, on the other hand, will weaken the wood.

The thermal expansion of wood is smaller than swelling due to absorption of water (see *Thermal Properties*). Because of this, thermal expansion can be neglected in cases where wood is subject to considerable swelling and shrinking. It may be of importance only in assemblies with other materials where moisture content is maintained at a relative constant level. The effect of temperature on plywood dimensions is related to the percentage of panel thickness in plies having grain perpendicular to the direction of expansion or contraction. The average coefficient of linear thermal expansion is about $6.1 \times 10^{-6}\,\mathrm{m\,m^{-1}\,°C^{-1}}$ for a plywood panel with 60% of the plies or less running perpendicular to the direction of expansion. The coefficient of thermal expansion in panel thickness is approximately $28.8 \times 10^{-6}\,\mathrm{m\,m^{-1}\,°C^{-1}}$.

The thermal conductivity k of plywood is about $0.11-0.15\,\mathrm{W\,m^{-1}\,K^{-1}}$, depending on species. This compares to values (in $\mathrm{W\,m^{-1}\,K^{-1}}$) of 391 for copper (heat conductor), 60 for window glass and 0.04 for glass wool (heat insulator).

From an appearance and structural standpoint, unprotected plywood should not be used in temperatures exceeding 100°C. Exposure to sustained temperatures higher than 100°C will result in charring, weight loss and permanent strength loss.

3.3 Permeability

The permeability of plywood is different from solid wood in several ways. The veneers from which plywood is made generally contain lathe checks from the manufacturing process. These small cracks provide pathways for fluids to pass by entering through the panel edge. Typical values for untreated or uncoated plywood of $\frac{3}{8}$ in. (9.5 mm) thickness lie in the range $0.014-0.038\,\mathrm{g\,m^{-2}\,h^{-1}\,mmHg^{-1}}$.

Exterior-type plywood is a relatively efficient gas barrier. Gas transmission ($\mathrm{cm^3\,s^{-1}\,cm^{-2}\,cm^{-1}}$) for $\frac{3}{8}$ in. exterior-type plywood is as follows:

oxygen	0.000 029
carbon dioxide	0.000 026
nitrogen	0.000 021

4. Mechanical Properties

The current design document for plywood is published by the American Plywood Association (1986), and entitled *Plywood Design Specification*. The woods used to manufacture plywood under US Product Standard PS 1 are classified into five groups based on elastic modulus, and bending and other important strength properties. The 70 species are grouped according to procedures set forth in ASTM D2555 (*Establishing Clearwood Strength Values*, American Society for Testing and Materials 1987). Design stresses are presently published only for groups 1–4 since group 5 is a provisional group with little or no actual production. Currently the *Plywood Design Specification* provides for development of plywood sectional properties based on the geometry of the layup and the species, and combining those with the design stresses for the appropriate species group.

4.1 Section Properties

Plywood section properties are computed according to the concept of transformed sections to account for the difference in stiffness parallel and perpendicular to the grain of any given ply. Published data take into account all possible manufacturing options under the appropriate standard, and consequently the resulting published value tends to reflect the minimum configuration. These "effective" section properties computed by the transformed section technique take into account the orthotropic nature of wood, the species group used in the outer and inner plies, and the manufacturing variables provided for each grade. The section properties presented are generally the minimums that can be expected.

Because of the philosophy of using minimums, section properties perpendicular to the face grain direction are usually based on a different configuration than those along the face grain direction. This compounding of minimum sections typically results in conservative designs. Information is available for optimum designs where required.

4.2 Design Stresses

Design stresses include values for each of four species groups and one of three grade stress levels. Grade stress levels are based on the fact that bending, tension and compression design stresses depend on the grade of the veneers. Since veneer grades A and natural C are the strongest, panels composed entirely of these grades are allowed higher design stresses than those of veneer grades B, C-plugged or D. Although grades B and C-plugged are superior in appearance to C, they rate a lower stress level because the "plugs" and "patches" which improve their appearance reduce their strength somewhat. Panel type (interior or exterior) can be important for bending, tension and compression stresses, since panel type determines the grade of the inner plies.

Stiffness and bearing strengths do not depend on either glue or veneer grade but on species group alone. Shear stresses, on the other hand, do not depend on grade, but vary with the type of glue. In addition to grade stress level, service moisture conditions are also typically presented—for dry conditions, typically moisture contents less than 16%, and for wet conditions at higher moisture contents.

Allowable stresses for plywood typically fall in the same range as for common softwood lumber, and when combined with the appropriate section property, result in an effective section capacity. Some comments on the major mechanical properties with special consideration for the nature of plywood are given below.

Bending modulus of elasticity values include an allowance for an average shear deflection of about 10%. Values for plywood bending stress assume flat panel bending as opposed to bending on edge which may be considered in a different manner. For tension or compression parallel or perpendicular to the face grain, section properties are usually adjusted so that allowable stress for the species group may be applied to the given cross-sectional area. Adjustments must be made in tension or compression when the stress is applied at an angle to the face grain.

Shear-through-the-thickness stresses are based on common structural applications such as plywood mechanically fastened to framing. Additional options include plywood panels used as the webs of I-beams. Another unique shear property is that termed rolling shear (see *Strength*). Since all of the plies in plywood are at right angles to their neighbors, certain types of loads subject them to stresses which tend to make them roll, as a rolling shear stress is induced. For instance, a three-layer panel with framing glued on both faces could cause a cross-ply to roll across the lathe checks. This property must be taken into account with such applications as stressed-skin panels.

See also: Laminated Veneer Lumber; Structural-Use Panels

Bibliography

American Plywood Association 1986 *Plywood Design Specification*, Form Y510. APA, Tacoma, Washington

American Plywood Association 1987a *Grades and Specifications*, Form J20. APA, Tacoma, Washington

American Plywood Association 1987b *303 Plywood Siding*, Form E300. APA, Tacoma, Washington

American Society for Testing and Materials 1987 *Annual Book of ASTM Standards*, Vol. 4.09: *Wood*. ASTM, Philadelphia, Pennsylvania

O'Halloran M R 1975 *Plywood in Hostile Environments*, Form Z820G. APA, Tacoma, Washington

Sellers T Jr 1985 *Plywood and Adhesive Technology*. Dekker, New York

US Department of Commerce 1983 *Product Standard for Construction and Industrial Plywood*, PS 1. USDC, Washington, DC (available from the American Plywood Association, Tacoma, Washington)

Wood A D, Johnston W, Johnston A K, Bacon G W 1963 *Plywoods of the World: Their Development, Manufacture and Application*. Morrison and Gibb, London

M. R. O'Halloran
[American Plywood Association, Tacoma, Washington, USA]

Preservative-Treated Wood

Wood has many unique characteristics which make it a valuable material for construction. However, in applications where it is exposed to wet conditions it is subject to degradation by microorganisms. In addition, it is vulnerable to insect attack in areas where the wood-destroying species exist, and it is also susceptible to fire. Fortunately, wood preservative systems have been developed which provide effective protection against microorganisms, insects and fire. These preservatives, treatment processes, specifications and general properties of the treated wood are described in detail in this article.

1. Wood Preservatives

Wood preservatives are generally classified into two broad groups; namely, oilborne and waterborne. The major preservatives in the oilborne group are creosote, which is an oily liquid that has a satisfactory viscosity for wood treatment, and pentachlorophenol (penta) which is oil-soluble. Copper naphthenate, tributyltin oxide (TBTO®) and copper-8-quinolinolate (Cu-8) are also oil-soluble compounds which have minor uses for special applications. A new biocide, propynyl butylcarbamate (3-10d0-2-), has been evaluated as a wood preservative and appears to have potential for above-ground applications (Hansen 1984). The waterborne preservatives are mainly inorganic salts and include the fire retardants.

1.1 Oilborne Preservatives

The hydrocarbon solvents used to dissolve the oilborne preservatives for the treatment of wood range from heavy oils to very light solvents, and have significant effects on the end product. The solvents currently being used are described by American Wood-Preservers' Association (AWPA) Standard P9. A brief description of the various types of solvents is as follows:

Type A—heavy-oil-type hydrocarbon
Type B—liquid petroleum gas (volatile solvent)
Type C—light petroleum solvent (mineral spirits)
Type D—methylene chloride (volatile solvent)

The Type A solvent is sufficiently heavy that it essentially remains as an integral part of the preservative system for the lifetime of the product and imparts a brownish, oily characteristic to the wood. The other three solvents are volatile and evaporate from the wood. Types B and D are extremely volatile, providing a dry, clean product. The Type C solvent evaporates at a slower rate and auxiliary solvents and water repellents are often added, so the resulting product is intermediate between the other two systems.

1.2 Waterborne Preservatives

The major waterborne preservatives are inorganic salts. These include chromated copper arsenate (CCA), ammoniacal copper arsenite (ACA), acid copper chromate (ACC), chromated zinc chloride (CZC) and fire retardants. CCA and ACA are by far the most widely used preservatives in this group because of their superior effectiveness against wood-destroying organisms and the fact that they undergo chemical reactions in the wood which makes them nonleachable.

In some cases TBTO and Cu-8 are formulated as aqueous solutions, dispersions or emulsions and so they could also be classified as waterborne preservatives. Furthermore, because of the recent surge in cost of petroleum solvents, several waterborne penta formulations are being developed and these systems may well be used extensively in the future.

Although wood preservatives are normally thought of as providing protection against decay by microorganisms, fire retardants can also be classified as wood preservatives for protection against thermal decomposition. There are two basic types of fire retardants and these are generally classified as interior and exterior types.

The interior fire retardants contain mixtures of inorganic salts such as ammonium phosphates, ammonium sulfate, borax and zinc chloride (Goldstein 1973). The actual formulations used by the various manufacturers are proprietary but are known to contain some or all of the chemicals listed above. Some of these salts are hygroscopic and impart additional hygroscopicity to the treated wood. Consequently, serious corrosion and blooming problems can develop if the treated wood is used in buildings that have relative humidities in the region of 80% or greater. Furthermore, these salts are highly water-soluble and cannot be used in applications where liquid water can cause leaching.

Recently, at least one manufacturer has developed a proprietary interior fire-retardant formulation that is relatively nonhygroscopic. Such formulations will help to eliminate many of the problems such as corrosion, blooming and lack of paintability that have been experienced with the earlier products.

The commercially available exterior fire retardants utilize reactive components that form insoluble compounds in the wood when they are cured at high temperatures (Goldstein 1973). This treatment results in a nonhygroscopic, noncorrosive, paintable product that can be used for such items as cedar shakes and shingles and exterior-grade lumber and plywood.

Accelerated laboratory tests have shown that the currently available exterior fire-retardant treated wood provides a reasonable degree of resistance to both decay and termites. Hence, this material can be used for above-ground applications without additional preservative treatment.

2. Treating Methods

In order for preservatives to be effective they must be applied to the wood in a manner that provides uniform distribution and sufficient penetration to control the wood-destroying agents. This is carried out by dissolving and diluting the preservative with a suitable solvent which is then impregnated into the wood by pressure methods.

Pressure treatment of wood is by far the most effective way to treat wood with preservatives because deeper penetration and closer control over preservative retentions and penetration can be obtained. This process is carried out in large pressure vessels by completely flooding the wood with preservative solution and applying pressure. Following this, the solution is returned to storage tanks for use in subsequent charges (Henry 1973).

3. Properties of Treated Wood Affecting Use

Since a number of preservatives and preservative carriers are available which impart different characteristics to the wood, it is important that care be exercised in selecting the right treatment for a given end use. Recommendations for various products are presented in Table 1. In addition, the following discussion briefly covers the considerations that are involved in selection of preservative systems.

3.1 Health and Phytotoxic Hazards

In general, once the common wood preservatives are impregnated into the wood and the solvent has evaporated, these products are safe to handle and use. However, because of their relatively high vapor pressure, penta and creosote should never be employed in totally enclosed structures that are to be used for housing people, plants or animals. Under such circumstances it is possible that the chemical vapors can build up sufficiently to create health hazards and undesirable odors or kill plant life.

The waterborne salts have no detectable vapor pressure, so they can safely be used in enclosed structures without concern for health or phytotoxic problems. However, in cases where foodstuffs will be in contact with the salt-treated wood, some contamination could occur. Hence, they should not be used for these purposes. Instead, Cu-8 should be used since it is fully approved for applications which require contact with foodstuff.

3.2 Type of Exposure

The type of exposure that wood is subjected to during use must be considered in the selection of an appropriate preservative. Wood that is exposed to ground contact must be treated with preservatives that are fixed in the wood so that they will not leach out under the high-moisture conditions and also have sufficient potency to control the broad range of microorganisms

Table 1
Preservatives recommended for various products and end uses[a]

		Pentachlorophenol								
Commodity	Creosote	In heavy oil	In light petroleum	In volatile solvent	CCA and ACA	ACC	CZC	Cu-8	Copper naphthenate	TBTO
Lumber and plywood										
Residential										
above ground	NR	NR	R	R	R	R	R	R	NR	R
ground contact	NR	NR	R	R	R	NR	NR	NR	NR	NR
House foundation	NR	NR	NR	NR	R	NR	NR	NR	NR	NR
Agricultural (above ground)										
nonfood contact	NR	NR	NR	NR	R	R	R	R	R	NR
food contact	NR	NR	NR	NR	NR	NR	NR	R	NR	NR
Agricultural (ground contact)	NR	NR	NR	NR	R	R	NR	NR	R	NR
Laminated for gluing	NR	NR	NR	R	NR	NR	NR	R	NR	R
Poles										
Utility	R	R	NR	R	R	NR	NR	NR	R	NR
Building, residential	NR	NR[b]	NR	NR	R	NR	NR	NR	NR	NR
Building, farm and commercial	R	R	NR	R	R	NR	NR	NR	R	NR
Playground equipment	NR	NR	NR	NR	R	NR	NR	NR	NR	NR
Log cabins	NR	NR	NR	NR	R	R	NR	R	NR	NR
Piling										
Salt water	R	NR	NR	NR	R	NR	NR	NR	NR	NR
Fresh water	R	·R	NR	NR	R	NR	NR	NR	R	NR
Posts										
Farm	R	R	R	R	R	NR	NR	NR	R	NR
Residential	NR	NR	R	R	R	R	NR	NR	NR	NR
Highway guardrail	R	R	NR	R	R	NR	NR	NR	R	NR
Crossties	R	R	NR	NR	NR	NR	NR	NR	R	NR
Crossarms	R	R	R	R	R	R	NR	NR	R	NR
Landscaping timbers	NR	NR	R	R	R	R	NR	NR	NR	NR

[a] R signifies recommended and NR signifies not recommended [b] Acceptable for exterior poles

that are present in the soil. In general, creosote, penta, CCA and ACA are satisfactory for these applications. At high retention levels, ACC and copper naphthenate are satisfactory for some products used in ground contact. The other three preservatives, namely, CZC, Cu-8 and TBTO, are satisfactory only for aboveground applications because of either leachability or lack of sufficient efficacy in ground-contact situations.

3.3 *Paintability and Appearance*

Depending on its end use, the paintability of wood may be of importance. Wood treated with waterborne salts, penta in volatile solvents, Cu-8 and TBTO can be painted without difficulty as long as the water or solvent has evaporated from the surface. Wood pressure-treated with penta in light petroleum (mineral spirits) cannot normally be painted without problems, but it is possible to paint material that has been dip-treated with this preservative as long as the solvent has dissipated.

The appearance and color of the treated wood can also be of importance for some uses. These factors vary with the different preservatives. The only preservatives which leave the wood virtually in its natural state are penta in a volatile solvent and TBTO. All of the others impart various colors to the wood.

It should be pointed out that in some cases wood treated with creosote and penta in heavy oil has a tendency to bleed after exposure to atmospheric conditions. This results in a dark, sticky surface which may be objectionable for some uses. Also, the treatment of wood with copper naphthenate results in an oily, tacky surface.

3.4 *Gluability*

In some cases it is desirable to glue treated wood for the production of glulam beams and other products. The normal preservative used for such applications is penta in a volatile solvent, as it provides the best gluability. It is possible to glue wood treated with waterborne salts, but this often requires special adhesives and surface preparation, so care must be exercised in this instance. Gluing wood treated with creosote and penta in heavy oil is extremely difficult and, consequently, is normally avoided.

3.5 *Corrosiveness*

Metallic hardware and fasteners are often used when structures are built with treated wood and this brings up the question of whether or not the preservatives will cause excessive corrosion of metals. In this regard, there is no evidence to show that the oilborne preservatives contribute to corrosion. In fact, if anything, creosote and heavy-oil treatments tend to inhibit corrosion.

There is no evidence to suggest that any of the preservatives leads to corrosion problems in the absence of excessive moisture. However, there is some indication that under high-moisture conditions some of the waterborne salt treatments can lead to excessive corrosion (Barnes et al. 1984). This is particularly true of the fire-retardant salts, which tend to be hygroscopic. In such cases, it is recommended that hot-dipped galvanized fasteners be used. In extreme cases, such as the All Weather Wood Foundation, it may be advisable to use stainless steel fasteners.

3.6 *Strength*

When lumber and plywood are treated with waterborne salts, penta, creosote, TBTO, Cu-8 or copper naphthenate, the strength properties are not affected, so no reductions of allowable stresses are required. However, when lumber is treated with CCA preservatives and kiln dried afterwards the modulus of rupture is reduced by 10% (Barnes and Mitchell 1984) and this should be taken into consideration in structural designs. There is very little information available on the effect of fire retardant treatments on the strength properties of wood products, so if this question arises it is advisable that the user contact the chemical supplier.

Quite often timber piling is subjected to prolonged periods of steaming or boultonizing (boiling in oil under reduced pressure) to condition them prior to pressure treatment with creosote or penta. These processes affect the strength of wood and, consequently, the allowable stresses are reduced by 10 and 15%, respectively.

4. *Standards, Specifications and Quality Control*

In the USA there are four major organizations concerned with the wood-preserving industry. These are the American Wood-Preservers' Association (AWPA), the American Wood-Preservers' Bureau (AWPB), the American Wood Preservers' Institute (AWPI), and the Society of American Wood Preservers (SAWP). A brief review of these organizations and their functions is presented below.

One of the main objectives of the AWPA is to develop and periodically update requirements and specifications for treating wood products. This is carried out by highly qualified technical committees and the resulting standards serve as a reference for nearly all of the production and sale of treated wood products in the USA. A list of the commodities represented by AWPA can be found in its book of standards. In general, the standards cover wood species, conditioning and treating requirements, preservative recommendations, retentions and penetration requirements for all of the commonly used commodities.

The AWPB was established in 1964 in order to provide an independent, nationwide quality-control program for treated wood products. This agency has its own standards—based on the AWPA standards—which are simplified for more effective consumer use.

Wood-preservation companies that elect to participate in the AWPB program must first pass a qualifying inspection of their plant, personnel and treated products. After qualification, a plant is authorized to use the AWPB quality mark which assures the consumer that these particular products meet the minimum requirements. Periodic follow-up inspections are made at each plant to ensure that the standards are being met. There is no question that this AWPB program has helped maintain the quality of treated wood products. Consequently, anyone specifying treated wood products should be aware of this program and take advantage of the AWPB quality mark.

AWPI and SAWP are nonprofit-making associations dedicated to disseminating information about the proper use and technical and environmental aspects of pressure-treated wood.

Information on wood preservation in countries other than the USA is to be found in a series of publications by Styrelson for Teknisk Utveckling, Stockholm (e.g., Cockcroft 1979).

See also: Decay During Use; Deterioration by Insects and Other Animals During Use; Fire and Wood; Protective Finishings and Coatings; Weathering

Bibliography

American Wood-Preservers' Association *Book of Standards.* American Wood-Preservers' Association, Washington, DC

Barnes H M, Mitchel P H 1984 Effects of posttreatment drying schedule on the strength of CCA-treated southern pine dimension lumber. *For. Prod. J.* 34(6): 29–33

Barnes H M, Nicholas D D, Landers R W 1984 *Corrosion of Metals in Contact with Wood Treated Water-Borne Preservatives.* American Wood-Preservers' Association, Washington, DC

Cockcroft R 1979 *Wood Preservation in the United Kingdom,* Information No. 153. Styrelson for Teknisk Utveckling, Stockholm

Goldstein I S 1973 Degradation and protection of wood from thermal attack. In: Nicholas D D (ed.) 1973 *Wood Deterioration and its Prevention by Preservative Treatments,* Vol. I: *Degradation and Protection of Wood.* Syracuse University Press, New York, pp. 307–39

Hansen J 1984 *IPBC—A New Fungicide for Wood Protection,* IRG/wp/3295. International Research Group on Wood Preservation, Stockholm

Henry W T 1973 Treating processes and equipment. In: Nicholas D D (ed.) 1973 *Wood Deterioration and its Prevention by Preservative Treatments,* Vol. II: *Preservatives and Preservative Systems.* Syracuse University Press, New York, pp. 279–98

Richardson B A 1978 *Wood Preservation.* Construction Press, Lancaster

Wilkinson J G 1979 *Industrial Timber Preservation.* Associated Business Press, London

D. D. Nicholas
[Mississippi State University, Mississippi State, Mississippi, USA]

Protective Finishes and Coatings

Protective finishes and coatings for wood used indoors can perform for many years without refinishing or severe deterioration. The durability of finishes on wood exposed to natural weathering processes, however, depends primarily on the wood itself. Other factors which contribute are the nature and the quality of the finish used, application techniques, the time between refinishings, the extent to which the surfaces are sheltered from the weather, and climatic and local weather conditions. Wood properties that are important in finishing are moisture content; density and texture; resin and oil content; growth pattern and orientation; and defects such as knots, reaction wood and diseased wood (Browne 1962).

The primary function of any wood finish is to protect the wood surface from natural weathering processes (sunlight and water), and help maintain appearance. Weathering erodes and roughens unfinished wood (see *Weathering*). Wood can be left unfinished to weather naturally, and such wood can often provide for extended protection of the structure (Feist and Hon 1984). Different finishes give varying degrees of protection from the weather.

The protection that surface treatment provides against light and water will be affected by the weather resistance of the bonding agents used in the finish (e.g., drying oils, synthetic resins and latexes), as these agents are subject to some degree of photolytic degradation. The mechanism of failure of paints and other finishes has been described by Hamburg and Morgans (1979) and will not be discussed further here. Protection of wood exposed outdoors by various finishes, by construction practices and by design factors to compensate for effects of weather has been addressed in great detail (Cassens and Feist 1986).

Two basic types of finishes (or treatments) are used to protect wood surfaces during outdoor weathering: (a) those that form a film, layer or coating on the wood surface, and (b) those that penetrate the wood surface leaving no distinct layer or coating. Film-forming materials include paints of all descriptions, varnishes, lacquers and also overlays bonded to the wood surface. Penetrating finishes include preservatives, water repellents, pigmented semitransparent stains and chemical treatments (Banov 1973).

1. Film-Forming Finishes

1.1 Paints

Film-forming finishes such as paint have long been used to protect wood surfaces. Of all the finishes, paints provide the most protection against erosion by weathering, and offer the widest selection of colors. A nonporous paint film will retard penetration of moisture and reduce problems of paint discoloration by wood extractives, paint peeling and checking, and warping of the wood. Proper pigments will essentially eliminate uv degradation of the wood surface. Paint, however, is not a preservative; it will not prevent decay if conditions are favorable for fungal growth. The durability of paint coatings on exterior wood is affected by the wood surface and the type of paint.

Paints are commonly divided into oil-based or solvent-borne systems and the latex or waterborne systems. Oil-based paints are essentially a suspension of inorganic pigments in an oleoresinous vehicle that binds the pigment particles and the bonding agent to the wood surface. Latex paints are suspensions of inorganic pigments and various latex resins in water, and form porous coatings. Acrylic latex resins are particularly durable, versatile materials for finishing wood and wood-related materials. Latex paints are used to a greater extent than oil-based paints for finishing wood, particularly for exterior use.

1.2 Varnishes and Lacquers

Clear varnishes or lacquers give wood an attractive initial appearance. Other treatments either change wood color or cover it completely. Unfortunately, clear varnish finishes used on wood exposed to sun and rain require frequent maintenance to retain a satisfactory appearance. The addition of colorless uv light absorbers to clear finishes has found only moderate success in aiding retention of natural color and original surface structure. It is generally accepted that opaque pigments found in paints and stains provide the most effective and long-lasting protection against light. Even using relatively durable clear synthetic resin varnishes, the weatherproof qualities of the wood–varnish system are still limited because uv light, which penetrates the transparent varnish film, gradually attacks the underlying wood. Eventually, the varnish begins to flake and crack off, taking with it fibers of the wood which have been degraded photochemically. Durability of varnish on wood to weathering is limited, and many initial coats are necessary for reasonable performance. Maintenance of the varnish surface must be carried out as soon as signs of breakdown occur. This may be as little as one year in severe exposures. Lacquers and shellacs are not suitable as exterior clear finishes for wood.

2. Penetrating Finishes

2.1 Water Repellents

A large proportion of the damage done to exterior woodwork (e.g., paint defects, deformation and decay) is a direct result of moisture changes in the wood and subsequent dimensional instability. Water repellents and water-repellent preservative treatments are used to protect wood from decay and moisture. Such treatments reduce absorption of water and retard growth of decay microorganisms. These penetrating treatments can also be used as natural finishes for wood (Cassens and Feist 1986). Pretreatment of wood with water repellents or water-repellent preservatives is very important in the finishing of wood (such as millwork) for exterior uses.

2.2 Stains

When pigments are added to water-repellent preservative solutions or to similar transparent wood finishes, the mixture is classified as a pigmented penetrating stain (sometimes referred to as an impregnating paint). Addition of pigment provides color and greatly increases the durability of the finish. The semitransparent pigmented penetrating stains permit much of the wood grain to show through; they penetrate into the wood without forming a continuous layer. Therefore, they will not blister and peel even if excessive moisture enters the wood. The durability of any stain system is a function of pigment content, resin content, preservative, water repellent and quantity of material applied to the wood surface.

Penetrating stains are suitable for both smooth and rough-textured surfaces; however, their performance is markedly improved if applied to rough-sawn, weathered or rough-textured wood. They are especially effective on lumber and plywood that does not hold paint well, such as flat-grained surfaces or dense species. Penetrating stains can be used effectively to finish exterior surfaces such as siding, trim, exposed decking and fences. Stains can be prepared from both solvent-based resin systems and latex systems; however, latex systems do not penetrate wood surfaces. Commercial finishes known as heavy-bodied, solid color or opaque stains are also available, but these products are essentially similar to paint because of their film-forming characteristics. Such stains, which can be oil- or latex-based, find wide success on textured surfaces and panel products such as hardboard.

2.3 Preservatives

Although not generally classified as wood finishes, preservatives do protect wood against weathering and decay, a great quantity of preservative-treated wood being exposed without any additional finish. There are three main types of preservative: (a) preservative oils (e.g., coal-tar creosote), (b) organic solvent solutions (e.g., pentachlorophenol), and (c) waterborne salts (e.g., chromated copper arsenate). These preservatives can be applied in several ways, but pressure treatment generally gives the greatest protection against decay (see *Preservative-Treated Wood*). Greater preservative content of pressure-treated wood generally results in greater resistance to weathering and improved surface durability. The chromium-containing preservatives also protect against uv degradation.

See also: Decay During Use; Radiation Effects; Surface Properties

Bibliography

Banov A 1973 *Paints and Coatings Handbook for Contractors, Architects, Builders and Engineers*. Structures Publishing, Farmington, Michigan

Browne F L 1962 *Wood Properties and Paint Durability*, USDA Miscellaneous Publication No. 629. US Forest Products Laboratory, Madison, Wisconsin

Cassens D L, Feist W C 1986 *Finishing Wood Exteriors: Selection, Application and Maintenance*, USDA Forest Service Agriculture Handbook No. 647. Forest Products Laboratory, Madison, Wisconsin. US Government Printing Office No. 1986-483-399

Feist W C, Hon D N-S 1984 Chemistry of weathering and protection. In: Rowell R M (ed.) 1984 *The Chemistry of Solid Wood*, Advances in Chemistry Series No. 207. American Chemical Society, Washington, DC

Hamburg H R, Morgans W M 1979 *Hess's Paint Film Defects: Their Causes and Cures*, 3rd edn. Chapman and Hall, London

US Forest Products Laboratory 1987 *Wood Handbook: Wood as an Engineering Material*. USDA Handbook No. 72. US Government Printing Office, Washington, DC

W. C. Feist
[US Forest Products Laboratory, Madison, Wisconsin, USA]

R

Radiation Effects

Radiation effects on wood are conventionally separated into those resulting from photooxidation and those from ionizing radiation (Fengel and Wegener 1984). The ultraviolet (uv) region of the electromagnetic spectrum provides the energy source for photoreactions, whereas other regions (x ray and γ ray) are responsible for the high-energy reactions induced by ionizing radiation. These two types of radiation differ fundamentally in energy and in the way they interact with a molecular chain. The quanta of uv radiation have energies in the range 10^{18}–2×10^{19} MeV, whereas high-energy radiation may have energies up to 2×10^{25} MeV. Degradation of a polymer chain in wood by uv light is caused by the absorption of energy in discrete units by specific functional groups (chromophores) that may be present in the chain. Typical uv absorption spectra of wood, cellulose and lignin are illustrated in Fig. 1. However, these specific light-absorbing groups are not needed for the absorption of high radiant energy, since the energy is transferred directly to the electrons that are in the path of the high-energy photons. Moreover, uv-light radiation merely tends to excite an electron within a specific functional group to a higher energy state, whereas high radiant energy tends to remove an electron from a molecule completely. Nonetheless, the ultimate effects of uv and ionizing radiation on wood are often quite similar.

Irradiation of wood has been used industrially to promote degradation of wood materials for saccharification; to produce pulp fibers with a low degree of polymerization; to modify wood per se; to induce graft copolymerization to improve the mechanical and chemical properties of wood; and to cure adhesives or coatings on wood surfaces (Fengel and Wegener 1984).

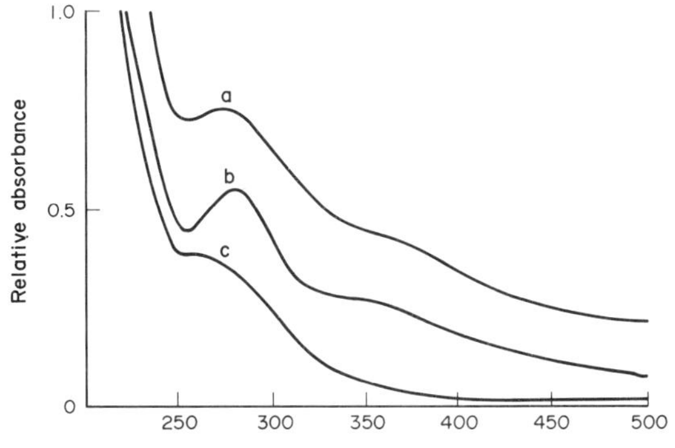

Figure 1
Ultraviolet absorption spectra: a, wood; b, cellulose; c, lignin

Wood absorbs a wide range of frequencies of electromagnetic radiation, from γ ray and x ray to uv and visible light. Energy absorption by wood causes a sequence of physicochemical changes. High-energy radiation can penetrate wood readily, whereas photoirradiation of wood is merely a surface reaction. This is because uv light cannot penetrate deeper than 75 μm. Visible light, on the other hand, can penetrate up to 200 μm into wood surfaces (Hon and Ifju 1978). However, visible light of 400–700 nm is insufficient in energy ($< 1.8 \times 10^{18}$ MeV mol^{-1}) to cleave chemical bonds in any of the wood constituents.

1. Formation of Free Radicals

The constituent molecules of wood are easily excited by electromagnetic energy. Certain frequencies are specifically absorbed by cellulose, hemicelluloses, lignin and extractives in wood. If the absorbed energy is high enough to break chemical bonds, the primary products are free radicals, and it is the way in which these radicals react which determines the nature of the overall degradation reaction (Hon et al. 1980). Ions and electrons may be formed if wood is exposed to high-energy radiation. The electron spin resonance (ESR) spectrometer can be used to detect free-radical formation and determine the structures of free radicals (Ranby and Rabek 1977). Most of the free radicals are usually unstable, and they tend to react with each other or with other chemical elements that are available in the environment to achieve a stable state. For example, free radicals may react with oxygen to form peroxy radicals which in turn will abstract protons to form hydroperoxide. (Hydroperoxide is unstable with respect to heat and light. It decomposes rapidly, causing free-radical chain reactions.)

2. Discoloration of Wood

The free-radical reactions in wood normally lead to the formation of coloring groups, that is, chromophoric and auxochromic groups. The chromophoric groups are carbonyl groups in cellulose and hemicelluloses; and quinones, carbonyl groups and conjugated double bonds in lignin (Hon and Glasser 1979). Generally, discoloration proceeds from yellow to brown to gray after irradiation but this depends greatly on the species of wood. Woods that are rich in extractives may become bleached before the browning becomes observable. The change in color of wood due to radiation can be monitored by means of a tristimulus colorimeter (Hon and Shiraishi 1987, Billmeyer 1981).

3. Anatomical Changes

Low doses of gamma radiation up to about 500 kGy (50 Mrad) do not cause a visible change in the wood structure; however, at higher doses, changes in the cell dimensions and an increase in the friability of wood are readily observed. It has been reported that the tangential vessel diameter, ray cell length and length and width of intervessel pits increase on exposure, whereas tangential vessel-wall thickness, ray cell double-wall thickness and latewood fiber double-wall thickness decrease (Tabirih et al. 1977). High-energy irradiated wood is also unstable in acidic or alkaline conditions. It degrades more rapidly than nonirradiated wood.

Ultraviolet light also affects the structure of wood. Degradation of the surface of the wood begins even at a relatively low irradiation intensity with an attack on the compound middle lamella which is rich in lignin. Higher intensities and longer exposure degrade the secondary walls (Hon 1983). The uv radiation also creates a contraction in the cell walls which results in microchecks along the compound middle lamellae. When wood is irradiated with wavelengths greater than 254 nm for 500 h, the transverse surfaces become disarticulated at the middle lamella region, and simple and bordered pits on exposed radial surfaces are severely damaged. Enlargement of pit apertures as well as loss of pit borders are also observed (Chang et al. 1982). Wood samples from 1200-year-old Japanese temples show a slow disintegration of the outer layer of the fibers. The primary walls and the secondary walls are found to be partly flaked off or completely missing on the exposed side of the fibers.

4. Chemical Changes

Tables 1 and 2 demonstrate the change in chemical composition in gamma- and uv-irradiated wood. Wood irradiated by the higher energy gamma rays suffers more degradation than wood subjected to lower energy uv radiation. Depolymerization and oxidation of cellulose, hemicelluloses and lignin are ordinarily observed. Liquids such as methanol, acetic acid and water, and gases such as carbon dioxide, carbon monoxide and hydrogen are generated during irradiation and lead to weight loss. The change in weight for wood irradiated with gamma rays is shown in Fig. 2.

5. Changes in Physical and Mechanical Properties

The changes in the chemical properties are reflected in the physical and mechanical behavior of wood. Exposure of pine, spruce and beechwood to varying amounts of ^{60}Co radiation results in changes in water absorption, compressive and bending strength and impact energy of rupture. A small reduction in equilibrium moisture content at all relative humidities is also observed with 1 MGy exposure. The change in mechanical properties is also dependent upon the dosages

Table 1
Effect of γ-radiation dosages on extractives, holocellulose and lignin contents of irradiated white oak wood (after Tabirih et al. 1977)

	Content[a] (%)			
Chemical constituent	Wood	Wood irradiated to 650 Mrad	Wood irradiated to 950 Mrad	Wood irradiated to 1900 Mrad
Extractives	8.98	52.8	58.0	69.1
Holocellulose	58.1	14.5	11.0	3.70
Lignin	32.6	30.2	29.6	27.0

[a] Approximate deviation based on handling is 1%

Table 2
Effect of uv light on holocellulose, cellulose, hemicelluloses, lignin, and extractives of southern pine which was irradiated in air for 1000 hours

	Content (%)			
Chemical constituent	Wood	$\lambda > 340$ nm	$\lambda > 280$ nm	$\lambda > 253$ nm
Holocellulose	68.1	61.1	58.2	42.8
Cellulose	45.3	44.9	43.7	41.2
Lignin	27.8	25.7	18.5	11.3
Extractives	10.5	13.5	21.3	45.7

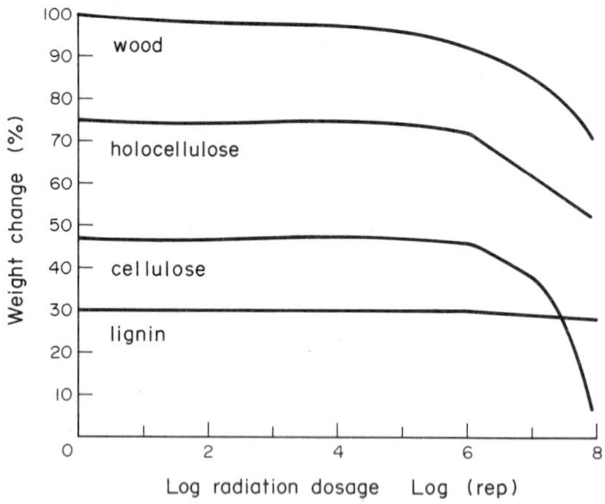

Figure 2
Changes of weight of a southern pine irradiated with gamma rays as a function of radiation dosage

applied. Generally, at dosages below 100 kGy either small increases (due to cross-linking) or no noticeable change in strength properties are observed. Beyond the 1 Mrad level, however, most of these properties decrease owing to degradation. The degree of crystallinity of wood remains unchanged up to 30 Mrad, but begins to decrease rapidly at about 1 kGy. Wood strength decreases with increasing dosage; tensile strength is affected most, compression strength least, and bending strength to an intermediate extent. A loss of elasticity and toughness of wood has also been observed.

See also: Decay During Use; Surface Properties

Bibliography

Billmeyer F W Jr 1981 *Principles of Color Technology*, 2nd edn. Wiley, New York

Chang S T, Hon D N-S, Feist W C 1982 Photodegradation and photoprotection of wood surfaces. *Wood Fiber* 14: 104–17

Fengel D, Wegener G 1984 *Wood: Chemistry, Ultrastructure, Reactions.* Gruyter, Berlin, Chap. 13, pp. 345–70

Hon D N-S 1983 Weathering reactions and protection of wood surfaces. *J. Appl. Polym. Sci.* 37: 845–64

Hon D N-S, Glasser W 1979 On possible chromophoric structures in wood and pulp—A survey of the present state of knowledge. *Polym.-Plast. Technol. Eng.* 12: 159–79

Hon D N-S, Ifju G 1978 Measuring penetration of light into wood by detection of photo-induced free radicals. *Wood Sci.* 11: 118–27

Hon D N-S, Ifju G, Feist W C 1980 Characteristics of free radicals in wood. *Wood Fiber* 12: 121–30

Hon D N-S, Shiraishi N 1987 *Handbook of Wood and Cellulosic Materials.* Dekker, New York, Chap. 10

Ranby B, Rabek J F 1977 *ESR Spectroscopy in Polymer Research.* Springer, Berlin

Tabirih P K, McGinnes E A Jr, Kay M A, Harlow C A 1977 A note on anatomical changes of white oak wood upon exposure to gamma radiation. *Wood Fiber* 9: 211–15

D. N.-S. Hon
[Clemson University, Clemson, South Carolina, USA]

Rattan and Cane

Rattan and cane are terms that can be used interchangeably for the stems of climbing palms belonging to the subfamily Calamoideae of the family Palmae or Arecaceae (Uhl and Dransfield 1987). They enter world trade as raw material for furniture manufacture, and are used locally in producing countries for a wide range of purposes, most importantly for weaving basketware. In common usage there is frequently confusion between rattan or cane and bamboo (see *Bamboo*), objects referred to as "bamboo" furniture in reality being made from rattan, and some so-called rattan furniture being constructed from the leaf stalks of palms such as raphia (*Raphia* spp.) or buri (*Corypha* spp.) or from bamboo. All rattan stems are solid and enter trade either whole or variously split and cored. The center of the natural distribution of rattans lies in India, Sri Lanka, Southeast Asia and the Malay archipelago, although there are outliers in West Africa, southern China, the west Pacific and Australia. In tropical America there is a group of climbing palms belonging to the genus *Desmoncus*, analogous with the Old World rattans but belonging to a quite different group of palms (subfamily Arecoideae: tribe Cocoeae). Although these New World climbing palms may be used locally for basketware or even cheap furniture, they are not used extensively as yet and seem to have limited potential. Most rattan entering world trade is collected from wild stands from tropical rain forests. With widescale forest clearance and overexploitation, present wild stocks of rattan are being rapidly depleted and trade protectionist measures have been introduced by some rattan-producing countries to protect local supplies and to increase the value of the exported product. However, these measures have tended to increase the pressure on wild stocks elsewhere. Forest departments and other agencies within the rattan-producing countries are now investigating the possibilities of rattan cultivation, by building on the experience of villagers in Central Kalimantan, Indonesia, where two species of rattan have been successfully cultivated for over a century (Dransfield 1979).

1. Botany and Distribution

There are at least 568 different species of rattan, belonging to the following 13 genera (in decreasing order of size): *Calamus* (about 370 species; West Africa, India and China to Australia and Fiji, with the greatest number of species in Borneo), *Daemonorops* (about 115 species; South China to New Guinea), *Korthalsia* (about 26 species; Indochina to New Guinea), *Plectocomia* (about 16 species; Himalayas to Borneo and the Philippines), *Eremospatha* (about twelve species; West Africa), *Laccosperma* (about seven species; West Africa), *Ceratolobus* (six species; West Malaysia to Java), *Plectocomiopsis* (five species); Thailand to Borneo), *Oncocalamus* (five species; West Africa), *Pogonotium* (three species; West Malaysia and Borneo), *Myrialepis* (one species; Indochina to Sumatra), *Calospatha* (one species; West Malaysia) and *Retispatha* (one species; Borneo) (Uhl and Dransfield 1987). Most of the important, good-quality canes entering world trade belong to the genus *Calamus*.

Rattans are conveniently divided into two major size classes: large-diameter canes, with stems greater than 18 mm in diameter, and small-diameter canes, with stems less than 18 mm in diameter. Large-diameter canes are used for the frames of items of furniture, while small-diameter canes are used, usually in the split or cored state, for binding frameworks and weaving seats and backs. Of the large diameter canes a few species are preeminent and because of this have been grossly overexploited in the past few years. In West Malaysia, Sumatra, South Thailand and South Kalimantan, *Calamus manan* (rotan manau or manau cane) is the premier quality large species, which, because of scarcity, is now being replaced in the trade by species of poorer quality such as *C. ornatus* and *C. scipionum*. In the Philippines, the most favored large cane is *Calamus merrillii* (palasan), but stocks have been virtually exhausted. Elsewhere there are fine large-diameter canes in Sulawesi, New Guinea and Sri Lanka which may be amenable to cultivation, but their botany and in some instances even their scientific identity are not known. Amongst the small diameter canes, *Calamus caesius* (sega) (South Thailand, West Malaysia, Sumatra, Borneo and Palawan) and *C. trachycoleus* (irit) (Kalimantan) are preeminent, but many more species enter the trade.

Rattans are mostly high climbing, many reaching the forest canopy. After an initial period of establishment growth during which the full stem diameter is built up, the stem grows upwards, not increasing further in diameter with age. The plant may consist of a single unbranched stem, which when harvested is killed, or it may produce suckers at the base, in which case the clump can be harvested continually. Very rarely the stems branch aerially in the forest canopy. Some species, e.g., *C. manan*, can grow to immense lengths; the longest ever recorded was about 185 m.

No reliable studies have been made on the actual or projected production of cane from wild stands of rattan, and exploitation has never been carefully controlled despite the existence of regulations. Because they are classed as minor forest products, there has been a tendency for rattans to receive rather little research attention from forest departments, yet as a forest product they have great significance to people living near forests as a supplementary source of cash, particularly in periods of agricultural scarcity.

2. Physical Properties

Rattan stems usually have elongate internodes and the nodes are clearly marked. The diameter of the internodes and indeed the stem itself does not usually vary significantly along their lengths. The cross section is usually more-or-less circular; however in several species of *Calamus* (e.g., *C. scipionum*), the cross-sectional outline is circular except for a subtriangular protrusion on one side, marking the position of the vascular supply to the climbing whips or inflorescences. The position of this ridge changes from one internode to the next, corresponding to the phyllotactic spiral of the living plant. This unevenness of the bare cane has implications for utilization, such canes being less favored than those with an even cross-sectional outline. The surface of the bare cane is usually smooth and frequently lustrous, pale yellow-brown to ivory-colored. In species of *Korthalsia*, however, the cane surface is reddish-brown and rough with incompletely separated remains of the leafsheaths. Such red canes, despite being extremely durable, are not favored by the trade, although they are extensively used for varying the texture and appearance of fine local basketware. The outer layer may be two or more cells thick. The cortex of the rattan stem includes scattered vascular bundles embedded in ground tissue. The anatomy of the stem can show both generic and specific differences. The structure of a typical vascular bundle is shown in Fig. 1.

The value of rattan in the furniture industry depends on the high degree of flexibility coupled with great strength and fine appearance. Not all canes have the same properties. The finest canes have a rather uniform cross-sectional structure with an even distribution of vascular bundles and even and not excessive lignification of the ground tissue. In some species of *Plectocomia*, starch is accumulated in the central part of the stem which is also rather soft. Cane from *Plectocomia* is generally worthless, being difficult to bend without causing distortion and being particularly prone to fungal and insect attack because of the accumulations of starch. Some species of *Plectocomiopsis* and *Eremospatha* have stems which are

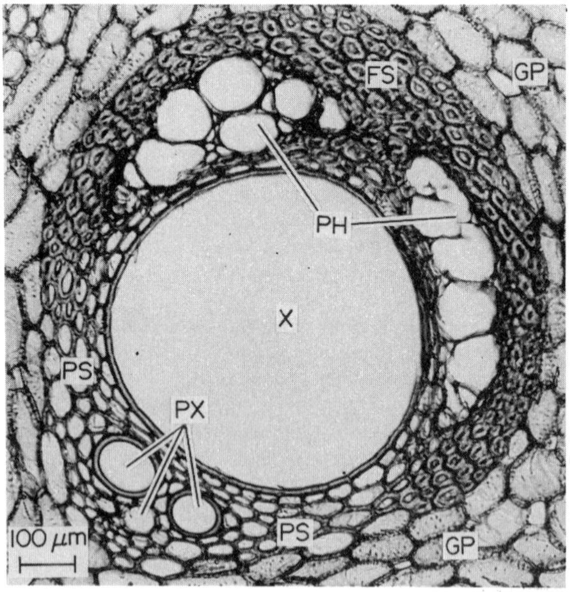

Figure 1
Vascular bundle of *Daemonorops angustifolia*, representing a common bundle type (*Calamus*, *Calospatha*, *Ceratobolus*, *Daemonorops*, *Korthalsia*). GP is the ground tissue, FS the fiber sheath, PH the phloem fields, X the metaxylem, PX the protoxylem and PS the parenchyma sheath (courtesy of Professor Walter Liese, University of Hamburg)

more-or-less triangular in cross section. These are also worthless, being impossible to bend without the cane splitting. In many of the better species of *Calamus* the outer layer of the cane, or epidermis, is heavily impregnated with silica. This silica layer gives a high luster to the surface of the cane. For some purposes such as the production of fine split cane (chair cane) the silica layer is removed by a process called "runti" or "lunti" which involves twisting the harvested stems in such a way that the silica layer snaps off in thin flakes. For other purposes the silica layer is carefully preserved and protected.

Studies on the physical properties of rattans are in their infancy. Tensile strength and other mechanical properties have been investigated for some species but in some instances the results of the studies are of little value because the scientific identity of the canes was not critically established, a problem which occurs in many aspects of rattan research. Goh (1982) investigated the mechanical properties of Rotan manau (*Calamus manan*) and found that, as in timber, the strength of green material increased upon drying. Air-dry material (14.4% moisture content) had a density of 750 kg m^{-3} and compression strength parallel to grain of 30 MPa, about one-half the value one would expect of timber of the same density (see *Strength*).

3. Processing

Processing of rattan is usually rather simple. Small-diameter canes newly harvested are cut into lengths of about 9 m and bent into two; they are bundled into loads sufficient for one man and carried out of the forest. If they are not to be processed immediately they may be stored under water for a short period. Initial processing of such small canes involves cleaning the cane surface of remains of the leaf sheath and the removal of dirt. This is often carried out manually by rubbing the cane surface with a handful of sand while continually washing the cane. Small canes may also be "runtied" at this point, the silica layer being removed by twisting the cane through closely placed rollers or bamboos. Canes are then sun-dried and additionally may be smoked over burning sulfur to prevent fungal attack. Fungi, particularly those responsible for "blue stain," may have a marked deleterious effect on harvested cane if it is not processed quickly. After drying and smoking, the canes are graded and packaged for shipment to a secondary processor or manufacturer. Secondary processing involves the splitting of the cane to produce skin (chair cane) and core. A wide range of machines are available for such processing. Usually they are designed in such a way that when a cane is passed into the machine over rollers, a die peels off the skin in four to eight strips while at the same time removing circular sectioned lengths of core from the remaining central portion, the number and diameter of the core strips being dependent on the die which is selected for the diameter range of the rattan itself. Strips of skin (chair-cane) are extensively used for weaving the backs of both cane and wooden furniture. Core is used in usually cheaper quality woven furniture and in handicrafts.

Large-diameter cane is usually cut into lengths of about 3 m as it is harvested, the resulting "sticks" being bundled and carried out of the forest. Processing of such large-diameter rattans is not usually carried out by the collectors themselves, but by a rattan merchant. The problem of blue-stain fungus attack is a particular problem with large-diameter canes, as the time which elapses between harvesting and delivery to or collection by the rattan merchant may be prolonged. Processing of large canes usually involves treating the sticks in heated oil or kerosene. Diesel, coconut or other oil may be used on its own or in combination with kerosene. The oil is often heated in tanks made from old oil drums and the cane immersed in the heated oil for short periods. The treatment seems to act by removing excess water, by bringing natural gums in the rattan to the cane surface and by effectively killing the rattan itself together with any fungi on the cut surfaces. After treatment with oil the sticks are further sun-dried before being graded, bundled and packaged ready for shipment to the manufacturer.

4. Trading

Traditionally, rattan harvested from the forest is bought from the collectors by middlemen who sell to small rattan merchants. The rattan merchants are responsible for grading and processing and finally export. In the past rattan was for the most part exported to Singapore, Hong Kong or Cebu (Philippines), all three ports acting in an entrepreneurial role with some small-scale manufacturing, resulting in the reexport of manufactured items or of processed cane. China has been a major importer of rattan and also a major exporter of rattan handicrafts. Recently, there have been changes to this age-old pattern. A few rattan cooperatives have been organized, removing middlemen from the local trade and thus increasing the price payable to the collectors, while governments have become increasingly aware of the advantages of adding to the value of the exported cane by carrying out processing and manufacture locally. To encourage this export, duties on raw cane have been raised or the export of raw cane banned. If such bans and prohibitive taxes are effective, then the future of rattan manufacturing businesses outside the producing countries will be bleak.

5. Commercial Uses

Almost all large-diameter rattan in trade enters the furniture industry. In contrast, some small-diameter rattan is used commercially in the production of high-quality matting, mostly for Japanese markets, some for handicraft manufacture (the value of which is considerable), and some for the furniture industry. The use of small-diameter canes for reinforcing concrete has been investigated but the potential for this use seems limited, despite the apparent resilience of rattan when thus used (Baharin 1978).

6. *The Future for Rattan*

Rattan appears at the moment to have an uncertain future. The most serious problem facing the industry is the supply of raw material. With widescale destruction of forest and overexploitation of the canes themselves, a world shortage of rattan seems unavoidable. Techniques for cultivating small-diameter rattan are available for two species, *Calamus caesius* and *C. trachycoleus*, but very few new plantations have actually been established. Furthermore, cultivation of large-diameter canes has not progressed beyond the experimental phase. Even if plantations were established now they would be unlikely to come into bearing before ten years have elapsed, so that the shortage of rattan is likely to become more severe. In the past, manufacturing has usually taken place in importing countries rather than the producing countries themselves. Producers are now erecting trade barriers with the intention of encouraging the manufacture of rattan furniture by the producing country. However, insufficient expertise in furniture production exists in the producing countries. Considerable changes in the rattan trade worldwide can be expected as all these factors take effect. One major concern is that disruptions to the traditional trading patterns may have the effect of destroying the market entirely, before cultivation can be initiated and the long-term supply assured.

Rattan has been the subject of two recent international symposia, in Malaysia in 1984 (Wong and Manokaran 1985) and in Thailand in 1987, besides numerous local symposia. There is a rattan information center at the Forest Research Institute Malaysia, Kepong, which acts as a clearing house for the Asian region, publishing a bimonthly bulletin and a bibliography of research papers (Kong-On and Manokaran 1986). There is thus much research interest in rattan but, so far, few results of significance to the silviculture of rattans and the long term supply.

See also: Bamboo, Coconut Wood

Bibliography

Baharin bin Puteh 1978 A study of rattan and its mechanical properties. Thesis, Mara Institute of Technology, Kuala Lumpur

Dransfield J 1979 A manual of the rattans of the Malay Peninsula. *Malay. For. Rec.* 29

Goh S C 1982 Testing of Rotan manau: strength and machining properties. *Malay. For.* 45(2): 275–77

Kong-On, H K, Manokaran N 1986 *Rattan: A Bibliography.* Rattan Information Centre, Forest Research Institute Malaysia, Kepong

Uhl N W, Dransfield J 1987 *Genera Palmarum, A Classification of Palms based on the work of H E Moore Jr.* Bailey Hortorium and International Palm Society, Kansas

Wong K M, Manokaran N 1985 *Proceedings of the Rattan Seminar 2–4 October 1984, Kuala Lumpur, Malaysia.* Rattan Information Centre, Forest Research Institute Malaysia, Kepong

J. Dransfield
[Royal Botanic Gardens, Kew, Richmond, UK]

Resources of Timber Worldwide

As a result of displacement of forests by agricultural, grazing and other land uses, and of timber harvesting for industrial and fuelwood uses, the timber resources of the world are declining in area and in total volume of wood. As of 1980, forest lands of all types totalled some 4×10^7 km^2 or about 32% of the world's total land area. The distribution of forests among the major geographic regions is highly uneven. The coniferous or softwood forests, which are of particular value for industrial uses, occur mainly in the temperate and boreal regions of the northern hemisphere. In contrast, more than 80% of the world's volume of broad-leaved or hardwood forests are in the tropics, while the temperate-zone broad-leaved forests are divided between the northern and southern hemispheres.

1. *Forest Area*

Data on forest areas, timber volumes and rates of timber harvesting and forest removal are neither complete nor accurate. Reasonably reliable data are available for not more than half of the world's forests, with estimates for the balance being derived from extensive reconnaissance, vegetation-type maps and other generalized data sources. Differences in classification as well as in units of measurement also contribute to wide differences in data cited by various authorities.

Table 1 summarizes available data as to the forest areas of the world. Forest lands with sufficient productivity to support timber stands with an appreciable degree of canopy closure represent two-thirds of the total forest area. These closed forests are the world's primary source of industrial wood, and some of them are also subject to extensive cutting for fuelwood purposes. The remaining one-third of the forest land consists mainly of open woodland or savanna types, but also includes some brushlands and temporarily deforested areas. These forests are not significant for industrial wood, but many of the areas are subject to heavy use for fuelwood production.

North American forests are predominantly coniferous, with locations largely in the temperate zone but

Table 1
Estimated forest areas of the world by major regions (various dates 1975–85)

Region	Total forest area (10^4 km^2)	Closed forest area[a] (10^4 km^2)	Total land area (10^4 km^2)	Closed forest as proportion of total land (%)	Closed forest area per capita (km^2)
North America	734	459	1829	25	0.02
Central America	65	60	272	22	0.005
South America	730	530	1760	30	0.024
Africa	800	190	2970	6	0.004
Europe	181	145	552	26	0.003
USSR	930	792	2227	36	0.03
Asia	530	400	2700	15	0.002
Pacific area[b]	190	80	842	10	0.036
World	4160	2656	13152[c]	20	0.007

[a] Forest land with 20% or more canopy cover [b] Australia, New Zealand, Papua New Guinea and the South Pacific Islands [c] Excluding Antarctica and Greenland

also including extensive boreal forest areas. North America is also a major source of temperate-zone broad-leaved forests, with such stands being concentrated in the eastern part of the continent. Central America mainly has tropical broad-leaved forests, but also includes some coniferous stands at higher elevations. South America has the world's most extensive tropical broad-leaved forests as well as some temperate broad-leaved stands, but coniferous species occur only in a few limited areas. European forests are generally similar to those of North America, with coniferous species dominating, an extensive boreal forest zone in addition to the greater area of temperate zone forests and a significant broad-leaved forest component, especially in central Europe. More than three-quarters of the African forests are of the open savanna type, with the closed forest area consisting almost entirely of tropical broad-leaved stands. The USSR includes more than half of the world's coniferous forest, with much of this forest area being in the boreal zone. In addition, the USSR has significant areas of temperate broad-leaved forests. Asian forests consist primarily of tropical broad-leaved species, but also include both temperate broad-leaved forests and coniferous forests in the more northern and mountainous zones. The natural forests of Australia, New Zealand, Papua New Guinea and the South Pacific Islands consist predominantly of broad-leaved stands, mainly in the temperate zone.

The temperate-zone coniferous forests, located mainly in the northern hemisphere, have almost entirely been brought under utilization. They serve as the world's major source of industrial wood supplies, and include a high proportion of the total forest area which is under forest management for continuous production of timber supplies. The boreal forests of Europe are also almost entirely in use and generally subject to forest management.

The major undeveloped coniferous forests of the world are in the boreal zone of the USSR, largely east of the Urals. Although the USSR has been developing major forest product complexes within its Siberian zone, these developments are concentrated in the relatively limited areas with a high proportion of pine and close to connections with the Trans-Siberian railroad. The greater part of the Siberian forests have only limited access, include extensive areas of permafrost, are believed to be generally low in productivity and consist dominantly of larch. The only other extensive undeveloped coniferous forest belt consists of the more northern of the Canadian forests. Again, limitations of economic access, the nature of the existing stands and generally low productivity serve to limit the possibilities of early development.

Although the USSR and northern Canada include practically all of the world's remaining undeveloped coniferous forests, it is the southern forests of the USA which have the greatest potential for expansion of coniferous industrial wood supplies during the 1990s. This region, with its extensive area of highly productive coniferous forests, a current surplus of growth over removals, a fully developed infrastructure and good access to markets, is a major center for expansion of the forest products industries.

The area of temperate forests is relatively stable. The forest area in Europe is actually increasing as a result of afforestation programs. In North America there is a limited net displacement of forests by other land uses, as well as some shift from coniferous forests to broad-leaved forests as a result of logging without subsequent management to favor the regeneration of the coniferous species. In New Zealand, Australia, Japan, Chile and some other nations, active programs of forest plantation establishment and management are leading to appreciable increases in the area and volumes of coniferous forest stands. In addition, in China,

Mongolia, North Korea and South Korea, current afforestation programs appear to be slightly more than offsetting the decrease in natural forest areas.

The reductions in the world's forest area are largely concentrated in the tropical broad-leaved forests. Estimates made for the period 1981–85 indicated that the closed forests of the tropics are being cleared at a rate of 7.5×10^4 km^2 annually, or about 0.6% of their total area, while the open woodlands are being cleared at an annual rate of 3.8×10^4 km^2, or about 0.5% of their total area (Lanly 1983). The heaviest impacts appear to be in the Amazon basin, but all tropical regions are experiencing some reduction in forest area. On a world scale, the principal causes are population pressure and the ever expanding need for food, with shifting cultivation and conversion of land to agriculture being the primary source of the reduction in forest areas. Fuelwood gathering and commercial logging without forest management are also important causes.

In contrast, the area of high-yielding industrial forest plantations in the tropics, including both coniferous and broad-leaved species, is estimated to have increased from 3.1×10^4 km^2 in 1975 to 4.8×10^4 km^2 in 1980. In addition, there were 2.2×10^4 km^2 of industrial plantations stocked with hardwoods not classified as fast growing and 4.4×10^4 km^2 of non-industrial tree plantations in 1980. Although these areas are limited compared with the decrease in the area of natural tropical forests, these plantations represent the beginnings of an important new source of wood supplies.

2. *Timber Volumes and Cutting Levels*

Overall, estimates of timber volume are appreciably more uncertain than the estimates of forest area, as they are based on the area estimates and have the additional uncertainties which surround the estimates of volumes per unit area. While the volume estimates for the more developed nations are generally based on national forest surveys, the volume estimates for tropical forests, open woodland areas and the boreal forests of limited access are highly approximate. Estimates of cutting levels or removals are also very uncertain. Estimates of the timber cut for industrial use are generally derived from census-type reports on forest products output, although in some instances data on forest removals are collected directly. However, data relating to the volumes of fuelwood cut in the forests are rarely collected regularly and any estimates of such volumes are highly approximate.

The estimates of timber volume and of cutting levels shown in Table 2 must thus be recognized as being subject to wide margins of error. According to these estimates, in the mid-1970s tropical broad-leaved forests included 47% of the volume in the closed forests of the world, coniferous forests included 39%

Table 2
Estimated timber volumes and cutting levels by major regions of the world (various dates 1975–1985)

Species group	Region	Timber volume in closed forests (10^9 m^3)	Industrial wood removals (10^9 m^3)	Industrial removals as a ratio of timber volume (%)	Timber volume in closed forest and open woodlands (10^9 m^3)	Total wood removals (10^9 m^3)	Total removals as a ratio of a total volume (%)
Conifers	North America	33	0.38	1.2	37	0.38	1.6
	Europe	10	0.20	2.0	10	0.21	2.5
	USSR	67	0.26	0.4	67	0.32	0.5
	Japan	1	0.03	3.0	1	0.03	2.5
	Other	8	0.08	1.0	8	0.19	2.5
	World	119	0.95	0.8	123	1.12	1.1
Temperate broad-leaved	North America	10	0.09	0.9	14	0.10	1.1
	Europe	6	0.07	1.2	6	0.11	2.0
	USSR	17	0.03	0.2	17	0.06	0.5
	Japan	1	0.02	2.0	1	0.02	2.0
	Other	8	0.04	0.5	10	0.19	1.9
	World	42	0.25	0.6	48	0.48	1.3
Tropical broad-leaved	Latin America	78	0.03		85	0.23	0.3
	Africa	42	0.03	0.1	60	0.27	0.4
	Asia and Far East	25	0.09	0.3	27	0.41	0.5
	World	145	0.15	0.1	172	0.90	0.5
All types	World	306	1.35	0.5	343	2.50	0.8

and temperate broad-leaved forests 14%. The distribution of industrial wood removals is strikingly different, with the closed coniferous forests supplying 70% of the cut, temperate broad-leaved forest 19% and tropical forests only 11%. North America, Europe and the USSR together hold 92% of the estimated volume of the coniferous forests of the world as well as 79% of the estimated volume of temperate broad-leaved forests.

Inclusion of the estimated volume in open woodlands as well as the volume in closed forests increases the total estimated volume by only 12%, with the greater part of this increase being in savanna-type forests in the tropics. However, inclusion of the estimated volume of fuelwood cut in the forests as well as the volume of industrial wood shows a total cut from all forests which is 85% greater than the cut of industrial wood from the closed forests. The fuelwood removals are largely concentrated in the less developed nations, with 85% of the estimated fuelwood removals coming from broad-leaved forests.

On a world basis the estimated level of removal of industrial wood and fuelwood from all forest areas amounts to 0.7% of the estimated total volume. This, however, reflects only removals for use and does not include volumes removed and destroyed in the process of converting forest land to other uses. Further, the effects of forest removals on the ability of forests to sustain future production depends not only on the volume of removals but also on their nature, the age and size class distribution of the trees in the forests from which they are obtained, and the level of forest management which is practised. Thus, the forests of Europe are subject to one of the highest ratios of removals to total volume of any of the forest regions of the world, yet these generally managed forests can sustain further increases in cut in the future while keeping the volume of the cut within the limits of growth. At the other extreme, industrial wood removals are estimated to amount to less than 0.05% of the volume of closed forests in Latin America, and removals including fuelwood amount to only 0.3% of total volume, yet these forests are rapidly being reduced in area and volume.

3. Utilization of the Timber Resources

The utilization of timber resources during the period 1979–81 reported by the Food and Agriculture Organization of the United Nations is summarized in Table 3. Sawlogs and veneer logs, used for the production of lumber, veneer and plywood, constituted 60% of the industrial wood production, with pulpwood constituting another 26%. Other industrial uses appear substantial in the world total, but actually are reported mainly for the USSR and China and are relatively minor in most parts of the world.

During this period North America produced 42% of the coniferous sawlog volume, Europe 20% and the USSR 26%. North America produced 47% of the total coniferous pulpwood, Europe 29% and the USSR only 17%. Production of sawlogs from broad-leaved forests was less concentrated, with Asia producing 37% of the volume, North America 16%, Europe 15%, the USSR 11% and Latin America 10%. However, pulpwood from broad-leaved forests was produced primarily in North America with 43% of the total and in Europe with 31% of the total. Thus, with the exception of certain high-value tropical species, the industrial wood of the world is produced largely within the regions that consume the manufactured forest products.

Fuelwood cut in the forests, which constitutes nearly half of the total removals for use, is used primarily in the less developed regions of the world. Such wood is used primarily in cooking and for domestic heat and is usually burned inefficiently, with low recovery of the latent energy. The use of industrial wood residues for fuel is not included. Such use is now a major source of power for the forest products industries. With the development of more efficient combustion technology and the rising costs of alternative power sources, the use of wood for industrial power could emerge as a

Table 3
Annual world wood production (1979–81 average)

Type of wood	Coniferous species (10^6 m^3)	Broad-leaved species (10^6 m^3)	All species (10^6 m^3)
Industrial			
sawlogs and veneer logs	603	249	852
pulpwood	255	111	366
pit props	23	10	34
other	99	73	172
total	980	443	1424
Fuelwood and charcoal	226	1483	1709
All types	1206	1927	3132

significant new element in the utilization of timber resources.

As a result of the continued rapid population growth in many of the less developed nations, the pressures for the cutting of fuelwood from both closed forests and open woodlands continue to increase. Critical shortages in the availability of timber are in prospect in the late 1980s and 1990s in many regions with broad-leaved forests of low productivity which are heavily used for fuelwood. Indeed, as of 1980 acute shortages of fuelwood already affected some 250 million people in arid regions of Africa, Asia and South America.

In terms of industrial wood uses in the more developed regions, timber resources generally appear adequate to enable a continuing increase in the production and consumption of forest products until the year 2000, but with major imbalances between domestic production and consumption in western Europe and Japan. The outlook for the more distant future is uncertain, depending upon assumptions as to the magnitude of increases in demand, achievements in forest management and further progress in utilization leading to more forest products from a given volume of timber cut in the forest.

Within this general framework, there are many issues at the specific level in terms of the availability of particular species, particular sizes and characteristics of logs and location relative to particular wood utilization centers. These are matters which are critical to specific operations and possibly to specific nations.

See also: Future Availability; Timbers of Africa; Timbers of Australia and New Zealand; Timbers of Canada and the USA; Timbers of Central and South America; Timbers of Europe; Timbers of Southeast Asia; Timbers of the Far East

Bibliography

Kallio M, Dykstra D P, Binkley C S (eds.) 1988 *The Global Forest Sector: An Analytical Perspective.* Wiley, Chichester.

Lanly J P 1983 Assessment of the forest resources of the tropics. *For. Abstr.* 44(6): 287–318

Persson T R 1974 *World Forest Resources*, Research Notes No. 17. Royal College of Forestry, Stockholm

Pringle S L 1976 Tropical moist forests in world demand, supply and trade. *Unasylva* 28: 106–18

United Nations Food and Agriculture Organization 1976 European timber trends and prospects, 1950 to 2000. *Timber Bull. Eur.* 29 (Suppl. 3): 1–309.

United Nations Food and Agriculture Organization 1982 *World Forest Products Demand and Supply 1990 and 2000*, FAO Forest Paper 29. United Nations Food and Agriculture Organization, Rome

United Nations Food and Agriculture Organization 1983 *1981 Yearbook of Forest Products.* United Nations Food and Agriculture Organization, Rome

United Nations Food and Agriculture Organization and Economic Commission for Europe 1985 *The Forest Resources of the ECE Region (Europe, the USSR, North America).* United Nations Economic Commission for Europe, Geneva

United Nations Food and Agriculture Organization and Economic Commission for Europe 1986 *European Timber Trends and Prospects to the Year 2000 and Beyond.* United Nations, New York

US Department of Agriculture, Forest Service 1982 *An Analysis of the Timber Situation in the United States 1952–2030*, Forest Resource Report No. 23. US Department of Agriculture, Forest Service, Washington, DC

J. A. Zivnuska
[Orinda, California, USA]

S

Shrinking and Swelling

One of the important characteristics of wood as a material is the shrinking and swelling associated with moisture content changes during processing and use. This article considers the theoretical and practical aspects of shrinking and swelling, and methods for increasing the dimensional stability of wood. Since shrinking and swelling are complementary properties, the shrinkage S is used to represent both phenomena. It is defined as the percentage reduction in the swollen dimension associated with loss of moisture, usually to the dry condition.

1. Cell Wall Shrinkage

Before discussing the shrinkage of wood, the shrinkage of the cell wall itself will be discussed, since removal of bound water from the cell wall is the primary cause of normal shrinkage. For the purposes of this discussion, moisture content M is defined as the mass of moisture per unit mass of dry wood, expressed as a percentage. The cell wall of green wood is completely saturated, and this moisture content is designated as the fiber-saturation point M_f. It is fully swollen at this point and has a specific gravity G_f^* of approximately 1.035, defined as the ratio of its dry mass to the mass of water displaced by its water-swollen volume. It should be noted that this definition of specific gravity is peculiar to wood since the weight and volume are measured at different moisture contents (see *Density and Porosity*).

The volumetric shrinkage of the cell wall as it dries is nearly linear with loss of moisture to the completely dry condition. The percentage shrinkage S^* from M_f to any lower value of M is, therefore, directly proportional to $M_f - M$. Thus, the total percentage volumetric shrinkage of the cell wall from M_f to the oven-dry condition, designated as S_f^*, is

$$S_f^* = M_f G_f^* \qquad (1)$$

The value of M_f varies among different woods, 27% representing an approximate mean for woods of the USA. Substituting this value and 1.035 for G_f^* into Eqn. (1) gives a total volumetric shrinkage S_f^* of about 28% for the cell wall from the green to oven-dry condition. Thus, the percentage volumetric shrinkage S^* of the cell wall for each percent reduction in M, based on fully water-swollen volume, is essentially equal to G_f^*, or 1.035% $\%^{-1}$.

Essentially all of the volumetric shrinkage in the cell wall of normal wood occurs at right angles to the cell axis or grain direction of the wood. This is because the microfibril orientation is predominantly parallel to the cell axis, and shrinkage occurs perpendicular to the microfibril direction.

2. Shrinkage of Wood

If small samples of green wood are dried slowly, under conditions such that there are no appreciable moisture gradients, shrinkage does not normally begin until all of the "free" water in the cell cavities has been removed, and the cell wall itself begins to lose moisture. This transition moisture content is designated as the fiber-saturation point M_f. As wood dries below M_f, normal shrinkage begins. However, the volumetric shrinkage is less than that of the cell wall itself. Experimental data of volumetric shrinkage from the green to oven-dry condition for woods of the USA show a nearly linear relationship between shrinkage S_f and the green volume specific gravity G_g, such that the mean ratio S_f/G_g is approximately 27, that is, it closely approximates M_f. This is remarkable, for it indicates that the cell cavity, on the average, remains constant in size as wood dries. The shrinkage S_g from the green to oven-dry condition is generally taken to be equal to S_f, since normal shrinkage does not occur above M_f.

The reason for the tendency toward a constant cell cavity size in wood during shrinkage is believed to be associated with the microfibril alignments in the cell wall. Although the bulk of these are aligned parallel to the cell axis, there is a narrow sheath on both the inside and outside of the cell wall where the alignment is more nearly horizontal. These layers act to reduce the shrinkage which might otherwise occur (see *Ultrastructure*).

It is sometimes convenient to express shrinkage in terms of shrinkage per unit change in wood moisture content, or in terms of unit change in relative humidity. The first term, designated as the volumetric hygroexpansion coefficient X_v, is defined as

$$X_v = (100/V)(dV/dM) \qquad (2)$$

where V is the wood volume. The second coefficient, termed the volumetric humidity expansion coefficient Y_v, is defined similarly as

$$Y_v = (100/V)(dV/dH) \qquad (3)$$

where dV/dH is the change in wood volume per percent change in relative humidity H. The term Y_v is probably more important than X_v since it indicates the dimensional changes associated with humidity changes to which wood is exposed in use.

Wood is orthotropic with respect to its shrinkage. Longitudinal shrinkage S_L in normal mature wood is small, of the order of 0.1–0.2% from M_f to oven-dry. However, in certain cases such as in wood from the center of trees (i.e., juvenile wood) S_L may be as much as several percent, because of the high microfibril angle with respect to the grain direction found in such cases. The same high S_L may be found in reaction wood, that is, compression wood in softwoods and tension wood in hardwoods (see *Macroscopic Anatomy*).

Tangential shrinkage S_T in wood is the largest directional component of shrinkage. It is 5–8% for many softwoods and may be 12% or more for dense hardwoods. The ratio of tangential to radial shrinkage over a given moisture range, often called the T/R ratio, is generally close to two for most woods. However, it may vary from slightly greater than unity for some woods to as high as three or more for others.

Several theories have been proposed for explaining anisotropic shrinkage in wood. There is general agreement to the effect that the low normal value of S_L is caused by the low fibril angle in the cell wall of wood, as has already been mentioned, and that excessive longitudinal shrinkage is related to higher angles. However, the greater tangential than radial shrinkage is more difficult to explain in terms of a single mechanism.

At least three groups of mechanisms have been proposed for explaining the high T/R shrinkage ratio. The first group is based on the effects of the gross wood structure such as ray tissue alignment and volume, or earlywood–latewood interactions. The second group is based on microfibril angle modifications in the radial and tangential walls of cells. The third group is based on differences in the structure and shrinkage characteristics of the tangential and radial cell walls. One (or more) of each group of mechanisms appears to operate for certain kinds of wood, but no single mechanism has been shown to operate to the exclusion of the others.

3. *Warping of Lumber during Drying and in Use*

Although there are disagreements among wood specialists as to the mechanism of differential shrinkage, there is no disagreement as to its effects in causing warping of lumber during drying and in subsequent use. The warping due to initial drying is generally the most severe. However, warping during use due to changing exposure conditions, although less severe, may also cause problems. The latter kind of warping, sometimes called "movement," can be minimized by drying the lumber initially to a moisture content equivalent to its average value during use (see *Drying Processes*). Probably the most common warping problem in lumber is the cupping which occurs in flat sawn boards as they dry. This is due to the differential tangential and radial shrinkage. The warping caused by excessive longitudinal shrinkage in parts of a board are most commonly designated by the terms crooking, bowing and twisting.

4. *Dimensional Stabilization of Wood*

Wood can be dimensionally stabilized, that is, its natural hygroexpansion tendencies can be minimized, by one or more of several methods. These can be divided into two groups: use of mechanical restraint and physicochemical treatment.

Mechanical restraint is probably the most commonly used technique, best exemplified by cross-laminating veneers into plywood. In this familiar technique, veneer laminae are glued together so that the grain orientations of alternate plies are mutually perpendicular. The low-shrinkage and high-strength characteristics of the plies in the parallel-to-grain direction minimize the shrinkage of the alternate laminae so that shrinkage in either of the width directions of properly balanced plywood is greatly reduced. Some increase of shrinkage in the thickness direction results, however, although this is generally not an important factor in use.

Stamm (1964) lists four categories of physicochemical treatments of wood which may be used to increase its dimensional stability, in addition to the cross-laminating method just described. These are: use of coatings and sealers, reduction of hygroscopicity, introduction of bulking chemicals into the cell walls, and cross-linking the cellulose chains within the cell walls to prevent water sorption.

The use of coatings and sealers to retard the rate of moisture movement into and out of wood is probably the most common and oldest method of minimizing shrinkage of wood. Paints and other coatings on exposed wood surfaces prevent rapid wetting and drying of the surface. This minimizes the development of steep moisture gradients near the surface which are responsible for surface failures such as checks and cracks, and for the subsequent entry of biological agents of deterioration. It should be emphasized that coatings do not prevent moisture movement but only retard it. Immersion of wood in water repellants or painting its surface with water-repellant coatings also helps to reduce surface moisture gradients because such treatment reduces the wetting of the wood surface by rain.

Reducing the hygroscopicity of wood by heating at temperatures above 100 °C also reduces its shrinkage characteristics. The reduction in hygroscopicity increases with both temperature and duration of heating. However, the mechanical properties of the wood are also degraded by heat treatment, although at a lower rate, and so this method has limitations.

Treatment of wood with bulking chemicals involves replacement of some of the water in the cell walls of green wood with various chemicals. These keep the wood in a partially swollen condition so that shrinkage is reduced in subsequent drying. Various methods have been used, ranging from soaking the green wood prior to drying in salt or sugar solutions or in polyethylene glycol, to application of more complex chemicals which react in the wood to form permanent insoluble bulking agents. The latter method includes liquid-phase treatment with thermosetting resins such as phenol formaldehyde and methyl methacrylate, as well as vapor-phase acetylation with acetic anhydride.

Cross-linking the cellulose chains by formaldehyde, particularly in the vapor phase, has been shown to reduce significantly the swelling of treated wood upon exposure to high humidity or to liquid water. Unfortunately, there is a corresponding loss in the toughness and abrasion resistance of the wood.

The high cost of chemical treatments of wood for dimensional stabilization generally precludes their use except for wood intended for specialized high-value purposes.

See also: Hygroscopicity and Water Sorption

Bibliography

Skaar C 1988 *Wood–Water Relations.* Springer, Berlin

Stamm A J 1964 *Wood and Cellulose Science.* Ronald Press, New York

C. Skaar

[Virginia Polytechnic Institute and State University, Blacksburg, Virginia, USA]

Strength

In the living tree, wood has several functions, one of which is structural support. The wood of the tree trunk has to withstand the compression loads from the weight of the crown above, and it has to resist bending moments resulting from wind forces acting on the entire tree. The ability of wood to resist loads (i.e., its strength) depends on a host of factors. These factors include the type of load (tension, compression, shear), its direction, and wood species. Ambient conditions of moisture content and temperature are important factors, as are past histories of load and temperature. Strength of a wood sample also depends on whether it is a small, clear piece free of defects or a piece of lumber with knots, splits and the like. This article will be concerned with the strength of clear wood and the major factors that influence it. Resistance and design values of commercial lumber are discussed elsewhere (see *Lumber: Behaviour Under Load*).

1. The Anisotropic Nature of Wood Strength

The cellular structure of wood and the physical organization of the cellulose-chain molecules within the cell wall (see *Macroscopic Anatomy*; *Ultrastructure*) make wood highly anisotropic in its strength properties. On the basis of the structural organization of wood, three orthogonal directions of symmetry can be distinguished. The longitudinal (L) direction, also referred to as the direction parallel to grain, is parallel to the cylindrical axis of the tree trunk and also to the long axis of the majority of the constituent cells. This direction has the highest proportion of primary bonds resisting applied loads, and is therefore the direction of greatest strength. The other two directions are the radial (R) and tangential (T) directions, which are perpendicular and parallel, respectively, to the circumference of the tree trunk. The latter two directions are also known collectively as directions perpendicular to grain. The strength perpendicular to grain is low because loads are resisted predominantly by secondary bonds.

The tensile strength of air-dry (~12% moisture content) softwoods parallel to grain is of the order of 70–140 MPa. For hardwoods it is often greater but may be less, depending on the particular species. The tensile strengths in the tangential and radial directions are about 3–5% and 5–8%, respectively, of the strength parallel to grain. The tensile strength perpendicular to grain of wood is thus only a small fraction of what it is parallel to grain, and in practical applications tensile stresses perpendicular to grain are avoided to the largest extent possible.

The compression strength (strength as a short column with a slenderness ratio of 11 or less) of air-dry softwoods parallel to grain is of the order of 30–60 MPa, which is much less than the tensile strength. The degree of anisotropy is also less, the compression strength perpendicular to grain being about 8–25% of the value parallel to grain. In contrast to tensile strength, which is always higher radially than tangentially, the compression strength may be either higher or lower, depending on species. Furthermore, compressive strength perpendicular to grain is often a minimum at 45° to the growth rings, that is, intermediate to the radial and tangential directions. The degree of anisotropy in compression depends on species, and in general is less in species of higher density. This is because compression failure of wood as a porous material occurs principally by instability at several levels. In compression parallel to grain, failure of cellulose microfibrils by folding can occur at stresses as low as one half of the ultimate failure stress. At higher stresses, a similar pattern of folding takes place on the cell-wall level, and these eventually aggregate into massive compression failures. In compression perpendicular to grain, failure occurs by collapse and flattening of the cells. Species with higher density have less porosity and thicker cell walls, which makes the cells much more resistant to collapse.

The shear strength of wood is characterized by six principal modes of shear failure, as illustrated in Fig. 1. The shear plane or failure plane may be in one of the three principal planes of wood (the LT, LR and TR planes), and failure in each shear plane may occur by sliding in one of two principal directions. These six

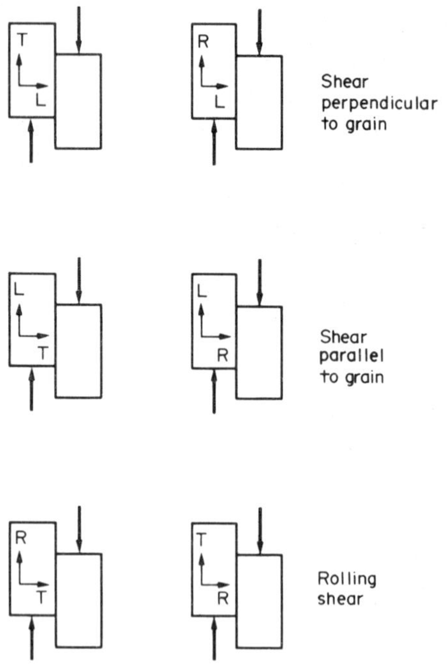

Figure 1
Schematic representation of the six principal modes of shear failure in wood, divided into three groups as shown

modes may be divided into three groups: shear perpendicular to grain, shear parallel to grain, and rolling shear. In shear parallel to grain, both the shear plane and the sliding direction are parallel to grain. In air-dry softwoods, the shear strength parallel to grain is of the order of 5–12 MPa. In shear perpendicular to grain, where the shear plane and the sliding direction are both perpendicular to grain, wood is most resistant, but because of its high degree of anisotropy often fails first in some other mode such as compression perpendicular to grain. Limited data indicate that the shear strength perpendicular to grain is of the order of 2.5–3 times the shear strength parallel to grain. In rolling shear, so-called because the wood fibers can be thought of as rolling over each other as failure occurs in this mode, the strength is least, amounting to 10–30% of the shear strength parallel to grain. There are no consistent or major differences between modes within each of the three groups of shear modes.

Certain problems relating to the strength of wood members can be dealt with by using linear elastic fracture mechanics. The required material parameters, the values of the critical stress intensity factor K_c, also vary because of the anisotropic nature of wood. In the opening mode of crack loading (mode I), there are six principal systems which are somewhat analogous to the six modes of shear failure. The crack plane may be located in one of three principal planes of wood structure, and in each of these the crack may propagate in one of two principal directions. Similarly, six systems are found in the other two modes of crack loading, the forward shear mode (mode II) and the transverse shear mode (mode III), and thus 18 fracture parameters are required to completely characterize wood. Since mode I is the most important, most of the available data are for K_{Ic}. Particular systems are denoted by two subscripts, the first representing the direction normal to the crack plane, and the second the direction of crack propagation. The RT, TR, RL and TL systems (i.e., the four systems where the crack plane is parallel to grain) have very similar K_{Ic} values. Those for the remaining systems, namely LT and LR, where the crack plane is perpendicular to grain, are again similar to each other but almost an order of magnitude greater than the other four. Thus, "weak" and "strong" systems can be distinguished. K_{Ic} values in the weak systems of air-dry softwoods are of the order of 150–500 $\mathrm{kPa\,m^{1/2}}$, and those in the strong systems are 6–7 times greater. Weak system K_{IIc} values are of the order of 1000–2500 $\mathrm{kPa\,m^{1/2}}$.

2. *Methods of Determining Wood Strength*

Many countries have established national standards for testing the mechanical properties of wood. By and large, the objectives and methods are similar. In the USA the methods are given in ASTM Designation D143-83, *Standard Methods of Testing Small Clear Specimens of Timber.* This standard has provisions both for sampling and for the tests themselves. The strength tests included are static bending, two types of impact bending, compression parallel and perpendicular to grain, tension parallel and perpendicular to grain, shear parallel to grain, hardness, and cleavage. All tests must be made at controlled (room) temperature and moisture content, which is usually either green (moisture content above the fiber saturation point (see *Hygroscopicity and Water Sorption*)) or 12%, and at prescribed speeds of testing. Test specimens are cut from sticks 50 × 50 mm in cross section with the grain parallel to the long dimension. European systems are usually based on sticks 20 × 20 mm in cross section.

The static-bending test is the most important single test. The test specimen is 50 × 50 × 760 mm and is center loaded as a simply supported beam over a span of 710 mm. Load-deflection data are taken so that an elastic modulus of bending may be calculated. The maximum bending moment is used to calculate the modulus of rupture, which is the presumed stress at the extreme fiber of the beam assuming that stress varies linearly through the depth of the beam. Actually, since the compression strength of wood is much less than its tensile strength, there is initial yielding on the compression side, followed by development of visible compression failures and enlargement of the compression zone. The neutral surface shifts toward the tension side as tensile stresses continue to increase. Maximum moment is reached when there is failure in tension. The modulus of rupture is the characteristic bending strength of wood, and its value is intermediate between the tensile and compressive strengths.

The dimensions of compression parallel to grain specimens are 50 × 50 × 200 mm. The maximum load is used to calculate the stress at failure, which is usually referred to as maximum crushing strength. For compression perpendicular to grain specimens, the dimensions are 50 × 50 × 150 mm. The specimen is laid flat, and the load applied through a metal loading block to the middle third of its upper surface (Fig. 2d). This is intended to simulate the type of loading as found in a stud or column bearing on a wood sill, so that the edges of the area under load are laterally supported by the adjacent unloaded area. In this test, a clearly defined maximum load cannot usually be obtained and only a fiber stress at proportional limit is computed.

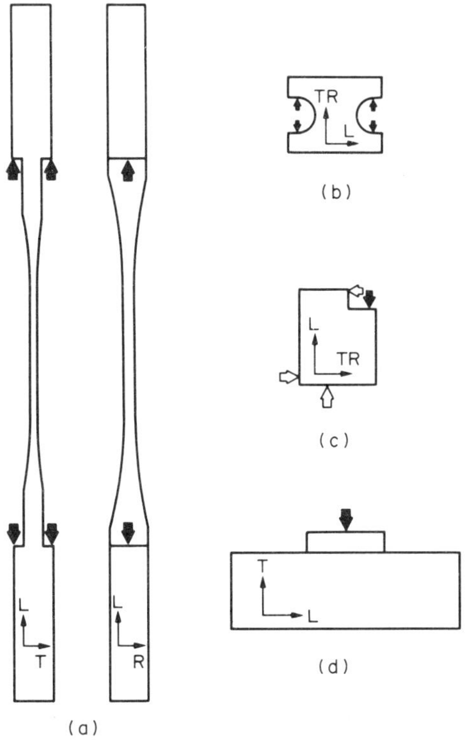

Figure 2
Standard test specimens for testing clear wood: (a) tension parallel to grain; (b) tension perpendicular to grain; (c) shear parallel to grain; and (d) compression perpendicular to grain. Application of loads is indicated schematically

Shear tests are generally made only parallel to grain. The specimen is 50 × 50 × 63 mm, with a 13 × 20 mm notch (Fig. 2c). The notch is cut so that the failure plane is the LT or LR plane in alternate specimens, and the results are averaged.

The four tests described so far are the most significant tests, since allowable stress values for visually graded structural lumber are based on their results. Data obtained from the remaining tests are useful for special purposes and for comparing relative strength of different species for particular applications. This includes tensile tests, because allowable stresses in tension parallel to grain are based on modulus of rupture data and there are no allowable stresses in tension perpendicular to grain. Tensile tests parallel to grain are difficult to make because of the high degree of anisotropy. Specimens are 460 mm long, have a 25 × 25 mm cross section at the ends, and are necked down to a minimum cross section of 4.8 × 9.5 mm (Fig. 2a). Load is applied to the shoulders of notches introduced near the ends. Load transfer from machine to specimen is therefore through shear over an area of 50 cm^2 at each end. Data for tensile strength parallel to grain are not available for most species and are not usually included in tables of strength data. Tensile tests perpendicular to grain are made on specimens with circular notches at each end, and loads are applied (through suitable fixtures) to the inside surface of the notches (Fig. 2b). Because of the geometry of the specimen, the tensile stresses perpendicular to grain are distributed very unevenly over the minimum cross section, so that the tensile strength is underestimated by about one-third.

The two types of impact bending tests are impact bending and toughness. The impact bending test involves dropping a hammer from successively greater heights, the maximum height to cause failure being recorded. This test is now considered to be obsolete. The toughness test is a single-blow impact test using a pendulum to break a 20 × 20 × 280 mm specimen by center-loading over a 240 mm span. The energy required to break the specimen is recorded. Toughness test data are not available for more than a few species.

The cleavage test, which is intended to measure the splitting resistance of wood, is made with a specimen which resembles the tension perpendicular to grain specimen. It has the same circular notch, but only at one end, and is 95 mm long overall. The load per unit of width, as applied at the notch, which causes splitting is recorded.

Hardness in wood is measured by embedding a spherical indentor of 11.3 mm diameter to a depth of 5.6 mm, and recording the load required to do so. Indentations are made on all three principal wood planes. Values obtained on the RT plane are referred to as end hardness and those on the LT and LR planes are known collectively as side hardness. Strength values for a few species as obtained from the various tests described are shown in Table 1.

Table 1
Strength data for four wood species at 12% moisture content

Property	Eastern white pine	Douglas fir	White oak (overcup oak)	Yellow poplar
Static bending: modulus of rupture (MPa)	59	85	87	70
Compression parallel to grain: maximum crushing strength (MPa)	33	50	43	38
Compression perpendicular to grain: fiber stress at proportional limit (MPa)	3.0	5.5	5.6	3.4
Shear parallel to grain: maximum shear strength (MPa)	6.2	7.8	13.8	8.2
Tension parallel to grain: maximum tensile strength (MPa)	78	130	101	154
Tension perpendicular to grain: maximum tensile strength (MPa)	2.1	2.3	6.5	3.7
Impact bending: height of drop causing complete failure (m)	0.46	0.79	0.97	0.61
Toughness: work to complete failure (J)	13	32	37	24
Cleavage: splitting force per unit width ($kN\,m^{-1}$)	28	32		49
Hardness: force to cause 5.6 mm indentation (kN)				
side grain	1.7	3.2	5.3	2.4
end grain	2.1	4.0	6.3	3.0

3. *Factors Affecting Wood Strength*

Various factors can strongly affect the strength of wood. The compression strength parallel to grain, for example, can more than double when drying wood from the green condition to 12% moisture content. The major factors can be divided into three groups as shown below.

3.1 Factors Related to Wood Structure

Although the nature and composition of wood substance is basically the same for all wood, the types of cells, their proportions and arrangements differ greatly from one species to the next. Balsa, a well-known lightweight wood, has an air-dry modulus of rupture of 19.3 MPa, as compared with values of over 200 MPa for a few African and South American tropical species.

Since wood is a biological material, it is subject to environmental and genetic factors that influence its formation. As a result, wood strength is very variable, not only from one species to the next, but also within a species. It will even vary depending on location within a single tree stem. Within a species, values of wood strength follow approximately a normal distribution. The variability of a particular species, as measured by the standard deviation, is generally proportional to the mean value. Therefore, it is convenient to express the standard deviation as a percentage of the mean, which is referred to as the coefficient of variation. Some strength properties tend to be more variable than others, but for most the coefficient of variation is of the order of 20%. This means that 95% of the values will fall in the range from about 60–140% of the mean value.

Much of the variation between and within species can be attributed to differences in wood density. The density of wood substance is constant at 1500 $kg\,m^{-3}$, so that wood density is in effect a measure of porosity (see *Density and Porosity*). As a result, there is a moderate to high degree of correlation between wood density and strength. It can be expressed as

$$S = k\rho^n$$

where S is a strength property, ρ is the wood density, and k and n are constants depending on the particular property. Values of n range from 1.00 to 2.25. If n is larger than unity it suggests that as density increases there is not only an increase in the amount of wood substance (which would imply $n = 1$) but qualitative changes in wood structure as well.

Grain direction has a major effect on strength as already discussed in detail. In structural lumber, which is used as an essentially linear load-bearing element (tension or compression member, or beam), it is desirable that the direction of greatest strength—the direction parallel to grain—should coincide with the long geometric axis. However, in practice the grain direction is often at an angle to the edge of the piece, a condition which is referred to as cross grain. In such a case, uniaxial stress along the geometric axis will have components both parallel and perpendicular to grain. The effective strength could be predicted if a suitable criterion for failure under combined stresses were available. For wood, such a general theory has not yet been found. An empirical relation which is often used is the Hankinson equation:

$$N \doteq PQ/(P \sin^2\theta + Q \cos^2\theta)$$

where N is the strength at an angle θ to the grain, and P and Q are the strengths parallel and perpendicular to the grain, respectively. The variation of strength with grain angle is similar to the variation in Young's modulus (see *Deformation Under Load*), so that even small angles can have a major effect on strength.

In addition to cross-grain, the major strength-reducing characteristics of structural lumber are knots. They are portions of branches which have become part of the stem through normal growth. Although generally harder and denser than the surrounding wood, their grain is more or less perpendicular, so that they contribute little to the strength parallel to the geometric axis of the piece. An additional factor is that the grain of the wood around the knot becomes distorted, representing localized cross-grain which also reduces strength.

3.2 Factors Related to the Environment

Equilibrium moisture content of wood is a function of ambient conditions of relative humidity and temperature (see *Hygroscopicity and Water Sorption*). Above the fiber saturation point moisture content has no effect on wood strength, but below it strength increases upon drying for most properties. A notable exception is toughness, which does not change much and may even decrease upon drying. The amount of increase depends on the particular property. Shear strength parallel to grain increases by about 3% for each 1% decrease in moisture content below the fiber saturation point; for maximum crushing strength in compression parallel to grain the increase is 6% per 1% moisture-content decrease. Most tabulated data give strength values when green (S_g) and at 12% moisture content (S_{12}). The strength (S_m) at moisture content M, in the range from 8% moisture content to the fiber saturation point (for strength adjustments commonly assumed to be 25%), can be estimated by the equation

$$S_m = S_{12}(S_g/S_{12})^{[(M-12)/13]}$$

Below 8% moisture content this exponential relation no longer holds as the rate of strength increase becomes less, and in the case of some properties there may even be a maximum at 6 or 7% moisture content. For example, tensile strength both parallel and perpendicular to grain decreases somewhat as the moisture content is decreased below this limit. Since the moisture content of wood in use is rarely much below 6%, this is of little practical significance.

Temperature has both immediate, reversible effects on wood strength and time-dependent effects in the form of thermal degradation (see *Thermal Degradation*). One immediate effect is that strength decreases as temperature is increased. There is an interaction with moisture content, because dry wood is much less sensitive to temperature than green wood. Some properties, such as maximum crushing strength in compression parallel to grain and modulus of rupture, are more sensitive to temperature than others, such as tensile strength parallel to grain. As a general rule, the effect amounts to a decrease in strength of 0.5–1% for each 1 °C increase in temperature. This is applicable in the temperature range of approximately −20 to 65 °C. Temperature effects over a greater range for several properties are illustrated in Fig. 3.

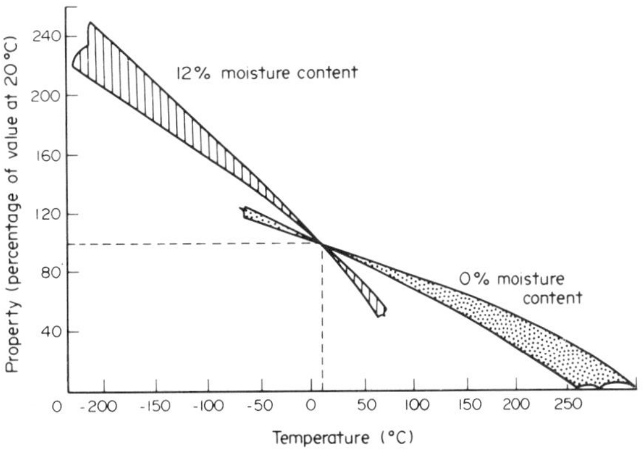

Figure 3
Immediate effect of temperature on bending strength, tensile strength perpendicular to grain, and compression strength parallel to grain. Variability is indicated by the width of the bands (adapted from *Wood Handbook*)

3.3 Agents Causing Deterioration

Deterioration or degradation of wood can be caused by biological agents, by chemicals, and by certain forms of energy. The principal biological agents of concern are decay fungi and wood-destroying insects, although damage may also be sustained because of bacteria, marine borers, and woodpeckers (see *Deterioration by Insects and Other Animals During Use*; *Decay During Use*). Damage by wood-destroying insects is by removal of wood substance, so that the effect on strength could be estimated by an assessment of the volume of material removed. In the case of wood decay, such estimates are more difficult because in the early stages of decay wood may appear sound but may have already suffered significant strength loss. To date, there are no satisfactory methods for estimating the residual strength of partially decayed wood members.

Chemicals which do not swell wood may have little or no effect on wood strength. Swelling chemicals, such as alcohols and some other organic solvents, may affect strength the same way water does, in proportion to the amount of swelling. The effect of such chemicals disappears if the chemicals are removed from the wood. Other chemicals may cause degradation of wood by such mechanisms as acid hydrolysis of cellulose. In general, wood is relatively resistant to mild acids but is degraded by strong acids or alkali. Heartwood tends to be more resistant to chemical degradation than sapwood, and softwoods are generally more resistant than hardwoods. Some chemicals used for preservative treatments do affect strength, and also the temperatures used in the treating process can have a permanent effect. Fire-retardant treatments usually result in some loss of strength. Common practice is to take a reduction of 10% in modulus of rupture for design purposes, but available data indicate that the effect on bending strength may be greater.

Wood under sustained loads is subject to static fatigue, commonly referred to as the duration of load factor. Available data show that for clear wood load capacity is linearly related to the logarithm of load duration. Load capacity decreases by 7–8% for each decade of increase on the logarithmic time scale. In structural lumber, the duration of load factor may be less pronounced, but current practice is to make adjustments for duration of loading in design. Dynamic fatigue in wood has not been studied in great detail. Results of tests on clear wood have shown that for fully reversed loading the fatigue strength after 30×10^6 cycles is about 30% of the static strength.

Wood can be degraded with large doses of nuclear radiation. Doses of γ rays in excess of 1 Mrad will cause a measurable effect on tensile strength; at a dose of 300 Mrad the residual tensile strength will be about 10%.

A more common form of degradation is by heat. The amount of degradation is a function not only of temperature but also time of exposure. Measurable effects on strength can be observed at temperatures as low as 65 °C. Degradation is more severe if the heating medium supplies water than if it does not. Softwoods are somewhat more resistant to thermal degradation than hardwoods. As in the case of other forms of degradation, including decay, toughness or shock resistance is affected first, followed by modulus of rupture, while stiffness is affected least. For softwood heated in water, a 10% reduction in modulus of rupture takes 280 days at 65 °C. At 93 °C the same reduction takes 8 days, and at 150 °C it takes only about 3 hours.

See also: Structure, Stiffness and Strength

Bibliography

American Society for Testing and Materials 1987 *Annual Book of ASTM Standards*, Vol. 4.09: *Wood; Adhesives*. American Society for Testing and Materials, Philadelphia, Pennsylvania

Bodig J, Jayne B A 1982 *Mechanics of Wood and Wood Composites*. Van Nostrand Reinhold, New York.

Dinwoodie J M 1981 *Timber, Its Nature and Behavior*. Van Nostrand Reinhold, New York

Forest Products Laboratory 1987 *Wood Handbook: Wood as an Engineering Material*. US Government Printing Office, Washington, DC

Gerhards C C 1982 Effect of moisture content and temperature on the mechanical properties of wood: An analysis of immediate effects. *Wood Fiber* 14: 4–36

Kollmann F F P, Côté W A Jr 1968 *Principles of Wood Science and Technology*, Vol. 1, *Solid Wood*. Springer, Berlin.

Patton-Mallory M, Cramer S M 1987 Fracture mechanics: A tool for predicting wood component strength. *For. Prod. J.* 37(7/8): 39–47

Schniewind A P 1981 Mechanical behavior and properties of wood. In: Wangaard F F (ed.) 1981 *Wood: Its Structure and Properties*. Pennsylvania State University, University Park, Pennsylvania

A. P. Schniewind
[University of California, Berkeley, California, USA]

Stringed Instruments: Wood Selection

Wood used for stringed instruments, such as pianos, members of the violin family and guitars, has been restricted to particular species. These are different for each instrument and for each of its parts, selection being based on long-standing experience. Furthermore, only wood of the highest quality and appearance is chosen. In other words, wood for stringed instruments has been selected empirically. The yield per log is presently very low; it is forecast that in the near future acquisition of suitable wood will become more difficult. Therefore, it is necessary to change the empirical selection method into a scientific one. Much research has been done to clarify scientific aspects of wood used for these instruments. In this article, basic principles and selection methods are described for wood for soundboards of pianos and sound boxes of violin-family instruments.

1. Wood Used for Stringed Instruments

Species of wood used for the soundboards of pianos and instruments of the violin family have been restricted to the genus *Picea*, namely European spruce (*Picea abies*), Sitka spruce (*Picea sitchensis*), and Akaezomatsu or Glehn's spruce (*Picea glehnii*). Similarly, backs of violins and related instruments are restricted to the genus *Acer*, namely sugar maple (*Acer saccharum*), Sycamore maple (*Acer pseudoplatanus*), Norway maple (*Acer platanoides*), and Itaya maple (*Acer mono*). In addition to species, wood used for instruments has also been restricted by tree age and trunk position, that is, only the lower part of trunks with a tree age of 100–300 years is acceptable. Boards are usually quarter-sawn, i.e., their face is in the longitudinal/radial (LR) plane. Soundboards are graded empirically, based on the extent of various defects, i.e., sloping grain, annual ring width, indented rings, color, reaction wood, sapwood, decay, knots, resin streaks, resin pockets, checks and injuries. Boards of acceptable grade have grain which is parallel to their length, a straight, uniform annual ring width of not more than 3 mm and only slight defects of the type mentioned above. The backs have grain which is parallel to their length and usually with a flame or curl pattern.

2. Quality Evaluation and Physical Properties

Wood is selected on the basis of the physical properties which are important for use. For the functions of soundboards, the physical properties most desirable are, generally, lightness, high elasticity, speed of propagation of vibration, length of duration of sound, and bigness of sound. These properties are related to physical quantities, namely specific gravity γ, Young's modulus E, sound velocity v which is proportional to

$(E\rho^{-1})^{0.5}$ (where ρ is density), internal friction Q^{-1} which is proportional to internal loss energy, and sound radiation damping v which is proportional to $v\rho^{-1}$. Consequently, it can be summarized that wood having low γ, high $E\gamma^{-1}$ and low Q^{-1} is suitable for soundboards. Backs participate in the vibration of sound boxes of violins, and thus they must also have suitable physical properties. As flexural vibration is excited in soundboards and backs, their physical properties are important not only in the longitudinal (L) but also in the radial (R) direction.

Characteristically, spruce has higher $E\gamma^{-1}$, lower Q^{-1} and higher $v\rho^{-1}$ than wood of other species in the L direction. Furthermore, it has been found that wood for higher-graded instruments has higher $E\gamma^{-1}$ and lower Q^{-1}. Thus, spruce wood for soundboards has the most desirable physical properties in the L direction. As frequency increases, the proportion of deformation caused by shear as compared to bending increases, so that the value of Q^{-1} increases in the high-frequency range. In the L direction, the value of spruce wood is lower in the low-frequency range and higher in the high-frequency range when compared to hardwoods. Spruce wood shows a higher dependence of Q^{-1} on frequency because its $E_L G_{LT}^{-1}$ value (where G_{LT} is the shear modulus in the LT plane), which determines the contribution of shear relative to flexural deformation, is higher. In wood the values of $E\gamma^{-1}$ are lower and values of Q^{-1} are higher in the R than in the L direction. Nothing noteworthy can be observed in these values in the R direction for spruce wood. Maple and other hardwoods are fairly inferior to spruce with respect to physical properties in the L direction, but slightly superior in the R direction. Furthermore, maple does not have any special characteristics in its physical properties compared with other hardwoods. Consequently, maple wood has been used for violin backs because of its beautiful appearance and its characteristic figure, rather than its physical properties.

3. Physical Properties and Structure

The relationship between physical properties which are important to soundboards and wood structure has been investigated by several researchers. In conifers, tracheids make up most of the constituent cells. From this it can be shown theoretically that $E\gamma^{-1}$ of wood in the L direction is approximately proportional to the average Young's modulus of the cell wall in the L direction. Therefore, differences in $E\gamma^{-1}$ values of coniferous woods in the L direction can be attributed to differences in cell-wall structure. For conifers it has been shown experimentally that wood of species having higher $E\gamma^{-1}$ values have smaller microfibril angles ϕ (which is a measure of the orientation of semicrystalline cellulose) and a higher degree of crystallinity α. These relationships can also be shown by numerical analysis of cell-wall models. Also, for spruce wood it has been shown experimentally that $E\gamma^{-1}$ decreases and Q^{-1} increases with increasing grain angle which corresponds to increasing values of ϕ. This explains why spruce boards having grain which is straight and parallel over their length have been used for soundboards. Reaction wood, which forms on the lower side of leaning stems in conifers, characteristically has darker color, a value of ϕ of the middle layer in the secondary wall, S_2, that is higher, and a higher lignin content in comparison with normal wood. In the L direction severe reaction wood has lower $E\gamma^{-1}$ and higher Q^{-1} than acceptable wood. However, light reaction wood is comparable to acceptable wood in its physical properties. As annual rings become wider, that is, as the proportion of latewood becomes lower, $E\gamma^{-1}$ decreases and Q^{-1} increases in the L direction. Therefore, wood having wide annual rings is inferior. The color of wood has hardly any influence on its physical properties, and therefore colored wood traditionally was not used only because of its appearance. Sapwood, which also has not been used, has high values of both $E\gamma^{-1}$ and Q^{-1}. Q^{-1} of sapwood is high because of its high equilibrium moisture content. Wood containing severely indented rings and resin streaks has inferior physical properties. However, wood containing slightly indented rings, and small knots and pitch pockets, is almost comparable to acceptable wood. Therefore, these kinds of wood have been regarded as inferior not because of their physical properties but because of their appearance.

4. Nondestructive Selection Methods

4.1 Soundboards of Pianos

In order to select wood suitable for piano soundboards more easily, physical quantities which can be used as selection criteria have been investigated. The parameters v, $E\gamma^{-1}$, Q^{-1} and $v\rho^{-1}$ discussed above can also be used as criteria. It has been found that there is high correlation between $Q^{-1}(E\gamma^{-1})^{-1}$ and $E\gamma^{-1}$ at any moisture content regardless of species, the three principal grain directions or grain angle. There is also a high correlation between $(QE)^{-1}$ and $E\gamma^{-1}$. If a material is subjected to sinusoidal stress with constant amplitude, the parameter $(QE)^{-1}$ is proportional to the energy per cycle dissipated as heat. Consequently, by measuring v or both E and γ, wood for piano soundboards can be selected easily. By selecting wood having higher $E\gamma^{-1}$ and low γ, wood having high $v\rho^{-1}$, i.e., high efficiency of sound radiation, can be obtained. In addition, the parameter $v\rho^{-1}Q$, $\bar{R}_{mn}$ and EG^{-1} have been proposed as criteria. $v\rho^{-1}Q$ is a parameter combining $v\rho^{-1}$ and Q^{-1}. $\bar{R}_{mn}$ is the average value of the sound pressure level in each mode produced by a board subjected to an oscillating force with constant amplitude. Therefore, $\bar{R}_{mn}$ is also a parameter related to the efficiency of sound radiation. Spruce wood has an especially high value of this quantity. EG^{-1} is related to the slope of the envelope curve of a sound spectrum and to the magnitude of a resonance point. Spruce wood has higher values of this quantity in the L and R directions than wood of other species.

4.2. Sound Boxes of Violin Family Instruments

The relationships between the vibrational characteristics of a finished violin and those of its parts as free plates are very complex. Therefore, the selection of wood for the soundboards and backs is also complex. According to the studies done so far, in order to make violins with fine tone and playing qualities, it is most desirable (but often quite difficult) to have the free-plate modes 1, 2 and 5 of a finished violin soundboard lie in a harmonic series, with mode 5 having a large amplitude and a frequency near 370 Hz, and to have the frequencies of modes 2 and 5 of the soundboard match those of the back. The shape, arching contours and thickness distributions of the soundboard and back plates are crucial in achieving these relations. The values of E, G, Q^{-1}, γ and v of the wood for soundboards and backs are very important to the sound of fine instruments. For the soundboards it is necessary to select wood having high $E\gamma^{-1}$ not only in the L but also in the R direction. The above may also be said of the other members of the violin family.

In order to determine these wood parameters instantly and automatically, measuring methods which pass the sound of wood being tapped through a FFT (fast Fourier transform) analyzer connected to a personal computer are being developed at present. Further improvements in selection methods of wood for instruments will require clarification of the relationship between the quality of finished stringed instruments and the physical properties of their components as free plates.

See also: Acoustic Properties

Bibliography

Holz D 1973 Untersuchungen an Resonanzholz. *Holztechnologie* 14: 195–202

Hutchins C M 1981 The acoustics of violin plates. *Sci. Am.* 245: 126–35

Kollmann F P, Côté W A 1968 *Principles of Wood Science and Technology 1.* Springer, Berlin

Meinel H 1957 Regarding the sound quality of violins and a scientific basis for violin construction. *J. Acoust. Soc. Am.* 29: 817–22

Ono T, Norimoto M 1983 Study on Young's modulus and internal friction of wood in relation to the evaluation of wood for musical instruments. *Jpn. J. Appl. Phys.* 22: 611–14

Ono T, Norimoto M 1984 On physical criteria for the selection of wood for soundboards of musical instruments. *Rheol. Acta* 23: 652–56

Ono T, Norimoto M 1985 Anisotropy of dynamic Young's modulus and internal friction in wood. *Jpn. J. Appl. Phys.* 24: 960–64

Yankovskii B A 1967 Dissimilarity of the acoustic parameters of unseasoned and aged wood. *Sov. Phys. Acoust. (Engl. Transl.)* 13: 125–27

T. Ono
[Gifu University, Gifu City, Japan]

Structural-Use Panels

Structural-use panels is a term used to identify a class of utilitarian wood-based panel products used for construction and industrial applications. The primary market for structural wood panel products is the light-frame construction industry. Panel products are shipped to suppliers and contractors for on-site assembly of residential and commercial structures. These wood panels are commodities manufactured and distributed in standard sizes and grades.

The structural panel products include plywood, oriented strandboard and waferboard. Plywood is not discussed here since it is adequately covered in a separate article (see *Plywood*).

1. Manufacturing

Waferboard and oriented strandboard manufacturing technologies are identical except for two differences: the shape of the flake and the technique called orientation. These differences are elaborated below.

1.1 Log Preparation

Logs are received at the plants and are either stored in water or stacked in tiers in the mill yard. From there they are transported to a deck where they are debarked and cut into the lengths suitable to accommodate the plant processing equipment. Debarking a log is most often accomplished by conveying it through either power-driven "ring" scraper knives or a very large drum with internal scrapers which remove the bark as the drum rotates.

Before beginning the process of transforming the blocks into flakes, nearly all manufacturers find it necessary to condition the logs. This is most often accomplished by soaking the short logs in a log pond for a period of time. This is important for good flake quality, especially in cold climates.

1.2 Conversion into Wafers or Strands

Flakes are produced by a variety of cutting techniques directly from the logs (often called roundwood). This differentiates waferboard or oriented strandboard from industrial particleboard which is produced from mill residue and much smaller particles.

There are two primary systems used to create flakes or strands. One is a drum-type machine where knives are aligned along the length of a cylinder which rotates and cuts the flakes as the logs are pressed against it.

The second common system is a disk machine in which the knives are located on the radii of the disk. The flakes are cut as the logs are pressed against the rotating disk. Flakes are usually 0.7–1.3 mm thick. Wafers are flakes which are squarish in size, the side length ranging from 20 to 50 mm, whereas strands are rectangular in shape typically with the length along the grain of the flake being three to five times longer than the width. Strands are typically 30–60 mm long and 10–25 mm wide.

The green flakes are transported by conveyors to the drying station. A typical flake dryer consists of concentric rotating drums over which the wet material makes several passes through the system before it leaves the dryer. Dryers of this type have been made since the 1940s and are commonly used in the agricultural industry for drying alfalfa and similar crops. The dryer is a concurrent multiple-cylinder rotary drum (direct-fired type) which utilizes heated air as the drying medium. Other types of dryers include single-pass jet dryers and tube dryers (Maloney 1977).

Wafers and strands are normally classified and sorted by size at least once. These classification systems depend on the process, and there are typically between three and five categories. The smallest category is not usually used in the final product but is used for fuel or other purposes. Two typical systems include screens which mechanically sort the material or air classifiers which pass the flakes through a moving air stream for stratification. Both processes can cause some mechanical damage to the flakes.

1.3 Adhesives and Additives

Adhesives currently used in structural-use panels are basically phenolic resins or the newer class of isocyanate resins (see *Adhesives and Adhesion*). Phenolic resins are applied to the flakes in either liquid or powder form. The resulting bonds have similar durability and exhibit moisture resistance. Liquid phenolic resins are usually applied by sprays in rotating drums or other systems which allow the flakes to pass by the resin application stations. Resin is applied in fine droplets on the flakes rather than an entire coating as would be typical for veneer plywood. Phenolic resins in powder form are manufactured by spray-drying the phenolic resin. The resin is delivered in bags and applied by introducing the resin with liquid additives such as wax to help in bonding the powder to the flakes prior to pressing.

The use of isocyanates within the waferboard and oriented strandboard industry is increasing. This class of resin utilizes a slightly different chemistry, but has the advantage that no water is contained in the system. Accordingly, flakes of higher moisture content can be used, thus reducing drying costs.

The most common additive in these products is a wax emulsion. The wax is added as a sizing agent to provide the finished product with resistance to water penetration. In addition to improving finished product properties with respect to water, wax also is of great assistance in distributing dry powdered resin during its application.

1.4 Assembly and Pressing

Felting, forming and spreading the furnish are terms referring to the formation of a uniform mat for consolidation of the final board. The preparation of a consistent mat of flakes is probably one of the most important parts of board manufacturing, secondary only to flake quality. Inconsistent distribution of flakes will result in changes in physical properties, density variations and abnormal responses to mechanical or physical loads. All forming machines meter the flakes uniformly across the width of the press to which it will be transferred. A typical operation involves moving carriers passing under the forming machine with segregated face flakes being laid on the mat first, followed by core flakes and then flakes for the top surface of the panel which becomes its other exterior surface.

The difference between waferboard and oriented strandboard becomes evident with the utilization of orientation devices. These are typically mechanically operated disks or veins which cause the rectangular strands to fall on the moving mat in a specific direction with respect to the long axis of the strand. With oriented strandboard, alternate layers of the product are aligned at right angles to each other, much as the veneers of plywood, although some versions of oriented panels do not orient the core and allow the flakes to fall in a random fashion much as an entire panel of waferboard would be manufactured. Following formation, the mats are then transported to the hot press. Press platens are heated with oil or steam and are closed usually with hydraulic rams. Press sizes of up to 1.2×15.2 m with 20 or more openings are becoming common. One recent development involves the injection of live steam during the pressing process to shorten press times even further. Press temperatures reach as high as 200 °C with press times from one to five minutes. With such pressures and temperatures, mat moisture content is quite important to prevent the occurence of steam pockets which cause the board to rupture upon release of pressure. This phenomenon is termed a "blow."

2. *Performance Standards*

Historical standards against which wood-based panel products have been manufactured have been product standards such as PS-1 for construction and industrial plywood and ANSI A208.1 for mat-formed wood particleboard. National concensus or product standards are most frequently promulgated through organizations such as the American National Standards Institute (ANSI), or in the case of US Product Standards, the Department of Commerce. Product or commodity standards are essentially manufacturing prescriptions for minimum-quality products. Requirements are not centered around the specific manufacturing process, and typically do not address the question of product application or how well the product is suited for any particular application.

A performance standard reverses the standard process by defining the end-use of the product and does not prescribe how the product will be manufactured. The objective of a performance standard is to ensure that the product will satisfy the requirements of the application for which it is intended. To do this the performance standard must define performance criteria and test methods. Performance standards were pioneered by the American Plywood Association (APA) in 1980, and have since been adopted by other trademarking organizations in the USA and most recently by the Canadian Standards Association in a standard expected to be adopted in 1988. The standards developed by the APA include one for single-layer floors and for sheathing. The intended end-use of these products is for light-frame wood construction as practised in the USA. As the name implies, panels are intended for combination subfloor underlayment and sheathing intended for roofs, walls and subfloors. The products are rated for their particular end-use. The standards cover panels for supports spaced 400–1220 mm apart for floors and roofs. Provisions are also made for wall bracing when the studs (vertical framing elements) are spaced 400 or 600 mm apart. A recent development includes a similar standard for siding where the panels (or panels cut into strips for lap) are to be the exterior building surface.

Panel performance is evaluated through testing under a qualification procedure for three basic areas of investigation: structural adequacy, dimensional stability and bond performance.

Testing for structural adequacy includes:

(a) verification of the ability to sustain concentrated loads both static and by impact;

(b) uniform-load testing to demonstrate the ability of the product to sustain wind or snow load;

(c) testing of fastener-holding ability to ensure that the panel has the ability to hold covering materials in place and to develop lateral loads in mechanically fastened connections to lumber or other framing;

(d) testing to demonstrate that wall sheathing panels can supply resistance for wall bracing.

Tests of physical properties include measuring the expansion of the panels when exposed to elevated moisture conditions. Two methods for measuring linear expansion have been developed. One is the oven-dry to vacuum-pressure-soak technique which is adequate for quality control purposes, but does not reflect field conditions. Accordingly, a performance test was developed which measures linear expansion when wetting on one side for a period of fourteen days. The test approximates severe exposure conditions as might occur during construction and should serve as a relative measure of linear expansion conditions that might be expected in service.

Finally, the bond durability of the panel must be demonstrated for site-built construction since panels will be called upon to resist wetting and drying and still be able to perform following exposure to these conditions. Results of extensive research have indicated that large-scale specimens exposed to hot-water soaking and oven drying and then subjected to the same structural performance criteria is adequate evidence of the ability of the panel to remain intact during exposure to weathering. These tests are equivalent to several years permanent exposure to the weather. Retention of strength following moisture cycling is also used as a measure of bond performance.

Certification of performance is a straightforward process under a performance-based system. First a sampling of the candidate product is tested according to the individual performance criteria discussed above. After demonstrating adequate structural performance, physical properties and bond durability, a span rating can then be assigned. Concurrent with performance oriented tests, a product evaluation is also performed in which properties unique to the product and suitable for quality assurance are measured. Such properties include bending strength and stiffness, thickness, density and others which may be useful in monitoring quality under a third-party basis. A mill specification is drawn up following this intensive product evaluation. This individualized mill specification is unique to the mill and product qualified under the standard.

3. *Physical Properties*

Physical property data have recently been collected for commercial products and this will be continuing as the industry matures.

3.1 Effects of Moisture

Many of the physical properties of structural panels are affected by the amount of moisture present in the wood. Wood is a hygroscopic material which nearly always contains a certain amount of water (see *Hygroscopicity and Water Sorption*). When a structural panel is exposed to a constant relative humidity, it will eventually reach an equilibrium moisture content (EMC). The EMC of plywood is highly dependent upon the relative humidity, but is essentially independent of temperature between 0 °C and 85 °C. Examples of EMC values for structural panels at 25 °C include 4% at 40% relative humidity (RH), 7% at 70% RH and 23% at 100% RH.

The expansion in length or width of structural panels from dry (less than 1% moisture content) to saturation is less than 0.5%. Linear expansion when exposed to a change in relative humidity from 50% to 90% RH is typically less than 0.3%. The dimensional stability of panels exposed to liquid water also varies. Tests were conducted with panels exposed to wetting on one side as would be typical of a rain-delayed construction site. Such exposure to continuous wetting on one side of the panel for 14 days resulted in less than 0.35% expansion across or along the major axis of the panel.

3.2 Permeability

The permeability of structural panels is different from solid wood in several ways. The overlapping flakes from which panels are made create many small gaps which provide pathways for fluids to pass when entering through the panel edge. Typical permeance for untreated or uncoated $\frac{3}{8}$-inch (9.5 mm) structural panels is about 0.23 $g\,m^{-2}\,h^{-1}\,mmHg^{-1}$.

4. Mechanical Properties

The current design document for plywood is published by the American Plywood Association and is entitled the *Plywood Design Specification.* As the title implies, it is directed toward plywood and does not apply to all structural panels. To correct this gap in information, the American Plywood Association is conducting an extensive testing program to determine strength properties for all structural panels. This work is expected to be completed in 1988. Preliminary data indicates that the resulting capacities of all structural panels will be very similar to those published in the *Plywood Design Specification*

Bibliography

American National Standards Institute 1979 *Mat-Formed Wood Particleboard,* ANSI A208, 1-1979. National Particleboard Association, Silver Spring, Maryland

American Plywood Association 1986 *Plywood Design Specification,* Form Y510. APA, Tacoma, Washington

American Plywood Association 1987 *Grades and Specification,* Form J20. APA, Tacoma, Washington

American Plywood Association 1987 303 *Plywood Siding,* Form E300. APA, Tacoma, Washington

Bengelsdorf M F 1981 *Linear Expansion and Thickness Swell of Wood Based Panel Products after One-side Wetting,* Report PT81-25. American Plywood Association, Tacoma, Washington

Canadian Standards Association 1985 *Waferboard and Strandboard,* CAN 3-0437.0-M85. Canadian Standards Association, Rexdale, Ontario

Maloney T M 1977 *Modern Particleboard and Dry Process Fiberboard Manufacturing.* Miller Freeman, San Francisco, California

O'Halloran M R, Youngquist J A 1983 An overview of structural panels and structural composite products. In: *Proc. Structural Wood Research State-of-the-Art and Research Needs Workshop.* American Society of Civil Engineers, New York

US Department of Commerce 1983 *Product Standard for Construction and Industrial Plywood,* PS 1. US Department of Commerce, Washington, DC

M. R. O'Halloran
[American Plywood Association, Tacoma, Washington, USA]

Structure, Stiffness and Strength

Wood exhibits unique structural features at virtually every dimensional scale—from the molecular to tree-size (see *Constituents of Wood: Physical Nature and Structural Function*; *Macroscopic Anatomy*; *Chemical Composition*; *Ultrastructure*). Glucose units are strung together to form molecular chains that are either amorphously tangled or assembled into crystalline regions; phenylpropane units are randomly linked at various locations to form an encrusting, cross-linked binding substance. Microfibrillar strands, composed of macromolecules, are helically wound in ribbon-like clusters to form layers of cell walls. The cell wall itself is composed of characteristic laminations: primary wall (P), secondary wall (S) and middle lamella (M). Growth rings are distinguishable by changes in various cell geometries and the ring itself is packed with cell types arranged in unique patterns specific to the tree species. Within the cross section of the tree, zones of sapwood and heartwood can be distinguished. Reactionwood and juvenile wood zones are defined by the abnormal characteristics of the cell. Knots and associated grain deviation patterns result from the attachment of branches to the stem of the tree. The shape of the stem itself is tapered and roughly cylindrically symmetric. What is the reason for this compounded complexity? How do these features, cascading down through the dimensional scales, affect the mechanical properties?

The quest of the materials scientist is to relate structural features to physical and mechanical properties. Once the structure–property relations are known for man-made materials, processing can be altered to change structure, thereby obtaining desired properties. The situation regarding structure–property relations for biological materials is different. The production process for manufacturing the material has evolved in conjunction with the evolution of the organism producing the material. The "design" of the material is intimately related to the requirements of the organism for survival. The design procedure is adaptive. Discovery of the structure–property relations for biological materials can involve finding reasons why the structure exists in relation to the function it performs. While the technologist may not be able to change the essential "production process" (genetic manipulation is possible but this changes only the parameters and not the overall blueprint), he can at least be able to utilize the material properly, exploiting its strengths and avoiding or compensating for its weaknesses.

In effect, the tree—the "manufacturer" of the wood—might be considered a factory for producing the material of which the factory is composed. The factory survives if the material performs its function under environmental duress. In addition to other functions (e.g., solar energy collection, water transport, sugar production), the tree must survive mechanical loading. It is from this perspective that scientists can understand the relationships between structure, stiffness and strength. This perspective can also carry over to the other functions, e.g., the structure of wood in relation to water transport, and a good discussion of form related to function in a tree can be found in Zimmermann and Brown (1971). A more general treatment of form versus function in biological systems is found in Thompson (1942).

1. Models

To study the relationships between structure and mechanical properties of wood, mathematical models are employed. Some are quite simple; others more complex. They all extract idealizations of the features of concern and deduce relationships that otherwise might be difficult to obtain by other means. The models are virtually indispensible tools for gaining insight into the reasons for the existence of the complex structure found in wood. On the other hand, the models are only as good as the assumptions made to employ them. Consequently, if experimental data cannot be obtained to validate model results, there is always the possibility for error.

The general purpose of most models is to establish the relationship between properties of a particular dimensional level to the properties at another scale. In all cases, "smearing of the microstructure" occurs. In other words, for some characteristic region the material within is assumed to be a continuum. The continuum is characterized by a single parameter, or set of parameters, that remain constant within the region or vary smoothly in a specified manner. Given values for these parameters (assumed or deduced in some way) and a well defined geometry, an analysis can be performed that will yield either (a) an effective value for a large scale parameter, such as an effective Young's modulus, or (b) a distribution of some variable, a stress component, say, over some larger region. Strength can be inferred from the maximum stress and the location of failure can be identified.

In the following sections the influence of structure on stiffness and strength of wood will be discussed for three scales of dimension: the scales of the tree (macro), the cell aggregate level (meso) and the cell wall (micro). The distinction in scales is arbitrary but is in fact useful. For solid wood products (lumber, plywood, particleboard), the strength and stiffness of relatively large units is important. These depend on properties at the cell aggregate level, which in turn are linked to the cell wall level. Thus, through a chain of structure–property relationships the properties of common wood products can be related to the ultrastructural features of the woody substance.

2. Macroscale

Banks (1973) considered the design of the tree from an engineering point of view. Among other examples illustrating performance efficiency, the taper of the stem is shown to be compatible with the idea of economical use of material. The tree, a spruce in this case, is considered to be a cantilevered beam built into the ground. The crown is assumed to have the shape of a triangle which is responsible for transverse loading caused by wind. Wind direction, being a random variable, is considered to be the reason why the tree trunk has a rotationally symmetric moment of inertia about the axis of the stem. Assuming the tree to be composed of a homogeneous, elastic material (a continuum hypothesis) which has a unique breaking strength in bending, then the taper of the trunk (diameter versus tree height) can be deduced from another assumption: the load-bearing capacity of stem is the same at any height level. The stem is divided into two regions: the lower and longest (ground to beginning of crown) and the upper (beginning of crown to top). The moment distribution in the lower segment is computed from a point load acting at the centroid of the crown and in the upper region it is computed from a ramp load (no load at the tip). It is found that the diameter will be proportional to the cube root of the difference between centroid height and vertical location in the lower region and will vary linearly with height in the upper zone (Fig. 1). Some measurements of tree taper agree with this analysis.

Schniewind (1962) reviewed the older literature and found that the mechanical theory of stem form based on equal resistance to bending moments along the height dates back to 1874. A number of authors have refined the theory to take into account the nonhomogeneous nature of wood in the stem such as the change in elastic modulus with height. In fact, based on empirical evidence that wood strength in bending is exponentially related to wood density, Schniewind provides a justification for horizontal density gradients as found in many species. If a given amount of wood is added to the tree stem each year, it is more efficient from the tree's point of view to strengthen each increment by "densifying" it, than to spread the mass out in a homogeneous manner. Moreover, the density response is apparently linked to crown development as radial density gradients clearly follow silvicultural treatments, i.e., thinning, fertilizer application and tree spacing (Megraw 1985).

While it is tempting to attribute a rationale to all patterns of organization and all features found in trees,

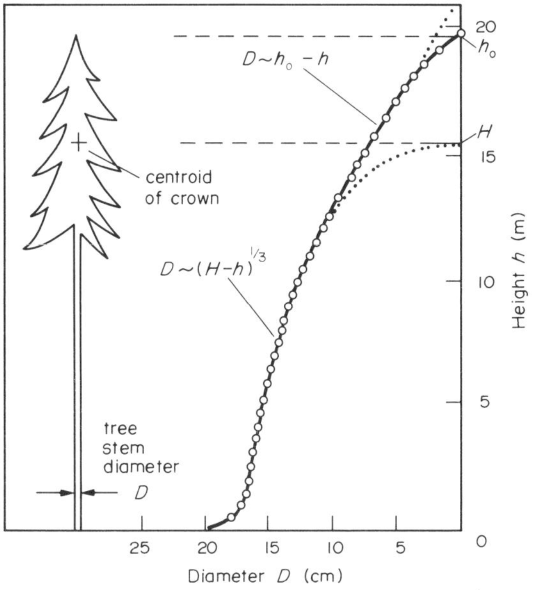

Figure 1
Taper in a 76-year-old spruce (adapted from Banks 1973)

it could also be that a structure or phenomenon is not optimal but simply does not appreciably reduce the survival capability of the tree. Growth stresses may be an example. In an earlier paper Archer and Byrnes (1974) developed a model to account for the axial tension in the outer radial zones of the stem and axial compression in the center. Extreme forms of residual stresses can cause splitting in logs when the trees are harvested. Treating the stem as a cylindrically orthotropic pole and using the "Boyd hypothesis" that each new growth increment is accompanied with a fixed growth strain state, Archer and Byrnes show that a residual stress pattern can be found which matches experimental data. The novel feature introduced in the analysis is the use of growth strains to produce the growth stresses. The growth strains are analogous to thermal or hygroelastic strains and are apparently the result of lignification of the walls in developing cells (Boyd 1973). Radial distributions of stresses, σ, in this type of material, under an axisymmetric assumption, are found to be a function of the radial coordinate r, raised to a power involving a material constant k, i.e., $\sigma = Cr^{\pm(k\pm1)}$, where $k=(E_T/E_R)^{1/2}$, E_T and E_R are the Young's moduli in the tangential and radial directions, respectively, and C is a constant. Thus, the radial position where the axial stress distribution changes from compression to tension is a function of the elastic constants. For a more detailed discussion of growth stresses see Archer (1987a).

In some species, spiral grain exists in which the wood fibers are not aligned with the axis of the stem but wind around the axis in a helical arrangement. The topic of spiral grain has been treated at length by Noskowiak (1963). Archer (1979) has modified his theory to account for this inclined principal direction of the material and has found that growth stresses are sensitive to the spiral grain angle. In addition, torsional shear stresses are also very sensitive to this parameter.

The crown of the tree is composed of branches. McMahon and Kronaurer (1976) have considered the design of the branches from a mechanical perspective. They were conceived to be rectangular in cross section with width b and depth h obeying power-law tapering with branch length position s, i.e., $b=b_0 s^{-\alpha}$ and $h=h_0 s^{-\beta}$. The exponents α and β are parameters whose values are predicted by two alternative design strategies. One is based on the concept of equal breaking resistance and the other on elastic similarity. In the latter case, the branch deflection under its own weight has the same relative shape for any segment. The value of β which, theoretically, produces this condition is 3/2. In contrast, the equal strength criterion yields $\beta=2$ on theoretical grounds. A plot of measured length versus diameter of oak branches asymptotically approaches an empirical relationship in which $\beta=1.50 \pm 0.13$. In addition, experiments involving the natural frequencies of shaking trees lead to an average value of $\beta=1.50$.

The connection between branch and trunk has not received much attention. Shigo (1985) has studied the grain configuration in the vicinity of this connection. He has found, for example, that cell orientation in the upper part of the junction is perpendicular to both stem axis and branch axis. A collar typically develops and is built up annually. The cellular arrangement appears to be related to sap and water transport as well as injury protection but must be compatible with supporting cantilevered loads acting on the branches. Lumber and veneer, whose cut surfaces sever these grain patterns, suffer significant strength losses resulting from the point-to-point change of principal material directions. Stress grades of lumber are essentially determined by the strength reducing aspects of knots and cross grain (see *Lumber: Behavior Under Load*).

An attempt has been made by Phillips et al. (1981) to characterize the grain pattern around the knot as an analog to flow around a circular object. Using this grain pattern as a model for the material property organization of an orthotropic anisotropic elastic solid, Cramer and Goodman (1983) attempted to predict the strength of individual pieces of lumber. A finite element approach combined with the principles of linear fracture mechanics was employed in the analysis.

3. *Mesoscale*

Using balsa wood to help develop a representative aggregate wood cell model, Easterling et al. (1982)

have developed a model to predict the crushing strength and elastic moduli in the three principal material directions for wood. The model essentially consists of tesselated hexagonal tubes packed in stories with embedded layers of ray cells (Fig. 2). Wood cells are represented as hexagonal tubes capped at the ends with prismatic caps. The tube walls are treated as continuous material with a density of $\rho_s = 1.5\ \mathrm{g\,cm^{-3}}$. The material is also considered as a transversely orthotropic material with respect to stiffness and strength. For example, the elastic modulus in the axial direction of the tubes is considered to be 35 GPa and in the transverse direction 10 GPa.

Due to the variation of gross wood density, ρ, with cell wall dimensions (i.e., ρ/ρ_s is proportional to cell wall thickness to hexagonal plate length, t/l) and to the stiffness of the tube structure being proportional to $(t/l)^3$, obtained from a structural analysis, the elastic modulus in the tangential direction is found to be related to the cube of gross density: $E_T = E_S\ (\rho/\rho_s)^3$.

Based on similar arguments involving geometry, but also including ray cells, the elastic modulus in the radial direction is found to be about twice the modulus in the tangential direction. The axial stiffness of the complex was shown to be related linearly to density. A plot of normalized stiffness against normalized density reveals that the trends in the data are consistent with the theory (Fig. 3).

Easterling et al. (1982) have also considered the crushing strength of wood using theoretical arguments based on geometry and continuous cell wall material. In this case, radial and tangential crushing strength are

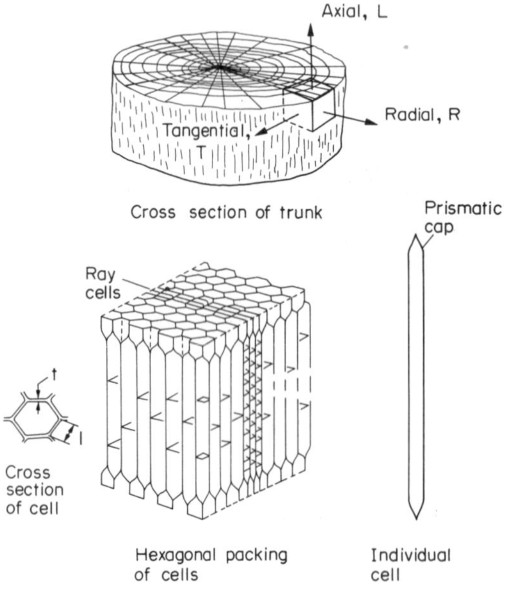

Figure 2
Cell aggregate model (adapted from Easterling et al. 1982)

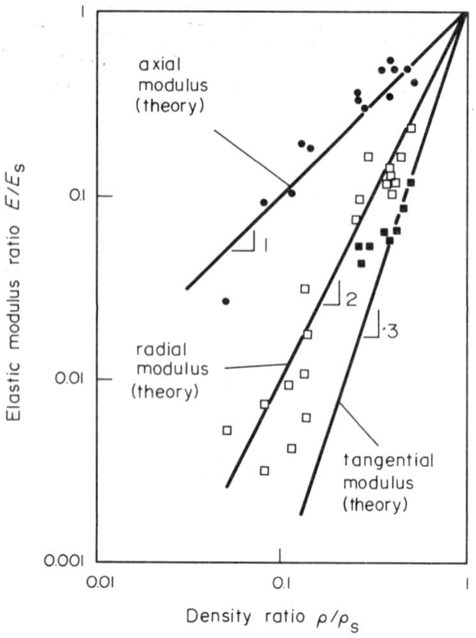

Figure 3
Normalized stiffness versus normalized density in the three orthogonal directions (adapted from Easterling et al. 1982)

proportional to the square of the cell wall thickness to hexagonal wall length ratio $(t/l)^2$. The axial crushing strength is linearly related to this ratio. Again, theoretical and experimental results were consistent. Thus, structural arrangements of the units that make up wood are seen to influence the properties of the gross wood.

While the typical arrangement of tracheids in softwoods is not a perfectly hexagonal honeycomb structure, Gillis (1972) has noted that there are a very large number of junctures involving the intersection of three double cell walls. He was lead, therefore, to develop a "triple-point element." In essence, the element is considered to be a portion of a rigid frame structure with one double wall parallel to the tangential direction and the other two inclined away from the tangential direction. The thickness of the inclined walls are the same but different from the tangentially oriented wall. All walls are considered to be composed of an isotropic material.

A strength of materials approach was taken to compute direct and bending strains resulting from forces acting through the three ends of the element. Equivalent stresses and strains were deduced from the response of the element and consequently effective stiffness moduli were determined for wood. On the basis of the analysis, Gillis was able to predict the ordering of the anisotropic elastic moduli for wood.

The ordering, for the most part, was the same as that found experimentally for the few species in which all the elastic constants have been measured. Hence, the ordering of the gross anisotropic elastic moduli for wood can be deduced by using a microstructural feature that is commonly found in aggregates of cells.

Other models on the mesoscale have involved accounting for differences in earlywood and latewood by introduction of a layered system. The density of latewood may be three or more times greater than that of the earlywood in some species, which implies similar differences in mechanical properties (see *Strength*). Similar layered models have also been used to model the reinforcing effect of rays in the radial direction of wood. In one case the two types of layering have been incorporated into a single model (Schniewind 1959).

4. Microscale

Mark (1967) has reviewed the early literature on cell wall models prior to development of his own models. Starting from the molecular level, Mark (1967) used a particular cellulose crystal structure to compute elastic modulus values for a fictitious continuum material, the framework, that deforms in a manner similar to the crystal structure. The stiffnesses of various chemical bonds (e.g., C–C, C–O and hydrogen bonding), combined with geometrical information concerning the crystal, were used to compute equivalent elastic constants. In the direction parallel to the cellulose chains, the elastic modulus E_{FL} was found to be a little less than a measured value of 134 GPa from the literature and the experimental value was therefore adopted. Other elastic constants for the framework were calculated from crystal structure while the constants for the lignin matrix were based on experimental values from the literature. For complete sets of elastic constants based on more recent methods of calculation see *Constituents of Wood: Physical Nature and Structural Function*.

Mark then treated each cell wall layer as a filament-wound composite of framework and matrix, calculating elastic properties of each layer according to its composition. He then proceeded to construct a model for a single fiber, making use of the geometry and proportions of the helically wound cell wall layers, i.e., M + P, Sl, S2 and S3 (Fig. 4). The purpose of this model was to compute internal stresses in the cell wall in order to find regions where the fiber is likely to break. His results, including a large shear stress in the S1 layer, supported experimental observations that fibers appear to rupture in the S1 layer when tested in tension.

Following Mark, there have been a series of publications dealing with the stiffness and strength of the cell wall. Cave (1968, 1969) predicted and measured the elastic modulus of the fiber in the long-axis direction as a function of microfibril angle. The microfibril wrappings were assumed to be distributed about a mean angle and the distribution function was considered to be Gaussian. In the model the concept of a balanced laminate was introduced. The adjacent walls of neighboring fibers having the same angle, but opposite inclination, prevents cell rotation when pulled axially. Only the S2 layer was considered since it is the thickest layer and tends to dominate cell behavior.

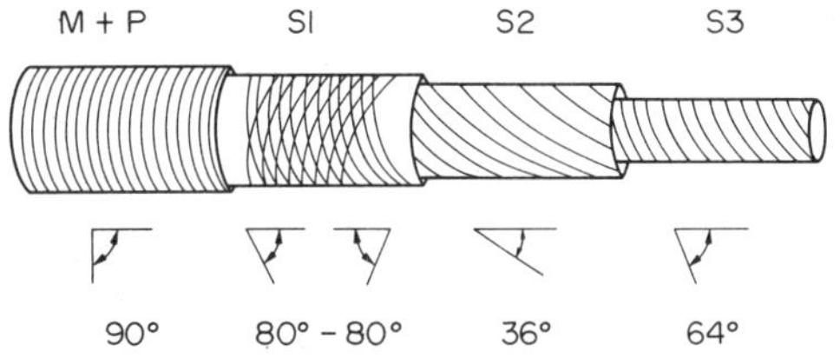

Figure 4
Microfibril angles of the layers of the radial cell wall (adapted from Mark 1967)

Quite independently, but similar to Cave's approach, Schniewind and Barrett (1969) developed a model for the cell wall based on the idea of a balanced laminate. The wood cell is considered to be a thin-walled tube subjected to a force along the cell axis. The wall can be broken down into many approximately flat, infinitesimal rectangular segments (Fig. 5). Each segment is layered with orthotropic materials having appropriately inclined principal directions to represent the S1, S2, S3, and M + P layers. A shear restraint was imposed on each layer. This approach differed from Mark. The inclusion of all layers differed from Cave. The overall stiffness of the cell, when compared to Mark's work, was increased. In addition, stresses were computed in the various layers of the cell wall and it was found that shear stresses in the S1 were small. Compressive stresses were higher, however, and

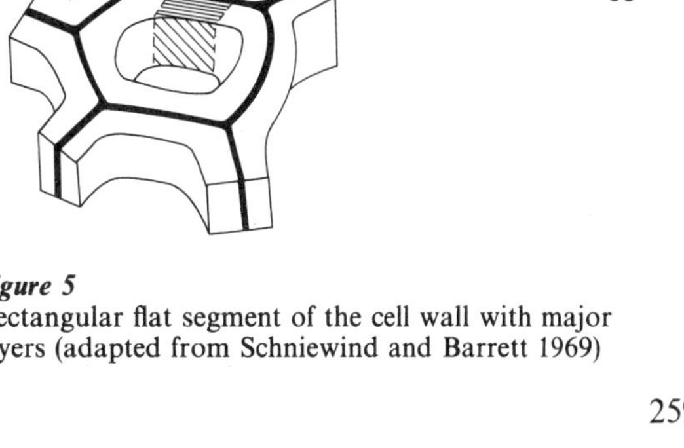

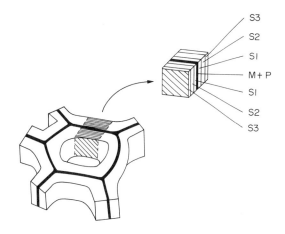

Figure 5
Rectangular flat segment of the cell wall with major layers (adapted from Schniewind and Barrett 1969)

it was postulated that failure in the S1 might be due to elastic instability of the microfibrils.

Schniewind (1970) has followed up with more work on internal stresses in various hypothetical cell types using the shear restraint assumption and Mark and Gillis (1970) have modified Mark's earlier work to include the ideas of shear restraint as well as improvements to account for different behavior in radial and tangential wall segments.

A three-dimensional model of a wood fiber was next developed by Tang (1972). The fiber was modelled as a laminated cylinder with each layer (S1, S2, S3 and M+P) possessing cylindrical anisotropy. A stress function approach was used to calculate stresses in the layers. Rather than using shear restraint, or variations of strain restriction themes, Tang chose to match normal and shear stresses across the layer interfaces. With this boundary condition, he found large tangential stresses in the S3 layer. In more detailed analysis, Tang and Hsu (1973) have explored the influence of the constituents of the layers themselves on the internal stress distributions in the wood fiber under tension. A large number of combinations of variables related to this microstructure were explored theoretically and it was concluded that the spacings between the microfibrils were very important with respect to the elastic properties of the cell wall.

Barrett and Schniewind (1973) next used a finite-element approach to solve for stresses in the cell walls of wood. A concentric, multilayered, orthotropic cylinder model was used, with each layer of the model being represented by a group of annular elements with appropriately transformed elastic compliance tensors. A generalized plane strain condition was imposed which allowed for the overall stiffness of the fiber to be computed. The finite element method naturally allows displacement boundary conditions, and consequently a complete shear restraint condition was imposed by setting tangential displacements at the outer surface to zero. Alternatively, an unattached, single fiber was considered by removing the constraint on outer nodal displacements. Their stress distributions differed from those of Tang (1972) substantially, a result not to be unexpected due to the differences in boundary conditions. The results, however, indicated that the two-dimensional approach utilized by the authors previously was quite adequate. The three-dimensional finite element analysis did allow a computation of radial stresses, which can not be done with a two-dimensional analysis, but these stresses were small and considered to be inconsequential. Shear restraint, the natural condition of a fiber surrounded by its neighbors, favoured the mechanical efficiency of the fiber.

Models used to explore the linkage between structure and mechanical properties of wood at any dimensional scale ultimately reflect on macroscopic mechanical behavior. Archer (1987b) has turned his attention to the cell and has used the idea of growth strain to model stress evolution in the wall of the developing cell. In this model, wood substance is laid down on the inside of a hollow cylinder, the fiber, rather than on the outside surface of a cylinder used to represent the trunk of the growing tree. The computed residual stresses based on this model indicate that the mechanical efficiency of the cell as it is developing appears to be optimized when the microfibril angle of the S2 layer is between 26° and 42°. Macroscopic properties are thus related to microstructural arrangements and a microscopic analysis reveals a rational for macroscopic survivability.

See also: Deformation Under Load; Strength

Bibliography

Archer R R 1979 on the distribution of tree growth stresses, Part III: The case of inclined grain. *Wood Sci. Technol.* 13: 67–78

Archer R R 1987a *Growth Stresses and Strains in Trees.* Springer, New York

Archer R R 1987b On the origin of growth stresses in trees, Part 1: Micromechanics of the developing cambial cell wall. *Wood Sci. Technol.* 21: 139–54

Archer R R, Byrnes F E 1974 On the distribution of growth stresses, Part I: An anisotropic plane strain theory. *Wood Sci. Technol.* 8: 184–96

Banks C C 1973 The strength of trees. *J. Inst. Wood Sci.* 6(2): 44–50

Barrett J D, Schniewind A P 1973 Three-dimensional finite-element models of cylindrical wood fibers. *Wood Fiber* 5(3): 215–25

Boyd J P 1973 Tree growth stresses, Part V: Evidence of the origin in differentiation and lignification. *Wood Sci. Technol.* 7: 92–111

Cave I D 1968 The anisotropic elasticity of the plant cell wall. *Wood Sci. Technol.* 2(4): 268–78

Cave I D 1969 The longitudinal Young's modulus of Pinus radiata. *Wood Sci. Technol.* 3(1): 40–48

Cramer S M, Goodman J R 1983 Model for stress analysis and strength prediction of lumber. *Wood Fiber* 15(4): 338–49

Easterling K E, Harrysson R, Gibson L J, Ashby F R S 1982 On the mechanics of balsa and other woods. *Proc. R. Soc. Lond., Ser. A* 383: 31–41

Gillis P P 1972 Orthotropic elastic constants of wood. *Wood Sci. Technol.* 6: 138–56

McMahon T A and Kronauer R E 1976 Tree structures: Deducing the principle of mechanical design. *J. Theor. Biol.* 59: 443–66

Mark R E 1967 *Cell Wall Mechanics of Tracheids.* Yale University Press, New Haven, Connecticut

Mark R E, Gillis P P 1970 New models of cell-wall mechanics. *Wood Fiber* 2: 79–95

Megraw R A 1985 *Wood Quality Factors in Loblolly Pine.* TAPPI Press, Atlanta, Georgia

Noskowiak A F 1963 Spiral grain in trees, a review. *For. Prod. J.* 13: 266–75

Phillips G E, Bodig J, Goodman J R 1981 Flow-grain analogy. *Wood Sci.* 14(2): 55–64

Schniewind A P 1959 Transverse anisotropy of wood: A function of gross anatomic structure. *For. Prod. J.* 9(10): 350–59

Schniewind A P 1962 Horizontal specific gravity variation in tree stems in relation to their support function. *For. Sci.* 8(2): 111–18
Schniewind A P 1970 Elastic behavior of the wood fiber. In: Jayne B A (ed.) *Theory and Design of Wood and Fiber Composites*. University of Washington Press, Seattle, Washington
Schniewind A P, Barrett J D 1969 Cell wall model with complete shear restraint. *Wood Fiber* 1: 205–14
Shigo A L 1985 How tree branches are attached to trunks. *Can. J. Bot.* 63(8): 1391–401
Tang R C 1972 Three-dimensional analysis of elastic behavior of wood fiber. *Wood Fiber* 3(4): 210–19
Tang R C, Hsu N N 1973 Analysis of the relationship between microstructure and elastic properties of the cell wall. *Wood Fiber* 5(2): 139–51
Thompson D 1942 *On Growth and Form*. Cambridge University Press, Cambridge.
Zimmermann M H 1977 *Trees Structure and Function*. Springer, New York

J. A. Johnson
[University of Washington, Seattle, Washington, USA]

Surface Chemistry

Chemical interactions at the wood surface are of great importance for wood utilization, in such areas as gluing and painting. Weathering of wood is essentially a surface phenomenon. Much of the value of wood in furniture and panelling relates to the appearance of the wood surface. This article deals with the chemical nature of wood surfaces and the chemistry of changes taking place in them.

1. Nature of Wood Surfaces

The concept of wood surface is difficult to define in a way which is both sufficiently precise and practical at the same time. Definition of the surface as a two-dimensional area is impractical since chemical surface interactions take place between three-dimensional atoms and molecules; definition of the surface as a layer of a certain specified thickness, such as length of the anhydroglucose repeat unit of cellulosic macromolecules, is not entirely satisfactory since various chemical and physical surface interactions involve surface layers of widely varying thickness. A more practical definition would be that of a surface layer of thickness involved in producing a certain specific surface effect. This definition is relative, however, since the thickness of such a layer would vary with the type of interaction in question.

The variation of the depth involved in various physical and chemical interactions represents an important consideration in studying the surface chemistry of wood as it makes it commonly impossible to arrive at the same results using different analytical methods.

Wood is an open-porous, composite, cellular biopolymer. Consequently the surfaces of wood can be either external, i.e., artificially created and comprising the interfaces between wood and its surroundings, or internal. The internal surfaces can be either permanent and comprising the interfaces between cell walls and cell lumens, or transient, opening in response to penetration of polar liquids into cell walls and comprising the interfaces between cell-wall material and such liquids.

The area of the external wood surface is generally much smaller than the areas of the permanent and particularly of the transient internal wood surfaces. The areas of the permanent and transient internal wood surfaces have been estimated by Stamm and Millett (1941). According to their data a cube of *Pinus lambertiana* Dougl. (sugar pine) wood with an edge of 1 cm and an external surface area of 6 cm^2 (density 0.36 $g\,cm^{-3}$) will have a permanent internal surface area of about 0.11 m^2 and a transient internal surface area of about 170 m^2.

1.1 Chemical Nature of Internal Wood Surfaces

The internal wood surface is composed of the surfaces of cell lumens, including the surfaces within pit openings, and of the surfaces of transient openings in cell walls. Very little is known about the chemical make-up of the transient surfaces; most likely they are predominantly composed of amorphous carbohydrates.

We have, however, more information on the chemical composition of the tertiary cell-wall layer, or warty layer, that for all practical purposes is identical with that of the tracheid surfaces in conifers. Various scanning electron microscopy studies indicate that this layer is a combination of lignin and amorphous carbohydrates encrusted to a varying extent with extractives. The nature of the encrusting extractives is not very well known. In some cases the extractives appeared to represent sodium-hydroxyde-soluble carbohydrates; in other cases they seemed to be at least partly nonpolar in nature. Most likely the nature of the extractives varies with the species.

1.2 Chemical Nature of External Wood Surfaces

Wood is composed of cellulose (linear polyanhydroglucose, 41–52% in softwoods, 37–52% in hardwoods), hemicelluloses (predominantly branched polyanhydromonosaccharides, 14–29% in softwoods, 24–39% in hardwoods) and lignin (cross-linked polymer of *p*-n-propylphenol-related units, 27–37% in softwoods, 18–32% in hardwoods), all three constituting the cellular structure of wood (see *Chemical Composition*). Wood also contains varying amounts of materials of chemically diverse nature deposited in cell lumens and cell-wall interstices (extractives, extraneous or associated materials), as well as small amounts of inorganics (generally below 1.0%). Thus the chemical composition of external wood surfaces will be influenced mainly by the factors determining the amounts, percentages and composition of the above five groups of materials in the total wood sample. These would include such factors as the taxonomic status of the wood species, conditions of tree growth, age of the tree, proportion of normal wood to reaction wood, location of the wood sample within the tree (e.g., extractive-rich heartwood vs. sapwood), proportion of springwood to summerwood and density of wood.

A number of additional factors dealing specifically with the surface can substantially modify the direct relationship between the chemical composition of bulk wood and that of surface layers.

2. Conditions and Methods of Wood-Surface Formation

Cell walls of wood are constructed of cellulose microfibrils embedded in hemicelluloses and lignin. The softening points of hemicelluloses and lignin are 50–60 °C and 90–100 °C, respectively. Because of the lower mechanical strength of hemicelluloses and lignin, particularly under temperature conditions above their softening points, the formation of the surface could take place preferentially in parts of the cell wall containing more of these materials, i.e., the compound middle lamella. Thus it has been demonstrated with wood of *Picea mariana* (Mill) B.S.P. (Koran 1968) that the percentage of surface created by tangential or radial failure across cell walls decreased from 40–50% to nearly zero with a temperature increase from 0 °C to >200 °C. Furthermore, in the case of tangential failures at >150 °C, the fiber faces revealed mainly the primary wall structure, heavily embedded in an amorphous matrix of lignin and hemicelluloses. This suggested that under some conditions hemicelluloses and lignin might become substantially enriched at wood surfaces.

The composition of wood surfaces has been studied by ESCA (electron spectroscopy for chemical analysis) (Young et al. 1982). The results express the ratio of oxygen-to-carbon atoms at the surface. Calculated values of this ratio are 0.83 for cellulose, 0.37 for softwood lignin and 0.10 or 0.11 for nonpolar extractives such as resin or fatty acids, respectively. In case of pinewood an oxygen-to-carbon ratio of 0.26 was obtained, which increased to 0.42 after removal of extractives with acetone. The problem of the lower-than-expected oxygen-to-carbon ratios obtained after extraction has not been satisfactorily solved as yet. While the results could be interpreted by an enrichment of lignin at the surface, the presence of nonpolar extractives covalently bound to the surface and thus not removable with solvents represents another possibility.

3. Redistribution of Extractives

Redistribution of extractives during or following formation of the surface represents an important aspect of the chemistry of wood surfaces. If the surface was formed prior to the removal of water from wood, e.g., in sawmills, or if wood was wetted after drying, evaporation of the moisture at the wood surface causes the movement of water to the surface, where it evaporates leaving water soluble extractives behind. This often results in undesirable discolorations of wood, particularly in darker woods that are rich in phenolic materials, such as *Sequoia sempervirens* (D. Don) Endl. (California redwood). The deposition of these materials can also lower the pH of the surface and interfere with gluing.

It has been well established that the surface energy and gluability of wood decrease with time of storage. This has been related to the formation of a nonpolar, lipophylic layer at the wood surface. The deposition of surprisingly small amounts of material is sufficient to substantially alter some properties of the wood surface, and even a monolayer of organic material can influence the wettability of wood (Baier et al. 1968). The studies by ESCA mentioned before indicated the presence of nonpolar extractives on the surface of pinewood as the oxygen-to-carbon ratio at the surface of wood was well below the ratios expected for either cellulose or lignin and approached those of nonpolar extractives. Extraction with acetone increased this ratio which was interpreted as removal of nonpolar extractives from the wood surface. Deposition of nonpolar materials at the surface of wood can negatively influence wood bonding, particularly in the case of neutral or acidic bonding agents. Thus the Swedish regulations for the production of laminated beams of *Pinus sylvestris* L. require gluing within 24 hours after formation of the wood surfaces to be joined.

The mechanism of wood-surface inactivation is controversial, however. Some evidence indicates that the loss of gluability is related to the migration of nonpolar extractives from the interior of the wood and their deposition on the surface, while other evidence points toward the deposition of foreign materials from the environment. Experimental results demonstrate that nonpolar, water insoluble extractives such as fatty acids, resin acids and steroids are well capable of migrating and becoming deposited on the wood surface at temperatures as low as ambient and also in kiln drying. Hemingway (1969) concluded that the transport of nonpolar extractives from the inside of wood was very unlikely, except for the region closest to the surface, since the amount of the free (i.e., not glyceridically bound) saturated fatty acids such as stearic and palmitic acids was not high enough in that region to interfere with gluing by surface contamination. Hemingway favored instead air-oxidation of unsaturated linoleic acid and deposition of the oxidation products on the surface.

4. Deposition of Foreign Materials

Foreign materials may be deposited on the surface of wood following its formation; such deposition may be purposeful or it may be incidental contamination which has been the subject of a considerable amount of work. Deposition of some materials is connected with the environment (dust, water of condensation, rainwater, organic vapors, grime, acids, aerosols); others originate from methods of wood-surface formation, while yet others constitute intentional treatments.

4.1 Incidental Deposition of Foreign Materials

As mentioned, some theories of wood-surface inactivation favor the deposition of nonpolar materials from the environment over the migration of such materials from the wood interior. Thus Nguyen and Johns (1979), using contact-angle methodology, wood of *Sequoia sempervirens* containing mainly polar extractives and wood of *Pseudotsuga menziesii* containing mainly nonpolar extractives, concluded that the contribution of the polar and dispersive force components to the total surface free energy of wood is directly related to the nature of wood extractives. At the same time they concluded that inactivation of wood surfaces over time is related to environmental rather than to wood factors (i.e., not to the movement of nonpolar extractives towards the surface) since the surface free energy due to dispersive forces decreased over time to slightly less than one half, regardless of the starting proportion of the dispersive and polar force components. Still another proposed mechanistic alternative involves the sorption of atmospheric gases (Marian 1967).

Another adverse effect of surface contamination of wood are discolorations of the wood surface. Contamination of wood with compounds of iron during storage or in use (rusty nails, rusty water, dust, flying metal particles) is a common cause of surface discoloration of woods high in phenolics and similar compounds, as iron reacts with most phenolic (e.g., tannins) and tropolonic extractives under formation of dark-colored complexes. In some cases such discolorations can be eliminated by treating the discolored wood with a solution of oxalic acid in water. Other common polyvalent metal ions are generally less tinctorial, however.

Contamination of the wood surface during surface preparation is apparently less serious. While small amounts of iron from cutting surfaces can be expected in principle to become embedded in wood during sawing and related operations, such contaminations become troublesome only in special cases.

4.2 Purposeful Deposition of Foreign Materials

In order to optimize certain wood properties wood is often treated with a variety of chemicals. These are deposited in the interior wood cavities and include wood preservatives against decay, such as (a) inorganic compounds of copper, zinc, arsenic and chromium, or organic compounds such as creosote and pentachlorophenol; (b) fire retardants such as compounds of phosphorus and boron; and (c) polymeric materials. Polymeric materials can be introduced as such or can be allowed to form in situ from introduced monomers. All of the above materials change the chemistry of the wood surface according to the nature of treatment. Other treatments are connected with the deposition of such films as paints, varnishes, or adhesives on the wood surface (see *Protective Finishes and Coatings*; *Adhesives and Adhesion*).

5. Changes Due to Heat and Oxidation

The chemistry of wood surfaces can become modified during surface formation, wood drying, storage and use by chemical reactions triggered by heat (pyrolysis) and/or atmospheric oxygen (air oxidation). During sawing and related operations, particularly in the case of excessive saw vibrations, the temperature of the wood surface can reach levels where wood begins to decompose. Although the magnitude of the wood temperatures incurred during sawing are not well known, the temperatures of the circular saws were found to depend upon the distance from the saw teeth and were generally 40–60 °C, but increasing to 100 °C and even 160 °C towards the saw teeth. The temperatures of the saw teeth are particularly high, however, reaching as high as 774 °C (Mote 1977, Zaitsev 1968). Although the times of contact between wood and the hot metal are very short, the high temperature levels can occasionally, particularly in case of saw malfunction, result in substantial pyrolytic and oxidative changes on the surface of wood. Drying of wood particles (particleboard) or wood veneer (plywood, laminates) at elevated temperatures represents another avenue where pyrolytic and oxidative changes at the wood surface can take place.

Wood degradation at moderately elevated temperatures or shorter exposures to higher temperatures in the presence of air or oxygen includes pyrolytic and oxidative changes, i.e., transformations initiated by increased temperature and transformations due to reaction with oxygen. In these reactions the wood components (carbohydrates and lignin) change independently from each other, i.e., wood behaves like a mixture of these materials. At longer exposures to higher temperatures the combustion process sets in consisting of autocatalytic pyrolytic decomposition coupled with the oxidation of the volatiles produced, the final products consisting of char, water and carbon dioxide.

Pyrolysis of cellulose has been the subject of intensive investigations and has been reviewed many times (Tillman 1981, Shafizadeh 1984). It begins with depolymerization by transglycosylation to yield levoglucosan and other monomeric and oligomeric sugar derivatives. Concurrently the dehydration reaction leads to the formation of the unsaturated materials. The same reactions most likely dominate the pyrolysis of hemicelluloses. Lignin pyrolysis is dominated by condensation reactions leading to the formation of ether linkages between the n-propyl sidechains and of alkyl–aryl bonds. As in case of cellulosics, this is paralleled by dehydration reactions leading to the formation of double bonds in the sidechains (Domburg et al. 1982). Oxidation of cellulose by atmospheric oxygen apparently begins to take place at about 140 °C. It is accompanied by depolymerization and results in the formation of carbonyl and carboxyl groups, some of which decarboxylate. The process is strongly catalyzed by moisture (Tryon et al. 1966, El-Rafie et al. 1983).

The amount of information on pyrolytic and oxidative changes on wood surfaces resulting from the history of surface preparation is meager, however. The information available is connected mainly with studies of surface inactivation in gluing processes and with dimensional stabilization of wood by exposure to moderately elevated temperatures.

Exposure of wood to temperatures moderately above 100 °C for long time periods results in loss of hygroscopicity connected with the loss of hydroxyl groups. A quantitative correlation was obtained between loss of hygroscopicity and loss of weight after heating wood samples of loblolly pine (*Pinus taeda* L.) and yellow poplar (*Liriodendron tulipifera* L.) to 200 °C for 5 min. This was explained by the formation of intramolecular epoxy groups between hydroxyls 2 and 3 of the anhydroglucose units of cellulose. Intermolecular ether linkages apparently do not form (Salehuddin 1970, Seborg et al. 1953). Heating of wood to 300 °C results in increased dimensional stabilization. This has been explained by the decomposition of hygroscopic hemicelluloses and other carbohydrates, followed by condensation and polymerization of the resulting furan-type compounds (Mitchell et al. 1953). Changes in the chemistry of wood surfaces due to increased temperatures were studied by Chow and Mukai (1972) by exposing microsections of white spruce (*Picea glauca* (Moench) Voss) to temperatures between 100 and 240 °C in air and in nitrogen. Below 180 °C the changes consisted mainly of oxidation, while above 180 °C they were of mixed pyrolytic and oxidative nature. The absorption of hydroxyl in infrared spectra decreased with time at 180 °C, the color of wood darkened, and crystallinity and the degree of polymerization (DP) of cellulose decreased. Carbonyl absorption of esters and carboxyls in infrared spectra decreased first and then increased with the temperature rise. Extractives were found to catalyze the rate of the oxidation (Chow and Mukai 1972).

Changes on the wood surface due to an increase in temperature also affect the extractives. Thus phenolic extractives, particularly tannins, are likely to undergo condensation reactions and to polymerize to the water-insoluble "synthetic phlobaphenes." Volatile extractives such as monoterpenoids are likely to volatilize and resin acids are likely to isomerize by double-bond migration within their structures. Additional changes should involve unsaturated fatty acids such as linolenic acid, which is likely to transform by intramolecular double-bond migration to conjugated positions, and could ultimately oxidatively cleave into lower-molecular-weight fragments (Hemingway 1969).

6. Changes Due to Exposure to Light

Electromagnetic radiation of the visible and ultraviolet regions such as daylight and light from incandescent and fluorescent lamps changes in time the appearance of wood by interacting with wood constituents. Such changes are commonly noticed on wood panelling and other wooden objects by a difference in wood color between areas exposed to light and areas protected from light, such as areas covered by paintings hung on a wall. The nature of the effect (darkening or lightening, and the kind of color change) is difficult to predict, however, as it depends upon the composition of the electromagnetic radiation, its intensity, temperature, moisture content of the wood, length of exposure and on the nature of the wood, particularly on the kind of extractives it contains (Kringstad 1973, Feist et al. 1982, Hon et al. 1982). It has been noted that exposure of lignin to light of wavelength <385 nm results in darkening, while exposure to light of wavelength >480 nm results in lightening of the color. Similar effect has been reported also in case of solid wood. Chemical changes involve initial formation of free radicals, and include chain scission, dehydrogenation, and dehydroxymethylation in the case of cellulose and hemicelluloses, and splitting of double bonds, formation of quinone structures, demethoxylation under formation of methanol, increased solubility (loss in Klason lignin content of wood) and polymerization in case of lignin. The formation of free radicals most likely involves the hydroxy groups of lignin since esterification or etherification of lignin increases the light stability of wood. In the presence of oxygen and water, hydrogen peroxide and peroxy groups also form. With solid wood at 45–50 °C and 50% relative humidity, and xenon arc as a light source, a loss of lignin and hemicelluloses from the surface was noted after 75 days of exposure. Water increased strongly the rate of material loss.

Extractives are particularly prone to color changes. It has been demonstrated that the colorless flavonols taxifolin and aromadendrin change by interaction with visible-ultraviolet light to the yellowish flavanonols quercetin and kaempferol by a photooxidative reaction sequence. Concurrently a general decrease in flavonoids and an increase in vanillin-related compounds was noticed (Minemura and Umehara 1979).

7. Changes in Wood Surface Due to Weathering

Weathering of wood is a complex process, dependent upon simultaneous action of several factors, such as solar radiation, moisture, temperature, air-oxygen and fungal microorganisms (mildew). Over time these agents give wood surfaces a characteristic gray color, they acquire a rough texture, checks and cracks, and become friable. The upper gray surface layer is about 125 nm thick. It is generally composed of degraded, disordered and loosely matted cellulosic fibers and contains very little if any lignin. This top layer forms mainly by direct interaction of ultraviolet radiation with wood substance, primarily with lignin under formation of free radicals. The transformations that follow occur under participation of other weathering factors and result in solubilization and disappearance of lignin. Other changes include oxidation and most likely depolymerization of cellulose; the former was demonstrated by ESCA and by infrared spectroscopy.

Below the gray layer is a brown layer (500–2500 nm) containing intermediate amounts of lignin (40–60% of normal). Since the brown layer can not form as a result of direct interaction of wood with light, as ultraviolet light cannot penetrate that deep, it must arise as the result of either energy transfer from the surface, or as the result of migration of free radicals from the surface into the wood interior. The brown layer is underlain in turn by normal, nonweathered wood (Feist et al. 1983).

8. Wood-Surface Modification

Certain reagents are purposely allowed to react with wood, primarily with the hydroxy groups of its constituents, in order to modify the properties of wood, particularly at its surface. Some of these reagents react with wood without the introduction of foreign molecular structures (heat, oxidation agents); in other cases foreign structures are covalently attached to the wood surface. These can include relatively small chemical units (methylation, acetylation), or larger groups including even the chains of molecular units (polymer grafting).

Treatment with physical agents includes treatment with heat, as well as with various forms of electromagnetic radiation, such as infrared, visible and ultraviolet light, and α, β and γ radiation. The respective chemistry of some of these has been discussed earlier. Such treatments, particularly in presence of air, result generally in oxidation of the surface under formation of carboxylic groups and cross-linking, an increase of the surface energy of wood, and improvements in adhesive properties.

A certain amount of information is available on the modification of wood surfaces by various types of plasma. Under plasma one generally understands an activated gas which includes various types of reactive chemical species, such as electrons, photons, positive and negative ions, free radicals and metastables. Plasmas can be generated in a variety of ways, using any gaseous material. Reaction of plasmas with wood surfaces can radically modify surface properties. Thus oxygen plasmas, such as corona plasma, oxidize the wood surface under formation of carbonyl and carboxyl groups, and drastically increase its surface energy, while other plasmas, such as radio-frequency (rf) acetylene plasma drastically decrease the surface energy by attaching nonpolar groups to the wood surface. Other chemical groups that can be attached to wood surfaces by various plasmas include amino groups (nitrogen or ammonia plasmas), acrylic acid and styrene derivatives. The reacting wood layer is generally very thin, making it often difficult to detect the presence of these groups by spectroscopic methods.

Oxidation has been used for a long time to alter the properties of wood surfaces. The most common oxidation reagents include hydrogen peroxide and other compounds with peroxy linkages, ozone, nitric acid, nitrates, chlorates, elemental halogens, metal ions such as Fe^{3+}, metal oxides, the derived acids and their salts such as chromates. Oxidations tend generally to increase the surface energy of wood. Concurrently pH tends to change, going down in case of hydrogen peroxide, nitric acid, ozone and halogens due to the production of organic acids and occasionally also due to introduction (HNO_3) or formation (HCl, HBr) of inorganic acids. In other cases, such as oxidation by chromates and nitrates the pH increases due to formation of salts of weaker carboxylic acids and reduction of the nitrate and chromate ions to N_2 and Cr_2O_3. As a result of oxidation, carboxylic and carbonyl groups are generally introduced into the wood. In case of oxidations by halogens or nitric acid, lignin becomes additionally nitrated or halogenated to some extent. Treatment of wood surfaces with acids, particularly if followed by heating, results in partial hydrolysis of cellulose and hemicelluloses and condensation of lignin. Under more drastic conditions the hydrolytically liberated monosaccharides, primarily the pentoses undergo a chain of transformations leading to furfural which ultimately polymerizes to polyfurfural, a material of reduced polarity.

Treatment of wood surfaces with alkalis transforms the carboxylic groups on the wood surface into corresponding salts. This ensures a high pH of the wood surface following removal of the introduced alkali. Alkalis remove also the surface fatty acids increasing in this way the surface energy of wood.

Methylation and esterification of wood surfaces has been extensively studied in connection with attempts to dimensionally stabilize wood. The reagents included dimethylsulfate/alkali in case of methylation, and ketene, acetic anhydride, acetyl chloride and phthalic anhydride in case of esterifications. Both types of treatments generally tend to decrease polarity and surface energy of wood.

Additional surface treatments include reactions with isocyanates (methylisocyanate, phenylisocyanate, 2,4-toluene diisocyanate) forming urethane linkages, aldehydes such as formaldehyde yielding acetalic linkages and expoxides such as ethylene or propylene oxide, forming β-hydroxyether groups. The changes in the properties of the wood surface following these treatments can be easily deduced from the nature of the respective chemical transformations (see *Chemically Modified Wood*).

Still another group of surface transformations includes generally undesirable changes resulting from fire or biological decay (see *Thermal Degradation*; *Decay During Use*).

See also: Chemical Composition; Surface Properties; Radiation Effects; Weathering

Bibliography

Baier R E, Shafrin E G, Zisman W A 1968 Adhesion: Mechanisms that assist or impede it. *Science* 162(3860): 1360–68

Chow S-Z, Mukai H N 1972 Effect of thermal degradation of cellulose on wood–polymer bonding. *Wood Sci.* 4: 202–8

Domburg G E, Skripchenko T N 1982 Process of formation of intermediate structures during thermal transformations of lignins (in Russian). *Khim. Drev.* 5: 81–88

El-Rafie M H, Khalil E M, Abdel-Hafiz S A, Hebeish A 1983 Behavior of chemically modified cottons towards thermal treatment. II: Cyanoethylated cotton. *J. Appl. Polym. Sci.* 28: 311–26

Feist W C, Rowell R M 1982 UV degradation and accelerated weathering of chemically modified wood. In: Hon D N-S (ed.) 1982 *Graft Copolymerization of Cellulosic Fibers*, ACS Symposium Series No. 187. American Chemical Society, Washington, DC, pp. 349–70

Feist W C, Hon D N-S 1983 Chemistry of weathering and protection. In: Rowell R (ed.) 1983 *The Chemistry of Solid Wood*, ACS Advances in Chemistry Series No. 207. American Chemical Society, Washington, DC, pp. 401–51

Hemingway R W 1969 Thermal instability of fats relative to surface wettability of yellow birchwood (Betula Lutea). *Tappi* 52: 2149–55

Hon D N-S, Chan H Ch 1982 Photoinduced grafting reactions in cellulose and cellulose derivatives. In: Hon D N-S (ed.) 1982 *Graft Copolymerization of Cellulosic Fibers*, ACS Symposium Series No. 187. American Chemical Society, Washington, DC, pp. 101–18

Koran Z 1968 Electron microscopy of tangential tracheid surfaces of black spruce produced by tensile failure at various temperatures. *Svensk Papperstidning* 71: 567–76

Kringstad K P 1973 Some possible reactions in light-induced degrading of high-yield pulps rich in lignin (in German). *Das Papier* 27: 462–69

Marian J E 1967 Wood, reconstituted wood, and glued laminated structures. In: Houwink R, Salomon G (eds.) 1967 *Adhesion and Adhesives*. Elsevier, Amsterdam, Vol. 2, Chap. 14, pp. 167–280

Mitchell R L, Seborg R M, Millett M A 1953 Effect of heat on the properties and chemical composition of Douglas-Fir wood and its major components. *For. Prod. Res. Soc. J.* 3(4): 38–42

Minemure N, Umehara K 1979 Color improvement of wood. (1). Photo-induced discoloration and its control (in Japanese). *Rept. Hokkaido For. Prod. Res. Inst.* 68: 92–145

Mote C D, Szymani R 1977 Principal developments in thin circular saw vibration and control research. Part 1: Vibration of circular saws. *Holz als Roh- und Werkstoff* 35: 189–96

Nguyen T, Johns W E 1979 The effects of aging and extraction on the surface free energy of Douglas fir and redwood. *Wood. Sci. Technol.* 13: 29–40

Salehuddin A B M 1970 A unifying physico-chemical theory for cellulose and wood and its application in glueing. Ph.D. Thesis, North Carolina State University, pp. 1–89

Seborg R M, Tarkow H, Stamm A J 1953 Effect of heat upon the dimensional stabilization of wood. *For Prod. Res. Soc. J.* 3(3): 59–67

Shafizadeh F 1984 The chemistry of pyrolysis and combustion. In: Rowell R (ed.) 1984 *The Chemistry of Solid Wood*, ACS Advances in Chemistry Series No. 207. American Chemical Society, Washington, DC, pp. 489–529

Stamm A J, Millett M A 1941 Internal surface of cellulosic materials. *J. Phys. Chem.* 45: 43–54

Tillman D A, Rossi A J, Kitti W D 1981 The process of wood combustion. In: *Wood Combustion*. Academic Press, New York, Chap. 4, pp. 74–97

Tryon M, Wall L A 1966 Oxidation of high polymers. In: Lundberg WO (ed.) 1966 *Autoxidation and Autoxidants*. Interscience, New York, Vol. II, pp. 963–68

Young R A, Rammon R M, Kelley S S, Gillespie G H 1982 Bond formation by wood surface reactions: Part I—Surface analysis by ESCA. *Wood Sci.* 14: 110–19

Zaitsev N A 1968 Temperature measurement on the cutting edges of the circular saws (in Russian). *Derevoobrab. Promst.* 17(4): 15

E. Zavarin
[University of California, Berkeley, California, USA]

Surface Properties

Every wood tissue within a cylindrical tree stem has an equal likelihood of being manufactured into an exterior surface area. Some knowledge of the basic properties of wood is therefore essential for a better understanding of the surface response of wood to its many applications.

1. Surface Areas

Wood morphology can be simplified by considering it as a foam-cellular plastic. There are three basic voids of varying sizes in the internal wood surface: the lumens, the pit openings, and the micropores in the cell wall. The calculated internal surface area for lumens in softwood (specific gravity 0.4) is about $0.2\ m^2\,g^{-1}$ or $800\ cm^2$ per cubic centimeter of wood, based on an average lumen diameter of 0.00333 cm. The fractional void volumes of wood with specific gravity 1.3 (lignum vitae) and 0.1 (balsa) are 0.11 and 0.93, respectively (Stamm 1964).

Observation of the pores in cell walls indicates that their size ranges from 2–30 nm with a median of 8–10 nm. Experiments with mercury porosimeter and nitrogen absorption isotherm techniques on hinoki (*Chamaecyparis obtusa* Endl.) showed four ranges of pore size: 9 μm (tracheid lumen), 1.3 μm and 40 nm (pit and pit membrane) and 4.5 nm (micropores in cell wall). The estimated total volume and total surface area for the pores were 1.5 $cm^3 g^{-1}$ and 1.89 $m^2 g^{-1}$, respectively. Micropores smaller than 30 nm in diameter accounted for less than one-third of the volume and more than half the surface area. The total surface area of 1.89 $m^2 g^{-1}$ is almost ten times greater than the surface area of the lumens alone.

Surface area measurements by the deuterium exchange method and expressed on the basis of the coverage of a monolayer of water molecules yielded values for wood of approximately 221 $m^2 g^{-1}$ (Taniguchi et al. 1978).

The porosity and surface profile of wood will affect the visual and tactile impression of wood surfaces and its end-use value. The relationship between the surface properties of wood and its physical and sensory aspects was examined recently (Sadoh and Nakato 1987).

2. *Chemical Variations*

Wood cells are aggregated according to age (sapwood, heartwood and juvenile wood) and according to season (earlywood and latewood), and all these groups have different properties of the external wood surface. Minimum lignin content occurs in latewood, while the maximum is usually found after the first-formed earlywood (Wilson and Wellwood 1965). This pattern of lignin content is somewhat opposite to that of the cellulose content which follows a similar pattern to that of specific gravity (see *Density and Porosity*). Because of this difference in earlywood and latewood chemical content, the exposing of wood in any anatomical direction will result in differences in chemical content and thus differences in reactivity.

Infrared spectra of wood samples taken from different positions of coniferous woods showed that the spectra were essentially similar, but with some difference in intensities of the absorption bands (Chow 1972). The intensity of the 1730 cm^{-1} band (carbonyl of carboxyl and ester groups) showed that the wood from near the sapwood–heartwood boundary has the highest carbonyl content, decreasing towards the pith. A similar pattern of chemical variation was observed in the distribution of phenolic dihydroquercetin (a flavonone extractive) in softwood species. These chemical changes in the growth process of the tree will be reflected in the properties of its wood surface. It is therefore important to establish the source and selection of materials to be used in research.

Wood surface chemistry is not as stable as generally anticipated. Though cellulose, lignin and hemicellulose are stable with relatively fixed positions in wood, the extractives, on the other hand, exist in a low-melting-point solid or liquid phase and may move toward the wood surface with time or during drying. These migrated extractives may coat the wood surface, hindering wood–chemical or wood–polymer interaction, depending on the quality and quantity of extractives deposited and on the characteristics of the polymeric system used. Extractives have frequently been reported to interfere with the hardening of cement and adhesives. A study of white spruce (*Picea glauca* (Moench) Voss) veneer surface showed that of the 1.3% acetone-soluble fatty substances present in the wood, only about 10% migrated to the surface of heartwood veneer and about 20% to the surface of sapwood veneer.

In practical applications, the wide range of acidity in different wood species (pH: 2.6 to 9.5) can further complicate the surface phenomena for paint adhesion and preservative efficacy. The proper adjustment of the pH of chemicals for a specific wood species should therefore be emphasized.

3. *Accessibility and Hydrogen Bonding*

The absorptive nature of a wood surface is not simply a chemical dependence. Surface changes may occur during wood processing: cutting machines may either densify or loosen surface tissues; high-temperature wood drying can change both physical and chemical properties of surfaces. Further, wood-surface history, such as storage time after surface preparation, will change the accessibility and activity of a wood surface. Because of this complexity of wood, the traditional theory of wetting, such as the study of liquid–wood contact angles, gave rather inconsistent and controversial results. Much effort has been devoted to finding a more representative technique for expressing the potential reaction of a wood surface to liquid chemicals. Recently, an immersion technique (Casilla et al. 1984) was reported for wettability measurement which monitored the force exerted on a wood specimen as it was immersed at a controlled rate into a liquid. It was shown that wood surface wettability was strongly influenced by the pH of the solution. Cationic type surfactant gave better wettability than anionic and nonionic types.

Hydroxyl groups play an important role in the wetting, chemical activity and bonding of wood surfaces. The hydrogen bond resulting from the interaction of hydroxyl groups has an energy level of about 8.8 kJ mol^{-1} and a mean bond length of 0.27 nm for O . . . O and 0.1 nm for the O–H distance in cellulosic fibers (Nissan 1967). Although about 25% of the total number of hydroxyl groups in pulp were reported available for bonding with others, in a conventional cellulosic paper network, the ratio of the effective hydrogen bonds to the total number of hydrogen bonds was only 3–4%.

At the wood surface, the complication introduced by the presence of chemicals other than carbohydrates makes estimation of hydrogen bonds difficult. Lignin, a phenolic block polymer with less hydroxyl content, is recognized as a hydrophobic component in association with the hydrophilic carbohydrates. The hydroxyl accessibility was found to depend on the source of the wood tissues and the drying history of the wood (Chow 1972). Approximate average accessibility for never-dried earlywood and latewood of Douglas fir (*Pseudotsuga menziesii* Franco) was 24% and 43%, respectively, while the accessibility of similar woods which had been previously dried to 8% moisture content was 16% and 33%, respectively. The pattern of variation of the percent accessible and resistant hydroxyl groups from pith to cambium in a Douglas fir tree is shown in Fig. 1. Ohsawá and Nakato (1986) found that cell-wall porosity, like hydroxy accessibility, depended on the drying history.

4. *Aging and Thermal Treatment*

From the time that a log is cut and drying is in progress, the wood substances in both the internal and external surfaces are further exposed to atmospheric gases. Oxidation proceeds more rapidly in the newly excised exterior surfaces. Discoloration of the wood surface occurs and is accelerated by light radiation.

This discoloration, in most cases, reduces the aesthetic value of wood. In experiments with Japanese larch wood (*Larix leptolepis* Gord), it was found that ten days' exposure of a wood surface to sunlight corresponds to nearly 40 h exposure to a xenon lamp and 5 h exposure to a carbon arc lamp. The photoinduced discoloration was found to have direct correlation

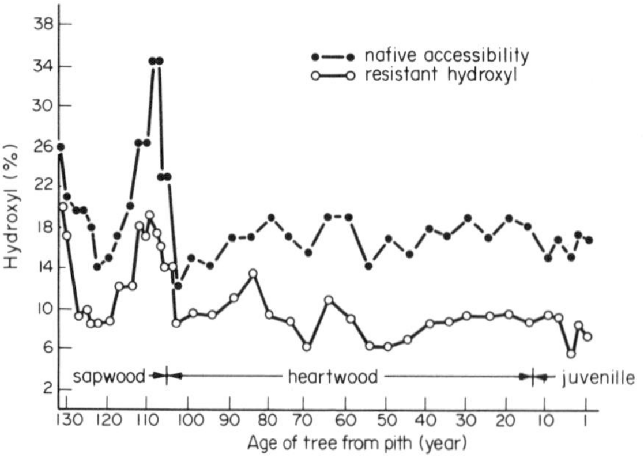

Figure 1
Variation of earlywood hydroxy accessibility in the radial direction of a Douglas-fir tree. The percentage difference between the native accessible and resistant hydroxyl is the remaining accessible hydroxy groups after drying of wood to 8% moisture content

with the electron-spin-resonance spectrum signal, suggesting the free-radical nature of the reaction (Minemura and Umehara 1979). The formation of phenoxy radicals is generally considered to be the intermediate product leading to color development (see *Radiation Effects*).

Ozone can react with a wood surface at a rapid rate under wet atmospheric conditions. The oxidation reaction subjects not only cellulosic substances but also lignin and extractives to severe degradation. Bonding of the ozone-treated surface of cellulose, polyethylene and wood strips shows a significant increase in dry bond strength (Goring 1971).

Wood surfaces are highly sensitive to thermal treatment. Measured by a thermal-mechanical method, the softing temperatures of isolated lignin, hemicellulose and cellulose were found to be 127–193 °C, 167–181 °C and 240 °C, respectively. Dry-powdered wood was found to soften at about 200 °C, and this temperature decreased with an increase in the moisture content.

There is a mass of data available on the thermal responses of wood. However, they are difficult to relate to wood-surface properties owing to the geometric and orientational effects in wood. Because the chemical substances in the exposed wood surface have a higher degree of freedom than embedded substances, the possible effect of softening temperatures of individual chemical components on the rheological behaviour of a wood surface cannot be ignored. This is particularly important since the majority of wood-composite bonds are made in the temperature range at which the softening temperature of wood chemicals have a significant effect.

A more detailed infrared spectral study of chemical changes of white spruce-wood surfaces on heating showed a significant increase in the 1730 cm^{-1} intensity after a certain period of heating, due to carboxylation or oxidation (Chow 1971). The time period to reach significant oxidation related curvilinearly to heating temperatures (100 °C to 240 °C). The reaction rates derived from the loss of 3350 cm^{-1} hydroxyl absorbance gave an approximate activation energy measure of 50 kJ mol^{-1} for unextracted native wood and about 84 kJ mol^{-1} for the extractive-free wood, suggesting a catalytic effect of extractives for oxidation of wood surfaces. The heat-induced color intensity (520 nm) showed a curvilinear relationship with a decrease in x-ray crystallinity and the degree of polymerization of cellulose.

Research on the physical and chemical properties of wood surfaces has become more and more important as engineered wood composite products find wider industrial acceptance. The use of wood fiber in manufacturing of automobile body parts is one of the best examples. Because the new application involves wood fiber, greater emphasis is needed on wood fiber surface characterization. Chemical instrumentation such as Fourier-transform infrared, Raman, ESR and ESCA spectroscopies and the new scanning technologies utilizing NMR or x-ray principles might help to accelerate the understanding of the surface properties of wood materials.

See also: Chemical Composition; Constituents of Wood: Physical Nature and Structural Function

Bibliography

Casilla R C, Chow S, Steiner P R, Warren S R 1984 Wettability of four Asian meranti species. *Wood Sci. Technol.* 18: 87–96

Chow S Z 1971 Infrared spectral characteristics and surface inactivation of wood at high temperatures. *Wood Sci. Technol* 5: 27–39

Chow S Z 1972 Hydroxyl accessibility, moisture content and biochemical activity in cell walls of Douglas fir trees. *Tappi* 55: 539–44

Goring D A I 1971 Polymer properties of lignin and lignin derivatives. In: Sarkanen K V, Ludwig C H (eds.) 1971 *Lignins: Occurrence, Formation, Structure and Reactions.* Wiley, New York, pp. 695–768

Hillis W E, Rozsa A N 1985 High temperature and chemical effects on wood stability. Part 2. The effect of heat on the softening of radiata pine. *Wood Sci. Technol.* 19: 57–66

Minemura N, Umehara K 1979 Color improvement of wood: Photo-induced discoloration and its control, Report No. 68. Hokkaido Forest Product Research Institute, Asahikawa, Japan, p. 145.

Nissan A H 1967 The significance of hydrogen bonding at the surface of cellulose network structure. In: Marchessault R H, Skaar C (eds.) 1967 *Surfaces and Coatings Related to Paper and Wood.* Syracuse University Press, New York, pp. 221–65

Ohsawa J, Nakato K 1986 Quantitative evaluation of micropore in Hinoki (*Chamaecyparis obtusa* Endl.) by nitrogen absorption. *J. Jpn. Wood Res. Soc.* 32(5): 373–77

Sadoh T, Nakato K 1987 Surface properties of wood in physical and sensory aspects. *Wood Sci. Technol.* 21: 111–20

Stamm A J 1964 *Wood and Cellulose Science.* Ronald Press, New York

Taniguchi T, Harada H, Nakato K 1978 Determination of water absorption sites in wood by a hydrogen-deuterium exchange. *Nature* (*London*) 272: 230–31

S. Chow
[Canadian Forest Products, Vancouver, British Columbia, Canada]

T

Thermal Degradation

When wood is exposed to elevated temperatures, changes can occur in its chemical structure that affect its performance. The extent of the changes depends on the temperature level and the length of time under exposure conditions. The changes in chemical structure may be manifested only as reduced strength, and hygroscopic water and volatile oil weight loss. In contrast, very drastic chemical changes may result in reduced strength and significant carbohydrate weight loss.

At temperatures below 100 °C, permanent reductions in strength can occur. The magnitude of the reduction depends on the moisture content, heating medium, exposure period and species. The strength degradation is usually not considered to result from the same thermal decomposition of the wood that occurs above 100 °C, since no significant carbohydrate weight loss occurs. The strength degradation is probably due to depolymerization reactions, although little research has been done on the chemical mechanism. Reviews by Gerhards (1979, 1982, 1983) and Koch (1985) summarize reduction in strength at temperatures below 100 °C (See *Strength*). If the wood has been treated with a chemical to reduce its flammability, more significant reductions in strength can occur at lower temperatures than for untreated wood. This is due to the presence of chemicals that catalyze the dehydration and depolymerization reactions. A review by Winandy (1987) summarizes the effects of elevated temperatures on strength properties of treated wood.

At temperatures above 100 °C, chemical bonds begin to break. The rate at which the bonds are broken increases as the temperature increases. Between 100 °C and 200 °C, noncombustible products, such as carbon dioxide, traces of organic compounds and water vapor, are produced. Above 200 °C the celluloses break down, producing tars and flammable volatiles that can diffuse into the surrounding environment. If the volatile compounds are mixed with air and heated to the ignition temperature, combustion reactions occur. The energy from these exothermic reactions radiates to the solid material, thereby propagating the combustion, or pyrolysis, reactions. If the burning mixture accumulates enough energy to emit radiation in the visible spectrum, the phenomenon is known as flaming combustion (see *Fire and Wood*). Above 450 °C all volatile material is gone. The residue that remains is an activated char that can be oxidized to carbon dioxide, carbon monoxide and water vapor. Oxidation of the char is referred to as afterglow.

The thermal degradation of wood can be represented by two pathways (Fig. 1), one occurring at high temperatures (> 300 °C), the other at lower temperatures. These two competing reactions occur simultaneously. Fire retardants work by shifting degradation to the low-temperature pathway.

1. Thermal Degradation of Wood Components

The thermal degradation of wood can be represented as the sum of the thermal degradation reactions of the individual components, namely cellulose, hemicellulose and lignin. However, the thermal degradation reactions of wood itself can vary from the sum of the individual-component reactions. Therefore, this discussion on thermal degradation includes analysis of the individual components and wood itself.

1.1 Cellulose

Cellulose is principally responsible for the production of flammable volatiles. Degradation occurs through dehydration, hydrolysis, oxidation, decarboxylation and transglycosylation.

By the low-temperature pathway, water is evolved from oven-dried cellulose, and the cellulose shows a large decrease in its degree of polymerization. The thermal degradation of cellulose can be accelerated in the presence of water, acids and oxygen. As the temperature increases, the degree of polymerization of cellulose decreases further, free radicals appear and carbonyl, carboxyl and hydroperoxide groups are formed. Thermal degradation rates increase as heating continues.

The primary reaction of the high-temperature pathway is depolymerization. This takes place when the cellulose structure has absorbed enough energy to activate the cleavage of the glycosidic linkage to produce glucose, which is then dehydrated to levoglucosan (1, 6-anhydro-β-D-glucopyranose) and oligosaccharides. The glycosidic linkages are hydrolyzable

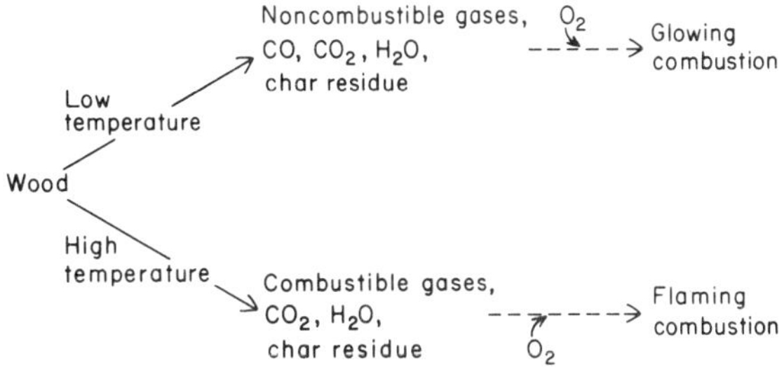

Figure 1
Degradation of wood by low-temperature and high-temperature pathways

at room temperature in the presence of strong acids. During pyrolysis, water and acids are produced from both the hemicellulose and the cellulose.

As temperature increases to around 450 °C, the production of volatile compounds is complete. The continuing weight loss is due to degradation of the remaining char.

1.2 Hemicelluloses

Hemicelluloses are less stable thermally than cellulose and evolve more noncombustible gases and less tar. Most hemicelluloses do not yield significant amounts of levoglucosan. Much of the acetic acid liberated from wood pyrolysis is attributed to deacetylation of the hemicellulose. Hardwood hemicelluloses are rich in xylan and contain a small amount of glucomannan. Softwood hemicelluloses contain a small amount of xylan and are rich in galactoglucomannan. Of the hemicelluloses, xylan is the least thermally stable, because pentosans are most susceptible to hydrolysis and dehydration reactions. The hemicelluloses degrade at temperatures from 200 °C to around 260 °C.

1.3 Lignin

Pyrolysis of lignin yields phenols from cleavage of ether and carbon–carbon linkages and produces more residual char than does pyrolysis of cellulose. The structure of lignin has been investigated using mass spectrometry to determine various lignin pyrolysis products. Dehydration reactions around 200 °C are primarily responsible for thermal degradation of lignin. Between 150 °C and 300 °C, cleavage of α- and β-aryl–alkyl–ether linkages occurs. Around 300 °C, aliphatic side chains start splitting off from the aromatic ring. Finally, the carbon–carbon linkage between lignin structural units is cleaved at 370–400 °C. The degradation reaction of lignin is an exothermic reaction, with peaks occurring between 225 °C and 450 °C; the temperatures and amplitudes of these peaks depend on whether the samples were pyrolyzed under nitrogen or air.

1.4 Wood

The influence of the individual components on the thermal degradation reactions of wood can be seen by plotting percentage weight loss as a function of temperature for the components and for wood itself (Fig. 2). The degradation of holocellulose, which consists of the alpha-cellulose plus the hemicelluloses, most closely follows that of wood. Lignin generally pyrolyzes at a slower rate than cellulose and holocellulose, although the degradation period begins somewhat earlier than for the holocellulose. Also, the presence of lignin increases the residual weight of the final char product. Alpha-cellulose and wood appear to degrade at similar rates, although wood begins to degrade at slightly lower temperatures than alpha-cellulose but higher temperatures than holocellulose. This lower degradation temperature of wood is primarily due to the hemicelluloses in the wood and holocellulose. The wood degradation resembles more closely the degradation pattern of the alpha-cellulose and holocellulose than the degradation pattern of the lignin. This is reasonable because cellulose and holocellulose account for approximately 50% and 75% of the wood, respectively.

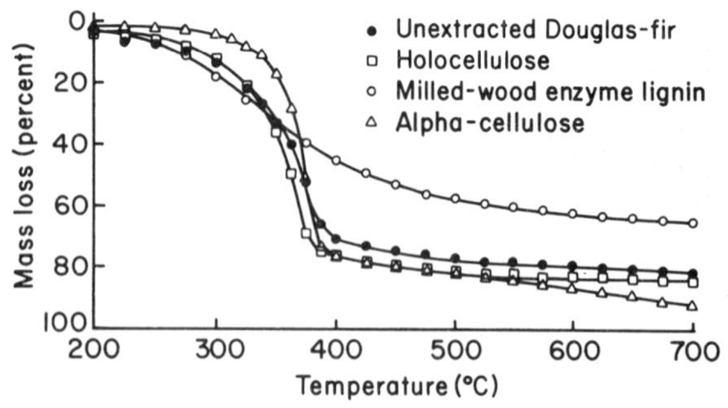

Figure 2
Mass losses of isolated Douglas-fir wood components heated in nitrogen at 5 °C min^{-1}

2. Kinetic Parameters of Thermal Degradation

The temperatures at which decomposition reactions of wood occur and the changes in specimen weight associated with the reactions can be found by thermogravimetric analysis. Quantitative methods can be applied to the thermogravimetric analysis curves (Fig. 2) to obtain kinetic parameters of thermal degradation, which govern the reaction rates. The kinetic parameters usually include the activation energy, the preexponential factor and the order of reaction. Typically, these describe the rate of mass loss and the heat of combustion of the volatiles. They are also useful in describing the temperature dependence of the mass loss rate. However, these kinetic parameters are highly dependent on experimental conditions such as heating rates, sample size and atmosphere and thus should only be considered as effective kinetic parameters having no physical significance. A wide range of values has been reported for both the activation energy and the preexponential factor, assuming a simple first-order reaction following the Arrhenius equation:

$$\dot{m} = A\exp(-E/RT)$$

where $\dot{m}$ is the mass loss rate, A the preexponential factor, E the activation energy, R the universal gas constant and T the temperature in kelvin. Independent measurements have been made on the individual wood components and wood itself. The following values are listed here only to indicate the range to be expected. These values were determined by different methods under different conditions and at different heating rates, and therefore they are not directly comparable.

The degradation of cellulose can be analyzed as a first-order Arrhenius equation. Effective activation energies for cellulose pyrolyzed in nitrogen have been found to range from 170 to 210 kJ mol^{-1} (Hirata 1979, Tang 1967). Effective activation energies for pyrolysis in air are lower, ranging from 109 to 151 kJ mol^{-1} (Akira 1979, Shafizadeh 1984, Stamm 1955). Some have reported that wood and alpha-cellulose degradation follow a two-step first-order decomposition.

The activation energy for hardwood xylan pyrolyzed in nitrogen ranges from 75 to 164 kJ mol^{-1}. Beall (1969) reported values for the activation energy around 13 kJ mol^{-1} for softwood xylan and around 34 kJ mol^{-1} for softwood glucomannan.

Activation energy values for lignin vary depending on the isolation procedures used to obtain the lignin. Beall (1969) and Tang (1967) determined activation energies of 46 and 88 kJ mol^{-1}, respectively, for lignin processed in sulfuric acid. Parker and LeVan (1988) determined an activation energy of 122 kJ mol^{-1} for milled-wood enzyme lignin. Ramiah (1970) found activation energies of 55 kJ mol^{-1} for periodate lignin and 80 kJ mol^{-1} for Klason lignin.

Activation energies for wood pyrolyzed in nitrogen range from 63 to 139 kJ mol^{-1} for pyrolysis temperatures less than 300 °C and from 109 and 227 kJ mol^{-1} for temperatures greater than 300 °C. For pyrolysis in air, the values range from 96 to 147 kJ mol^{-1}. Shafizadeh (1984) showed that pyrolysis proceeds faster in air than in an inert atmosphere and that this difference gradually diminishes around 310 °C. Atreya (1983) found this difference to disappear around 400 °C.

See also: Thermal Properties

Bibliography

Akira K 1979 A study on the carbonization process of wood. Japan Forestry and Forest Products Research Institute Bulletin No. 304, pp. 7–76

Atreya A 1983 Pyrolysis, ignition, and fire spread on horizontal surfaces of wood. Ph.D. thesis, Harvard University, Cambridge, Massachusetts

Beall F C 1969 Thermogravimetric analysis of wood, lignin and hemicelluloses. *Wood Fiber* 1(3): 215–26.

Beall F C 1971 Differential calometric analysis of wood and wood components. *Wood Sci. Technol.* 5: 159–75

Browne F L 1958 Theories of the combustion of wood and its control, FPL Report 2136. US Department of Agriculture Forest Service, Forest Products Laboratory, Madison, Wisconsin (reviewed and reaffirmed in 1965)

Gerhards C C 1979 Effect of high-temperature drying on tensile strength of Douglas-fir 2 × 4's. *For. Prod. J.* 29(3): 39–46

Gerhards C C 1982 Effect of moisture content and temperature on the mechanical properties of wood: an analysis of immediate effects. *Wood Fiber* 14(1): 4–36

Gerhards C C 1983 Effect of high-temperature drying on the bending strength of yellow-poplar 2 × 4's. *For. Prod. J.* 33(2): 61–67

Hirata T 1979 Changes in degree of polymerization and weight of cellulose untreated and treated with inorganic salts during pyrolysis. Japan Forestry and Forest Products Research Institute Bulletin No. 304, pp. 77–124

Koch P 1985 Drying southern pine at high temperature—a summary of research at Pineville, LA from 1963–1982. In: Mitchell P H (ed.) 1985 *Proc. North American Drying Symp.* Mississippi Forest Products Utilization Laboratory, Mississippi State University, pp. 1–38

McDermott J B, Klein M T, Obst J R 1986 Chemical modeling in the deduction of process concepts: A proposed novel process for lignin liquefaction. *Ind. Eng. Chem. Process Des. Dev.* 25: 885–89

Parker W, LeVan S 1988 Kinetic properties of the components of Douglas-fir and the heat of combustion of their volatile pyrolysis products. *Wood Fiber Sci.* (in press)

Ramiah M V 1970 Thermogravimetric and differential thermal analysis of cellulose, hemicellulose, and lignin. *J. Appl. Polym. Sci.* 14: 1323–37

Shafizadeh F 1984 The Chemistry of pyrolysis and combustion. In: Rowell R M (ed.) 1984 *The Chemistry of Solid Wood*, Advances in Chemistry Series 207. American Chemical Society, Washington, DC, pp. 489–530

Stamm A J 1955 Thermal degradation of wood and cellulose. Presented at the *Symp. on Degradation of Cellulose and Cellulose Derivatives.* Sponsored by the Division of Cellulose Chemistry, 127th National Meeting of the American Chemical Society, Cincinnati, Ohio, April 4–7

Tang W K 1967 Effect of inorganic salts on pyrolysis of wood, alpha-cellulose, and lignin determined by dynamic thermogravimetry, FPL-RP-71. US Department of Agriculture Forest Service, Forest Products Laboratory, Madison, Wisconsin

Winandy J E 1987 Effects of treatment and redrying in mechanical properties of wood. In: *Proc. FPRS Conference on Wood Protection and Techniques and the Use of Treated Wood in Construction.* Forest Products Research Society

S. L. LeVan
[US Forest Products Laboratory, Madison, Wisconsin, USA]

Thermal Properties

Temperature fluctuations which occur in wood both in nature and in buildings affect many wood properties, such as its mechanical properties. This article deals with these temperature changes, the conduction of heat and thermal expansion. How much and how fast wood heats up, how warm it feels to the touch, and how fast a heat wave passes through wood—all these depend on its thermal diffusivity, which in turn is related to thermal conductivity and to specific heat. The thermal properties also affect fire performance and combustion, but this article is only concerned with temperatures up to 375 K, at which wood begins to degrade thermally. The behavior of wood at higher temperatures is treated elsewhere (see *Fire and Wood*; *Thermal Degradation*).

Moisture influences the thermal properties of wood both as hygroscopic or bound moisture in the cell walls and as free water in the cell cavities. The two differ with regard to freezing. Bound moisture is in a kind of solid state at all temperatures and cannot freeze, whereas the free water freezes at 273 K. Water-saturated cell walls hold a maximum of about 30% bound moisture, the percentage being expressed on the basis of oven-dry weight, as is usual in mechanical wood technology. Free water is the moisture in wet wood in excess of 30% moisture content. Since ice and water differ in thermal properties, the thermal properties of wet wood abruptly change at 273 K, the change being greater the more free water the wood contains.

Most thermal properties vary with wood density ρ, a term which in this article stands for the ratio of mass to volume (g cm^{-3}), both at 12% moisture content, an average value for wood in use indoors and outdoors. Elsewhere, wood density is often expressed in terms of specific gravity S, the ratio of the oven-dry weight to the weight of a volume of water equal to the volume of the sample at a specified moisture content. The relationship between specific gravity on the basis of volume at 12% moisture content (S_{12}) and density is $\rho = 1.12S_{12}$.

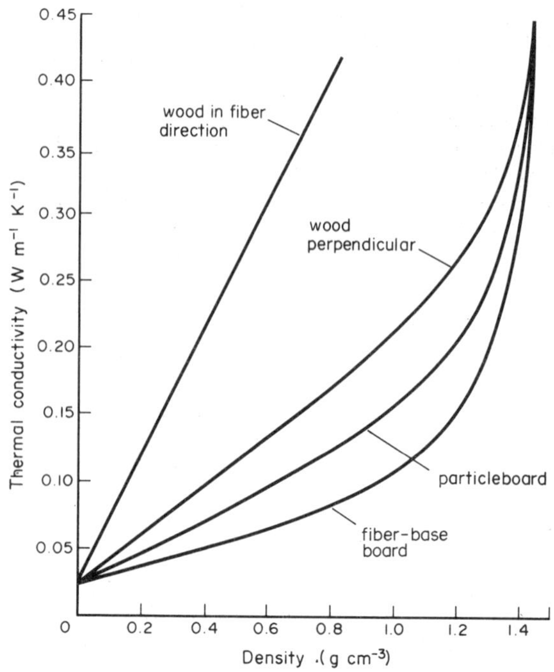

Figure 1
Relationship between thermal conductivity and density for wood and wood-base panel products at 12% moisture content and 300 K (after Kollmann and Côté 1968. © Springer, Berlin. Reproduced with permission)

1. Thermal Conductivity

Since wood lacks the free electrons which so rapidly transfer heat and electricity in metals, it conducts heat by the relatively inefficient transfer of vibrational energy from one particle to the next. For this reason, and because wood's hollow cells trap air, wood and wood-based panel products are low in thermal conductivity and so feel warm to the touch as the material conducts little heat out of the skin where the temperature-sensing nerves are located.

Wood is an insulator. Special insulating materials surpass it in this regard, but they lack the strength needed for many purposes in construction and elsewhere. The insulating value or resistance to heat flow naturally increases in proportion to the material's thickness t, and is inversely proportional to thermal conductivity k. For construction materials, the resistance is expressed in the USA as an R value, where $R = t/k$ (ft^2 h °F Btu^{-1}). The units as an integral part of the definition are omitted. Common insulating blankets of fiberglass 8.9 cm (3.5 in.) thick, for example, achieve R-11, whereas wood of the same thickness and density 0.5 g cm^{-3} at 12% moisture content has R-4.4.

Differences in thermal conductivity between wood species are largely explained by different densities (see *Density and Porosity*). The average thermal conductivity of a species deviates by less than 10% from the average of all species of similar density. In the density range 0.3–0.8 g cm^{-3} of most commercial woods, conductivity increases in proportion to density (see Fig. 1). The line of conductivity as a function of density intercepts the zero-density axis at the conductivity of still air (0.024 W m^{-1} K^{-1}). Wood cell cavities are too small (~0.005 cm diameter) to transfer a significant amount of heat by convection of air and by radiation.

In the longitudinal or fiber direction, which is lengthwise in most pieces, wood conducts heat 1.5–2.8 times faster than in the transverse direction, perpendicular to the fibers. This is due to the orientation of fibers and of cellulose-chain molecules. The higher ratios apply more to dry wood, the lower ratios more to wet wood in which water equalizes the difference between longitudinal and transverse conductivity. In dry wood the ratio averages 2.5. This article is mainly concerned with transverse conductivity, which determines heat flow in the direction of thickness and is of greater significance than longitudinal conductivity.

Considering the transverse directions, one may expect that wood conducts heat somewhat faster radially, due to radially oriented rays, than tangentially, along the growth rings. However, dense latewood bands of temperate-zone woods may compensate or even overcompensate for the effect of rays. Hence in some wood species the ratio of radial-to-tangential conductivity may reach 1.3, compared to only 0.9 in others, depending on the volume of ray tissue and on the earlywood–latewood contrast in wood density. In general, the difference between radial and tangential conductivity can be disregarded. In most wood items heat flows in intermediate transverse directions anyway.

The thermal conductivity of water (0.562 $\mathrm{W\,m^{-1}\,K^{-1}}$ at 273 K) is more than 20 times higher than that of air, and also exceeds the conductivity of cell-wall substance (about 0.42 $\mathrm{W\,m^{-1}\,K^{-1}}$). Therefore, thermal conductivity increases with increasing moisture content u ($\mathrm{g\,g^{-1}}$), the relationship being approximately linear: $k_u = k_{12}\,(1+u)/1.12$. This means thermal conductivity in the transverse direction increases by about 1% for each percent of added moisture. Moisture also contributes to heat conduction by diffusion as vapor from the warm to the cold side, where it condenses, releasing the large latent heat of the vapor. Some vapor diffuses to the cold side even under stationary conditions, because capillary forces in wet wood pull the condensed water back to the warm side. Transfer of heat by diffusing and condensing water vapor is probably the reason why thermal conductivities, measured by different authors using different methods, vary considerably. The averages determined in some investigations deviate from the values of Fig. 1 by as much as 50%. The true averages of all real values are probably several percent higher than those of the regression line in Fig. 1.

As with most insulating materials, wood's thermal conductivity increases with temperature, as shown in Fig. 2 for the density 0.6 $\mathrm{g\,cm^{-3}}$. In lighter woods temperature has a slightly stronger effect, in denser wood a somewhat weaker one. The thermal conductivity ratios in Fig. 2 apply to unsteady-state conditions and are therefore for moist wood at an elevated temperature, at which diffusing vapor transfers plenty of heat, higher than for steady-state conditions. The 60% moisture-content curve is an estimate based on extrapolation and on the thermal conductivity of ice (2.2 $\mathrm{W\,m^{-1}\,K^{-1}}$ at 273 K), which is four times higher than the thermal conductivity of water. The negative slope of the 60% curve below freezing reflects the fact that the thermal conductivity of ice decreases as temperature rises. For ice itself the curve would be twice as steep.

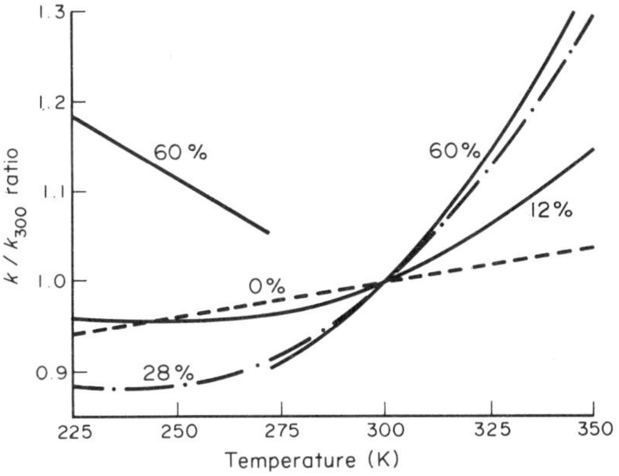

Figure 2
Relationship between thermal conductivity (expressed as a fraction of the value at 300 K) and temperature for wood of 0.6 $\mathrm{g\,cm^{-3}}$ density perpendicular to grain at various moisture contents (data from Kuhlmann 1962)

The thermal conductivity of water increases with temperature above 273 K to the extent shown in Fig. 2 for wood at 12% moisture content. Due to the transfer of heat by vapor, the effect of temperature on conductivity is greater in moist wood than in water. However, water-soaked wood in which all cell cavities are filled has no space for diffusing vapor; its thermal-conductivity–temperature relationship resembles that of water. Therefore, temperature has the strongest influence on thermal conductivity at a given moisture content, which is lower for dense wood than for light wood. Dense wood, after all, has relatively little cell cavity space for water and for diffusing vapor. Otherwise the relationships shown in Fig. 2 for wood of 0.6 $\mathrm{g\,cm^{-3}}$ apply approximately also for other densities; in lighter species the effect of temperature tends to be somewhat stronger, due to their ample cell-cavity space for diffusing vapor.

In wood-based panel products, fibers are usually oriented in the plane of the panel, so that heat flows across the panel perpendicular to the fibers, as in its flow across solid wood. In solid wood, however, all fibers lie parallel to each other and are grown together, whereas in the panels a number of fibers lie at angles and touch each other only over a part of their length. Areas of deficient contact restrict heat transfer from fiber to fiber and reduce conductivity to below that of solid wood (Fig. 1).

Contacts or heat bridges between fibers are at a minimum when all fibers cross each other at right angles. None of the wood-based panels is structured this way, but fiber-base board (that is, hardboard and fiberboard, formerly called insulating board) comes close, consisting mainly of reassembled fibers oriented at random in the panel plane. Fiberboard conducts heat about half as well as solid wood of the same density (Fig. 1). However, in very light fiberboard in which air occupies most of the volume, and in very dense board in which the fibers are pressed into close contact, conductivity approaches that of solid wood (Fig. 1).

Particleboard consists of discrete pieces such as long splinters and flat shavings, which like the fibers in fiber-base board lie randomly oriented in the plane of the panel. However, all fibers within the particles naturally run parallel to each other, so that only at the interface between particles does the flow of heat meet with significant resistance. An adhesive bonding the particles together is restricted to tiny spots, rather than being a continuous film, and so conducts little. Hence, thermal conductivity of particleboard lies in between the conductivities of solid wood and of fiber-base board (Fig. 1).

As discussed above, the conductivities of the boards apply to panels in which all fibers and particles lie in the plane. In reality, some fibers and particles point out of the plane and conduct heat across the panel relatively well. Therefore, the conductivities shown in Fig. 1 are minimal; most panels conduct 5–40% more heat. Due to the orientation, conductivity of particleboard depends on the kind of particle; conductivities tend to increase in the sequence of wafers, shavings, flakes, splinters and sawdust. In layered boards, low-density layers have a relatively strong influence, of course, but the effect of low density is frequently cancelled by coarseness of the particles. Particles of bark incorporated in particleboard may increase or decrease conduction of heat as compared with wood particles, because bark is less anisotropic on the one hand, but a better insulator on the other.

Moisture seems to have a stronger influence on transverse thermal conductivity of wood-base boards than on solid wood, in the order of over 2% for each added percent of moisture. The influence of temperature again tends to be considerably stronger than shown for solid wood in Fig. 2.

Plywood consists of layers, typically veneers, whose grain runs roughly at right angles to each other. The layers are much thicker than the particles of particleboard and are bonded together by continuous adhesive film, so that the thermal conductivity of plywood through its thickness equals the conductivity of solid wood.

2. *Specific Heat*

The specific heat is a measure, for example, of how much energy furniture and structural wood absorb when a cold room is heated up, and how much heat the items store. It also has a strong influence on the rate of cooling and of heating, as explained in the discussion of thermal diffusivity in Sect. 3.

The specific heat c of wood seems to be the same for all wood species. The value increases with temperature, as does c of other solids, particularly those of the nonmetallic kind. The specific heat of oven-dry wood, in the temperature range 200–400 K, is given by $c_0 = 4.86(T-44)$; accordingly, the mean value of c over the 293–373 K temperature range amounts to 1405 J kg^{-1} K^{-1}. This is a relatively high value compared with the c of other solids.

Since specific heats of water and ice (4187 and 2100 J kg^{-1} K^{-1}, respectively) are much higher, the specific heat of wood increases with increasing moisture content. From about 200 K to 280 K, and moisture contents up to 30% ($u = 0.3$), the following partly empirical equation applies:

$$c_u = \frac{4.86(T-44) + 4187u}{1+u} \qquad (1)$$

At higher temperatures, the c of bound water ($u \leqslant 0.3$) was found to be greater than that of ordinary free water, which occurs also in cell cavities above the fiber saturation point ($u = 0.3$). Therefore an empirical term $c_+ = (T-273)^{1.7}$ has to be added to Eqn. (1). The sum $c_u + c_+$ is realistic in the range 274–360 K, for $u \leqslant 0.3$. Specific heats measured for moist wood varied by several percent in the various investigations. The calculated values c_u and c_+ correspond to averages of the various measurements, and are therefore not exact for any given condition. By contrast, the value for c_0 deserves more confidence, as specific heats measured on oven-dry wood had standard deviations in the order of only $\pm 1\%$.

Wood-based panels are practically equal to solid wood in specific heat, as they contain at most only 10% of added binders, whose specific heat is about one-half that of wood.

3. *Thermal Diffusivity*

The thermal diffusivity of a material determines the rate of temperature change in the material when subjected to change in ambient temperature. The time required for any point inside the material to reach a given temperature is inversely proportional to diffusivity. Wood ranks low in thermal diffusivity among structural materials; it takes a long time for heat waves to penetrate timbers and, for example, to pass through solid wood doors exposed to fire on one side.

Thermal diffusivity α of a material depends on the amount of heat conducted into the material and on the temperature rise caused by the conducted heat. It is defined by $\alpha = k/(c\rho)$ and can be calculated on this basis. Thermal diffusivity can also be determined using measured temperatures and the differential equation for the variation of temperature with time.

For oven-dry wood, calculated and measured diffusivities agree fairly well, but for moist wood some measured diffusivities are up to three times higher than calculated diffusivities as a result of transfer of heat by diffusing vapor. Under the transient conditions of temperature-change measurements, the proportion of heat transferred by vapor is relatively large, particularly in permeable wood heated in steam or hot water.

Dense wood tends to be lower than light wood in thermal diffusivity, but the difference amounts to 20% at most between woods of density 0.7 and 0.3 g cm^{-3}. Generally, measured diffusivity increases with higher temperature, particularly in wet wood, whereas wood moisture reduces thermal diffusivity, especially in the low-moisture-content range (see Fig. 3). Compared with the values for water, ice is four times higher in thermal conductivity but has only one-half the specific heat. Therefore, the thermal diffusivity of wet wood is much higher below 273 K than above 273 K.

Longitudinal thermal diffusivities should exceed transverse diffusivities by a factor equal to the ratio of

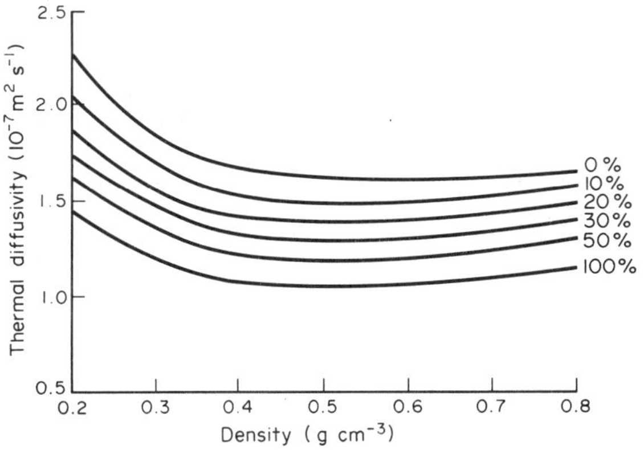

Figure 3
Relationship between thermal diffusivity at 333 K and density for wood in the transverse direction at various moisture contents (after Maku 1954. © Wood Research Institute, Kyoto University, Japan. Reproduced with permission)

thermal conductivities, that is 2.5. In reality, some measured longitudinal diffusivities are considerably higher, nearly six times higher in the case of wood heated in steam, the reason being heat transfer by diffusing vapor, and the fact that permeability for vapor in the fiber direction is more than 10 times higher than transverse permeability.

4. *Thermal Expansion*

At constant moisture content, wood like most substances expands when heated and contracts when cooled. However, rising temperatures are usually associated with decreasing relative humidities, which cause wood to dry and shrink, while at falling temperature humidity rises, so that wood absorbs moisture and swells. Shrinking and swelling tend to overshadow thermal expansion; under many conditions wood shrinks when heated and expands when cooled. Above 375 K the material is subject to thermal degradation; the resulting irreversible contractions are superimposed on the reversible dimensional changes. Thermal expansion of wood rarely occurs without associated effects, but it merits consideration nevertheless.

Thermal expansions vary considerably from piece to piece within the same type of wood, as may be expected for a biological material. Percentagewise, the variations are especially large in the fiber direction (longitudinally), in which the coefficients of linear expansion of oven-dry wood are only 4×10^{-6} K^{-1} on average. Some pieces expand twice that much. Longitudinal expansions differ little from species to species and seem to be independent of wood density, but they increase toward higher temperatures and near 375 K are roughly 20% higher than at 275 K. Moisture affects longitudinal expansion percentagewise as shown in Fig. 4 for transverse expansion; the coefficients reach a maximum of about 8×10^{-6} K^{-1} around 20%, dropping below 4×10^{-6} K^{-1} near 30% moisture content.

The low longitudinal expansion, less than one-half of that of steel, is explained by the longitudinal orientation of cellulose-chain molecules. Transversely the material expands much more, particularly in the tangential direction of annual rings, in which the coefficients are about 1.4 times larger than radially, the lower radial expansion being explained by the restraint of ray cells. Figure 4 shows averages of transverse coefficients a for wood of density 0.5 g cm^{-3}. For other densities the coefficients a_ρ at 0% moisture content are given approximately by

$$a_\rho = 29 + [36 \times 10^{-6}(\rho - 0.5)]$$

The low coefficients of wet wood (Fig. 4) result from internal drying and readsorption of moisture. At rising temperature, the equilibrium moisture content decreases and some moisture diffuses out of the cell walls into the cell cavities, so that the cell walls and the wood as a whole shrink. Above 350 K the equilibrium moisture content slightly increases, the cell-wall substance readsorbs moisture out of cell cavities, and the wood swells.

Freezing of free water in cell cavities has no consequences for wood dimensions as long as the cell cavities contain some air space into which the ice can expand. Only water-soaked wood increases in size as it freezes, the increase being largest for wood of low density in which cell cavities comprise a large percentage of the wood volume.

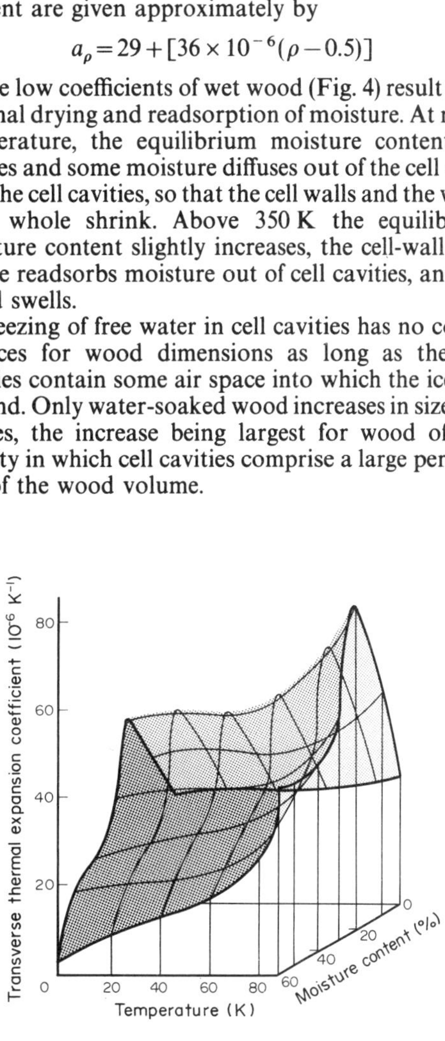

Figure 4
Coefficients of linear thermal expansion at various temperatures and moisture contents for wood of density 0.5 g cm^{-3} in the transverse direction (adapted from Kubler et al. 1973)

Frost temperatures also affect bound moisture and wood shrinkage. The vapor pressure of ice is lower than that of supercooled water and of bound moisture near fiber saturation. Therefore, some moisture diffuses out of the cell walls to freeze in cell cavities and causes wood to shrink, providing that cell cavities have space for the frozen-out moisture. The difference between vapor pressure of ice and of supercooled water grows as temperature decreases, so that the amount of frozen-out moisture and the resulting freeze shrinkage increase towards lower temperatures. In the tangential direction, in which wood shrinks most (see *Shrinking and Swelling*), freeze shrinkage reaches 2% in some species at 230 K, which is 15 times greater than true thermal contraction due to cooling from 273 to 230 K. Freeze shrinkage is reversible, of course, and takes several days at 230 K, while true thermal expansion is by its nature immediate at all temperatures.

Plywood, as a layered panel product in which the fiber directions of laminations are at right angles to each other, expands in directions of length and width more than solid wood in the fiber direction but less than solid wood transversely. However, since wood is much stiffer along the fibers than transversely, the expansion remains close to that of solid wood longitudinally; it averages $7\times10^{-6}\,K^{-1}$. Moisture levels and temperature have similar effects as in solid wood. In the direction of panel thickness, of course, plywood expands as much as solid wood transversely.

In particleboard and fiberboard the fibers are oriented in the panel plane as in plywood, rather than pointing in the direction of panel thickness. Hence, the coefficients are of same order of magnitude as in plywood.

See also: Thermal Degradation

Bibliography

Christoph N, Brettel G 1977 Investigation on thermal expansion of wood in relation to density and temperature (in German). *Holz Roh- Werkst.* 35: 99–108

Deliiski N 1977 Mathematical model for conductive heating of wood (in German). *Holz Roh- Werkst.* 35: 141–45

Hearmon R F S, Burchham J N 1956 The specific heat and heat of wetting of wood. *Chem. Ind.* 31: 807–8

Kollmann F F P, Côté W A 1968 *Principles of Wood Science and Technology*, Vol. 1, *Solid Wood*. Springer, Berlin, pp. 240–57

Kubler H, Liang L, Chang L S 1973 Thermal expansion of moist wood. *Wood Fiber* 5: 257–67

Kuhlmann G 1962 Study on thermal properties of wood and particleboard in relation to moisture and temperature in the hygroscopic range (in German). *Holz Roh- Werkst.* 20: 259–70

Lewis W C 1967 *Thermal conductivity of wood-base fiber and particle panel materials*, Research Paper FPL 77. US Forest Products Laboratory, Madison, Wisconsin

MacLean J D 1946 *Rate of Temperature Change in Short-Length Round Timbers.* US Forest Products Laboratory, Madison, Wisconsin

McMillin C W 1970 Specific heat of ovendry loblolly pine wood. *Wood Sci.* 2: 107–11

Maku T 1954 *Studies on the Heat Conduction in Wood*, Bulletin No. 13. Wood Research Institute, Kyoto University, Japan

Schneider A, Engelhardt F 1977 Comparative investigation on the thermal conductivity of particleboards and wood bark boards (in German). *Holz Roh- Werkst.* 35: 273–8

Steinhagen H P 1977 *Thermal Conductive Properties of Wood*, Forest Products Laboratory Report FPL-9. US Forest Products Laboratory, Madison, Wisconsin

H. Kubler
[University of Wisconsin, Madison, Wisconsin, USA]

Timbers of Africa

West Africa, with forests covering about 500 000 km^2, is one of the world's most important sources of hardwoods. In Central Africa, which is less accessible, the forest area approaches some 700 000 km^2. East African trees, many of which grow at a higher altitude and in a less humid climate, include some which are coniferous, and present efforts, including particularly the planting of *Pinus radiata*, a fast growing species similar to *Pinus sylvestris*, should eventually provide an alternative to the European softwood supply.

Some commercial species occur throughout tropical Africa and there are many species which, although for various reasons not generally available for export, are of value for local use.

This article describes the most important commercial woods. Timber names can be very confusing, as they vary in relation to the locality of origin. A few alternatives are given, but lack of space precludes anything like a full list. Scientific names provide the most reliable nomenclature.

Descriptions refer to the size of the tree and the country of origin, followed by notes on the obvious features of wood—color, texture and density. Brief reference is made to outstanding strength values, ease of seasoning and durability, and also working qualities and utilization.

ABURA (*Mitragyna ciliata*), also called bahia and subaha, is common in West Africa. Logs are exported up to 5.5 m long and 0.75 m in diameter. The wood is unusually plain and lusterless, pinkish brown in color, with fine and uniform texture. A fairly light wood, it has a density of 0.56 g cm^{-3} when seasoned. It dries well with little degradation and is stable in service. It is not a durable wood and is fairly resistant to impregnation. Abura works well with hand and machine tools but tends to blunt sharp edges rather quickly. It is used for construction work where strength is not important, and for flooring not exposed to heavy wear.

AFARA (*Terminalia superba*), also known as limba, ofram, akom and fraké, is a tall, buttressed tree, with a diameter of up to 1.5 m above the buttresses. The wood, in plentiful supply, is of a drab yellow color, but some trees have a greenish gray heartwood, often quite dark and sometimes streaked. Its texture is moderately coarse, and its density varies between 0.48 and 0.64 g cm^{-3} when seasoned. Afara dries rapidly with little degradation and is very stable, but is not a durable wood. It works well and is used for utilitarian plywood. In the solid it is useful for general furniture production. It is less popular in the UK than in the rest of Europe.

AFRICAN "WALNUT" (*Lovoa trichiliodes*), otherwise known as dibetou, tigerwood, congowood, alona wood and lovo wood, is related to the mahoganies and is widely distributed. Logs are up to 1 m in diameter and 3.5 m or more in length. The wood resembles the mahoganies in texture but is golden brown in color and sometimes marked with black lines. When quarter sawn it exhibits the characteristic stripe familiar in sapele. Its density is around 0.55 g cm^{-3} when seasoned. It is very resistant to impregnation but is moderately durable. It bears little resemblance to the true walnut but is a high class furniture wood and works well. Veneers are very decorative.

AFRORMOSIA (*Pericopsis elata*) or kokrodua has been recently introduced and the wood has proved to be popular. Logs are available up to 10 m long by 1.5 m in diameter. The wood is strong and fairly heavy, its density being on average about 0.70 g cm^{-3} when seasoned. It is of a rich brown color which varies considerably. It seasons slowly but with little tendency to split and warp, and is stable in service; it is resistant to decay and insect attack.

The timber is used for ship- and boatbuilding where durability and strength are required and is also regarded as a high class furniture wood both in the solid and as veneer.

AFZELIA (*Afzelia* spp.), also called doussie, apa, chanfuta and mkora, is widely distributed and logs are available up to 7.5 m long and 1 m in diameter. The wood is handsome in general appearance, reddish or golden brown, coarse and majestic in texture and with an average density of around 0.80 g cm^{-3}. It has to be dried slowly, but shrinkage is minimal. It is noted for its durability and its resistance to insect attack. It is not an easy wood to work with hand or machine tools, but the handsome finish repays the effort. An excellent structural wood for staircases and fittings on a large scale in public buildings, it is also used in veneer form.

AGBA (*Gossweilerodendron balsamiferum*) or tola branca is one of the largest trees in tropical Africa, up to 60 m in height, clear of branches for 30 m, unbuttressed and cylindrical. A useful wood of comparatively light weight, its density is about 0.51 g cm^{-3} when seasoned. It is of pale pinkish brown color and moderately fine texture. Exudation of gum is common but not sufficient to constitute a defect. The timber seasons easily with little degradation and is very stable. It is comparable in strength with mahogany and fairly resistant to insect and fungal attack. Agba is a good joinery wood especially for chairs, tables and cupboards. It is also manufactured into veneer and plywood.

ANTIARIS (*Antiaris africana* and *A. welwitschii*) from tropical Africa, also called oro, chen chen and ako, is up to 45 m in height and 1.5 m in diameter. The wood closely resembles obeche but is slightly heavier, its density averaging 0.40 g cm^{-3} when seasoned. It is pale yellow or straw colored. Antiaris is liable to warp during seasoning but is stable in service. It is low in strength and perishable; logs have to be extracted quickly to avoid insect attack and discoloration from fungi. It works easily with hand tools and its uses include interior parts of furniture and boxmaking. Selected logs provide decorative veneer.

AVODIRE (*Turraeanthus africanus*) is a small tree, often of indifferent shape for conversion into boards. Related to the mahoganies, the wood is very white when freshly cut but darkens to pale yellow on exposure. It is similar in texture to the mahoganies and is often wavy grained, which produces an attractive mottled figure. It seasons quickly but may warp, although it is stable in service. Strong for its weight, its density is about 0.5 g cm^{-3}. It is a delightful wood to work, but is most frequently met in veneer form, avodire curls being particularly attractive.

AYAN (*Distemonanthus benthamianus*), also called Nigerian satinwood and movingui, is a tall slender tree distributed in West Africa. Logs may be up to 1 m in diameter and 3–4 m in length. The wood is very attractive, being of a pale but bright yellow color and of medium texture. Of medium weight and strength, its density is about 0.65 g cm^{-3} when seasoned. It seasons without distortion and is fairly stable in service. It is used for indoor decorative work, flooring and panelling, particularly with woods of contrasting colors. Occasionally logs are figured, and veneers cut from them are very attractive. Satinwood is quite a descriptive name.

BERLINIA (*Berlinia* spp.) is well distributed in West Africa and logs up to 5.5 m long and 1 m in diameter are exported. The wood is pale brown in color with irregular purplish markings. It is of the same hardness and density as oak, around 0.70 g cm^{-3}, and rather coarse textured. Berlinia seasons rather slowly, shrinks appreciably and shows some movement when dry. The heartwood resists impregnation, but the wood is naturally fairly durable. Its working qualities are somewhat variable but on average are fairly good. It is useful for decorative fittings on a large scale as well as for heavy structural work.

AFRICAN BLACKWOOD (*Dalbergia melanoxylon*) is a small tree, 9 m tall by 20 cm in diameter, occurring widely in East and Central Africa. Small logs are exported. The wood is almost black or dark purplish brown, fine textured, oily and very heavy, with a density of 1.2 g cm^{-3} when seasoned. Seasoning needs care, as the wood is apt to split, but it has good working qualities and permits the cutting of screw threads. Its most important use is for woodwind orchestral instruments, for which it is considered to be unrivalled for appearance, working qualities and tone. The natural oil prevents absorption of water.

BUBINGA (*Guibourtia* spp.) is usually called Kevazingo when it is highly figured. Trees are fairly large with a clear bole of 18 m and up to 1.5 m diameter. The wood is deeply colored and heavy, with a density of 0.80–0.96 g cm^{-3} when seasoned. The color is attractive reddish or purplish brown with darker streaks, the texture being fairly coarse. Although it is used mostly in the form of rotary-cut veneer, working qualities in the solid are reasonably good. Bubinga is related to the rosewoods and has a certain similarity, being heavy and of decorative appearance.

CAMWOOD (*Pterocarpus soyauxii*) or African padauk is usually exported in limited quantities as billets up to 1.5 m long and 0.3 m in diameter or as logs up to 1 m in diameter. The heartwood is extremely deeply colored, bright dark red, very lustrous and of coarse texture. A moderately heavy wood, the density varies between 0.70 and 0.82 g cm^{-3} when seasoned. Like many leguminous timbers it is very stable and seasons slowly with little tendency to warp. With due regard for its weight and hardness it is reasonably responsive to hand working processes. A wood eminently suitable for imposing situations, it is recommended for parquet flooring, ornamental turnery and furniture, and is also available as veneer.

CEIBA (*Ceiba pentandra*) is common and widely distributed in tropical countries. It is closely related to bombax, with which it may be mixed in commercial consignments. The wood is very soft, coarse in texture and of variable light weight, its density being 0.20–0.40 g cm^{-3}, averaging 0.32 g cm^{-3} when seasoned. There is no distinction between heartwood and sapwood. Of pale color, almost white, it is very prone to discoloration. It is perishable but is easily impregnated; it has very low strength. Most suitable for veneer for inside laminations in plywood and blockboard, it is also used for insulation purposes. Its light weight and softness call for sharp tools.

CORDIA (*Cordia millenii* and *C. platythyrsa*) represents two species that are plentiful in tropical Africa and other species in central South America. Logs up to 3.5 m long and 1 m in diameter are exported. The wood is comparatively light, with a density of about 0.43 g cm^{-3} when seasoned. It is coarse textured and of pale to medium pinkish brown color. Quarter sawn surfaces may show a striped effect. Reported to season quickly, it is stable in use, and its strength properties are average for its weight. It is similar to South American freijo (*Cordia goeldiana*). Some of it is quite attractive and is used for furniture making, and it should prove suitable for boatbuilding.

DAHOMA (*Piptadeniastrum africanum*), also called ekhimi and dabema, is a fairly large tree up to 37 m in height with a diameter of 1.25 m above the buttresses. It is widely distributed in West and Central Africa but not exported in large quantities. The wood is of a rather variable brown, not particularly attractive color and when freshly cut possesses a disagreeable odor. It is similar in weight to oak, its density being around 0.70 g cm^{-3} when seasoned. Seasoning must be slow, as the timber shows a tendency to warp. It is strong when used in large sizes and is moderately durable. It is not a pleasant wood to work with and is used mostly for heavy construction.

DANTA (*Nesogordonia papaverifera*) or otutu is a moderately sized tree with a clear cylindrical bole up to 15 m long and 1 m in diameter. The wood is of a warm red color, the inner heartwood being darker than the outer heartwood. It is of a fine texture, often interlocked and showing a stripe on quarter sawn surfaces. It is moderately heavy, with a density of around 0.70 g cm^{-3} when seasoned, and has pronounced elastic qualities. It is moderately durable but resistant to preservative treatment. The best examples are quite attractive and the wood should be better known for furniture and general interior fittings and ballroom floors. It is effective for ornamental turnery.

EBONY (*Diospyros* spp.) is a trade name under which are marketed two principal African species of diospyros as well as several others. The trees cover a wide area and grow to a height of 21 m and a diameter of 0.75 m. The wood is well known for its black color, often with brown streaks. It is very fine textured and heavy, with an average density of 1.00–1.02 g cm^{-3}. Being very hard and resistant, ebony is not very easy to work, but the special character of the wood makes up for the difficulties. It turns well, takes a high polish and is used chiefly for musical instruments and inlay work. It is also manufactured into veneers.

EKKI (*Lophira alata*), also known as hendui, kaku and red ironwood, is one of the most beautiful trees in the world when in flower. A large rain-forest tree, it is up to 30 m in height and exceeds 1.5 m in diameter. The wood is very hard and heavy, its density being 0.96–1.12 g cm^{-3} when seasoned. It is dark purple brown in color and of coarse texture, and shows a white deposit in its large pores. It is difficult to season owing to splitting and warping and is very resistant to fungal attack, termites and marine borers. Difficult to work, ekki has a severe blunting effect on cutting edges. It is used for heavy marine constructions and for heavy duty flooring.

ESIA (*Combretodendron macrocarpum*), also called abale, owewe and minzu, is a common and large tree but of limited utility owing to difficulties encountered in seasoning. The wood is of reddish brown color, coarse textured and moderately heavy, its density being about 0.70 g cm^{-3}, similar to oak. The wet wood has an unpleasant odor. Esia dries slowly and warping, splitting and shrinkage are very extensive. It is moderately durable but resistant to impregnation. The sapwood is perishable but is easily treated. Although supplies are abundant, the wood is used only for construction work and railway sleepers. A saving grace appears to be decorative possibilities as veneer.

GABOON (*Aucoumea klaineana*) or okoume is a species almost confined to Gabon, where it is very plentiful. It is a large tree with a bole up to 21 m long above the buttresses and up to 2 m in diameter. The wood is a uniform pale gray or pinkish gray and is slightly lighter in weight than mahogany, with a density of about 0.43 g cm^{-3} when seasoned. It is easily dried, either naturally or artificially, and there is no distortion. It has long been an ideal wood for manufacture into plywood. Neither strong nor durable, it is hardly used in the solid.

GEDUNOHOR (*Entandrophragma angolense*), otherwise known as tiama or edinam, is a large tree distributed across tropical Africa, supplying logs up to 9 m long and 1.8 m in diameter; it is a member of the mahogany family, the *Meliaceae*. The wood is similar to mahogany but is somewhat plainer and not so lustrous. Slightly less heavy than sapele and utile with a density of 0.55 g cm^{-3}, it does not season without considerable distortion, but it is stable in service. It is moderately resistant to insect and fungal attack but does not respond to impregnation. Uses include furniture, superior joinery and block and strip flooring. Occasional figured logs are valuable as decorative veneers.

GUAREA (*Guarea cedrata* and *G. thompsonii*) includes two species marketed together under one trade name; it is also called Nigerian pearwood or bosse. Logs are available up to 4.5 m long and up to 1.25 m in diameter. The woods are of the mahogany family. *Guarea cedrata* has a scent like cedar. It frequently exudes gum, but not sufficient to constitute a defect. Both woods are pale pinkish brown in color and very slightly finer in texture than mahogany, and somewhat heavier, the density ranging from 0.55 to 0.62 g cm^{-3}. Guarea seasons readily and is moderately stable. The heartwood does not respond to impregnation but is durable. Uses include furniture and flooring, plywood and decorative veneers.

IDIGBO (*Terminalia ivorensis*), also called emeri and framire, occurs in West Africa, though it is not a common tree. Logs are available 3–10 m long and up to 1 m in diameter. The wood is of a pale yellow or yellow–brown color, with a medium to open texture, and its density varies considerably, averaging 0.54 g cm^{-3}. It is very stable in service and seasons readily without distortion, but strength values are rather low. Difficult to impregnate, it has a natural resistance to fungal attack. It works well with hand tools and is suitable for high class joinery and furniture. It is attractive when stained and polished.

ILOMBA (*Pycnanthus angolensis*), otherwise known as akomu, otie and walele, occurs over most of tropical Africa. and is exported in large quantities in long cylindrical logs 7.5–11 m long and 0.6–0.9 m in diameter. The wood is pale colored, pinkish white to brown, plain and featureless and of light to medium weight, with a density of about 0.51 g cm^{-3}. It needs slow seasoning, as it is prone to splitting and warping, but is moderately stable in service. The sapwood is liable to discoloration. The timber is perishable but responds to impregnation. It works easily with hand and machine tools and gives a good finish. It is used chiefly for plywood production, furniture and moldings.

IROKO (*Chlorophora excelsa*) or mvule is distributed over the whole of tropical Africa. The tree is tall and straight, up to 20 m clear of branches, and logs may be 1.8 m in diameter. The wood is yellow or brown, darkening to deep brown, coarse textured, and has a density of 0.64 g cm^{-3}. It seasons well, is stable in use, and has natural durability. Iroko "stone" is the result of damage to the tree and may be later occluded and undetectable. Otherwise, good working qualities, strength and pleasing appearance make iroko one of the world's most useful woods.

MAHOGANY (*Khaya* spp.) trees, also known as acajou or acajou d'Afrique, may supply logs 25 m long above the buttresses, and up to 2 m in diameter. The wood varies from pinkish brown to a warm reddish brown and is of light to medium weight, the average density being 0.53 g cm^{-3}. It is somewhat coarser in texture than the Central American mahogany (*Swietenia* spp.). It seasons well, is stable and is very resistant to methods of impregnation, but is moderately durable. It is a standard wood for high class furniture and decorative joinery and interior fittings, and as some trees produce figure in great variety, the veneer trade is considerable.

MAKORE (*Tieghemella heckelii*), also known as cherry mahogany and baku, is a large tree with a cylindrical bole yielding logs up to 7.5 m long and 1.25 m in diameter. It is unrelated to the mahoganies. The wood is somewhat heavier than mahogany, with a density of about 0.64 g cm^{-3} and is of a warm reddish color. It seasons moderately well, is stable in use and is somewhat stronger than mahogany. It is rated as very durable. As a decorative wood it is of value for craftwork, but the dust in working can be unpleasant. It is used for high class furniture and fittings, and for work where strength and durability are required.

MANSONIA (*Mansonia altissima*), despite its specific name, is a medium sized tree supplying logs 3.5 m long by 750 mm in diameter. The wood is similar in color to American black walnut but is somewhat finer in texture. Unfortunately the color fades quickly on exposure. Mansonia is moderately stable, very resistant to decay and seasons without difficulty; fairly strong for its weight, it has a density around 0.6 g cm^{-3}. It is a good working wood which does not blunt cutting edges. It is used as a substitute for walnut in furniture, superior joinery and domestic flooring. In any abrasive work it quickly causes severe nasal irritation; the use of masks is necessary.

MISSANDA (*Erythrophleum guineense* and *E. ivorense*) is a large tree up to 37 m in height and 1.5 m in diameter, not of good shape for conversion. It is also known as tali, erun, sasswood, potrodom, kassa and muave. The wood is very hard and heavy, with a density of around 0.90 g cm^{-3}, coarse textured and deep reddish brown in color. It is difficult to season, being liable to distortion, but is very strong and very durable. Not an easy wood to work, it is used locally for heavy construction, harbor installations and railway sleepers. It is exported mainly as flooring strips for floors subjected to heavy wear.

MTAMBARA (*Cephalosphaera usambarensis*) is a large tree of the same family as Ilomba occurring in Tanzania, 25 m in height and in diameter up to 1.8 m above the buttresses. The wood is related and similar to banak (*Virola* spp.) from South America. Fine to medium in texture, it is a plain wood, somewhat heavier than mahogany, with a density of about 0.60 g cm^{-3}. It seasons slowly and is moderately stable in service, with average strength for its weight, and is rated as nondurable. Working qualities are good and it is used for nondecorative furniture and interior joinery. Logs are suitable for rotary-cut veneers for plywood.

MUHIMBI (*Cynometra alexandri*) occurs in Uganda, Tanzania and Zaire, and has a clear bole up to 12 m long and 0.75 m in diameter. The wood is fine textured and heavy, with a density of about 0.90 g cm^{-3} and a dull reddish brown color. It seasons with some tendency to surface checking, but is very durable and possesses high strength values. Difficult to work, it has a severe effect on cutting edges. Muhimbi is used locally for heavy construction work where strength and durability are important. It is exported in the form of flooring strips for heavy pedestrian traffic and ballroom floors.

MUHUHU (*Brachylaena hutchinsii*) is a small tree, common in East Africa, supplying logs up to 6 m long and 450 mm in diameter. The wood is very hard and heavy, with a density of about 0.91 g cm^{-3} when seasoned, and is of a pleasing yellowish brown color. There is a degree of surface checking in seasoning, but the wood is exceptionally stable in service as flooring and is very durable. Not an easy wood to work, it is used mostly for flooring strips and blocks, for which it gives outstanding wear. It is used in East Africa for decorative and characteristic carving of graceful animals.

MUNINGA (*Pterocarpus angolensis*), also called ambila, mukwa and kajat, is a medium sized tree yielding logs up to 3 m long and 350 mm in diameter. The wood is very handsome in appearance, of golden or reddish brown color and moderate in texture and weight, with a density of 0.64 g cm^{-3} when seasoned, of excellent working qualities and exceptionally stable. It is available in limited quantities, since its unrivalled qualities are well known in the countries of origin in East Africa. It is eminently suitable for the highest class of work, such as furniture, decorative joinery and fitments, and is also manufactured into veneer. Its response to varying conditions of humidity has been found to be the lowest of any wood.

NIANGON (*Tarrietia utilis*) or nyankom, common in the coastal forests of West Africa, is a medium sized tree with a bole up to 18 m long and 0.9 m in diameter. The wood is mahogany colored, somewhat more open in texture than mahogany and heavier, with a density of about 0.64 g cm^{-3} when seasoned. It works well with hand and machine tools and feels distinctly greasy when handled. Seasoning is not difficult and the wood is moderately stable in use. It is fairly durable and very resistant to preservative treatment. It is suitable for all purposes for which mahogany is used, including boatbuilding and flooring not exposed to heavy wear.

OBECHE (*Triplochiton scleroxylon*) or wawa is a common tree in West Africa, supplying wood which is one of the most important in industry. It is available in large sizes up to 1.8 m in width. The wood is light in weight, with a density of 0.38 g cm^{-3} when seasoned, open textured and of a white to pale straw color. Rapid seasoning is necessary to avoid discoloration and the wood is not durable and not readily impregnated with preservatives. Obeche works easily with hand and machine tools, but cutting edges must be sharp in end-grain working. It is used in large quantities for general cabinet work, interior joinery and kitchen furniture and is generally of good appearance and free from defects.

OGEA (*Daniellia ogea*) also called oziya and faro, is a large tree, over 30 m in height, giving logs up to 2 m in diameter, and is common in southern Nigeria. The wood is light in weight, with a density of about 0.48 g cm^{-3}, pinkish brown in color with darker streaks and rather coarse textured with wide sapwood. It seasons readily and is moderately stable in service, but the wood is perishable. Machine-planed surfaces are apt to be rather woolly. It is a utilitarian wood, useful for temporary constructions, boxes and packing cases, and suitable for rotary-cut veneer for plywood manufacture. Figured logs are of value for decorative veneers.

OKAN (*Cylicodiscus gabunensis*) or denya, distributed in the rain forests of West Africa, is a large tree with a clear bole up to 24 m long and over 1 m in diameter. The wood is very hard and heavy, with a density of 0.96 g cm^{-3} when seasoned, and is of an unusual greenish brown color, darkening on exposure to deep brown. It is coarse textured with interlocked grain. It is rated as very durable, being resistant to termites and marine borers. It is used for heavy constructional work under exposed conditions such as marine piling, dock gates, piers and jetties, and is suitable for heavy duty flooring.

OKWEN (*Brachystegia nigerica* and *B. kennedyi*) is common in southern Nigeria and is up to 1.25 m in diameter. The proportion of sapwood is unusually large. The wood resembles mahogany but is less attractive, somewhat darker in color and heavier, with a density of around 0.70 g cm^{-3} when seasoned. It shows a striped figure on quarter sawn surfaces. Its durability is moderate and the wood is very resistant to methods of impregnation. Okwen seasons with difficulty and there is considerable distortion. It is not an easy wood to work and is used for rough construction and temporary work where resistance to insect and fungal attack is unimportant.

OLIVE (*Olea hochstetteri*) or musharagi is a large tree up to 28 m tall, different in size and shape from the olive trees of the Mediterranean, and yields logs up to 1 m in diameter. The wood is very outstanding in quality and very attractive in appearance, being light brown in color and heavily ornamented with dark brown lines and streaks. Fine textured, it is very hard and heavy, with a density of 0.88 g cm^{-3} when seasoned. Seasoning is slow with some splitting and distortion. Olive is very durable and is especially suitable for heavy duty parquet flooring in public buildings. It is also good for ornamental turnery, carving and decorative work in general.

OPEPE (*Nauclea diderrichii*) also known as kusia and badi, is a large unbuttressed tree yielding logs up to 1.25 m in diameter. The wood is very yellow in color, fairly coarse textured and characterized by a wavy figure. It is fairly heavy, with a density of 0.75 g cm^{-3} when seasoned. It is stable in service, but surface checking is common during seasoning. A strong durable wood, it is in demand for work in exposed situations such as piers, decking, bridges and piles. Its color and figure offer attractive possibilities for turnery and decorative work generally, and it makes hardwearing flooring, but the color is rather bright for large areas.

PTERYGOTA (*Pterygota bequaertii* and *P. macrocarpa*) or awari is a large tree in the rain forests of West Africa, yielding logs up to 1.25 m in diameter. The wood is uniform in color, pale yellow to almost white, similar to yellow sterculia of the same family. Fairly coarse textured, it is moderately light in weight, its density varying from 0.56 to 0.66 g cm^{-3} when seasoned. Its strength varies with the density. It is readily impregnated with preservatives but is naturally rather perishable and prone to discoloration. A rather plain wood, it is of value for interior joinery and utilitarian furniture. It is also useful for plywood manufacture.

PENCIL CEDAR (*Juniperus procera*) occurs mainly at high altitudes in Kenya. The tress are large, reaching a height of 36 m and a diameter of 1.5 m. The bole is often fluted. The wood is attractive in its red color and fine texture and is rather heavier than the American species, with a density of around 0.58 g cm^{-3} when seasoned. It is classed as durable but is not often used for exposed work. The characteristic odor is pronounced. It is a delightful wood to work with hand tools but is exported mainly as pencil slats. It is used locally for joinery and furniture manufacture and in the UK occasionally for greenhouses, and it is also manufactured into veneers for wardrobe linings.

PODO (*Podocarpus* spp.) is found at high altitudes in East Africa. The trees are up to 30 m in height and 0.65 m in diameter. The wood is indistinguishable by species and is exceptionally smooth, clean and uniform in appearance. The density is about 0.50 g cm^{-3} when seasoned. Podo is very pleasing to work with hand tools, but needs care and very sharp edges to cut cleanly across end grain. It is useful for high class joinery, food containers and in large sizes for interior fittings. The lack of growth rings renders podo very different in appearance from northern softwoods (conifers). It is not durable but can be readily impregnated with preservatives.

"RHODESIAN TEAK" (*Baikiaea plurijuga*) is a small tree 15 m in height and 0.75 m diameter from Zimbabwe and adjoining countries. The wood is exported chiefly as flooring blocks and is not related to true teak. Its handsome color, dark reddish brown with black markings, its hardness and weight (density 0.90 g cm^{-3}), its stability and resistance to wear and its durability all add up to its being a preeminent wood for decorative flooring. Used for this purpose with woods of different color, for instance olive and muhuhu, the wood could almost rival marble. Locally it is used for furniture, heavy construction work and railway sleepers.

SAPELE (*Entandrophragma cylindricum*) is a large tree with a straight cylindrical bole clear of branches to a height of 30 m and up to 1.25 m in diameter. The wood is very popular, of finer texture than African mahogany and appreciably heavier, with a density of 0.64 g cm^{-3} when seasoned, and the well known stripe on quarter sawn surfaces is characteristic. Some logs are highly figured and are valuable for decorative veneers. Sapele seasons slowly and there is often a measure of distortion; it is fairly durable but resistant to impregnation. It is used for furniture of the highest class and is strong enough for staircases and fittings in public buildings.

TCHITOLA (*Oxystigma oxyphyllum*) is a large tree yielding cylindrical logs up to 1 m in diameter. It is exported as logs and as veneer. The wood is also known as tola, not to be confused with agba or tola branca. It is a decorative wood of reddish brown color with darker markings. Moderately coarse in texture, it is of medium weight, with a density of 0.60 g cm^{-3} when seasoned. Reported to be variable in durability and strength, it seasons readily and is stable in use. It works without difficulty and is used for decorative woodwork, mostly in veneer form.

UTILE (*Entandrophragma utile*) or mufumbi, distributed over equatorial Africa, is a large tree clear of branches for 24 m and up to 1.8 m in diamter. Akin to sapele but very slightly more open in texture, somewhat deeper in color and less refractory in seasoning, it lacks the cedarlike scent of sapele. It is moderately durable and resistant to impregnation. It is used for similar purposes as sapele, the two species being alike. At its best utile is a handsome wood and selected logs are cut into veneer which is highly decorative. Plain logs are used for plywood and it is a very popular wood used for a wide range of work.

WENGE (*Millettia laurentii*) or dikela is a small tree from Zaire up to 0.6 m in bole diameter. The wood is dark brown and lighter brown in close and narrow bands. On flat sawn surfaces this is a decorative feature. It is hard, fairly coarse textured and heavy, with a density of around 0.80 g cm^{-3} when seasoned. Reported to season slowly, it is stable in service, has high strength and stiffness and is very durable. For a hardwood it works without difficulty. It is used for cabinet making, interior decorative joinery, parquet flooring and fancy goods. Panga panga, *Millettia stuhlmannii* is very similar. It is more popular in continental Europe than in the UK.

ZEBRANO (*Microberlinia brazzavillensis*), also called zebrawood, is exported from Gabon and Cameroun as short logs up to 1 m in diameter. The wood is almost unique in its pale brown color, handsomely marked with dark brown lines and stripes. This coloration varies considerably from one tree to another. Zebrano is used almost exclusively for decoration, chiefly as veneer. It does not season very readily but is usually cut into quarter sawn boards and flitches. The veneers need to be handled with care to avoid splitting. The density, around 0.70 g cm^{-3}, is similar to that of the allied genus *Berlinia*, which is sometimes called rose zebrano, but the two woods are quite different.

See also: Resources of Timber Worldwide

Bibliography

Bolza E, Keating W G 1972 *African Timbers—The Properties, Uses and Characteristics of 700 Species*. Division of Building Research, CSIRO, Melbourne

Brown W H 1978 *Timbers of the Wolld*, Vol. 1: *Timbers of Africa*. Timber Research and Development Association, High Wycombe, UK

Chudnoff M 1984 *Tropical Timbers of the World*, Agriculture Handbook No. 607. US Department of Agriculture, Washington, DC

Desch H E 1981 *Timber: Its Structure and Properties*, 6th edn. (revised by J M Dinwoodie). Macmillan, London

Lincoln W A 1986 *World Woods in Color*. Macmillan, New York

Rendle B J (ed.) 1969 *World Timbers*, Vol. 1: *Europe and Africa*. Benn, London

C. W. Bond
[Leamington Spa, UK]

Timbers of Australia and New Zealand

The distinctive features of the trees of Australia and New Zealand are the preponderance of hardwood species in the native forests and the rapid enlargement of pine plantations to provide future needs. The climatic conditions in the forested areas of both countries cover the whole range except for extreme cold.

1. Australia

In Australia the total area of productive or potentially productive native forests and plantations is estimated at around 43 million hectares, which is 5.6% of the total land area.

The *Eucalyptus* genus is by far the most important of Australian forest trees. Its species dominate most forest areas and are scattered over much of the open woodland. A wide variety of hardwoods is produced from these species—timbers which display a considerable range of characteristics such as color, weight, hardness, toughness, strength, elasticity, durability and fissibility. Because of this diversity of properties, eucalypt timbers have innumerable uses ranging from heavy structural purposes to paper manufacture and including those where the finest appearance is paramount. While there are over 450 known species of eucalypts, only about 40 are regarded as commercial varieties. Of these, the most plentiful are mountain ash, messmate, alpine ash (which three are included amongst the ash group and are also known collectively as Tasmanian oak), jarrah, karri, blackbutt and spotted gum.

Relatively few conifers occur naturally in Australia. The formerly extensive stands of hoop pine (*Araucaria cunninghamii*) are largely gone but plantations of this species are now being established. Cypress pine (*Callitris* spp.) is still relatively plentiful, but it grows sparsely over a very wide area. There are several noneucalypt hardwoods that are commercially important.

The most successful plantation species in Autralia is radiata pine (*Pinus radiata*) and about 800 000 hectares have been established over a wide geographical range, making up approximately 70% of the total coniferous plantations. Caribbean pine (*P. caribaea*), slash pine (*P. elliottii*) and hoop pine plantations are established mainly in Queensland, and maritime pine (*P. pinaster*) in Western Australia. Some eucalypt plantations have also been established, but the properties of wood from fast grown trees may differ from that of the slow grown.

It has been estimated that Australia could be self-sufficient in wood requirements around the year 2000, mainly through the continuation of the softwood plantation program. Currently about 25% of the total sawn wood consumption is made up of imports, mainly from Canada, the USA, New Zealand, and Malaysia. Australia is not a large exporter of forest products by world standards, but at times such markets have been important in the local regions. In recent years a significant export market in eucalypt woodchips to Japan has been developed.

The present article covers the major commercial species, for which Table 1 gives mean values for density, modulus of rupture and modulus of elasticity, all at 12% moisture content; where available, the 95% probability range for density is also included.

Table 1
Physical and mechanical properties of major commercial species[a]

	Density (g cm^{-3})		Modulus of elasticity (MPa)	Modulus of rupture (MPa)
	mean	range[b]		
Ash group of eucalypts	0.67	0.48–0.86	15500	115
Blackbutt	0.86	0.72–1.00	18800	145
Red River gum	0.855	0.735–0.975	11160	100
Spotted gum	0.97	0.745–1.08	18800	142
Red ironbark	1.06	0.91–1.22	16950	150
Jarrah	0.805	0.69–1.00	12950	112
Karri	0.89	0.79–0.985	19000	130
White cypress pine	0.67	0.55–0.76	8960	80
Radiata pine (Aust.)	0.53	0.410–0.660	11480	87
Podocarp group		0.45–0.63	6950–9650	50–80
Tawa	0.64		13900	95
Rimu	0.60		8700	70

[a] At 12% moisture content [b] 95% probability

MOUNTAIN ASH (*Eucalyptus regnans*), ALPINE ASH (*E. delegatensis*) and MESSMATE (*E. obliqua*) are the most important commercial species in the ash group of eucalypts. The timbers of the group are all pale brown with a pinkish tinge. They have an open texture and a straight grain and are of moderate durability and hardness. They may be dried readily but have a tendency to check and are subject to collapse. However, good recovery is achieved by reconditioning. Shrinkage is high. These timbers are used for a wide range of products including framing, general construction, flooring, joinery, cabinet work, furniture, wood wool and pulp for paper.

BLACKBUTT (*Eucalyptus pilularis*) yields timber with a light brown to brown color with an occasional pink tinge. The wood has an open texture and is usually straight grained. It seasons well with some collapse but responds to reconditioning. Fairly durable and strong, it is readily worked with hand or machine tools. It forms a most useful general purpose timber suitable for most structural applications, flooring, cladding, poles, crossarms and fencing.

RED RIVER GUM (*Eucalyptus camaldulensis*) is a red timber with its grain often interlocked and wavy. The texture is close and it may exhibit a pleasing figure. The timber is durable in contact with the ground, with a good resistance to termite attack. It is a somewhat difficult timber to season without excessive degradation caused by checking and distortion. It has a high shrinkage. It is most suitable for structural uses, especially where durability is an important consideration.

SPOTTED GUM (*Eucalyptus maculata*) is timber varying in color from light to dark brown. The sapwood is white, often exceeding 50 mm in width, and is highly susceptible to *Lyctus* borer. The grain may be straight or interlocked and occasionally wavy. The texture is open and coarse. The timber has a greasy feel. It is fairly durable and is one of the heavier eucalypts. Care is necessary in drying, to avoid checking and distortion. Spotted gum is a good structural timber used where strength, hardness and shock resistance are required such as in bridge construction, heavy framing, boatbuilding and agricultural machinery. It is the most satisfactory available Australian timber for tool handles.

RED IRONBARK (*Eucalyptus sideroxylon*) yields a dark red timber with a narrow, pale yellow sapwood. It has a moderately fine texture with interlocked grain. The timber is very hard, dense, strong and durable. It is slow drying but is generally used in the green condition in large sizes. It is employed for all forms of heavy construction where strength and durability are important, including piles and poles.

JARRAH (*Eucalyptus marginata*) produces timber which is reddish-brown in color but may vary from pink to dark red with a narrow, pale-colored sapwood. It is coarse in texture and the grain is commonly interlocked. Kino veins or pockets may be present. Hard and heavy, jarrah has a good reputation for durability in the ground and is resistant to termites. Care must be taken in seasoning to avoid distortion or checking. The timber finishes and polishes well. Its fire resistance is well known. Jarrah is an extremely versatile timber, finding applications ranging from all types of construction to high-class furniture. In particular, jarrah flooring is world-renowned for its beauty and durability.

KARRI (*Eucalyptus diversicolor*) has a reddish brown wood closely resembling jarrah in appearance but generally lighter in color. The texture is coarse and the grain is often interlocked. Care is necessary in seasoning to avoid checking. Shrinkage is high. It is a hard, heavy, stiff and tough timber with somewhat lower durability than jarrah. Owing to its high strength and availability in large sizes and long lengths, karri is well known in Australia and overseas as an excellent wharf and bridge timber. It is also used in house framing, flooring, agricultural implements, shipbuilding, mine guides, furniture and plywood.

WHITE CYPRESS PINE (*Callitris columellaris*) is the most plentiful of the small number of conifers native to Australia. The color of the wood varies from light yellow to dark brown. It has a characteristic resinous odor. Small firm knots are common. The timber has very good resistance to decay and termite attack. Its shrinkage is very low, permitting it to be used often in the green or partly seasoned condition. It machines to a fine finish and takes a high polish. White cypress pine is used extensively for framing, cladding and flooring, and also for posts and transmission poles.

RADIATA PINE (*Pinus radiata*), although a species exotic to Australia and New Zealand, it has, through extensive plantings, become the most plentiful single species currently produced in both countries. The heartwood is light brown with a paler sapwood which may be up to 100 mm wide. Growth rings are distinct. The timber works well with hand or machine tools. It has a low durability but is normally preservative-treated if used in situations where it would otherwise be subject to deterioration. Radiata pine can usually be dried quickly with little degradation, but wood close to the pith requires special drying techniques to prevent distortion due to the presence of spiral grain. Radiata pine is a general purpose timber with its uses ranging from railway sleepers (treated) to furniture. Structurally it is suitable for framing, particularly for housing. In larger applications it is often used in glue-laminated form. Many large industries such as paper, pulp, sawmilling particleboard and plywood are based on the plentiful supply of radiata pine.

2. *New Zealand*

The native forests of New Zealand can be grouped into three main classes, but there are naturally transitional and subgroups within this classification. The principal classes are the podocarp–broadleaved forest, the kauri (*Agathis* spp.) forest and the beech (*Nothofagus* spp.) forest. The distribution of the podocarp–broadleaved forest tends to be universal, but it dominates in the north, in warm wet lowlands and on the lower mountain slopes. The principal podocarp is rimu. Other commercially important podocarps include totara, matai, miro and a lowland species, kahikatea. Kauri, which formerly dominated the forest of the northern regions, is now carefully managed and protected. Beech forests are dominated by the southern beech species and are located mainly in the South Island, in the higher-altitude areas of both islands and in the dry lowlands. The commercial exploitation of native forests is being reduced.

Coniferous species form the major proportion of the exotic species, although eucalypts and other hardwoods have recently been established for specialist purposes. The fast growing conifers, natives of North America, have thrived in the New Zealand environment, particularly radiata pine. This species has become dominant, as it has proved to be a resilient tree to manage, useful for a wide range of forest products and with a significantly superior growth rate compared with other exotic conifers. Douglas fir (*Pseudotsuga menziesii*) continues to be planted in the colder, more severe environments.

The total area afforested by exotic species is currently estimated to be in excess of 1 000 000 hectares and expanding.

RADIATA PINE (*Pinus radiata*) grown in plantations is the major commercial species in New Zealand. Its properties are very similar to those described for the Australian grown material.

KAHIKATEA or NEW ZEALAND WHITE PINE (*Podocarpus dacryaioides*), MATAI (*P. spicatus*) and TOTARA (*P. totara*) are among the podocarps of New Zealand. Their wood is mostly pale yellow to yellow–brown, although totara may be reddish brown. The grain is straight with fine texture and the growth rings are indistinct. The durability of the podocarps varies from moderate to high. Totara is renowned for its durability, particularly resistance to marine borers. All species are worked readily by hand or machine tools. The timbers generally dry well but there is some tendency to split. The podocarps are used mainly for framing, cladding, flooring and joinery. Totara is used for wharf and bridge construction and shipbuilding.

TAWA (*Beilschmiedia tawa*) is a white to yellow colored wood with a straight grain and fine texture. The heartwood is nondurable and the sapwood is susceptible to *Lyctus* borer attack. Tawa kiln dries fairly readily but with some tendency to checking. The timber is used as flooring due to its good wearing properties. Other uses include interior joinery, furniture, panelling and plywood.

RIMU (*Dacrydium cupressinum*) has sapwood of pale yellowish color, gradually darkening through an intermediate zone to a heartwood which is a light brown color. The timber has a straight grain and a fine even texture, and is moderately durable. It is worked easily by hand and machine tools. It dries readily, with a slight tendency to surface checking. Uses include cladding, flooring, furniture, interior joinery, turnery and panelling.

See also: Resources of Timber Worldwide

Bibliography

Boland D J, Brooker M I H, Chippindale G M, Hall N, Hyland B P M, Johnston R D, Kleinig D A, Turner J D 1984 *Forest Trees of Australia*, 4th edn. Nelson/CSIRO, Melbourne

Bolza E, Kloot N H 1963 *The Mechanical Properties of 174 Australian Timber*, Technical Paper No. 25. Commonwealth Scientific and Industrial Research Organization, Division of Forest Products, Melbourne

Bootle K R 1983 *Wood in Australia: Types, Properties and Uses*. McGraw-Hill, Sydney
Brown W H 1978 *Timbers of the World*, Vol. 8: *Timbers of Australasia*. Timber Research and Development Association. High Wycombe
Building Research Establishment 1972 *A Handbook of Hardwoods*, 2nd edn. Her Majesty's Stationery Office, London
Building Research Establishment 1977 *A Handbook of Softwoods*, 2nd edn. Her Majesty's Stationery Office, London
Department of Primary Industry 1986 *Australian Forest Resources* 1985. Australian Government Publishing Service, Canberra
Ditchburne N, Kloot N H, Rumball B 1975 *The Mechanical Properties of Australian-Grown Pinus radiata*, Technical Paper, 2nd Series, No. 9. Commonwealth Scientific and Industrial Research Organization, Division of Building Research, Melbourne
Hillis W E, Brown A G (eds.) 1984 *Eucalypts for Wood Production*, 2nd edn. Commonwealth Scientific and Industrial Research Organization/Academic Press, Melbourne
Kloot N H, Bolza E 1977 *Properties of Timbers Imported into Australia*, Technical Paper, 2nd series, No. 17. Commonwealth Scientific and Industrial Research Organization, Division of Building Research, Melbourne
Standards Association of Australia 1986 *Timber Classification into Strength Groups*. AS2878–1986. Standards Association of Australia, Sydney

W. C. Keating and W. E. Hillis
[Commonwealth Scientific and Industrial Research Organization, Clayton, Victoria, Australia]

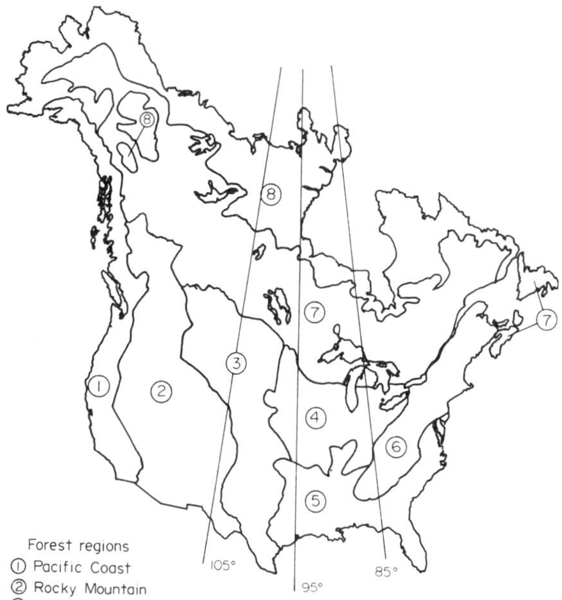

Figure 1
Major forest regions of Canada and the USA

Timbers of Canada and the USA

Canada and the USA have some of the largest and most diverse temperate and boreal forests in the world. Great softwood (gymnosperm) and hardwood (angiosperm) forests extend from the polar regions of the north to the Mexican border on the south. The timber volume and land area occupied by softwood forests in Canada and the USA are second only to those of the vast reserves in the USSR (*World Wood* 1980). The hardwood forests, although not as great as the softwood forests, are very diverse in species composition—only those of mainland China are more diverse among temperate hardwood forests.

The Canadian and US hardwood and softwood forests provide rich reserves of timber. The species and characteristics of the timber are molded by diverse climatic, topographic and edaphic conditions into a variety of raw materials for many products and uses.

This article considers 35 of the most important Canadian and US species or species groups, comparing their properties and uses. Other species are mentioned without detail.

Figure 1 shows a division of Canada and the USA into eight major forest regions of similar species and ecological conditions (Sargent 1933, Brockman 1968, Elias 1980). The distribution of species is referred to in terms of these major regions in most cases.

1. Softwoods

1.1 Cedars

Numerous species are commonly called cedar. These include trees of the genera *Chamaecyparis*, *Juniperus*, *Libocedrus* and *Thuja*. Woods of all these genera are aromatic, and their heartwoods have natural decay resistance. The most important species is Western red cedar, or Pacific red cedar (*Thuja plicata* Donn. ex D. Don). It is found in the Pacific Coast Region and the north-central portion of the Rocky Mountain Region. The sapwood is white and the heartwood is rusty-red to brown and very decay resistant. The wood is of straight, even grain and medium texture; it machines well, but is low in nail-holding ability. The average specific gravity is 0.32. The wood is low in strength. The principal uses are for transmission poles, shingles, siding, posts, laminated roof decking and panelling.

1.2 Douglas-Fir

Douglas-fir is found in the Pacific Coast and Rocky Mountain areas: Pacific Coast Douglas-fir (*Pseudotsuga menziesii* (Mirb.) Franco var. *menziesii*) and Rocky Mountain Douglas-fir (*P. menziesii* var. *glauca* (Beissn.) Franco). The sapwood is whitish to pale yellowish whereas the heartwood is pinkish to dark reddish. The wood is fairly difficult to work, paints poorly and is difficult to treat with preservatives. The average specific gravity is 0.45 for coastal and northern Rocky Mountain areas and 0.43 for the southern Rocky Mountain area. The wood is strong in relation to its weight and quite stiff. Douglas-fir is one of the most important timber species of Canada and the USA. The major uses are for structural lumber, plywood, laminated beams, piling and poles. It is also used for sash and doors, molding, flooring and pulpwood.

1.3 Pines

There are 36 native species of pines found in Canada and the USA (Critchfield and Little 1966).

(*a*) *Southern pine*. Four major southern pines are marketed together. They are loblolly pine (*Pinus taeda* L.) shortleaf pine (*P. echinata* Mill.), longleaf pine (*P. palustris* Mill.) and slash pine (*P. elliottii* Engelm.).

These species are found in the Southern Pine Region, with loblolly and shortleaf throughout the region and longleaf and slash on the coastal plain and piedmont. The woods of these species cannot be distinguished in lumber form. The sapwood is whitish, the heartwood reddish brown to brown. The average specific gravity varies with species from 0.50 to 0.55. The wood, strongest of the North American pines, is high in strength and stiffness. It does not hold paint well but has good nail-holding ability and glues well. It is hard to work but is easily treated with preservatives. The principal uses are structural lumber, plywood and pulpwood. Other uses are for posts, poles, piling, railroad ties, boxes, millwork and flooring.

Minor southern yellow pines are Virginia pine (*P. virginiana* Mill.), pitch pine (*P. rigida* Mill.) pond pine (*P. serotina* Michx.), spruce pine (*P. glabra* Walt.), sand pine (*P. clausa* (Chapm.) Vasey) and Table Mountain pine (*P. pungens* Lamb.).

(*b*) *Northern pines*. Three pines are found in Canada and the northeastern USA: eastern white pine (*P. strobus* L.), jack pine (*Pinus banksiana* Lamb.) and red pine (*P. resinosa* Ait.). As structural lumber these three species are grouped together as "northern pines" or in the spruce–pine–fir category. Eastern white pine is the most important species.

Eastern white pine is found in the southeastern portion of the Boreal Region and throughout the Appalachian Region. The sapwood is white to pale yellowish white, and the heartwood is cream to light brown. The wood is straight grained and of medium uniform texture. The average specific gravity is 0.32. It is moderately weak in strength. The wood dries well, works easily, holds paint well, glues well and is moderate in nail-holding ability. Major uses are for furniture, shop lumber, millwork, pattern stock, sash and doors, and structural lumber.

(*c*) *Western pines*. Of about twenty pine species found west of the 95th meridian, only four are of major commercial importance. Ponderosa pine is the most important species.

Ponderosa pine (western yellow pine (*P. ponderosa* Dougl. ex Laws.)) and jeffrey pine (western yellow pine (*P. jeffreyi* Grev. and Balf.)) are marketed together as ponderosa pine. Jeffrey pine is found in the Sierra Nevada Mountains of California. Ponderosa pine is widely distributed in the Pacific Coast Region, the Rocky Mountain Region and east into the Central Prairie Region. The sapwood is whitish and the heartwood is yellowish to light reddish to orange–brown. The wood works and dries easily, glues well and is average in nail-holding ability. The average specific gravity is 0.37. The wood is moderately weak in strength. Principal uses are for molding and millwork, shop lumber, sash and doors, and structural lumber. Other uses are for boxes, furniture and novelties.

Other commercial species of western pine are lodgepole pine (*P. contorta* var. *latifolia* Engelm.) sugar pine (*P. lambertiana* Dougl.), and western white pine, (Idaho white pine (*P. monticola* Dougl. ex D. Don.)).

1.4 Redwood

Coast redwood (*Sequoia sempervirens* (D. Don.) Endl.) is found in northwestern California in the Pacific Coast Region. Although redwood is limited in range, it is an important species. The sapwood is white, the heartwood is pinkish red to deep reddish brown and is decay resistant. The wood is straight and coarse grained. It works easily, holds paint well and is intermediate in nail-holding ability. The average specific gravity is 0.38. The wood is moderately low in strength and is moderately stiff. It is used for siding, decking, molding, structural lumber and tanks and vats.

1.5 Spruce

Five species of Canadian and US spruce are of commercial importance. Eastern spruces are white spruce (*Picea glauca* (Moench) Voss), black spruce (*P. mariana* (Mill.) B. S. D.), and red spruce (*P. rubens* Sarg.). Black spruce and white spruce are transcontinental in distribution in the Boreal Region and extend into the northern Appalachian Region as well. Red spruce is found in the Appalachian Region at high elevations. Western spruces include engelmann spruce (*Picea engelmannii* Parry ex Engelm.), found at high elevations throughout the Rocky Mountain Region; and sitka spruce (*Picea sitchensis* (Bong.) Carr.), found from southern Alaska to northern California in the Pacific Coast Region. The wood of all these spruces is very similar. The sapwood is white to light yellow; the heartwood is light yellow to brown. The wood is straight grained, fine textured and easily worked. It paints and glues well and dries easily. The average specific gravity is 0.35–0.41. It is moderate in strength but has a very high strength-to-weight ratio and is moderately stiff. The major uses are structural lumber and pulpwood. Other uses include aircraft parts, musical instruments, boat masts and spars, and factory lumber.

1.6 True Firs

Western species are grouped as "white fir" or in the spruce–pine–fir or Hem-fir categories for structural lumber. Eastern species are marketed primarily in the spruce–pine–fir category.

Western firs include: Pacific silver fir (*Abies amabilis* Dougl. ex Forbes), California red fir (*A. magnifica* A. Murr.) and noble fir (*A. procera* Rehd.), which are confined to the Pacific Coast Region; and white fir or concolor fir (*A. concolor* (Gord and Glend.) Lindl. ex Hildebr.), grand fir (*A. grandis* (Dougl. ex D. Don) Lindl.), and subalpine fir (*A. lasiocarpa* (Hook.) Nutt.), which are found in both the Rocky Mountain Region and the Pacific Coast Region. The wood is generally white to reddish brown, with little difference between sapwood and heartwood. It is straight grained, medium in texture and easy to work. The average specific gravity varies from 0.35 for grand fir to 0.42 for noble fir. The wood is moderate in strength and moderately stiff. Primary uses are for structural lumber, plywood and pulpwood. Other uses are boxes, sheathing, pallets and laminated products.

Eastern true firs include balsam fir (*Abies balsamea* (L.) Mill.) and fraser fir (*A. fraseri* (Purch) Poir.), and are similar to western firs in use.

1.7 Western Hemlock

Western hemlock or Pacific hemlock (*Tsuga heterophylla* (Raf.) Sarg.) is found primarily in the Pacific Coast Region and in the northcentral Rocky Mountain Region. The sapwood and heartwood are nearly the same color, whitish to light yellowish brown with a purplish to reddish brown hue. The wood is straight, even grained and has medium texture. It machines moderately well, glues easily and holds paint moderately well. The average specific gravity is 0.42. The wood is moderately strong and stiff. Principal uses are for structural lumber, plywood, molding and millwork, siding, ladder rails and pulpwood. Related species of commercial use are Eastern hemlock (*T. canadensis* (L.) Carr.) and Mountain hemlock (*T. mertensiana* (Bong.) Carr.).

1.8 Western Larch

Western larch (*Larix occidentalis* Nutt.) is found in the northern half of the Rocky Mountain Region at elevations between 600 and 2000 m. The sapwood is whitish to straw colored, the heartwood reddish brown. Average specific gravity is 0.48. The wood is strong in all mechanical properties and is often marketed with Douglas-fir. Major uses include structural lumber, transmission poles, plywood and pulpwood. A related species, Tamarack (*L. laricina* (DuRoi) K. Koch), found throughout the Boreal Region, is similar to western larch but is only moderately strong.

2. *Hardwoods*

2.1 Aspen

Two species are generally called aspen: trembling aspen, also called quaking aspen, or popple (*Populus tremuloides* Michx.), and bigtooth aspen or largetooth aspen (*P. grandidentata* Michx.). Trembling aspen is transcontinental in the Boreal Region, also extending into every other region except the Southern Pine Region. Bigtooth aspen is limited to southeastern Canada in the Boreal Region, the northern half of the Appalachian Region and the northwestern portion of the Central Hardwood Region. The sapwood is nearly white, the heartwood whitish to light brown. The wood has fine even texture and is generally straight grained. The average specific gravity is 0.39 for trembling aspen and 0.37 for bigtooth aspen. The wood is moderately low in strength and moderately stiff. It has moderate nail-holding ability. Principal uses are pulpwood, flakeboards and lumber.

Related species with similar appearance and properties are eastern cottonwood (*P. deltoides* Bartr. ex Marsh.) and black cottonwood (*P. trichocarpa* Torr. and Gray).

2.2 Black Cherry

Black cherry (*Prunus serotina* Ehrh.) is found throughout most of the eastern half of the USA and in the most southern portions of Ontario and Quebec. The sapwood is whitish to light reddish brown and is narrow, the heartwood is light to dark reddish brown. The wood is generally straight grained, works well, finishes smoothly, glues well and is stable. The average specific gravity is 0.47. Principal uses are lumber and veneer for furniture, panelling, musical instruments and trim. This beautiful wood is valued for its color and texture.

2.3 Black Walnut

Black walnut (*Juglans nigra* L.) is found primarily in the Central Hardwood Region, the southern portion of the Appalachian Region and into the Southern Pine Region. The sapwood is nearly white, the heartwood chocolate brown. The wood is straight grained to highly figured. The average specific gravity is 0.51. The wood is strong, stiff, machines well and finishes beautifully. It glues well and has good nail- and screw-holding ability. The major uses are for lumber and veneer used in furniture, cabinetry and panelling, and for turnery, gun stocks and novelties. Black walnut is one of the most valuable woods of the world.

2.4 Hard Maple

This category includes sugar maple, rock maple (*Acer saccharum* Marsh.) and black maple (*A. nigrum* Michx. f.) Both species are found throughout the Central Hardwood Region and sugar maple in the southeastern portion of the Boreal Region and northern portion of the Appalachian Region. The sapwood is white; the heartwood is light reddish brown. The wood is uniform in texture, and may be straight grained or highly figured. The average specific gravity is 0.64. The wood is very strong and stiff. It machines well, glues moderately well, finishes well, holds screws and nails well, and dries easily. The major use is for lumber, which is manufactured into furniture, flooring, toys and sporting goods. Other uses are for musical instruments, woodenware, butcher blocks and millwork.

2.5 Hickory

Four major species of true hickory are found in the eastern USA: shagbark (*Carya ovata* (Mill.) K. Koch), shellbark (*C. laciniosa* (Michx. f.) Loud.), pignut (*C. glabra* (Mill.) Sweet) and mockernut (*C. tomentosa* Nutt.). These species are found in the Central Hardwood, Appalachian and Southern Pine Regions. The sapwood is whitish, the heartwood pale brown to reddish brown. The wood is coarse textured and straight grained. The average specific gravity varies from 0.60 to 0.68. The wood is very strong and has extraordinarily high shock resistance. It machines well, finishes well, has good nail- and screw-holding ability, but does not glue well. The principal uses are lumber and veneer for furniture, cabinets, panelling, tool handles, sporting goods and ladder rungs. Other uses include railroad ties, flooring and charcoal. Similar species called pecan or pecan-hickory include bitternut hickory (*C. cordiformis* (Wangenh.) K. Koch), pecan (*C. illinoensis* (Wangenh.) K. Koch) and nutmeg hickory (*C. myristicaeformis* (Michx. f.) Nutt.).

2.6 Oaks

The oaks are the predominant hardwood tree species of Canada and the USA, making up 35% of the total timber volume. There are two groups of oaks, the red oaks (subgenus *Erythrobalanus*) and white oaks (*Leucobalanus*). The woods within each of these subgenera (see Table 1) cannot be separated and will be described as one wood.

(*a*) *Red Oaks.* The range of red oaks covers southern Ontario, Quebec and the eastern USA, except for Florida. The sapwood is whitish, whereas the heartwood is pinkish to reddish brown. The wood has few or no tyloses in the pores, and has broad multiseriate rays. It is straight grained, machines well, finishes well, holds nails and screws well and is difficult to dry. The average specific gravity varies from 0.50 to 0.65. It is strong and stiff. The major uses are lumber and veneer for furniture, and pulpwood. Other uses include millwork, flooring, railroad ties, pallets, boxes and piling.

(*b*) *White Oaks.* The distribution of the white oaks is the same as for the red oaks, except that bur oak extends into Manitoba. The sapwood is whitish to light brown, the heartwood light to dark brown. The wood has tyloses in the pores, and has broad multiseriate rays. The average specific gravity varies from 0.60 to 0.70. The wood has high strength, is stiff and is quite resistant to decay. The principal uses are lumber and veneer for furniture, and pulpwood. Other uses are railroad ties, tight cooperage, flooring, piling and pallets.

Table 1
Common and specific names of red and white oaks

Common name	Specific name
Red oaks	
Scarlet	*Quercus coccinea* Muenchh.
Southern red	*Q. falcata* Michx.
Cherrybark	*Q. falcata* var. *pagodifolia* Ell.
Laurel	*Q. laurifolia* Michx.
Blackjack	*Q. marilandica* Muenchh.
Water	*Q. nigra* L.
Nuttall	*Q. nuttallii* Palmer
Pin	*Q. palustris* Muenchh.
Willow	*Q. phellos* L.
Northern red	*Q. rubra* L.
Shumard	*Q. shumardii* Buckl.
Black	*Q. velutina* Lam.
White oaks	
White	*Quercus alba* L.
Swamp white	*Q. bicolor* Willd.
Overcup	*Q. lyrata* Walt.
Bur	*Q. macrocarpa* Michx.
Swamp chestnut	*Q. michauxii* Nutt.
Chinkapin	*Q. muehlenbergii* Engelm.
Chestnut	*Q. Prinus* L.
Post	*Q. stellata* Wangenh.
Live	*Q. virginiana* Mill.

2.7 Red Alder

Red alder (*Alnus rubra* Bong.) is found in the Pacific Coast Region. The sapwood and heartwood are indistinguishable—white to cream when cut, changing to reddish brown or yellow brown when dry. The wood has fine even texture, is generally straight grained, works easily and dries easily. It glues well, holds paint well and is average in nail- and screw-holding ability. The average specific gravity is 0.37. Its strength is moderate. The major uses are for lumber and pulpwood. The lumber is used primarily for furniture.

2.8 Soft Maple

This category includes red maple (*Acer rubrum* L.) and silver maple (*A. saccharinum* L.). They are found in the Central Hardwood, Appalachian and Southern Pine Regions. The sapwood is white and wide; the heartwood is light brown to grayish. The wood is generally straight grained, of uniform fine texture, machines well, is moderately good for gluing, finishes well and dries well. The average specific gravity is 0.47 for silver maple and 0.53 for red maple. The wood is moderately strong. Its major uses are for lumber and pulpwood. The lumber is used for upholstered furniture framing, panelling, and planing-mill products, boxes, pallets and veneer.

2.9 Sweetgum

Redgum (*Liquidambar styraciflua* L.) is found throughout the Southern Pine Region. The sapwood is whitish; the heartwood varies through shades of red, reddish brown, brown and gray. The color variations often result in color-figure patterns. The wood texture is uniform, but the grain is interlocked, which may cause warping during seasoning. The average specific gravity is 0.46. The wood is moderately strong and stiff. It machines fairly well and has good nail- and screw-holding ability. Principal uses are for lumber and veneer for furniture, and pulpwood. Other uses are for pallets, boxes, crates and slack cooperage.

2.10 White Ash

White ash (*Fraxinus americana* L.) is found in the Central Hardwood, Appalachian and Southern Pine Regions. The sapwood is whitish, the heartwood brown to grayish brown. The wood is generally straight grained, dries easily and machines well. The average specific gravity is 0.56. The wood is strong and high in shock resistance. It has high nail- and screw-holding ability. The principal uses are for tool handle stock, lumber and veneer for furniture, sporting goods and implements.

Other ash species are green ash (*F. pennsylvanica* Marsh.), black ash (*F. nigra* Marsh.), blue ash (*F. quadrangulata* Michx.) and Oregon ash (*F. latifolia* Benth.).

2.11 Yellow Birch

Yellow birch (*Betula alleghaniensis* Britton) is found in the southeastern Boreal Region and in the Appalachian Region at higher elevations. The sapwood is whitish, the heartwood light to reddish brown. The wood has fine even texture, dries moderately well, finishes well and machines well. It is average in gluing and has good nail- and screw-holding ability. The average specific gravity is 0.55, and the wood is strong. The principal uses are for lumber and veneer in furniture and cabinetry. It is also used for toys, turnery and woodenware. Closely allied species are sweet birch (*B. lenta* L.), which is found in the Appalachian Region and is similar to yellow birch in properties and use, and paper birch (*B. papyrifera* Marsh.), which is transcontinental in the Boreal Region but is weaker and used primarily for pulpwood, lumber, clothes pins and toothpicks.

2.12 Yellow Poplar

Yellow poplar or "poplar" (*Liriodendron tulipifera* L.) is found in the Central Hardwood, Appalachian and Southern Pine Regions. The sapwood is whitish, and the heartwood is usually greenish brown but may be dark green, purple, black, blue, pink or yellow. The wood is straight grained and of uniform fine texture; it machines well, dries easily, glues well, and has average nail- and screw-holding ability. The average specific gravity is 0.42, and the wood is moderate in strength. The major uses are for lumber, veneer and pulpwood. Furniture is the primary use of lumber and veneer. Other uses include boxes, musical instruments, toys and sporting goods.

See also: Resources of Timber Worldwide

Bibliography

Brockman C F 1968 *Trees of North America: A Field Guide to the Major Native and Introduced Species North of Mexico.* Golden Press, New York

Critchfield W B, Little E L Jr 1966 *Geographic Distribution of the Pines of the World*, USDA Forest Service Miscellaneous Publication 991. US Government Printing Office, Washington, DC

Elias T S 1980 *The Complete Trees of North America: Field Guide and Natural History.* Van Nostrand Reinhold, New York

Everett T H 1971 *Living Trees of the World.* Doubleday, New York

Hosie R C 1969 *Native Trees of Canada*, 7th edn. Canadian Forestry Service, Ottawa, Ontario

Little E R Jr 1971 *Atlas of United States Trees*, Vol. 1: *Conifers and Important Hardwoods*, USDA Forest Service Miscellaneous Publication 1146. US Government Printing Office, Washington, DC

Little E R Jr 1979 *Checklist of United States Trees (Native and Naturalized)*, USDA Forest Service Agricultural Handbook No. 541. US Government Printing Office, Washington, DC

Panshin A J, deZeeuw C 1970 *Textbook of Wood Technology*, 3rd edn., Vol. 1, McGraw-Hill, New York

Sargent C S 1933 *Manual of the Trees of North America (Exclusive of Mexico).* Houghton Mifflin, Cambridge, Massachusetts

US Forest Products Laboratory 1987 *Wood Handbook: Wood as an Engineering Material*, USDA Agricultural Handbook No. 72, rev. edn. US Government Printing Office, Washington, DC

World Wood 1980, 21: 17–46. World wood review

R. R. Maeglin
[US Forest Products Laboratory, Madison, Wisconsin, USA]

Timbers of Central and South America

The forests of Latin America cover a vast area encompassing a diversity of plants and ecological habitats. They stretch from the West Indies and Mexico southward through Central and South America. Forests vary from open stands of pine in the mountains of northern Mexico through dense jungles in the Amazon basin, to temperate hardwood forests in southern Argentina and Chile. With a forest area of over 1000 million hectares, Latin America contains almost a quarter of all the world's forests. South America alone has 950 million hectares of forests and is the largest timber reserve in the western hemisphere.

The botanical composition of these forests is quite heterogeneous. In the lowland wet tropics there are 100 or more tree species per hectare. Of the many thousands of tree species in the South American tropical forests about 500 are more or less known to the timber trade. The great majority of these are hardwoods.

Many of the obstacles which hinder development of the timber resource are common throughout the region. The most important ones are the low concentrations of presently commercially valuable tree species. Poor transportation systems, difficult terrain and seasonal weather conditions are also obstacles.

Large tracts of forest land remain virtually undisturbed even though some species have been prized on world markets for centuries. Mahogany (*Swietenia* spp.) was appreciated by the early Spaniard explorers. The first recorded use was in 1514 during the construction of the cathedral of Santo Domingo. Extensive use of mahogany for fine furniture began to develop in the mid-1700s.

The wood of lignum vitae (*Guaiacum officinale*) has been an article of trade since the early 1500s for its medicinal properties. It was not until the 1800s that this wood, with its exceptional strength, tenacity and self-lubricating properties, was used for bearings or bushing blocks to line the stern tubes of propeller shafts of steamships. Other important and valuable woods that were early recognized by the Europeans as a source of dye were log wood (*Haematoxylon campechianum*) and Brazilwood (*H. brasiletto*), from which the country Brazil got its name. A closely related dyewood called pernambuco (*Guilandina echinata*) is now highly prized for violin bows. It is to the bow what spruce is to the sounding board and maple is to the back of a violin.

Brazilian rosewood (*Dalbergia nigra*) and West Indian satinwood (*Zanthoxylum flavum*) are exquisite woods known to commerce for about 350 years. By the mid-1800s, Central American rosewood (*Dalbergia stevensonii*) and cocobolo (*D. retusa*) were being exported to world markets. Balsa (*Ochroma pyramidale*), the lightest commercial wood, came into prominence during World War I, and large quantities were consumed in manufacture of life preservers, rafts and mine buoys. Today its insulating properties are utilized in the manufacture of refrigerated containers.

In the following paragraphs the properties and uses of some Latin American woods are summarized. For each wood the common and scientific name, range, general description of the tree and wood, and uses are given. In general, specific gravity values are described as light (< 0.4), medium (0.4–0.8) and heavy (> 0.8). Specific gravity or density serves as an excellent indicator of other wood properties (see *Density and Porosity*).

ACAPU (*Vouacapoua americana*), wacapou, or partridgewood is native to Surinam, French Guiana and northeast Brazil. The medium sized trees have clear boles to 15–23 m but mostly not more than 0.6 m in diameter. The very durable heartwood is dark chocolate brown or reddish brown with light brown striping. This coarse textured, fairly straight grained wood is moderately difficult to work and dry because of its hardness and high density. It is used in heavy construction, flooring (strip and parquet), furniture, cabinetwork, panelling and railroad crossties.

ANDIROBA (*Carapa guianensis*) or cedro macho occurs in the West Indies and from Honduras to northern South America. These large trees commonly have diameters of 0.6–1.0 m and often occur in pure stands. The durable to nondurable heartwood is salmon to dark reddish brown, often not sharply demarcated from sapwood. This medium density, straight or roey grained wood must be dried with care or serious defects will occur. Its workability is good, but it is reported to be harder than mahogany. Andiroba is similar to Honduras mahogany in color and properties and they are often confused or substituted for each other. Furniture, flooring, millwork, turnery and all types of construction where durability is not a factor are some of its main uses.

ANGELIQUE (*Dicorynia guianensis*) or basralocus occurs in the Guianas and Amazon region but is most abundant in Surinam and French Guiana, where it may make up 10% of the forest stands in some locations. The well formed trees grow to heights of 45 m and diameters of 1.5 m. The medium textured heartwood is dark brown to purplish or reddish brown. This moderately heavy, usually straight-grained wood has a high silica content; it may be difficult to dry, with a tendency to checking and slight warping. Owing to its high decay resistance, angelique is used in marine construction, railroad crossties and ship decking. It is also used for furniture, flooring and planking.

ARARACANGA (*Aspidosperma* spp., araracanga group) or mylady is found from Mexico to the lower Amazon region including the Guianas. These large canopy trees attain heights of 35 m with diameters of 0.6–1.0 m. The durable, medium textured heartwood is bright orange–red to reddish brown when freshly cut but comes light pinkish brown or pale yellow–brown upon exposure. This hard, heavy wood with straight to irregular grain machines well and can be dried without difficulty at a moderate rate. Uses include furniture, flooring, interior work, panelling, turnery, railroad crossties and boat framing.

BALSA (*Ochroma pyramidale*) is widely distributed throughout tropical America, usually on bottom-land soils at low elevations. Native trees are 18–27 m high and 0.75–1.2 m in diameter. Ecuador is a large exporter of balsa, most of it from plantations. The heartwood is pale brown and the sapwood (comprising mostly commercial timber) is nearly white. Balsa is the lightest commercial hardwood (about 0.16 g cm^{-3}) and is very soft, usually straight grained and coarse to medium textured. It is easily dried and very easy to work with sharp thin-edged tools. The wood is very prone to insect and fungi attack and must be converted rapidly. Uses include toys, model airplanes, core stock, rafts and floats but mostly as insulation against heat, vibration and sound.

BANAK (*Virola* spp.) and CUANGARE (*Dialyanthera* spp.) are so similar that they are mixed in the lumber trade. Banak is distributed from Belize and Guatemala to Venezuela, the Guianas, Brazil, Peru and Bolivia. Cuangare is found only in Colombia and Ecuador. Trees may reach 45 m high and 1.5 m in diameter. The nondurable heartwood with a variable texture is pinkish golden brown to deep reddish brown. These are light to medium density woods with straight grain that generally machine well if sharp knives are used. Banak is reported to be difficult to dry, with a strong tendency to warp, check, honeycomb and collapse, but cuangare dries rapidly except when wet streaks cause collapse. Both woods are used for moldings, panelling, veneer and core stock, furniture components, particleboard and general carpentry.

BRAZILIAN ROSEWOOD (*Dalbergia nigra*) or jacaranda is scattered in the eastern forest of Brazil and has become very scarce in the more accessible regions due to long-term exploitation. Trees sometimes attain a height of 40 m with short irregular boles and trunk diameters of 0.9–1.2 m. The very durable and medium to rather coarse textured heartwood is various shades of brown to chocolate or violet and is irregularly and conspicuously streaked with black. This hard, heavy, straight grained wood has a fragrant roselike odor and needs to be dried slowly to prevent checking, but it has excellent working properties. Brazilian rosewood is highly prized for decorative veneers, fine furniture, musical instruments, knife handles, fancy turnery and marquetry.

BULLETWOOD (*Manilkara bidentata*), beefwood or balata is widely distributed throughout the West Indies, Central America and northern South America. Well-formed trees reach heights of 30–45 m with diameters of 0.6–1.2 m. The heartwood is light to dark reddish brown with straight to slightly interlocked grain. This hard, heavy wood is difficult to dry and often develops severe checking and warp. Its working properties are good despite its high density, and it is highly resistant to fungi and termite attack. Uses include heavy construction, furniture parts, turnery, tool handles, flooring, boat frames and other bent work, railroad crossties, violin bows, billiard cues and other specialty items. Tapped trees yield a latex called balata or gutta-percha which is used in products such as golf balls and insulated wires.

CANALETE (*Cordia* spp. of the *C. gerascanthus* type), bocote or siricote is distributed from Mexico to Argentina and the West Indies. The small to large trees, sometimes 30 m tall, grow in the tropical dry zones. The very durable heartwood is tobacco-colored to reddish brown with irregular dark brown or blackish streaks and variegations. This hard, heavy, fine-to-medium textured wood can be readily worked and finishes very smoothly but is difficult to dry. Often surface checks and end splits develop. This rather recently known wood is highly prized for veneer, fine furniture, turnery and small novelty items.

CAPOMO (*Brosimum* spp. of the *B. alicastrum* type), ojoche or ramon occurs from southern Mexico to the Peruvian Amazon. Trees reach a height of 35 m with straight cylindrical boles clear to 20–25 m; the diameter may range up to 0.75–1.0 m. The nondurable heartwood and sapwood are yellowish white with straight to irregular interlocked grain. This medium to heavy wood with fine-to-medium texture is easy to moderately difficult to air-dry, with a tendency to twist. Although it is easy to moderately difficult to machine, proper cutters should be used because of its density and silica content. Uses include general construction, flooring, furniture, veneers and tool handles. The cooked seeds are edible.

CARIBBEAN PINE (*Pinus caribaea*) is native to Belize, Honduras, Nicaragua, Guatemala, the Bahamas and Cuba but has been widely introduced as a plantation species throughout the world. The trees grow to a height of 30 m with trunk diameters of 0.75–1.0 m. The moderately durable, somewhat coarse textured heartwood is generally golden brown to red–brown. This low-to-medium-density wood with straight grain works well but dries rather slowly, with a tendency for end splitting in thick stock. Compression wood is often present in plantation trees. Uses include general light and heavy construction, carpentry, flooring, joinery, utility poles, railroad crossties (treated), utility plywood and pulp and paper products.

CATIVO (*Prioria copaifera*) is distributed in the lowland areas from Nicaragua to Colombia, often in nearly pure stands. Trees are usually 20–30 m high with clear boles to 12 m and 0.5–1.0 m in diameter. The nondurable and rather fine textured heartwood is medium to light brown, frequently attractively streaked, and straight grained. The sapwood is thick, pinkish to white, and often exudes a considerable amount of gum. This gum may cause problems with finishing and drying. The low-to-medium-density wood dries easily and machines well but with a slight tendency to fuzziness. Uses include interior trim, veneer and plywood, millwork and furniture and cabinet work.

CAVIUNA (*Machaerium* spp.) or pau ferro is widely distributed throughout the tropics but is most abundant in Bolivia and southeast Brazil. Trees are medium sized. The highly decay-resistant heartwood is brown to dark purple–brown and frequently streaked. This hard, heavy wood has a fine to coarse texture and straight to irregular grain. It is generally useful for the same purposes as Brazilian rosewood (*Dalbergia nigra*) which include fine furniture, decorative veneers, turnery and cabinet work.

CEIBA (*Ceiba pentandra*) or kapok tree occurs throughout the tropics. In the New World it grows from Mexico to Brazil and Ecuador. Very large trees reach heights of 45 m and diameters of 2 m above the buttresses. The generally nondurable coarse textured heartwood is pinkish white to ashy brown with straight grain. This soft, lightweight wood dries rapidly with little warp or checking. It is easy to machine, but sawn surfaces are often fuzzy and tears occur in shaping, boring and turning. Uses include plywood, packaging, lumber core stock and pulp and paper products. Floss on the seeds is marketed as kapok.

COCOBOLO (*Dalbergia retusa*) grows in the Pacific region of Central America, extending from southwestern Mexico to Panama. Small to medium sized trees grow to 14–18 m with diameters of 0.5–0.6 m. The highly durable heartwood is usually a deep rich orange–red with black striping or mottling. This hard, heavy, fine textured wood is oily and slightly pungent. It has excellent drying and machining properties, but the natural oils cause problems in gluing. The fine dust from machining may cause dermatitis. Cocobolo is a highly prized wood for inlay work, high quality turnery, knife handles, musical and scientific instruments, furniture and fancy cabinet work and other specialty items.

COURBARIL (*Hymenaea courbaril*) or jatoba is distributed from southern Mexico and the West Indies to northern Brazil, Bolivia and Peru. Trees grow to a height of 40 m with diameters of 1.5–1.8 m. The very durable heartwood, with medium to coarse texture, is salmon red to orange–brown when fresh, but upon exposure it becomes russet to reddish brown, often with dark streaks. This heavy wood dries with only slight checking and warping. It is moderately difficult to machine due to its high density and is difficult to plane due to interlocked grain. Uses include tool handles (good shock resistance), steam-bent parts, flooring, furniture, turnery, gear cogs, wheel rims and other specialty items. Trees exude a rosinlike gum (copal) used in varnishes. The seed pods contain an edible pulp.

FREIJO (*Cordia* spp. of the *C. alliodora* type) or laurel blanco occurs from southern Mexico to the southern edge of the South American tropics. The variable-sized tree is frequently 12–18 m in height and 0.45–0.6 m in diameter. The moderately durable heartwood is yellowish to brown and uniform or more or less streaked and variegated. This medium density, usually straight grained wood air-seasons rapidly with only slight warping and checking. It works well and finishes smoothly. Uses include general construction, millwork, fine cabinet and furniture components, flooring, decorative veneer and boat construction, and for some applications it is used as a substitute for teak, walnut or mahogany.

GONCALO ALVES (*Astronium graveolens*) or gateado is distributed from Mexico to Brazil and Ecuador. Trees attain heights of 35 m with diameters of 0.6–1.0 m. The very durable and fine to medium textured heartwood is brown, red or dark reddish brown with nearly black strips after exposure. This hard, heavy wood with straight to wavy grain is moderately difficult to season but is not difficult to work in spite of its high density. Uses include heavy durable construction timbers, fine furniture and cabinet wood, decorative veneers, knife handles, billiard cue butts and other specialty items.

GREENHEART (*Ocotea rodiaei*) or Demerara greenheart is found in the Guianas and northern Brazil. Trees attain heights of 40 m with diameters up to 1.0 m. The very durable, fine textured heartwood varies from light to dark olive green or blackish. This hard, heavy wood with straight to wavy grain dries very slowly with a marked tendency to check and end-split. It is moderately difficult to work and dulls cutting edges rather quickly. Greenheart is especially used in marine and ship construction, lock gates, docks and pilings. Other uses include industrial flooring, vats, heavy construction, turnery, billiard cue butts and fishing rods.

GUANACASTE (*Enterolobium cyclocarpum*), kelobra or conocaste, is distributed from Mexico southward through Central America to Venezuela, Guyana and Brazil. Trees grow to 18–30 m with a short trunk 1–2 m in diameter. The durable, coarse textured heartwood with typically interlocked grain is brown with various shadings and sometimes with a reddish tinge. This soft, light-to-medium weight wood dries and works easily, but raised and chipped grain is common in planing. Tension wood is also common, resulting in fuzzy grain in most machining operations. Uses include core stock, panelling, pattern wood, interior trim, furniture components and veneer.

HONDURAS MAHOGANY (*Swietenia macrophylla*) or caoba is native from Mexico to the upper Amazon and its tributaries in Peru, Bolivia and Brazil and is planted in many tropical countries for lumber or ornament. The tree sometimes reaches a height of 45 m and 1.8 m or more in diameter. The durable heartwood varies from light reddish or yellowish brown to a rich salmon color. This straight-to-roey-grained medium density wood is exceptionally stable and is easily dried, worked and finished; it is often sliced or rotary-cut into fine veneers. Mahogany is one of the finest and best known furniture and cabinet woods. It is highly prized for interior trim, fancy veneers, musical instruments, boatbuilding, patternmaking, turnery and carving.

HONDURAS ROSEWOOD (*Dalbergia stevensonii*) has been reported only in Belize. Trees attain heights of 15–30 m, but the trunks are short with diameters to 1 m. The very durable, medium textured and slightly aromatic heartwood is pinkish brown to purple with alternating dark and light zones. This hard, heavy and generally straight grained wood dries slowly with a marked tendency to check and is moderately difficult to machine. Uses include musical instruments, decorative veneers, fine furniture and cabinets, knife handles, fine turnery and many specialty items.

HURA (*Hura crepitans*), assacu or possumwood is distributed throughout the West Indies and from Central America to northern Brazil and Bolivia. Trees commonly reach heights of 27–40 m with clear boles and diameters up to 3 m. The generally nondurable heartwood with fine to medium texture is pale yellowish brown or pale olive gray. This rather soft, low density wood, with straight to interlocked grain, is moderately difficult to dry. Sometimes warping is severe. It works easily but may machine to fuzzy surfaces due to tension wood. Uses include general carpentry, boxes and crates, veneer and plywood, joinery, furniture, particleboard and fiberboard.

IMBUIA (*Phoebe porosa*) or Brazilian walnut grows mostly in the moist forest of Parana and Santa Catarina in southern Brazil. Trees attain a maximum height of 40 m with 2 m diameter. The reportedly durable heartwood is yellowish or olive to chocolate brown with straight to curly and wavy grain. This rather fine textured and medium density wood is easy to dry, but thick stock dries slowly and may develop honeycomb and collapse. It machines satisfactorily, but the fine dust may cause dermatitis. Uses include fine furniture and cabinet work, panelling, flooring, gunstocks, decorative veneer and joinery.

IPE (*Tabebuia* spp., lapacho group), lapacho or guayacan is distributed throughout continental tropical America. Trees may grow to 45 m with a trunk diameter of 2 m, but more frequently the height is 30 m with diameters of 0.6–0.9 m. The very durable heartwood with fine to medium texture is olive brown to blackish, often with lighter or darker striping. This hard, heavy wood with straight to irregular grain is rather easy to dry with only slight checking and warping. It is moderately difficult to work and has a blunting effect on cutting edges. The fine yellow dust (lapachol) produced in most machining operations and characteristics of this group of species may cause dermatitis in some workers. Uses include railroad crossties, heavy construction, tool handles, turnery, industrial flooring and decorative veneers.

LEMONWOOD (*Calycophyllum candidissimum*) or degame occurs in Cuba and is also found from southern Mexico to Colombia and Venezuela. The small to medium sized trees usually reach 12–15 m in height and up to 0.75 m in diameter. The generally nondurable heartwood ranges from light brown to oatmeal color and is sometimes grayish. This high-density fine textured wood is difficult to work and moderately difficult to dry, with some tendency to warp and check. Degame has been used for the manufacture of archery bows and fishing rods and is suitable for tool handles, turnery, shuttles and other textile manufacturing items.

LIGNUM VITAE (*Guaiacum* spp.) is distributed throughout the West Indies and from Mexico to the northern fringe of Colombia and adjacent areas in Venezuela. The small trees usually grow 6–9 m in height with diameters of 0.25–0.5 m. The very durable, very fine textured and slightly scented heartwood is dark greenish brown to almost black with strongly interlocked grain. This very hard, heavy wood is difficult to dry and work. It has an oily feel due to a high resin content. Owing to its capacity for self-lubrication, hardness and durability, lignum vitae is valuable for bearings and bushing blocks for propeller shafts of ships. Other uses include pulley sheaves, mallet heads and turnery.

MONTEREY PINE (*Pinus radiata*), radiata pine or insignis pine is native to the central coast of California but has been planted and is now being harvested in Chile, New Zealand, Australia and South Africa. Plantation-grown trees may reach a height of 24–27 m with diameters 0.6–0.9 m in 20 years. The straight grained heartwood, which is durable above ground, is light brown to pinkish brown. This low-density fine textured wood dries rapidly with little degradation and machines easily. Radiata pine is being used locally and exported as veneer and plywood, pulp and paper, particleboard and fiberboard, light construction, boxes and crates and millwork.

NARGUSTA (*Terminalia amazonia*) is distributed from southern Mexico to Brazil and Peru. Trees may reach a height of 40–45 m with diameters of 1.2–1.5 m. The medium textured, somewhat durable heartwood is yellowish olive to golden brown, sometimes with prominent reddish brown stripes. This medium to high density wood with roey grain has very variable seasoning characteristics. Some material is easy to dry, while other material dries with considerable checking and warping. It is generally fair to difficult to work. Uses include flooring, railroad crossties, furniture, boatbuilding, turnery and utility plywood. It has been suggested as a possible substitute for oak.

PARANA PINE (*Araucaria angustifolia*) is distributed in Paraguay, Argentina and the Brazilian plateau regions of Rio Grande do Sul, Santa Catarina and Parana. Mature trees reach heights of 25–35 m with long clear boles. Diameters up to 1.5 m are possible. The nondurable uniformly-textured heartwood is various shades of brown, often with bright red streaks. This low to medium density wood with straight grain is worked easily but is more difficult to dry than most softwoods. Principal uses include framing lumber, interior trim, sash and door stock, furniture, case goods and veneer.

PEROBA (*Aspidosperma* spp., peroba group) or peroba rosa occurs in southeast Brazil and la Selva Misionera of Argentina. Large trees reach a maximum height of 38 m with diameters of 1.2–1.5 m. The durable fine textured heartwood is rose red to yellowish and often variegated or streaked with purple or brown. Upon exposure the wood becomes brownish yellow to dark brown. This medium-to-heavy wood dries with little checking, but some warp may develop. It works with moderate ease although some difficulties occur with irregular grain. Peroba is favored for fine furniture and decorative veneers, but is also used for flooring, interior trim, sash and doors and turnery.

PRIMAVERA (*Cybistax donnell-smithii*) grows in southwest Mexico, on the Pacific coast of Guatemala and El Salvador and in north-central Honduras. Trees attain a height of 30 m with diameters 1.2–1.5 m. The nondurable to moderately durable heartwood with a medium to rather coarse texture is cream colored, yellowish white to pale yellowish brown with more or less striping. This light to medium density wood with straight to roey grain is easy to dry and work. Uses include fine furniture, decorative veneers and interior trim.

PURPLEHEART (*Peltogyne* spp.) or amaranth is native from Mexico to southern Brazil but is generally found in the north-middle part of the Brazilian Amazon region. The trees grow to heights of 50 m with diameters to 1.2 m. When freshly cut, the heartwood is dull brown, but it turns a deep purple upon exposure. With prolonged exposure, especially to the sun, the color becomes dark brown or almost black. This durable, hard, heavy, generally straight grained wood is easy to moderately difficult to dry and moderately difficult to work. Because of its high density, it is used for tool handles, heavy construction and shipbuilding, and because of its deep purple color, it is also used for turnery, marquetry, cabinets, fine furniture and many specialty items.

ROBLE (*Tabebuia* spp., roble group) or mayflower occurs from the West Indies and Southern Mexico to Venezuela and Ecuador. Trees commonly reach 12–18 m in height with diameters of 0.4–0.6 m. The moderately to very durable heartwood with medium to coarse texture is light brown to golden. This medium density wood with straight to roey grain dries and machines easily, but some care is required in planing to prevent torn and chipped grain. Uses include flooring, furniture, cabinetwork, interior trim, tool handles, decorative veneers and boatbuilding. For some applications, roble is suggested as a substitute for ash and oak.

SANTA MARIA (*Calophyllum brasiliense*) or Maria grows throughout the West Indies and from Mexico southward into northern South America. Trees attain a height of 30–45 m with a long straight bole 1–2 m in diameter. The durable heartwood, with generally interlocked grain, varies in color from pink or yellowish pink to brick red or rich reddish brown. This medium density wood is fairly easy to work but is moderately difficult to dry, with warp being moderate to severe. Maria is widely used in the tropics for general construction, flooring, furniture and boat construction and is a favored general utility timber.

SIMAROUBA (*Simarouba amara*), marupa or aceituno grows from Venezuela and the Guianas to the Amazon region of Brazil. These large trees reach a height of 40–45 m with diameters of 0.5–0.6 m. The nondurable medium textured heartwood cannot be differentiated from the whitish or straw colored sapwood. This light to medium density wood with usually straight grain dries and works easily. It is used in interior construction, furniture components, veneer and plywood, patternmaking, millwork, boxes and crates, particleboard and fiberboard.

SPANISH-CEDAR (*Cedrela* spp.) or cedro occurs from Mexico to Argentina. Under favorable conditions, trees will reach heights of over 30 m and diameters of 1–2 m. The usually straight grained heartwood is pinkish to reddish brown when freshly cut but becomes red to dark reddish brown after exposure. The distinctive cedary-smelling wood is soft, of medium weight and rated as durable. It has excellent weathering characteristics and is easy to dry and work. Uses include millwork, cabinets, fine furniture, musical instruments, boatbuilding, patterns, veneer and cigar wrappers and boxes.

SUCUPIRA (*Bowdichia* spp.) ranges from Venezuela and the Guianas to southeast Brazil. The tree is medium sized to large, being up to 45 m high with a diameter up to 1.2 m. The very durable heartwood with a coarse texture is dull chocolate to reddish brown with striping due to abundance of parenchyma. This hard, heavy wood is difficult to work due to its high density and interlocked and irregular grain. No information is available on drying. Suggested uses for sucupira are heavy, durable construction and railroad crossties.

TREBOL (*Platymiscium* spp.) or macawood is distributed from southern Mexico to the Brazilian Amazon region. Trees attain heights of 25 m with diameters of 0.7–1.1 m. The very durable heartwood is bright red to reddish or purplish brown and more or less distinctly striped. This hard heavy wood with straight to roey grain works fairly easily and is reported to dry slowly, with a slight tendency to warp and check. Trebol is used in fine furniture and cabinet work, decorative veneers, musical instruments, turnery, joinery and specialty items (violin bows, billiard cues).

TROPICAL WALNUT (*Juglans* spp.) or nogal is distributed from southern Mexico to Colombia, Ecuador and Peru. Trees grow mostly to 18 m in height with diameters to 1 m. The very durable coarse textured heartwood is chocolate brown with a purplish cast and generally darker than the North American black walnut. This medium density wood with straight to irregular grain dries very slowly and severe honeycombing and collapse may occur in thicker stock. It works well and peels and slices readily, but the veneers also dry slowly. Tropical walnut is used in the same applications as North American black walnut, especially decorative veneers and fine furniture.

WEST INDIAN SATINWOOD (*Zanthoxylum flavum*) grows in the lower Florida Keys and the West Indies. It is generally a small tree, sometimes reaching a height of 12 m with a trunk diameter up to 0.5 m. The nondurable fine textured heartwood is creamy or golden yellow, darkening with exposure. This high density wood with interlocked grain is reported to season without difficulty. It has a dulling effect on cutting edges but is an excellent turnery wood. The dust may cause dermatitis. West Indian satinwood is rare but highly prized for cabinetmaking, fine furniture, inlays, turnery, fancy veneers and specialty items.

See also: Resources of Timber Worldwide

Bibliography

Berni C A, Bolza E, Christensen F J 1979 *South American Timbers—The Characteristics, Properties and Uses of 190 Species.* Division of Building Research, CSIRO, Melbourne

Chudnoff M 1984 *Tropical Timbers of the World*, Agriculture Handbook No. 607. US Department of Agriculture, Forest Service, Washington, DC

Erfurth T, Rusche H 1976 *The Marketing of Tropical Wood, B: Wood Species from South American Tropical Moist Forests*, FO: MISC/75/29–1. Food and Agriculture Organization of the United Nations, Rome

Instituto de Pesquisas Tecnologicas 1971 *Fichas de Caracteristicas das Mateiras Brasileiras.* Instituto de Pesquisas Tecnologicas do Estado de Saõ Paulo, Saõ Paulo

Longwood F R 1972 *Present and Potential Commercial Timbers of the Carribean, with Special Reference to the West Indies, the Guianas, and British Honduras*, Agriculture Handbook No. 207. US Department of Agriculture, Forest Service, Washington, DC

Record S J, Hess R W 1943 *Timbers of the New World.* Yale University Press, New Haven (reissued by Arno Press, New York)

Rendle B J (ed.) 1969 *World Timbers*, Vol. 2: *North and South America.* Benn, London

Takahashi A 1975 *Compilation of Data on the Mechanical Properties of Foreign Woods*, Part II: *Central and South America*, Research Report of Foreign Wood No. 4. Foreign Wood Laboratory, Faculty of Agriculture, Shimane University, Matsue, Japan

US Forest Products Laboratory, 1987 *Wood Handbook: Wood as an Engineering Material*, Agriculture Handbook No. 72 (revised edn.). US Department of Agriculture, Forest Service, Washington, DC

R. B. Miller

[US Forest Products Laboratory, Madison, Wisconsin, USA]

Timbers of Europe

Europe, with the exception of the vast resources of coniferous trees in the north and east, cannot be classed as a timber exporting region. Generally, the eastern countries of Europe produce better-quality timber. This applies particularly to the softwoods, which need a colder climate than that of the milder western countries in order to produce slower grown, finer textured material.

This article presents descriptions of the major commercial species of Europe, with reference to tree size, country of origin, and wood properties including color, texture and density. Where appropriate, strength values, ease of seasoning, durability, working qualities and utilization are also discussed.

ALDER (*Alnus glutinosa* and *A. incana*) is a small tree, up to 14 m high, growing chiefly in low-lying places and on the banks of streams and rivers. The wood is pale colored, slightly pinkish brown, soft and light, with a density of 0.53 g cm^{-3} when seasoned, and with no distinction between heartwood and sapwood. It is generally straight grained, the texture being fine and uniform. It machines easily and is manufactured into utilitarian plywood, chiefly in the USSR, and into boxes and crates. It is the usual wood for clog soles. It turns well and is used for cheap handles, brush backs and broom heads.

ASH (*Fraxinus excelsior*) grows in favourable conditions to a height of 30 m. Ash is one of the most useful of European timbers because of its elasticity and toughness. The density varies from about 0.5 to 0.8 g cm^{-3} when seasoned, slow-grown trees producing the lightest wood. The wood is sometimes attractively figured. The name olive ash refers to wood of a brownish color, often streaked, and hence of high value for decorative veneers. The wood is available in a variety of sizes but the strongest, heaviest and most-elastic wood is from young coppice-grown trees. Older trees occasionally show black heart. Chief uses are for bent work, sports equipment and tool handles.

BEECH (*Fagus sylvatica*) is one of the most useful of European hardwoods, in plentiful supply, the tree growing to a height of more than 30 m. In Europe the practice of steaming immediately after conversion renders the wood a warm pinkish brown color. Fully quarter-sawn surfaces show an attractive figure. It is a hard and close grained wood, with a density of around 0.72 g cm^{-3} when seasoned. It is not resistant to insect and fungal attack but can be readily impregnated with preservatives. The timber is used extensively for chair making, mass production of furniture generally, flooring and laminated sports equipment. It works well with machine tools and the logs are suitable for rotary-cut veneer for plywood manufacture.

BIRCH (*Betula verrucosa* and *B. pubescens*) is a small tree, up to 18 m in height, widely distributed in northern Europe and the USSR, where it is regarded as important. The wood is plain white to pale straw in color, and fine textured, showing no distinction between heartwood and sapwood. The average density is around 0.66 g cm^{-3} when seasoned. The timber is used for a variety of purposes such as turnery, brush backs, broom heads and chair parts, but is best known in good quality plywood. Wood with various types of figure is produced by trees showing rough bark, and such examples are manufactured into veneer of high decorative value.

BOX (*Buxus sempervirens*) is a small tree up to 6 m in height and usually not more than 150 mm in diameter, attaining larger sizes in southeast Europe, where logs may be up to 400 mm in diameter. An outstanding wood, extremely hard and fine textured, pale but bright yellow in color, box has a density varying from 0.85 to 1.15 g cm^{-3} when seasoned. It should be dried slowly because of its tendency to split. It is used extensively for engravers' blocks, draftmen's instruments, mallet heads and rollers where resistance to wear is important. There are various tropical woods, botanically unrelated, which are very similar to box in texture and color and which are available in larger sizes.

CEDAR (*Cedrus* spp.) Three species are regarded as true cedars. They are planted as ornamental trees and may grow to imposing specimens, up to 30 m high and 2.5 m at the base. Cedar wood has a distinctive odor when freshly cut, is pale brown in color, fairly light in weight, and has a density of about 0.56 g cm^{-3} when seasoned. It is plain except for the growth rings, which are sufficiently prominent to give the wood a decorative appearance. The timber is very resistant to fungal and insect attack and is suitable for garden furniture, ornamental gates and fences. It works well and is also used for chests and linings of wardrobes.

CHERRY (*Prunus avium*) is in restricted supply, but the tree may grow up to 25 m in height, producing wood of medium weight, the density averaging 0.6 g cm^{-3} when seasoned. It is quite an attractive wood, pale pinkish brown with some variation in appearance owing to the growth rings, which are quite distinct. It is not resistant to insect and fungal attack, but it can be impregnated and it is a good working timber with hand and machine tools, being attractive for turnery. Classed as a high-grade furniture wood, it is also manufactured into veneer.

CHESTNUT (*Castanea sativa*) may grow up to 30 m in height and sawn timber is available in widths up to 200 mm. The growth of mature trees is often markedly spiral and this constitutes a defect in sawn material. The wood is similar to oak (to which it is botanically related) in both color and texture, except for the large rays present in oak. Chestnut, therefore, has no figure comparable with oak and is less strong, softer and lighter, with a density of about 0.55 g cm^{-3} when seasoned. This accounts for the ease of working compared with oak, and the two timbers are often mixed in furniture manufacture. Coppice-grown chestnut, which splits easily, is used extensively for fencing.

ELM (*Ulmus* spp.) may grow up to 35 m in height. Various species are converted into boards and they are not normally distinguishable. The fatal elm disease has been responsible for the destruction of most trees in Britain and the countryside is the poorer for this denudation. Wood grown in Britain is "wild" in texture, while that from Europe tends to be of straighter grain. Wych elm (*Ulmus glabra*) is considered the best. The average density is 0.56 g cm^{-3} when seasoned. Elm works well in machining, but care is necessary in planing. It is not durable and needs preservative treatment. The timber is commonly used for garden furniture. The better grades are effective for high-grade furniture with pleasing characteristics. Elm burr veneers are very decorative.

HOLLY (*Ilex aquifolium*) is a white wood with a greenish or grayish tinge, dense and heavy, its density being around 0.8 g cm^{-3} when seasoned. Not often straight grained, it is available only in small sizes, as the tree rarely exceeds 9 m in height and 0.3 m in diameter. The wood dries slowly and is apt to split. It is not easy to work, but it can be turned to an excellent finish. It is used chiefly for inlaying, wood engraving and for carvings such as chess pieces, both white and black; when stained it is an effective substitute for ebony. Being fine textured it is useful for fretwork and intricate examples such as house furniture in miniature.

HORNBEAM (*Carpinus betulus*) is widely distributed in Europe, of medium height (18 m) and up to 0.6 m in diameter. Its density is around 0.75 g cm^{-3} when seasoned, and being typically cross grained, the wood is very tough and difficult to split. Seasoning is not difficult and the wood is comparatively stable; for this reason it is useful for piano action parts. It can be worked to fine measurements and can be drilled without splitting. Other uses are for mallet heads, wood screws and the moving parts in wooden machinery. It is sometimes stained to resemble ebony. Related to beech, it is somewhat heavier and less easy to work.

HORSE CHESTNUT (*Aesculus hippocastanum*) grows to a height of 30 m and up to 1.5 m in diameter. The wood is ivory colored, fine textured, soft and light, with a density of about 0.5 g cm^{-3} when seasoned. It seasons well and is stable in service. It is readily attacked by fungi and insects but takes preservatives. The clean, smooth appearance and good working qualities render the wood useful for kitchenware and toy making, and it is effective for carving. It is also manufactured into rotary-cut veneers for fruit baskets and trays for fruit storage, and has been used for cheaper grade violins.

LABURNUM (*Laburnum anagyroides*) is widely cultivated as an ornamental tree 9 m in height and 0.8 m in diameter. The wood is one of the harder and heavier European hardwoods, with a density of around 0.85 g cm^{-3} when seasoned. The heartwood is deep brown in color, the narrow sapwood being pale yellow. It is reported to season well and to be durable. Botanically related to the tropical rosewoods, laburnum is a highly decorative wood used for a variety of purposes. Not available in large sizes, it is particularly useful for inlaid work and musical instruments (flutes and recorders). It turns well for ornamental work, and small-section end-grain veneers (oysters) are used in special examples of cabinet work.

LARCH (*Larix decidua*) is grown in Britain almost exclusively in plantations but is widely distributed in northern Europe and Siberia. It is a large tree, one of the few deciduous conifers, and will grow to 30 m in height and up to 1 m in diameter. The wood is reddish brown, harder than most conifers and somewhat heavier, with a density of around 0.64 g cm^{-3} when seasoned. Larch seasons slowly but is stable and strong in service. It is durable, but when used in the round the narrow light-colored sapwood needs preservative treatment. It is used for outdoor work in buildings, gates and fencing, rustic work and pergolas. Slow grown wood for colder climates is of value for superior joinery, the growth rings providing an attractive feature.

LIME (*Tilia vulgaris*) is a free-growing tree up to 30 m or more in height and 1.2 m in diameter, often planted in avenues. The wood is very plain in appearance, fine textured and whitish or pale brown in color. It is not durable but is easily impregnated. A light, soft hardwood, with a density of around 0.55 g cm^{-3} when seasoned, it is used for a variety of purposes such as toy making, brush handles and nondecorative turnery; the best grades are used for piano keys. Much of the finest carved work in the past has been done in lime, the easily worked wood being ideal for fine detail.

OAK (*Quercus robur* and *Q. petraea*) encompasses two main species with little difference between them. The density varies, with an average of 0.7 g cm^{-3} when seasoned; that grown in the north being somewhat heavier, while further south the wood is more even and easier to work. Oak has long been a favorite for furniture, staircases and panelling, flooring and decorative work generally. It is also durable and of value for fencing, posts, gates and heavy structural work. Sapwood, however, is prone to insect attack and should not be included. The large rays form the well-known figure on quartersawn surfaces, and plain and figured veneers are in constant demand. So-called brown oak is from older trees, the color being due to a fungus which is not classed as a defect. Brown oak is somewhat lighter in weight and shrinks considerably during seasoning.

PEAR (*Pyrus communis*) is a small tree up to 12 m in height and 0.4 m in diameter. The wood is fine textured, light red–brown in color and moderately heavy, with a density of around 0.72 g cm^{-3} when seasoned. It is reported to season very slowly, but when dry the wood is stable. Because of this it has been used in the past for precision work such as T squares, set squares and laboratory apparatus. Stained black it is useful as a substitute for ebony. It is also used for wood engraving, musical instruments and saw handles.

PLANE (*Platanus acerifolia*) is a large tree up to 30 m in height and more than 1 m in diameter. London plane, so called because the tree flourishes in the atmosphere of towns and cities, is conspicuous for shedding its bark in large flakes. The timber is whitish in color with numerous large rays which are reddish brown and which form an attractive figure on quartersawn surfaces. Consequently the wood is manufactured into veneer to display this figure under the name lacewood. It is useful for decorative panelling, cabinet making and turnery. It is similar to beech but is slightly softer and less heavy, with a density of 0.64 g cm^{-3} when seasoned, and is somewhat whiter in color.

POPLAR (*Populus* spp.) includes aspen, a tree rarely exceeding 18 m in height and 0.6 m in diameter, black poplar, a fast growing tree up to 30 m and 1.4 m in diameter, and Lombardy popular, which is a well-known ornamental tree but of no use as timber. The wood is soft and light, with a density of about 0.45 g cm^{-3} when seasoned, and is nearly white or very pale brown in color and fine textured. A utility wood, of no decorative value, it has been found very suitable for the floors of railway wagons and lorries, being resistant to abrasion. It is also used for crates and boxes, chip baskets, toys, matches and match boxes, and is manufactured into plywood.

ROBINIA (*Robinia pseudoacacia*) is usually grown as an ornamental tree, the flowers being white and fragrant. It is also known as acacia and black locust. The wood is ring-porous, greenish when first cut, mellowing to a golden brown, and fairly coarse in texture. It is strong, tough, and resistant to decay, moderately hard and heavy, with a density of around 0.7 g cm^{-3} when seasoned. It works well with hand and machine tools and is a useful and attractive furniture wood. Comparatively plentiful in France, it is used as a substitute for ash in bent work, wheels, ladders and agricultural implements. It needs care in seasoning owing to a tendency to warp.

SCOTS PINE (*Pinus sylvestris*) is a tree very widely distributed in Europe and northern Asia, also known as red or yellow deal, often with a geographical prefix. It may grow to a height of 30 m, averaging 0.75 m in diameter, but much of the wood used is from smaller trees. Its quality varies considerably and is related to the size and frequency of knots. The average density is around 0.48 g cm^{-3} when seasoned. The wood seasons well but is rather prone to blue stain in the sapwood, which is present in a large proportion. Growth rings are conspicuous. It is classed as nondurable, but the sapwood responds to preservative treatment. It is used in vast quantities for construction and general carpentry, poles and pit props. Scots pine is the commonest of all woods in north Europe. This, coupled with its ready response to working with all hand and machine tools, renders it the most useful and the most valuable of northern forest products.

SPRUCE (*Picea abies*) is a tree up to 45 m tall and 1.5 m in diameter, and is widespread in northern Europe. Numerous young trees are sold each year as Christmas trees. Exported as whitewood or white deal with a geographical prefix, the wood is almost white and shows no difference between sapwood and heartwood. It is somewhat lighter in weight than Scots pine, its density being about 0.42 g m^{-3} when seasoned. In Britain, with its milder winters, wood of inferior quality is produced. It seasons well and is more stable than Scots pine, though less strong. British grown wood, being wider ringed, tends to be of coarser quality than that grown in colder climates. The timber is not very durable and is not easily impregnated, but it works well with hand and machine tools. Uses include interior joinery and structural work, food containers, packing cases and wood wool.

SYCAMORE (*Acer pseudoplatanus*) is a large tree up to 30 m in height and 1.5 m in diameter. The wood is one of the most valuable of European hardwoods. White and fine textured, it is lighter than beech, with a density of 0.6 $g\,cm^{-3}$ when seasoned. It is susceptible to fungal decay but is too valuable for use in exposed situations. It requires care in seasoning, as it discolors readily. The uses of the timber are many and varied. It is used in the textile industry and for ornamental turnery. Its clean appearance and nontainting character make it especially suitable for kitchen, bakery and dairy utensils. Some trees produce highly figured wood of value for decorative veneers, sometimes with a close wave giving rise to the name fiddle-back sycamore, this being in demand for violins.

WALNUT (*Juglans regia*) is another extremely valuable hardwood, a native of eastern Europe and cultivated in western Europe since early times. It may grow to 25 m in height. The bole, usually rather short, may be 1 m in diameter. The wood is of moderate density, about 0.64 $g\,cm^{-3}$ when seasoned, and medium textured. It is of dark purplish brown color, variable in three distinct zones: sapwood, outer heartwood and inner heartwood which is often streaked with dark markings and hence very decorative. Walnut seasons slowly, shows high strength values, is tough and is difficult to split, and hence is used for the best gun stocks. Most of it is manufactured into veneers and these are much sought after for the best class of furniture and for fittings and panelling in the most lavish public buildings.

WILLOW (*Salix* spp.) is found in several species and varieties which are used for various purposes such as wicker baskets and furniture. White willow is used exclusively for cricket bats. Pollarded trees are specially cut to produce long straight branches for fence posts and traditional hurdle making. The wood is straight grained, fine textured and light in weight, with a density of around 0.38 $g\,cm^{-3}$ when seasoned. It is plain, whitish or pale pinkish brown and is resistant to impact, tending to dent rather than split. It seasons readily but is not a durable wood. It is used principally for toys, floors for railway wagons and trucks, clog soles and wood pulp.

YEW (*Taxus baccata*) is a slow growing tree planted more for ornament than timber production in Britain but widely distributed in Europe and north temperate Asia. The tree is usually of poor shape for conversion into good timber. The narrow sapwood is white while the heartwood is a warm orange–brown color of value for decorative work. For a softwood (conifer) it is hard and heavy; its density is about 0.67 $g\,cm^{-3}$ when seasoned, and it is extremely durable. Yew is especially good for turning, for which it is highly prized, as it gives a fine finish and its color is almost unique. For high quality furniture it is much in demand by individual craftsmen for its striking beauty. Suitable material is manufactured into veneer. In medieval times yew was considered the wood par excellence for bows used in weaponry and sporting events.

See also: Resources of Timber Worldwide

Bibliography

Brown W H 1978 *Timbers of the World*, Vol. 6: *Timbers of Europe*. Timber Research and Development Association, High Wycombe

Desch H E 1981 *Timber, Its Structure and Properties*, 6th edn (revised by J M Dinwoodie). Macmillan, London

Lincoln W A 1986 *World Woods in Color*. Macmillan, New York

Panshin A J, de Zeeuw C, Brown H P 1980 *Textbook of Wood Technology*, 4th edn. McGraw-Hill, New York

Rendle B J (ed.) 1969 *World Timbers*, Vol. 1: *Europe and Africa*. Benn, London

C. W. Bond
[Leamington Spa, UK]

Timbers of Southeast Asia

Southeast Asia is the source of many world-famous woods such as meranti, teak, rosewood, padauk and ebony. Using a broad definition of the region, the floras can be divided into four groups: the Indian flora (India, Sri Lanka, Bangladesh), the Indochinese flora (Burma, Thailand, Cambodia, Laos, Vietnam), the Malaysian flora (Malaysia, Indonesia, the Philippines) and the Papua New Guinea flora (Sulawesi, Papua New Guinea). Of these, the Malaysian flora in particular with its jungles and its majestic trees of the *Dipterocarpaceae* family that grow to over 60 m forms a treasurehouse of woods for world trade.

In this article, 38 selected species or species groups representing particularly important commercial woods are described. The information given includes nomenclature, distribution (origin) and descriptions of characteristics. Physical and mechanical properties are shown in Table 1.

BALAU (*Shorea* spp.), also called thitya (Burma), teng (Thailand), phchek (Cambodia), chik (Laos), ca-chac (Vietnam), yakal (Philippines), selangan batu (Borneo) and bangkirai (Indonesia), is composed of about 50 species distributed from India to Southeast Asia, and there are 30 species in Borneo alone. Balau comprises the heavier species of *Shorea* (see also MERANTI). The sapwood is dark yellow and the heartwood deep brown, often with purple or pale brown streaks. Balau is fairly difficult to work with tools and dries with much shrinkage and a tendency to split. It weathers very well but is liable to insect attacks. It cannot be glued with resorcinol, PVA or phenol adhesives. It is

Table 1
Physical and mechanical properties of timbers of Southeast Asia

	Density (g cm^{-3})	Shrinkage coefficient (%)[a]		Bending modulus of elasticity[b] (MPa)	Bending strength[b] (MPa)
		Radial	Tangential		
Balau	0.75–1.02	0.19	0.44	19600	157
Batai	0.26–0.48	0.13	0.23	7550	37
Belian	1.05	0.22	0.40	18340	185
Bintangor	0.595	0.18	0.22	10000	94
Ceylon ebony[c]	0.73	0.17	0.28	9800	77
Champak	0.54	0.18	0.25	11080	87
Chengal	0.88	0.10	0.26	19580	148
Cottonwood tree	0.25	0.08	0.17	6600	42
Durian	0.53	0.14	0.24	11720	74
Eaglewood	0.33–0.40	0.14	0.28	7550	53
Eng	0.85	0.15	0.30	13500	96
Giam	0.88–1.04	0.13	0.27	14200	132
Indian laurel	0.75	0.16	0.26	10300	74
Indian rosewood	0.79	0.09	0.20	10790	62
Jelutong	0.40–0.47	0.09	0.22	7840	51
Jongkong	0.56	0.16	0.28	11380	84
Kapur	0.57–0.76	0.16	0.35	13440	105
Kauri	0.40–0.67	0.18	0.36	11500	100
Kelat	0.76	0.21	0.34	13140	102
Keledang	0.67	0.18	0.26	11800	84
Kempas	0.87	0.08	0.10	18530	122
Keranji	1.09	0.18	0.33	25530	270
Keruing	0.64–1.18	0.24	0.37	20000	78
Khasya pine	0.44	0.10	0.27	8430	68
Meranti: dark red	0.48–0.63	0.08	0.25	12750	87
light red	0.40–0.54	0.12	0.26	7850	62
yellow	0.47–0.52	0.10	0.27	10400	78
white	0.53–0.63	0.16	0.32	12550	90
Merawan	0.63–0.90	0.18	0.43	15400	162
Merkus pine	0.63	—	—	17160	126
Mersawa	0.58–0.73	0.22	0.39	9900	84
Nyatoh	0.64–0.71	0.21	0.30	12260	110
Padauk	0.75	0.11	0.17	14300	142
Punah	0.72	0.20	0.36	12840	67
Ramin	0.54–0.70	0.21	0.39	15000	117
Rengas[d]	0.65	0.10	0.18	14950	111
Sengkuang	0.58	0.14	0.29	12550	101
Sepetir[e]	0.60	0.19	0.26	11380	87
Teak	0.60–0.68	0.10	0.19	12750	98
Terap	0.42	0.11	0.26	10290	69
White lauan	0.46–0.68	0.09	0.25	9810	78

[a] Per 1% change in moisture content [b] At 15% moisture content [c] Data are actually for Indian ebony (*Diospyros melanoxylon*) [d] *Melanorrhoea torquata* [e] Swamp sepetir (*Pseudosindora palustris*)

used for structural timber, flooring, marine piling, shipbuilding, sleepers and barrels.

BATAI (*Albizia falcataria*), also called miracle tree or Molucca albizia, moluccan sau (Philippines), jinling or sengum (Malaya), kayu machis (Sarawak), sengon laut or jeungjing (Indonesia) and white albizia (Papua New Guinea), is indigenous from Molucca to New Guinea but is now planted throughout Southeast Asia. Batai is noted as one of the fastest-growing trees in the world, and can reach a height of 15 m in three and 30 m in ten years. The wood is pale pinkish brown. Its texture is rather coarse and it works easily. Batai is used for plywood cores, particleboard, fiberboard, matches, clogs, packing cases and pulp.

BELIAN (*Eusideroxylon zwageri*), also known as Borneo ironwood, tambulian (Sabah, Philippines) and billian or ulin (Indonesia), is found in Borneo, Sumatra and the Philippines. The sapwood is pale yellow and the heartwood pale brown or pale yellow with occasional red hues, darkening on exposure. Belian is fairly difficult to work with tools. It is resistant to marine borers and termites, but its high extractive content interferes with some finishes. Belian is used for heavy timber construction, wharf decking, marine piling, bridge construction, shipbuilding and flooring.

BINTANGOR (*Calophyllum* spp.) is generally known by that name in Malaysia, but *C. inophyllum* is also called penaga laut (Malaysia) and Alexandrian laurel (India). It is distributed from India to Oceania. The sapwood is pale yellow–brown, and the heartwood pinkish or reddish brown. The wood is rather difficult to work but takes a lustrous, smooth finish. The fruits are edible and the seeds contain 55% of an oil which is used for lamps and soap. The sap, leaves and roots are used in medicine. The wood is used for general construction, flooring, boat building, furniture, musical instruments, gun stocks and plywood.

CEYLON EBONY (*Diospyros ebenum*), also called Bombay ebony, ebans (India) and kayu malam (Malaysia, Indonesia), is found in India and Sri Lanka. It was the original ebony of commerce, which can now include any of over 200 species of *Diospyros* in the tropical zone. The color varies with the species, the heartwood being generally black, sometimes with irregular brown streaks, and the sapwood yellowish white with black streaks. The texture is fine and even, and the wood is easy to turn and to carve but must be dried slowly. It is high in strength but brittle, and highly resistant to insect and fungus attack. Ebony is used for fancy goods, notably piano keys, fingerboards for violins or cellos, brush backs and high class furniture.

CHAMPAK (*Michelia champaca*), other names for which are champa (India), sagah (Burma) and champar (Thailand), is found in India, Burma, Thailand, Indochina, Malaysia and Kalimantan. The sapwood is pale brown and the heartwood dark brown with a green hue on exposure. The texture is medium, and the wood is lustrous and slightly oily. It works well, except for turning, but must be seasoned carefully to avoid checking. Champak is used for boatbuilding, furniture, joinery, plywood and pulp.

CHENGAL (*Balanocarpus heimii*) is found only in the Malay peninsula. The heartwood turns dark brown on exposure, the sapwood being pale yellow. The wood planes well to a lustrous surface, and is hard and durable. Uses are in heavy construction, flooring, barrels and tanks. The wood also yields a resin, "dammar penak," used in varnish.

COTTONWOOD TREE (*Bombax malabaricum*), also called cotton wood, semul or letpan (Burma), ngui (Thailand), simur (Malaysia), kapok or randoe allas (Indonesia), and bān-zhi (China), is distributed from India to Taiwan and northern Australia. With sapwood and heartwood not demarcated, it is pale yellowish white when freshly cut, turning to a yellowish brown or grayish brown on exposure. The grain is straight and the texture coarse. The timber is light and lustrous and machines and dries well but is not durable. Fibers enveloping seeds are used like kapok or silk or cotton fibers. The wood is used for matches, containers, musical instruments, toys, pencils, canoes, plywood, pulp and fiberboard.

DURIAN (*Durio* spp. and *Neesia* spp.) is distributed from Burma to Southeast Aisa, and is famous as a fruit tree (*D. zibenthinus*) more than for its wood. The sapwood is paler in color than the heartwood, which is pinkish or reddish brown. A coarse textured wood, durian is easy to work and dry but is not durable. It is used for interior trim, furniture, ceiling boards, flooring and plywood.

EAGLEWOOD (*Aquilaria* spp.), also called agar wood (India), chan krasna (Cambodia), kayu garu (Malaysia), akyau (Burma) and karas (Indonesia), is distributed from India to the Philippines. Eaglewood is white, rather homogenous, soft and not durable. It is fairly easy to work with hand and machine tools and easy to dry. Richly resinous and odorous wooden blocks are called agar wood. Dark wood rich in resin is used as incense. The wood contains 0.75–2.5% of an oil which is used to blend perfume. Other uses are in gemboxes, rosaries, beads, crucifixes, interior fittings, furniture and plywood.

ENG (*Dipterocarpus tuberculatus*), also known as pluang (Thailand), in (Burma) and khlong (Cambodia), is found in Bangladesh, Burma, Thailand and Cambodia. The sapwood is pale gray, distinct from the dark reddish brown heartwood. The wood is coarse textured, hard, tough and moderately durable, but it should be treated with preservative. Eng is used for heavy construction, framing, sleepers and vehicle flooring.

GIAM (*Hopea* spp.) is the name for the heavier species of *Hopea*, while the lighter species are called merawan. It is also sometimes confused with balau (*Shorea* spp.). Giam is distributed from southern India to the Philippines and New Guinea. The sapwood is light brown and the heartwood a darker brown, with generally darker tones than merawan. The wood is fine textured, hard and durable. It is used for construction, sleepers, boatbuilding, texture rollers and cooperage.

INDIAN LAUREL (*Terminalia alata*, *T. coriacea* and *T. crenulata*), also called amari or sain (India) and taukkyan (Burma), grows in India, Bangladesh, Burma, West Pakistan and southwest Thailand. It is valued as a figured and handsome wood. The sapwood is reddish white, distinct from the heartwood, which is dark reddish, dark yellowish brown or pinkish beige, sometimes with dark streaks. The texture is rather coarse, and it can be worked smoothly with machine tools but not with hand tools. It must be seasoned with care, as there is a marked tendency to split and distort. It has excellent weathering properties. Nailing is not possible without predrilling, but it can be glued readily and looks good with a wax finish. It is used for high class interior trim, furniture, brushbacks, handles, large marine piles, boatbuilding, sleepers, panelling and sliced veneer.

INDIAN ROSEWOOD (*Dalbergia latifolia*), also called Bombay blackwood (India) and sono keling (Indonesia), is found in India, Java and Burma. Related species are Asiatic rosewood (*D. bariensis*), Burma blackwood (*D. cultrata*), Burma tulipwood (*D. oliveri*), Thailand rosewood (*D. cochinchinensis*) and sissoo (*D. sissoo*). Indian rosewood has pale yellow–white sapwood and heartwood that is dark red–purple, grayish red–purple and dark red with black streaks. The texture is coarse, but the surface is lustrous and takes a good wax finish. However, the wood contains extractives that interfere with the cure of unsaturated polyester-based paint and make the sawdust irritating to workers. It is very durable and is used for high class furniture, parquet, cabinetry, musical instruments, interior trim and fittings and handles.

JELUTONG (*Dyera costulata* and *D. lowii*), also known as yelutong (Thailand) and jelutung (Indonesia), grows in Malaysia, Sumatra and Borneo. It is pale yellow, turning pale beige after exposure. Its texture is fine and even. It is soft and fairly easy to work and to season. It finishes, stains, varnishes, nails and glues well but is not durable. It is used for patternmaking, carvings, drawing boards, blackboards, matches, wooden clogs, furniture, plywood, particleboard, fiberboard and pulp. The sap contains a latex used for making chewing gum.

JONGKONG (*Dactylocladus stenostachys*), also called merebong, is found in Borneo. The sapwood is not demarcated from the pale yellowish white or pale orange–brown heartwood, turning to dark or reddish brown on exposure. It is fine textured wood, often with pinholes on tangential surfaces due to included phloem in radial strands. Jongkong is fairly easy to work but is liable to attack by fungi and insects, especially termites. It is used for general construction, flooring, interior trim and plywood cores.

KAPUR (*Dryobalanops* spp.), also termed kapor or Borneo camphorwood, is found in Borneo, Malaysia and Sumatra. The sapwood is pale yellowish brown with a red tinge, and the heartwood pale red or deep reddish brown. Kapur has excellent resistance to fungi but is liable to insect attack. It is easily worked and takes a smooth finish. It splits easily in nailing and is stained black by iron nails or screws under wet conditions. Kapur can be glued well with urea-based but not with phenolic adhesives. It has an odor resembling camphor and contains much silica. Uses are in heavy construction where insect attack is not a problem, flooring, vehicle bodies, furniture and plywood.

KAURI (*Agathis alba*), also called agathis, East Indian kauri, dammar or Borneo kauri, almaciga (Philippines), damar putih (Indonesia), damar minyak (Malaysia), bindang (Sarawak) and menghilan (Sabah), is distributed from India to New Zealand. It is a softwood, with sapwood not demarcated from the heartwood, which is light yellow, straw-colored, pale beige and pale yellowish pink with a pink tinge, turning brown on exposure. Its growth rings are not clearly demarcated and its texture is fine and even. It is worked smoothly by hand and machine tools and seasons rapidly and well. It splits easily in nailing and is liable to termite attacks. Kauri yields a resin, copal, that is used in varnish. It is used for house construction, flooring, panelling, matches, pencils, interior joinery, patternmaking, kitchen furniture and plywood.

KELAT (*Eugenia* spp.) is a genus widely distributed in the tropical zone, and there are about twenty species used for timber in Malaysia. The sapwood is lighter in color than the heartwood, which is grayish, golden or reddish brown. Fine textured, it takes a lustrous finish but is rather difficult to work and to dry and is liable to split in nailing. Kelat is used for building construction, furniture, musical instruments, sleepers and bridge timbers.

KELEDANG (*Artocarpus* spp.), also called nangka or jack tree, includes the heavy species of *Artocarpus* and is distributed from India to Southeast Asia. The light species are described under TERAP. The heartwood is gold brown, dark yellowish brown or russet brown according to species. Keledang is a coarse textured wood that splits easily in nailing. It has excellent resistance to weathering and to insect attack. It is used in house construction, interior trim and furniture and for pulp. Its seeds and fruit are edible.

KEMPAS (*Koompassia malaccensis*), also known as tong bueng (Thailand), kayu raja (Sarawak) and mengaris (Borneo), is found in Malaysia, Indonesia and Thailand. The sapwood is pale yellow in color, distinct from the heartwood, which is reddish brown with pale brown streaks or lines due to the confluent or aliform parenchyma which surrounds the vessels. The texture is rather coarse and the wood is hard and fairly difficult to work with hand and machine tools. It is prone to surface checking and to splitting in nailing, and it is liable to insect attacks. It is used for sleepers (treated with preservative), structural timber, poles, plywood cores, parquetry and high class flooring.

KERANJI (*Dialium* spp.) is distributed widely in Southeast Asia. The sapwood is pale brown and the heartwood turns dark gold or reddish brown according to species. Keranji is a hard and tough wood that is fairly difficult to work and to dry without checking. It is used for underwater construction, decorative work and flooring and makes attractive panelling. Its fruits are edible.

KERUING (*Dipterocarpus* spp.), also called gurjun, hora (Sri Lanka), kanyin (Burma), yang (Thailand), chhoeuteal (Cambodia), dau (Vietnam), apitong (Philippines) and lagan (Indonesia), is distributed from India to Southeast Asia. The sapwood is pale yellowish or grayish brown, distinct from the heartwood, which is reddish brown, purplish red and orange. The texture is coarse and the wood contains 1% of silica. Keruing is hard but fairly easy to work with hand and machine tools. It splits easily in nailing and is not durable. Some species exude resin in service even after drying. It is used for plywood, particleboard, fiberboard, pulp, structural timber, flooring, sleepers, poles, marine piles and wagon sides.

KHASYA PINE (*Pinus khasya*), also called Benguet pine, northern Burma pine or Langbian three-leaved pine, is found in India, Burma, Thailand, Indochina and the Philippines. The sapwood is whitish or cream-colored and not distinct from the heartwood, which is pale yellow or reddish yellow. The wood works smoothly by hand and machine tools, except for gumming caused by its high resin content.Its uses are in house construction, mine timbers, poles and boxes.

MERANTI (*Shorea* spp.) is divided into the following four groups depending on wood color: (a) dark red meranti, also called dark red seraya (Sabah), meranti ketuko (Indonesia), nemesu (Malaysia) and red lauan (Philippines); (b) light red meranti, also called light red seraya (Sabah), meranti merah (Indonesia), meranti rambai (Malaysia, Sumatra, Kalimantan), tembaga (Malaysia), saya (Thailand) and almon or mayapis (Philippines); (c) yellow meranti, also called yellow seraya (Sabah), meranti kuning (Indonesia) and yellow lauan (Philippines); and (d) white meranti, also called melapi (Sabah), meranti puteh (Indonesia), lun pateh (Sarawak), manggasinoro (Philippines), bo-bo (Vietnam), lumbor (Cambodia) and pa-nong (Thailand). The red meranti group consists of about 70 species, the yellow meranti group about 39 species and the white meranti group about 25 species, the largest number in each group being found in Malaysia and Borneo. The color of the heartwood of meranti varies from whitish to dark red, as indicated by the subgroup names. The sapwood is generally paler but not always well defined. The wood is coarse textured, varies from soft to medium hard and tends to be brittle. It dries easily but often with distortion and can be nailed readily. Meranti is not resistant to fungi and insects. It machines and works well except for white meranti, which dulls tools because it contains silica but can be peeled for veneer. It is used for plywood, fiberboard, particleboard, pulp, building construction, interior trim and fittings, furniture, boxes and crates, boat- and shipbuilding and vehicle bodies.

MERAWAN (*Hopea* spp.), also called thengan (Burma), takien (Thailand), sao (Vietnam), koki (Cambodia), manggachapui (Philippines) and selangan (Indonesia), is the group of lighter species of *Hopea* (see GIAM for the heavier species). Merawan is found in Burma, Thailand, Indonesia, Malaysia and the Philippines, much of it in continental Southeast Asia. The sapwood is pale yellow or pale yellowish brown and the heartwood yellowish brown, sometimes olive brown. The texture is rather fine and even, but the wood is fairly difficult to saw and work. Merawan dries slowly, without distortion or surface checking. Planed surfaces are lustrous. The wood stains well but glues poorly and is not suitable for plywood. It is durable in contact with the ground. Merawan is used for construction, flooring, ceiling boards, panelling, doors, handrails, interior trim and fittings, vehicle bodies, ship- and boatbuilding, barrels and vats. The wood also yields a resin, dammar, used to make varnish.

MERKUS PINE (*Pinus merkusii*), also called Indochina pine, Sumatra pine and Mindoro pine, is found in Burma, Thailand, Java, Sumatra and the Philippines. The wood is pale beige with growth-ring markings of dark reddish brown or brownish gold. It contains much transparent resin, is liable to blue staining and is relatively nondurable. Uses are in interior trim and fittings and general construction.

MERSAWA (*Anisoptera* spp.), also known as krabak, thing-kadoo (Burma), phdiek (Cambodia), bac (Laos), venven (Vietnam), pengiran (Sabah) and palosapis (Philippines), is found from Bangladesh to the Philippines. The timber may be pale yellow or pale yellowish brown in color, with pale pink or pale brown streaks. Its texture is coarse and even, and the blunting effect of the silica content on cutting edges is severe when it is worked by hand and machine tools. It is used in light construction, flooring, interior fittings, furniture and plywood.

NYATOH (*Palaquium* spp. and *Payena* spp.) also called mernaki, gutta-percha tree or padang, kirihiriya (India), phan-sat (Laos), svakom (Cambodia), viet (Vietnam), nato (Philippines), njatoh (Indonesia) and white or red planchonella (Papua), is distributed from the Seychelles and India to Australia, New Zealand and Polynesia. The sapwood is lighter in shade than the heartwood, which varies from grayish pink and pinkish beige to dark reddish brown. The texture is moderately fine and even. Depending on silica content, nyatoh is more or less readily worked with hand and machine tools. It seasons fairly well, with a tendency to collapse depending on the species. Liable to insect and termite attacks and extremely resistant to impregnation, it is used for building construction, ceiling boards, doors, furniture and plywood. Some species yield gutta-percha.

PADAUK (*Pterocarpus* spp.), other names for which are narra (Philippines), sena (Malaysia), angsana (Sabah), pradoo (Thailand), sonokembang (Indonesia), vermilion wood (USA) and amboyna (the name for figured veneer), is distributed throughout Southeast and southern Asia. Padauk is a remarkably handsome wood. Its sapwood is white or pale yellow, the heartwood orange–brown, red–brown or pinkish brown with purple streaks. It has fine texture and works readily with hand and machine tools. It dries, finishes, glues and coats well and is durable and resistant to impregnation. Padauk is used for decorative veneer, fine furniture, cabinetry and panelling and is also a source of dye.

PUNAH (*Tetramerista glabra*), also called entuyut (Sarawak), is found in Indonesia, Malaysia, Sarawak and Brunei. The heartwood is pinkish brown with an orange–brown tinge after seasoning; the sapwood is lighter but not distinct, and is prone to blue stain. Punah is used for building construction and furniture.

RAMIN (*Gonystylus bancanus*), also called melawis (Malaysia) and ramin telur (Sarawak), includes other species of *Gonystylus* on occasion, and is found from Malaysia to the Philippines. Fresh-cut wood is pale yellow and emits an offensive odor but turns whitish yellow and loses the odor as it dries. The wood has a fine texture and is moderately hard with good resistance to abrasion due to excellent side grain. It machines and seasons well, but large shrinkage and splitting may occur. The timber splits easily in nailing but has good nailholding power. It is used for interior trim, flooring, precision instruments and furniture.

RENGAS (*Gluta* spp. and *Melanorrhoea* spp.), also known as gluta (India), thit-say-pen (Burma), rak (Thailand), kroeul (Cambodia) and son (Vietnam), is distributed from India to Southeast Asia. Rengas is composed of various genera and species of the *Anacardiaceae* family. The sapwood is very wide and white–yellow to straw in color, clearly demarcated from the heartwood, which is vivid red with dark streaks. The wood is hard and dulls tools quickly due to the silica content. The sap is a strong irritant and even dry wood may irritate the skin. Rengas is used for cabinetry, decorative interior fittings, sliced veneer and sleepers. The famous japan lacquer is produced from the sap of Burmese species.

SENGKUANG (*Dracontomelum* spp.), also called dào or lamio (Philippines), dahu (Indonesia) and New Guinea walnut is distributed throughout Southeast Asia. The wood varies from green–yellow with irregular dark brown streaks to grayish brown like walnut. It machines, works and finishes well but is not durable. It is used for furniture, cabinetry, plywood and sliced veneer.

SEPETIR (*Sindora* spp.), also called makata (Thailand), go or gu (Vietnam), kayu galu (Philippines) and sindur (Indonesia) is found throughout Southeast Asia. A closely related species is swamp sepetir or sepeti paya (*Pseudosindora palustris*), grown in Borneo only. The wood is brown and has a rather coarse texture with an oily feel and a fine odor. Sepetir is rather difficult to work, as tools gum up quickly. It is used for cabinetry, furniture, joinery, musical instruments and veneer.

TEAK (*Tectona grandis*), also known as kyun (Burma), sak (Thailand), may sak (Laos, Cambodia, Thailand) and djati (Indonesia), is distributed in the monsoon zone in tropical Asia. It is also planted in other tropical and subtropical zones of the world. The sapwood is narrow and yellow–brown and the heartwood yellowish brown or dark brown, sometimes figured with darker markings. Its texture is fine to coarse and it has a handsome color and figure. The wood has a waxy feel and strong odor reminiscent of old leather. It is fairly easy to work and to season. It finishes well but prevents the hardening of unsaturated polyester coatings. It is resistant to insects and fungi; temples and houses built of teak in the hot and moist climate of Thailand have kept well for centuries. Uses of teak are for high class interior fittings and trim, furniture, cabinetry, panelling, flooring, boat- and shipbuilding, carvings, sliced veneer and sleepers.

TERAP (*Artocarpus* spp.) consists principally of the light species of *Artocarpus* in Malaya and Sadah, while the heavy species are called keledang. Terap also includes species from other genera: *Antiaris* spp. and *Paratocarpus* spp. The wood is yellowish brown and fairly easy to work and to dry. It is prone to termite and marine borer attacks. Terap is used for plywood, boxes and joinery.

WHITE LAUAN (*Pentacme contorta*), also called lamao, is found in the Philippines. The wood is pale gray at first, turning pale pinkish brown after seasoning. Its texture is coarse and it machines, works, dries, glues and finishes well, but it is not durable. It is used for house construction, interior trim, flooring, cabinetry, furniture, boxes, fiberboard and plywood. (See MERANTI for red lauan.)

See also: Resources of Timber Worldwide; Timbers of the Far East

Bibliography

Brown W H 1978 *Timbers of the World*, Vols. 3–5,8: *Timbers of Southern Asia; Timbers of South East Asia; Timbers of the Philippines and Japan; Timbers of Australasia.* Timber Research and Development Association, High Wycombe, UK

Chudnoff M 1984 *Tropical Timbers of the World*, Agriculture Handbook No. 607. US Department of Agriculture, Washington, DC.

Keating W G, Bolza E 1982 *Characteristics, Properties and uses of Timber*, Vol. 1: *South-east Asia, Northern Australia and the Pacific.* Texas A & M University Press, College Station, Texas

Lincoln W A 1986 *World Woods in Color.* Macmillan, New York

Rendle B J (ed.) 1970 *World Timbers*, Vol. 3: *Asia and Australasia and New Zealand.* Benn, London

A. Takahashi
[Shimane University, Matsue City, Japan]

Timbers of the Far East

The Far East—Japan, Korea, China and their surrounding regions—includes various climatic types ranging from subtropical to subpolar zones. The number of tree species is very large, with 840 in Taiwan alone and 2500 in Japan, but this article is confined to six species of softwood and eight of hardwood representing the commercially most important timbers, though related species with similar properties and uses are mentioned. The information given includes nomenclature, distribution (origin), and descriptions of tree and wood characteristics. Physical and mechanical properties are shown in Table 1.

AKAMATSU (*Pinus densiflora*), also known as Japanese red pine, sonamu (Korea) and rìběn chisōng (China), is found in Japan, Korea and northeast China. Akamatsu is the most widely distributed and abundant conifer in Japan and is an important species for planting. It is a large tree, attaining a height of 30 m and a diameter of 0.5 m or more. The sapwood is pale yellow and distinct from the pale beige heartwood. The growth rings are distinct and wide with prominent summerwood bands. The grain is usually straight and the texture is coarse. The wood has a high resin content and the surface feels greasy. It may also contain pitch streaks. The wood is moderately hard and heavy and moderately easy to work and to dry. It has excellent weathering properties, but is liable to blue stain before drying. The timber is used in house construction, joinery, furniture, musical instruments, carvings, sliced veneer, bridge timbers and railway sleepers, and for pulp, fiberboard, particleboard and activated carbon. Species with similar properties and uses are kuromatsu or Japanese black pine (*P. thunbergii*), Taiwan red pine (*P. taiwanensis*), Masson's pine or ma wei son (*P. massoniana*) and Chinese pine (*P. tabulaeformis*). Masson's pine is an important plantation timber in China.

CAMPHOR TREE (*Cinnamomum camphora*), also called Cinnamon wood, kusunoki (Japan) and zhāngshù (China), is found in Japan, Taiwan, China and Indochina. It is an evergreen broadleaved tree up to 25 m high with a diameter of 2 m and occasionally 3 m. The wood is diffuse porous with pores solitary or in pairs. The growth rings are conspicuous, delineated by a narrow and dense band of summerwood. The heartwood is grayish pink, pale beige or pale yellow–pink and is not clearly distinguishable from the sapwood. The grain is generally interlocked but occasionally straight, and the texture is coarse. Planed surfaces are lustrous and attractive, and occasionally finely figured. The vessels contain abundant tyloses. The wood has a characteristic camphor odor but no taste; it contains about 2–10% camphor oil consisting of camphor, safrole, cineole, camphene and other constituents. The wood is moderately light in weight and soft. It seasons fairly well with some tendency to warp and collapse. It is moderately hard to split and moderately strong in impact loading and in endwise compression. It is durable when exposed to conditions favorable to decay. Uses are in building construction, flooring, decorative posts, cabinets, interior trim, doors, furniture, precision equipment, carvings, fancy goods, decorative veneer and plywood, and last but not least the production of camphor oil for use in insecticides and pharmaceutical products. There is also a famous species, Borneo camphor wood (*Dryobalanops aromatica*), in southeast Asia that is used for producing camphor.

HINOKI (*Chamaecyparis obtusa*), also known as Japanese cypress, pyeonbaek (Korea) and rìběn bianbăi (China), is found in Japan, where it is one of the principal plantation trees. A conifer, it reaches a height of 30 m and a diameter of 1 m. The sapwood is not demarcated from the pale yellow–pink to pink heartwood. The grain is usually straight, giving an attractive, fine and regular figure. The texture is fine and even. The wood contains γ-cadinene, α-terpineol and borneol and is scented and oily; the scent is prized by the

Table 1
Physical and mechanical properties of Far East timbers

	Density ($g\,cm^{-3}$)			Shrinkage coefficient (%)[a]		Bending modulus of elasticity[b] (MPa)	Bending strength[b] (Mpa)
	min.	max.	mean	radial	tangential		
Akamatsu	0.42	0.62	0.52	0.18	0.29	11300	88
Camphor tree	0.39	0.67	0.49	0.19	0.32	8800	69
Hinoki	0.30	0.49	0.40	0.12	0.23	8820	74
Japanese beech	0.50	0.75	0.60	0.18	0.41	11800	98
Japanese larch	0.37	0.56	0.46	0.18	0.28	9800	79
Japanese oak	0.45	0.90	—	0.19	0.35	9800	98
Katsura tree	0.37	0.63	0.47	0.17	0.28	8300	74
Keyaki	0.47	0.84	0.69	0.16	0.28	12000	98
Korean pine	0.35	0.57	0.49	0.08	0.21	7260	39
Manchurian ash	0.43	0.74	0.55	0.17	0.31	9300	93
Royal paulownia	0.17	0.37	0.27	0.09	0.23	4900	34
Sen	0.40	0.69	—	0.17	0.34	8330	68
Sugi	0.27	0.41	0.35	0.10	0.25	7350	64
Yezo spruce	0.32	0.48	0.40	0.15	0.29	8820	69

[a] Per 1% change in moisture content [b] At 15% moisture content

Japanese but generally disliked by Americans and Europeans. The wood is light, soft, resilient and tough. It is worked smoothly by hand and machine tools and dries rapidly and easily, but its resistance to splitting is low. It planes well, producing a lustrous surface. The wood has excellent durability, with high resistance to attack by insects and fungi. The timber is used in house construction, foundations, siding, interior trim, joinery, precision equipment, handles, carvings, bridge construction, poles, piles and sliced veneer, and for pulp, fiberboard and particleboard. Species of similar appearance, properties and uses are Taiwan red cypress or tai wǎn gui (*C. formosensis*) and Port Orford cedar (*C. lawsoniana*).

JAPANESE BEECH (*Fagus crenata*), also known as Siebold's beech, Buna (Japan), neodobam namu (Korea) and shuǐ gīng gǎng (China), occurs in Japan, where it is the most abundant hardwood. It is a deciduous broadleaved tree that often attains a height of 30 m and a diameter of 1.5 m. The wood is diffuse porous and the vessels occupy about 40% of the wood volume. Sapwood and heartwood are not demarcated, both being a pale beige or light brown. False heartwood is common and is beige or pale grayish yellow. The rays are conspicuous on all surfaces, forming a silvery grain on the radial surface. The grain is usually straight or roey and the texture fine. The wood is moderately hard and heavy but is not difficult to work and finishes smoothly; it is very suitable for steam bending. It must be seasoned carefully to avoid warping and splitting, and is liable to various fungal and insect attacks in the green condition. Uses are in furniture, flooring, tool handles, slack cooperage, woodenware, vehicle bodies, railway sleepers, plywood, fiberboard, particleboard and pulp. This wood is similar to European beech (*F. sylvatica*), Taiwan beech (*F. hayatae*), Korean beech (*F. multinervis*) and Chinese beech (*F. longipetiolata*).

JAPANESE LARCH (*Larix leptolepis*), also known as karamatsu (Japan), nagiopsong (Korea) and luò yè sōng (China), is found in Japan, where it is one of the principal plantation species. It is a deciduous conifer with an average height of 30 m and a diameter of up to 1 m. The sapwood is beige and distinct from the heartwood, which is light reddish brown. The distinctive growth rings are delineated by dark and wide bands of summerwood. The grain is usually straight and the texture coarse. The wood has a greasy feel, due to a high resin content, and often exudes resin or pitch when green. It is moderately heavy and hard, and contains much arabinogalactan, a hemicellulose fraction which can interfere with the setting of cement. The wood can be worked smoothly by hand and machine tools. It dries well with moderate distortion, but splits easily. The heartwood has high durability under water. The timber is used in building construction, foundations, shingles, piles, railway sleepers, bridges and shipbuilding, and for pulp, fiberboard and particleboard. The bark is a source of tannin, and the wood of terpineol. The wood is not suitable for concrete forms or excelsior-cement board. Species with similar properties and uses are Dahurian larch (*L. gmelini*) and Siberian larch (*L. sibirica*).

JAPANESE OAK (*Quercus crispula*), also known as kraft oak, mizunara (Japan), tokukasu namu and mulcham namu (Korea), tsuo-shu and shuǐ giu (China) and dub mongoruskii (USSR), is found in Japan, Korea, northern China, Siberia and Sakhalin. It is a representative Japanese tree and sometimes forms entire forests in Japan. The deciduous broadleaved tree is up to 30 m high with a diameter of 1.5 m. The sapwood is pale whitish yellow and clearly demarcated from the heartwood, which is beige with silver grain when quartersawn, or with grayish brown stripes. Vessels, with tyloses, 0.1–0.3 mm in diameter, from one to three cells wide, form a ring in the earlywood. The grain may be irregular or straight and the texture is coarse. The wood varies from light and soft when slowly grown to hard and heavy when grown fast, the hard and heavy wood being rather difficult to saw and plane. It must be seasoned with care, as there is a marked tendency to split and distort. It planes to a smooth surface and polishes or varnishes well. Uses are in interior trim, flooring, stairs, handrails, doors, panelling, furniture, tight cooperage and skis; the bark is used to extract tannin. Other species of oak of similar properties and uses in northeast Asia are Mongolian oak (*Q. mongolica*), kashiwa (*Q. dentata*), kunugi (*Q. acutissima*), konara (*Q. serrata*) and abemaki (*Q. variabilis*).

KATSURA TREE (*Cercidiphyllum japonicum*), also called katsura (Japan), giesu namu (Korea) and lián xiāng shù (China), occurs in Japan. Deciduous and broadleaved, katsura is a large tree that reaches a height of 30 m with a diameter of 1 m and occasionally 2 m. The tree is native to Japan, but there is a variety in China. The wood is diffuse porous and the sapwood is pale yellowish beige with occasional pale green streaks and clearly demarcated from the heartwood, which is pale reddish brown or light yellowish brown. The summerwood is narrow and distinctive, the grain fairly straight and the texture fine and even. Vessels occupy up to 52% of the wood volume. The wood is very light and soft, can be worked smoothly by hand and machine tools and seasons very well with little distortion. It planes to a very smooth lustrous surface and splits easily. The timber is not durable when used under conditions favoring decay. Katsura contains katsuranin, a phenolic compound, in the heartwood and tannin in the bark. Uses are in implements, furniture, interior trim, musical instruments, boxes, carvings, parquetry, inlaid work, patterns, woodenwear, sliced veneer and plywood.

KEYAKI (*Zelkova serrata*), also known as zelkova and keaki, neutinamu (Korea) and guāng re jǔ shu (China), occurs in Japan, China, Taiwan and Korea. It is one of the best timbers in Japan. The sapwood is a pale yellowish beige and the heartwood light yellowish brown or brownish gold. The wood is ring-porous with vessels 0.1–0.25 mm in diameter in groups of one to three in the earlywood. The texture is rather coarse, but the wood is highly prized for various kinds of beautiful figure. The wood is moderately heavy and hard but can be worked smoothly by hand and machine tools and has good steam bending properties. It contains keyakinin, a phenolic compound, in the heartwood and is resistant to termites and fungal decay. The timber is used in construction of temples and shrines, interior and exterior trim, flooring, ceilings, doors, implements, furniture, ship- and boatbuilding, railroad ties, carvings, crossarms and sliced veneer.

KOREAN PINE (*Pinus koraiensis*), also called benimatsu, chōsen-matsu, chōsen-goyō (Japan), chatsu namu (Korea) and hongsōng or hai sōng (China), grows in Korea, Japan, northeast China and Siberia. It is an important plantation tree in Korea and China. A conifer, it reaches a mean height of 30 m and a diameter of 1 m. The sapwood is pale beige and the heartwood pinkish beige or dull red. The grain is usually straight and the texture rather coarse with distinct growth rings delineated by a dark band of latewood. The surface is lustrous and greasy. Resin may exude on the surface of quartersawn heartwood, sometimes in heavy streaks. The wood is moderately light and soft, can be worked well with hand and machine tools and seasons rapidly and well. It has excellent weathering properties. Uses are in housing construction, siding, interior trim, joinery, furniture, musical instruments, precision equipment, carvings, crates, patterns, ship- and boatbuilding and vehicle bodies. The seeds are edible. Related pines of similar properties and uses are Siberian stone pine or Manchurian pine (*P. sibirica*) and Japanese white pine or himekomatsu (*P. pentaphylla*).

MANCHURIAN ASH (*Fraxinus mandshurica*), also known as Japanese ash and Asiatic ash, tamo and yachidamo (Japan), deulme namu (Korea), shui-chuiryu, howa chuiryu, riběn shuǐgu liŭ (China) and yasen mandsurushii (USSR), is found in northern Japan, northeast China, Korea, Siberia and Sakhalin. It is a deciduous broadleaved tree about 25 m high and 1 m in diameter with a long and straight bole. The sapwood is pale yellowish white and distinct from the heartwood, which is beige or grayish yellow. The vessels are arranged in groups of two or three in the earlywood and have a diameter of 0.1–0.4 mm. The grain is mostly straight and the texture coarse, producing a handsome figure. The wood is moderately hard and heavy and easy to work and to dry; it has good steam bending properties, and takes a smooth finish. It splits readily and makes excellent firewood. The timber is used in interior trim, furniture, general implements, boat- and shipbuilding, sporting goods, musical instruments and plywood and for pulp, fiberboard and particleboard. Species of similar properties and uses are shioji (Japan) or bai la shu (China) (*F. spaethiana*), Korean ash (*F. rhynchophylla*, toneriko (*F. japonica*), Chinese ash (*F. chinensis*) and harunire or Japanese elm (*Ulmus davidiana* var. *japonica*).

ROYAL PAULOWNIA (*Paulownia tomentosa*), also called empress tree and princess tree, kiri (Japan), odong namu (Korea) and riběn pào tóng (China), is found in Japan, Korea and China. For centuries it has been planted extensively in Japan and Korea for its timber. A deciduous broadleaved tree about 10 m high with a diameter of 0.5 m, it grows very rapidly. The sapwood is not demarcated from the beige or grayish-yellow heartwood, which often has wide streaks of grayish-brown hue. The growth rings are delineated by a somewhat darker band of latewood and are 5–10 mm wide. The grain is usually straight and the texture somewhat coarse. Planed surfaces are fairly lustrous. The wood is ring-porous with abundant tyloses. It is the lightest and softest of all Japanese woods and can be worked smoothly by hand and machine tools. It dries well with little degradation and distortion. The wood has a very low thermal conductivity and is difficult to burn. Uses are in ceiling boards, decorative interior trim, joinery, furniture, small boxes, woodenwear, musical instruments, sporting goods and carving. The bark is used to extract dye. Species of similar properties and uses are kawakami paulownia or Taiwan giri (*P. kawakamii*), fortune paulownia or kokonoe giri (*P. taiwaniana*) and Korean paulownia or chōsen giri (*P. coreana*). Fortune paulownia is the principal species planted in China and Taiwan. Recently it has also been planted in Central and South America.

SEN (*Acanthopanax ricinifolius* or *Kalopanax pictus*), other names for which are Japanese ash, castor arabia or sem, harigiri (Japan), womu namu or eom namu (Korea) and ci qiū or tsu-chiu (China), occurs in Japan, northeast China, Korea, Siberia and Sakhalin. On the world market sen is best known as a hardwood plywood species. It is a deciduous broadleaved tree up to 20 m high and 1 m in diameter. The wood is ring-porous, resembling a soft type of ash. The sapwood is pale yellowish white and not clearly demarcated from the heartwood, which is pale yellowish or grayish brown. Vessels in earlywood are mostly solitary, 0.2–0.4 mm in diameter. The grain is straight and the texture coarse. The wood is rather light and soft and has a prominent growth-ring figure on flat-sawn surfaces; it is nonresinous, odorless and tasteless. It machines, finishes and dries well. It is easy to split, has moderate resistance to decay and is fairly resistant to impregnation. Uses are in interior trim, furniture, cabinets, doors, musical instruments, sporting goods, carvings, turnery, vehicle bodies, ship- and boatbuilding, railway sleepers, bridge timbers, sliced veneer and plywood and for pulp, particleboard and fiberboard.

SUGI (*Cryptomeria japonica*), also called common cryptomeria, Japanese redwood (Japan), sam namu (Korea) and liŭ shā (China), is found in Japan, eastern and central China and Taiwan. Sugi is the most important plantation tree in Japan and Taiwan. It is a conifer that attains a height of 30–40 m with a diameter of 1 m or more. The sapwood is pale reddish yellow and clearly demarcated from the heartwood, which varies from dull reddish brown to deep red. Growth rings are distinct and relatively wide, but the summerwood bands are narrow. The grain is straight and the texture coarse and even. The wood is very light and soft and fairly easy to work with hand and machine tools. It dries well without degradation but is easy to cleave. The heartwood weathers well. The wood has a pleasant scent because it contains about 2% of hydroxysugiresinol, and is permeable to preservatives. The timber is used in building construction, panelling, ceiling boards, joinery, furniture, poles, piles, vats and boxes. Species with similar properties and uses are taiwania (*Taiwania cryptomerioides*), Chinese fir or shā mū (*Cunninghamia sinensis* or *C. lanceolata*) and Formosa fir (*C. konishii*). These species are important plantation trees in China and Taiwan.

YEZO SPRUCE (*Picea jezoensis*), also known as yeddo spruce, yezomatsu (Japan), gamunbi namu or kamunpi (Korea) and yūlin-sōng (China), grows in northeast Asia, northern Japan, Korea and Sakhalin. It is one of the most important plantation trees in Japan, and is a conifer that reaches a height of 30 m and a diameter of 1 m. The sapwood and heartwood are not demarcated, both being pale beige or pale yellow. Fine pitch streaks are found frequently. The grain is straight and the texture fine and even. Planed surfaces are lustrous. The wood is very light, soft and resilient. It machines, seasons and finishes well. The wood splits easily but has excellent weathering properties. Uses are in house construction, interior trim, siding, joinery, musical instruments, poles, ship- and boatbuilding, vehicle bodies and airplanes and for pulp and fiberboard. Species with similar properties and uses are Hokkaido spruce or akayezomatsu (*P. glehnii*), Siberian spruce (*P. obovata*), Japanese fir or todomatsu (*Abies sachalinensis*), Taiwan fir (*A. kawakamii*), Taiwan spruce (*P. morrisonicola*), euan sun (*P. asperata*) and others in China (*P. manshurica*, *P. pungsanensis*, *P. tonaiensis*, *P. aurantiaca* and *P. retroflexa*).

See also: Resources of Timber Worldwide; Timbers of Southeast Asia

Bibliography

Brown W H 1978 *Timbers of the World*, Vols. 3–5: *Timbers of Southern Asia; Timbers of South East Asia; Timbers of the Philippines and Japan.* Timber Research and Development Association, High Wycombe
Hirai S (ed.) 1980 *Encyclopedia of Woods*, Vols. 1–17. Kanae Shobō, Tokyo [in Japanese]
Lincoln W A 1986 *World Woods in Color.* Macmillan, New York
Rendle B J (ed.) 1970 *World Timbers*, Vol. 3: *Asia and Australia and New Zealand*, Benn, London

A. Takahashi
[Shimane University, Matsue City, Japan]

Trade, Prices and Consumption of Timber in the USA

Since the turn of the century, the annual consumption of timber products in the USA has consistently hovered between 311 and 368 million cubic meters of wood. The only major drop in the consumption level occurred during the depression of the 1930s. Physical scarcity of wood fiber has not, with limited exception, been of major concern. The store of growing stock in the USA has outpaced consumption and will probably continue to do so. Economic scarcity as measured by changes in resource and product prices is, however, readily apparent for a number of timber species and product categories—most notably sawtimber. Timber prices are expected to increase through the year 2030, and shifts in the mix of wood and nonwood products consumed will probably occur.

1. Prices, Production and Consumption: General Trends

For decades, wood has played an important role in the development of the economic and social fabric of the USA. As a material, it has found its way into a variety of uses. Since the turn of the century, however, notions of wood scarcity have surfaced at regular intervals. In fact, scarcity concepts applied to wood fiber supplies often serve as foundations for many of today's public and private forestry programs. These concepts are basically twofold, namely physical scarcity and economic scarcity. The distinction between the two is critical to the development of policies focused on wood supplies, and has been recognized by numerous authors (Barnett and Morse 1963, Smith 1979, Potter and Christy 1962, Manthy 1978). Physical concepts of scarcity are embodied in Malthusian doctrine which implies that wood supplies are insufficient to meet consumption requirements and needs, and that social standards decline as a result. Major concern is focused on the physical adequacy of wood fiber. In contrast, economic concepts of scarcity imply price as a device which signals changes in the condition of supply and demand for a resource. Prices rise because of supply and/or demand factors. Supply conditions leading to long-term price increases include reduced physical stock of wood as a raw material, declining quality of timber resources, rapid rises in the cost of labor or capital required to process timber, and the advantages of monopoly power. Demand changes resulting in long-term price increases include increases in expansion of timber consumption rates faster than the market can supply (Manthy 1977).

In general, physical scarcity of timber-growing stock is not readily apparent in the USA—at least nationwide and for all raw material categories combined (Table 1). Growing stock volumes increased nearly 18% between 1952 and 1976 and are expected to increase by 25% over 1976 levels by 2030. Consumption levels in the latter year are expected to be 424×10^6 m^3 greater than occurred in 1976—an increase that is substantially behind the 5.1×10^9 m^3 increases in growing stock that will occur during the same period. Interpreting aggregate figures of this sort must be done cautiously since the physical supply of some materials has not kept pace with "desirable" consumption levels (e.g., high-quality veneer logs). However, in absolute terms, growing stock volumes are expected to far outpace increases in consumption levels in the years ahead.

Economic scarcity concepts applied to timber as a raw material lead to somewhat different conclusions. In general, there has been a rising trend in the price of all forest products combined since 1870 (Fig. 1). The trend can be attributed almost exclusively to sawlog price increases before 1945 and rising veneer log prices in the late 1940s (Manthy 1977). Between 1950 and the early 1970s, real forest product prices remained relatively stable—a situation evident for nearly all major forest product categories (i.e., sawlogs, veneer logs and pulpwood). Price rises since 1970 can be attributed mainly to increased sawlog prices.

Table 1
US roundwood consumption, timber removals, net annual growth and growing stock inventory, 1952, 1962, 1970 and 1976 with projections to 2030

Item	Amount of wood (10^6 m^3)								
	1952	1962	1970	1976	1990	2000	2010	2020	2030
Roundwood consumption	337	328	360	377	575	643	714	762	801
Growth stock inventory	17086	18345	19268	20132	22204	23103	24662	24545	25189
Net annual growth	394	473	559	613	642	632	615	601	590
Removal from growing stock	336	339	398	403	510	523	554	586	603

Source: Forest Service 1980a

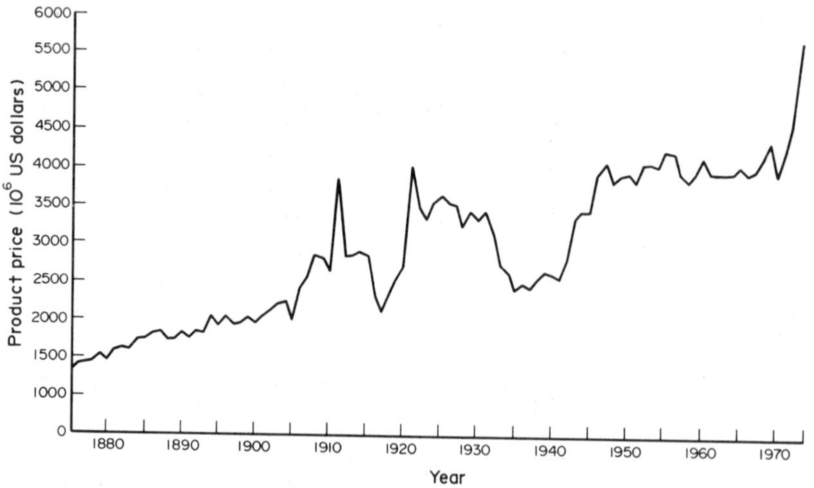

Figure 1
Trends in product price (deflated by the 1967 wholesale price index) for all US forest products (after Manthy 1978a)

2. *Price and Consumption Levels 1950–1985: Wood Product Classes*

Wood as a material is consumed in a variety of manufactured forms. When categorized according to product class, trends in price and consumption of wood become more apparent and meaningful to policy-setting organizations. In 1984, domestic wood fiber consumption (estimated at 198.2 Mt air-dry weight) was distributed as follows: lumber, 24%; plywood and veneer, 5%; panel products, 4%; woodpulp, 28%; fuelwood, 35%; and other industrial products, 4%.

2.1 Lumber

Nearly one-quarter of all wood fiber consumed in the USA is in the form of lumber (Table 2). Since 1950, total consumption has seen only moderate increases. In contrast, real prices for lumber have increased dramatically. Prices for all types of lumber in 1985 were nearly 4 times those in 1950, with the bulk of the increases occurring since the mid-1950s. The most dramatic increases in price have occurred in the softwood lumber market, although hardwood lumber prices have also experienced substantial increases.

2.2 Plywood

Consumption of plywood in the USA increased over 400% between 1951 and 1985 (Table 3). The bulk of this increase is attributed to rising demand for softwood plywood. Plywood prices declined between 1950 and the early 1960s, remained stable through 1979 and since then have declined. The most dramatic decrease

Table 2
US lumber consumption and producers prices by hardwood and softwood, 1950–1985

	Consumption (10^6 m^3)			Price (1967 US dollars)		
Year	Total	Softwood	Hardwood	All lumber	Softwood lumber	Hardwood lumber
1950	96.5	78.8	17.7	83.6	88.1	82.1
1951	91.3	72.9	18.4	93.7	95.6	88.2
1952	92.5	75.3	17.2	91.3	95.2	81.2
1953	91.8	74.6	17.2	90.5	93.2	82.8
1954	91.3	74.3	17.0	88.9	91.8	81.0
1955	94.6	76.7	17.9	94.5	97.7	85.7
1956	96.5	77.4	19.1	96.5	98.5	91.1
1957	82.6	68.9	13.7	90.9	92.6	86.3
1958	85.2	70.8	14.4	89.5	90.8	86.3
1959	95.6	79.5	16.1	96.6	98.7	89.9
1960	85.0	69.9	15.1	92.1	92.7	90.8
1961	83.8	69.6	14.2	87.4	87.9	86.2
1962	88.0	72.7	15.3	89.0	90.1	86.0
1963	92.5	75.0	17.5	91.2	92.1	88.8
1964	96.3	78.8	17.5	92.9	93.3	92.2
1965	97.0	78.8	18.2	94.0	93.1	97.4
1966	96.3	77.4	18.9	100.1	97.7	108.7
1967	91.5	73.5	18.0	100.0	100.0	100.0
1968	97.9	80.2	17.7	117.4	120.7	104.5
1969	96.8	78.4	18.4	131.6	134.5	120.1
1970	93.2	76.0	17.2	113.7	113.3	114.6
1971	102.7	85.7	17.0	136.0	141.6	113.4
1972	108.1	91.6	16.5	159.4	167.7	126.2
1973	109.0	91.8	17.2	205.2	214.3	169.0
1974	94.6	77.9	16.7	207.1	211.4	189.5
1975	87.3	73.4	13.9	192.5	200.6	160.3
1976	102.4	87.1	15.3	233.0	248.1	176.0
1977	114.2	98.2	16.0	276.5	297.4	200.3
1978	120.3	103.8	16.5	322.4	346.0	235.8
1979	118.0	100.2	17.2	354.3	380.0	260.0
1980	100.7	84.5	16.3	325.8	345.1	252.0
1981	91.6	77.2	14.4	325.1	343.0	255.2
1982	88.0	76.7	11.6	310.8	321.6	262.4
1983	105.0	92.5	12.7	352.6	369.8	283.7
1984	114.4	100.3	14.4	349.8	353.9	319.7
1985	116.6	102.4	14.2	340.5	345.3	307.2

Source: Ulrich 1987

Table 3
US plywood consumption and producer prices by hardwood and softwood, 1950–1985

Year	Consumption (10^6 m^2)			Price (1967 US dollars)		
	Total	Softwood	All hardwood	Softwood plywood	Hardwood plywood	Plywood
1950		248.2		148.5	180.9	121.0
1951	394.0	278.2	115.8	144.2	172.9	118.9
1952	413.4	294.1	119.3	135.2	162.0	111.6
1953	485.1	356.7	128.4	142.8	164.8	121.1
1954	402.1	370.0	132.1	134.4	159.0	111.9
1955	657.0	490.2	166.8	137.1	163.3	114.1
1956	674.6	503.3	171.3	127.9	144.7	112.8
1957	688.6	523.9	164.7	117.9	127.1	108.6
1958	768.0	601.5	166.5	117.2	126.3	107.8
1959	923.9	712.0	211.9	121.8	134.3	109.5
1960	889.1	720.6	168.5	115.5	119.3	110.9
1961	977.6	789.2	188.4	113.5	116.4	109.8
1962	1088.3	865.0	223.3	109.3	112.1	105.6
1963	1206.2	963.1	243.1	110.9	115.2	105.4
1964	1336.0	1062.0	274.0	109.3	111.5	106.4
1965	1439.3	1152.2	298.1	107.1	109.4	104.0
1966	1498.0	1189.5	308.5	104.2	106.3	101.5
1967	1478.1	1185.3	292.8	100.8	100.0	100.0
1968	1692.2	1331.5	360.7	112.9	126.0	98.0
1969	1608.5	1240.6	367.9	115.0	130.7	97.7
1970	1656.0	1304.2	351.8	98.2	102.9	92.8
1971	1923.8	1510.4	413.4	100.5	111.4	88.3
1972	2117.1	1637.8	479.3	109.7	130.1	87.6
1973	2027.1	1628.3	398.8	115.2	144.0	83.7
1974	1648.7	1372.1	276.6	100.6	116.7	81.3
1975	1655.9	1386.3	269.6	92.2	114.7	68.3
1976	1910.1	1598.1	312.1	102.2	135.2	66.9
1977	2042.0	1728.8	313.3	109.3	152.3	65.8
1978	2127.7	1789.0	338.7	112.6	155.9	67.0
1979	1952.8	1656.3	296.5	106.3	136.8	71.8
1980	1611.2	1407.2	204.1	91.7	114.9	65.7
1981	1605.0	1659.8	223.9	83.7	104.5	61.2
1982	1550.7	1365.9	184.9	77.5	94.3	60.5
1983	1890.6	1649.3	241.3	80.5	102.4	59.3
1984	1945.2	1724.9	220.4	77.9	97.8	58.1
1985	1996.4	1762.7	233.7	74.9	98.1	52.7

Source: Ulrich 1987

Table 4
US pulpwood consumption and prices, by selected regions and species, 1950–1985

Year	Consumption (10^6 m^3)	Price (1967 US dollars m^{-3})			
		Midsouth[a]		Wisconsin[b]	
		Southern pine	Hardwoods	Pine	Aspen
1950	116.2	4.10	3.59	4.80	3.55
1951	134.7	4.32	3.93	5.53	4.54
1952	130.7	4.46	4.01	5.92	4.33
1953	135.7	4.50	4.04	5.71	4.06
1954	134.3	4.50	4.04	5.59	3.90
1955	149.3	4.59	4.06	5.50	3.93
1956	168.5	4.75	4.11	5.55	4.11
1957	160.9	4.52	3.95	5.30	3.77
1958	154.5	4.46	3.82	5.03	3.57
1959	169.2	4.56	3.81	5.46	3.75
1960	178.1	4.66	3.81	5.13	3.78
1961	178.7	4.62	3.81	5.26	3.79
1962	190.4	4.59	3.83	4.95	3.71
1963	195.7	4.59	3.82	4.79	4.01
1964	209.8	4.63	3.81		
1965	226.1	4.65	3.95		
1966	242.8	4.66	4.17		
1967	242.4	4.73	4.19	5.59	4.32
1968	253.5	4.77	4.26	5.41	4.30
1969	272.2	4.80	4.36	5.08	4.14
1970	274.3	4.70	4.23	5.29	4.00
1971	274.5	4.62	4.19	4.99	3.88
1972	285.5	4.81	4.30	4.95	3.77
1973	308.5	4.88	4.36	5.09	3.79
1974	316.2	4.87	4.36		
1975	259.3	4.52	4.06		
1976	297.4	4.48	4.00	5.52	3.81
1977	308.4	4.46	3.99	5.45	3.66
1978	320.2	4.37	4.00	4.98	3.42
1979	337.5	4.69	3.97	4.83	3.52
1980	331.1	4.43	3.75	4.59	3.53
1981	322.9	4.73	3.70	4.67	3.42
1982	311.7	4.85	3.66	4.51	3.40
1983	339.3	4.81	3.64	4.58	3.24
1984	368.0	4.89	3.64	4.44	3.22
1985	350.8	4.84	3.71	4.88	3.34

[a] Weighted average for all stumpage points [b] Delivered to mill
Source: Ulrich 1987

in real price occurred in hardwood plywood markets, a decline of nearly 56% since 1950. The latter can be attributed in large measure to innovative production processes focused on eastern and western timber species.

2.3 Pulpwood

Domestic pulpwood consumption rose dramatically between 1950 and 1985, i.e., from 116×10^6 m^3 to 336×10^6 m^3 (Table 4). In general, this is consistent with slow but positive increases in per capita consumption of pulp products nationwide. For selected regions and species, dramatic changes in pulpwood prices are not evident. If anything, they are remarkably consistent through time, region and timber species. A major structural change in the pulpwood market has been the advent of chips produced from roundwood and byproducts of primary processing plants. In 1950, 6% of USA pulpwood consumption was in the form of chips (i.e., 4.5×10^6 m^3). By 1985, the percentage had risen to 41% (126.9×10^6 m^3).

2.4 Particleboard, Hardboard and Insulation Board

Hardboard and particleboard have seen major increases in consumption since 1950 (Table 5). Consumption increases have been especially sharp for particleboard, which between 1950 and 1979 increased by a factor of 200, only to decline in the early 1980s followed by a significant recovery in 1984 and 1985. Real prices for particleboard have declined since the

Table 5
US particleboard, hardboard and insulation board consumption and producer prices, 1950–1985

	Consumption[a]			Price (1967 US dollars)		
Year	Particle-board	Hardboard	Insulating board	Particle-board	Hardboard	Insulating board
1950	2					101.2
1951	3					95.8
1952	3					100.3
1953	4					106.9
1954	4	123	246			112.2
1955	7	146	279			114.7
1956	10	152	277			116.2
1957	17	158	248			116.7
1958	22	171	270		107.1	117.7
1959	27	213	293		107.8	120.6
1960	25	201	265		107.0	120.0
1961	30	225	263		107.9	119.2
1962	38	259	268		108.5	111.2
1963	46	291	284		109.8	108.7
1964	59	309	309		107.9	106.0
1965	75	321	315	107.8	105.7	101.7
1966	93	325	288	103.6	102.1	98.6
1967	103	317	300	100.0	100.0	100.0
1968	132	396	327	96.7	95.9	100.5
1969	159	447	340	96.6	93.7	102.2
1970	163	438	302	84.4	92.6	100.1
1971	221	527	361	82.0	88.7	100.4
1972	287	616	369	80.9	85.8	99.9
1973	319	647	369	78.5	78.1	90.3
1974	278	595	302	72.0	73.7	83.6
1975	250	483	271	65.0	67.3	82.3
1976	322	586	314	67.6	71.8	88.0
1977	384	651	324	72.6	73.5	91.6
1978	422	762	326	81.4	75.0	96.8
1979	391	721	315	70.3	69.9	84.4
1980	344	588	262	71.0	69.7	77.4
1981	334	554	197	72.4	74.3	82.6
1982	290		171	73.4	76.3	83.8
1983	368		218	75.6	77.3	86.7
1984	416		255	76.7	75.3	86.9
1985	428		267	75.5	76.2	92.4

[a] particleboard, 1.91 cm basis; hardboard, 0.32 cm basis; insulating board, 1.27 cm basis   Source: Ulrich 1987

mid-1960s, reflecting a rapid increase in available supply of the product. In general, all "board products" have exhibited declines in real prices.

2.5 Paper, Paperboard, Building Board and Woodpulp

Between 1950 and 1985, consumption increased by 167% for paper, 177% for paperboard and 236% for woodpulp (Table 6). Per capita consumption of these products has also increased, often markedly during the same period. In constant US dollars (1967 = 100), prices have been relatively stable, except for building board and woodpulp. The latter experienced substantial price rises in the 1970s. Building board prices have exhibited a general decline since 1959.

2.6 Sawtimber Stumpage

Some of the USA's most dramatic shifts in timber product prices have occurred at the stumpage level, notably sawtimber stumpage (Table 7). Increases were especially dramatic for western sawtimber species and most apparent in the 1950s and 1960s. For example, between 1950 and 1967, real prices for Douglas-fir sawtimber varied in the range of 18–50 US dollars per

Table 6
US paper, paperboard, building board and woodpulp consumption and producer prices 1950–1985

	Consumption (Mt)				Price (1967 US dollars)			
Year	Paper	Paper-board	Building board	Wood-pulp	Paper	Paper-board	Building board	Wood-pulp
1950	16.802	11.047	1.227	17.138	83.0	99.3	99.5	99.0
1951	17.756	11.627	1.273	18.683	83.4	111.9	94.3	106.4
1952	16.961	10.821	1.310	18.198	89.3	111.2	98.6	106.7
1953	17.639	12.418	1.377	19.533	91.6	110.0	105.1	105.7
1954	17.821	12.140	1.492	19.865	92.2	109.8	110.4	106.2
1955	19.341	13.796	1.667	22.323	94.3	111.8	112.9	109.0
1956	20.767	14.111	1.696	23.938	96.6	114.9	114.3	110.0
1957	19.835	13.905	1.609	23.278	97.0	113.0	114.9	107.9
1958	19.527	13.955	1.724	23.385	95.9	111.3	114.7	108.7
1959	21.577	15.226	2.018	26.162	96.5	111.0	116.9	108.4
1960	21.983	15.365	1.869	26.563	97.7	110.2	116.2	107.7
1961	22.403	16.053	1.933	27.812	98.3	103.1	116.1	102.5
1962	23.246	17.058	2.066	29.511	98.4	103.4	111.6	100.3
1963	23.927	17.782	2.255	31.474	98.5	105.5	110.5	99.0
1964	25.369	18.740	2.454	33.777	99.5	107.2	108.0	103.6
1965	26.769	19.885	2.565	35.721	97.9	105.1	104.5	103.6
1966	28.846	21.541	2.396	38.388	97.7	102.4	101.1	100.2
1967	28.801	20.833	2.407	38.126	100.0	100.0	100.0	100.0
1968	30.157	22.783	2.831	42.522	99.5	93.6	98.4	97.6
1969	31.794	24.212	3.000	44.751	99.1	93.3	99.1	93.9
1970	31.699	23.530	2.828	43.969	100.5	91.6	91.5	99.3
1971	32.347	23.916	3.366	45.243	100.2	89.8	90.2	98.3
1972	34.351	26.378	3.780	48.243	97.6	88.6	89.3	93.6
1973	35.704	27.307	3.875	49.986	90.1	85.4	83.7	95.2
1974	35.498	25.718	3.452	49.670	92.8	95.1	77.1	136.0
1975	30.137	22.765	3.052	43.380	98.9	97.4	72.7	162.0
1976	34.466	25.850	3.634	48.930	99.6	96.2	75.8	156.3
1977	36.490	27.039	3.800	50.363	100.1	90.7	80.8	144.7
1978	38.452	28.137	4.142	51.443	98.5	85.8	89.5	127.3
1979	39.703	28.942	3.910	52.559	97.5	85.8	77.4	133.4
1980	39.142	27.764		53.204	95.5	87.3	76.7	141.5
1981	39.034	28.918		53.199	95.4	88.0	79.0	135.3
1982	37.948	26.529		51.247	95.7	85.2	80.0	126.6
1983	41.511	29.333		54.505	93.0	82.8	82.5	114.5
1984	44.776	31.479		58.644	97.6	90.7	83.5	128.0
1985	44.847	30.610		57.533	98.3	89.0	83.3	112.3

Source: Ulrich 1987

Table 7
Average stumpage prices for sawtimber sold from National Forests, by species group, 1950–1985

	Price (1967 US dollars m^{-3})			
Year	Douglas-fir	Ponderosa pine	Southern pine	Eastern hardwoods
1950	8.70	9.74	14.17	
1951	12.13	16.04	16.52	
1952	12.65	13.43	18.91	
1953	10.04	12.87	17.00	
1954	8.04	13.52	14.74	
1955	14.30	12.91	15.83	
1956	18.09	13.04	17.91	
1957	12.22	11.26	14.70	
1958	10.00	8.78	14.30	
1959	16.87	9.43	16.13	9.70
1960	14.65	8.74	15.83	10.43
1961	12.70	5.57	12.35	8.26
1962	11.39	7.39	11.91	8.52
1963	12.83	7.26	11.57	9.91
1964	17.48	8.74	12.78	10.35
1965	19.17	8.91	14.26	11.26
1966	21.78	8.61	16.83	12.78
1967	18.13	9.65	16.65	11.74
1968	25.96	12.83	17.91	10.00
1969	35.57	29.00	21.09	12.35
1970	16.52	12.65	17.35	10.61
1971	18.70	14.35	19.91	9.39
1972	26.17	24.00	23.96	12.52
1973	44.57	29.78	30.13	14.83
1974	54.96	27.30	20.70	12.48
1975	42.13	17.69	14.17	8.43
1976	41.87	24.17	20.65	8.30
1977	50.57	29.43	22.43	8.48
1978	52.00	34.22	27.96	8.52
1979	72.81	44.09	28.65	8.65
1980	69.95	33.36	25.14	8.48
1981	51.94	28.93	25.49	7.53
1982	17.18	9.74	18.49	8.18
1983	23.19	14.92	20.18	8.61
1984	18.62	17.18	19.53	12.62
1985	17.79	14.28	12.79	9.22

Source: Ulrich 1987

thousand board feet. Beginning in 1973, prices increased substantially—often more than 25–50% per year. Similar although less dramatic increases occurred in ponderosa pine markets, especially in the years 1978 and 1979. Beginning in 1979, however, the real stumpage price for all species began a dramatic decline through 1985. This was especially noticeable for Douglas-fir stumpage.

2.7 Price Outlook for Hardwood and Softwoods

The prices of wood fiber and its products are almost certain to be higher in the future (Table 8). Softwood stumpage prices are likely to increase in all regions of the USA (e.g., 2.5% annual increases in the South, 3.8% annual increases in the Rocky Mountain region and 1.8% annual increases in the Douglas-fir region of the Pacific Northwest). In all regions, however, the rate of increase in softwood timber prices will be largest in the 1990s and rates of increase will decline in decades beyond this period. Hardwood stumpage prices are also expected to increase through the year 2030. The average rate of increase will range from 0.7% per year for hardwood plywood to 1.2% per year for hardwood lumber. Despite uncertainties involved in forecasting price levels, it remains probable that substantial increases in relative prices of most timber types and timber products will occur. Such increases are likely to be largest for softwood sawtimber, high-quality hardwood timber and the products made from this timber, primarily lumber and plywood.

3. Trade in Timber Products

The USA has been a major net importer of wood products since the early 1900s. In 1985, the net volume imported was 38.8×10^6 m^3 roundwood equivalent. In that year, 83.9×10^6 m^3 of roundwood equivalent was exported (17% of roundwood produced in the USA), while 122.7×10^6 m^3 of roundwood equivalent was imported (30% of total USA timber product consumption).

In 1986, softwood logs represented over 37% of the value of wood products exported from the USA, an increase of 13% since 1981 (Table 9). Well over 60% of the softwood log export volume in 1986 was shipped to Japan. Softwood lumber was the second most valuable USA wood product export in 1986. As for imports, nearly 55% of USA expenditures for imported wood products were incurred for purchase of softwood lumber. Almost all (2.8 billion US dollars) of this softwood lumber import total was paid to Canadian producers. In 1986, the USA incurred a trade deficit in wood products of over 2.1 billion US dollars.

The 3022.2 million US dollars of wood products exported by the USA in 1986 was sold to market in the following countries: Japan (43%), Canada (13%), China (6%), South Korea (5%), FRG (4%), Taiwan (4%), UK (3%), Italy (3%), Mexico (3%), Belgium/Luxembourg (2%), Egypt (2%), Australia (1%), Netherlands (1%), other countries (10%).

The future of USA trade in timber products will depend largely on the economic supply of timber in major forested regions of the world, and on wood product demands originating in major markets. Being a source of most US wood imports, the timber situation in Canada will have a major influence on future US timber-trade prospects.

See also: Future Availability; Industries Based on Wood in the USA: Economic Structure in a Worldwide Context; Resources of Timber Worldwide

Table 8
Stumpage price indexes for 1976 and projections of equilibrium price indexes to 2030, by hardwood and softwood timber and US regions

Timber type and region	1976 Price level (1967 US dollars)	Projected price indexes (1967 US dollars)				
		1990	2000	2010	2020	2030
Softwoods						
Northeast	100.0	166.1	185.1	213.6	245.3	275.5
North Central	100.0	154.0	180.9	207.3	238.9	279.0
Southeast	138.9	229.6	280.0	358.0	434.6	526.6
South Central	138.9	230.6	281.6	358.5	434.3	524.7
Rocky Mountain	138.7	473.0	514.4	704.1	859.7	1045.0
Pacific Northwest						
Douglas-fir subregion	164.2	275.0	228.2	287.4	355.8	430.3
Ponderosa pine subregion	113.8	300.5	330.6	425.1	500.8	608.1
Pacific Southwest	146.5	300.8	334.7	416.3	490.2	579.9
Hardwoods						
Northeast	100.0	104.1	92.1	93.0	98.8	105.1
North Central	100.0	99.7	93.1	97.9	109.8	123.3
Southeast	100.0	133.9	99.1	101.7	112.9	126.4
South Central	100.0	136.3	123.6	137.3	166.9	203.0

Source: Forest Service 1980b

Table 9
US trade balance in wood products (in thousand US dollars) by Commodity, 1986

Commodity	Exports	Imports	Net balance
Softwood logs	1129370	7219	1122151
Hardwood logs	97265	4054	93211
Poles, piles and posts	20511	12429	8082
Wood chips	170022	19607	150415
Softwood lumber	638945	2833804	(2194859)
Hardwood lumber	337137	130782	206355
Softwood flooring	102	469	(367)
Hardwood flooring	6784	29325	(22541)
Siding	963	120175	(119212)
Molding	10414	113036	(102622)
Treated lumber, flooring, siding and molding	4049	19876	(15827)
Railroad ties	6116	3337	2779
Softwood veneer	15176	15570	(394)
Hardwood veneer	81764	114328	(32564)
Softwood plywood	131313	34163	97150
Hardwood plywood	13355	484515	(471160)
Hardboard	23506	56099	(32593)
Particleboard	34366	136640	(102274)
Cellular wood panels	1527	170	1357
Gypsum and plasterboard	9299	99089	(89790)
Other panel products	30197	33307	(3110)
Miscellaneous wood products	260020	899447	(639427)
Total	3022201	5167441	(2145240)

Note: Totals may not add due to rounding Source: Foreign Agricultural Service 1987

Bibliography

Barnett H J, Morse C 1963 *Scarcity and Growth: The Economics of Natural Resource Availability*. Johns Hopkins University Press, Baltimore, Maryland

Clawson C 1978 Will there be enough timber? *J. For.* 76: 274–76

Foreign Agriculture Service 1987 *Wood Products: International Trade and Foreign Markets*, Circular Series WP-1-87. US Department of Agriculture, Washington, DC

Forest Products Research Society 1979 *Timber Supply: Issues and Options,* Proceedings No P-79-24. Forest Products Research Society, Madison, Wisconsin
Forest Products Research Society 1980 *Timber Demand: The Future is Now,* Proceedings No P-80-29. Forest Products Research Society, Madison, Wisconsin
Forest Service 1980a *An Analysis of the Timber Situation in the United States 1952–2030.* US Department of Agriculture, Washington, DC
Forest Service 1980b *An Assessment of the Forest and Rangeland Situation in the United States,* FS-345. US Department of Agriculture, Washington, DC
Hair D 1978 Does the US face a shortfall of timber? *J. For.* 76: 276–78
McKeever D B, Hatfield C A 1984 *Trends in Production and Consumption of Major Forest Products in the United States.* Forest Products Laboratory, Madison, Wisconsin
Manthy R S 1977 Scarcity, renewability, and forest policy. *J. For.* 75: 201–5
Manthy R S 1978a *Natural Resource Commodities: A Century of Statistics.* Johns Hopkins University Press, Baltimore, Maryland
Manthy R S 1978b Will timber be scarce? *J. For.* 76: 278–80
Potter N, Christy F T 1962 *Trends in Natural Resource Commodities: Statistics of Prices, Output, Consumption, Foreign Trade and Employment in the United States 1870–1957.* Johns Hopkins University Press, Baltimore, Maryland
Risbrudt C D, Stone R N 1980 *Trends and Patterns in Wood Products Consumption and Production.* Forest Products Laboratory, Madison, Wisconsin
Skog K, Risbrudt C D 1982 *Trends in Economic Scarcity of US Timber Commodities.* Forest Products Laboratory, Madison, Wisconsin
Smith V K 1970 *Scarcity and Growth Reconsidered.* Johns Hopkins University Press, Baltimore, Maryland
Spelter H, Phelps R S 1984 Changes in postwar US lumber consumption patterns. *For. Prod. J.* 34(2): 35–41
Ulrich A H 1981 *US Timber Production, Trade, Consumption, and Price Statistics: 1950–1980,* USDA Miscellaneous Publication No 1408. US Department of Agriculture, Washington, DC
Ulrich A H 1984 *US Timber Production, Trade, Consumption, and Price Statistics: 1950–1983,* USDA Miscellaneous Publication No 1442. US Department of Agriculture, Washington, DC
Ulrich A H 1987 *US Timber Production, Trade, Consumption, and Price Statistics: 1959–1985,* USDA Miscellaneous Publication No 1453. US Department of Agriculture, Washington, DC

P. V. Ellefson
[University of Minnesota, St Paul, Minnesota, USA]

U

Ultrastructure

Ultrastructure implies sub-light-microscopic structure; that is, the structure that cannot be resolved with a light microscope but requires the higher resolving power of electron microscopy. Light-microscope resolving power is limited to ~ 0.2 μm, while the scanning electron microscope (SEM) has a lower limit <10 nm, and the transmission electron microscope (TEM) can resolve structures separated by less than 0.5 nm.

As one moves from the relatively gross wood structure, where distinct differences can be observed between the wood of the angiosperms (hardwood; Fig. 1) and that of the gymnosperms (softwood; Fig. 2), toward finer structure, similarities increase. The chemical constituents are essentially the same, although the proportions and distribution of each may vary. The cell walls made up of these constituents (cellulose, hemicelluloses and lignin) are very similar in basic organization also, although some variations have been reported in normal wood. For reaction wood—i.e., compression wood in conifers and tension wood in hardwoods—both the ultrastructure and chemical constituent organization are quite different (see *Chemical Composition*).

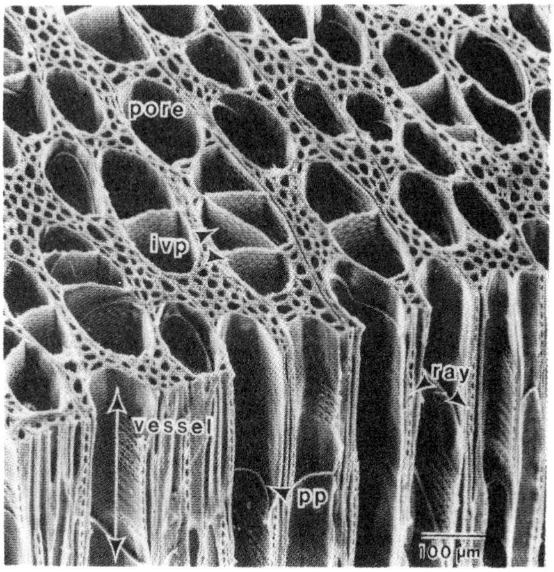

Figure 1
Scanning electron micrograph of cottonwood showing the complex nature of a typical hardwood. The large vessel elements are prominent. Intervessel pitting (ivp) and ray contact area pitting interrupt the inner walls. Simple perforation plates (pp) can be seen at the junction of two vessel elements

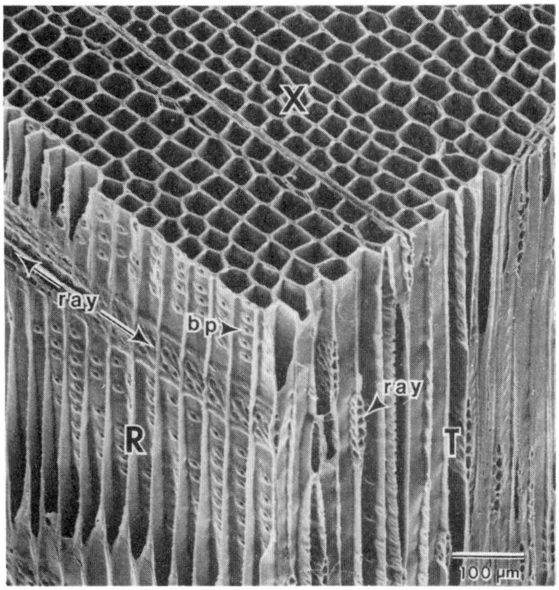

Figure 2
Scanning electron micrograph of a typical softwood, sugar pine, showing cross (X), radial (R) and tangential (T) surfaces. The longitudinal tracheid inner walls are punctuated by bordered pits (bp), and they also conform to the rays which contact every tracheid at least once along its length

1. The Wood Cell Wall

Although the morphological characteristics of coniferous tracheids, vessel elements, libriform fibers and other longitudinal elements are quite dissimilar, the walls of each cell type are very much alike. There are generally three layers in the secondary wall, the portion of the wall which the cell itself produces. At the time of cell division the thin primary wall encloses all of the living substance which makes it possible to support or promote biosynthesis of the three major chemical constituents of wood. The cellulose is formed as long strands called microfibrils. These become the framework of the secondary wall and are laid down in a closely packed array in alternating layers much like plywood.

A generalized model of typical cell-wall organization appears in Fig. 3. Note that the structure of the

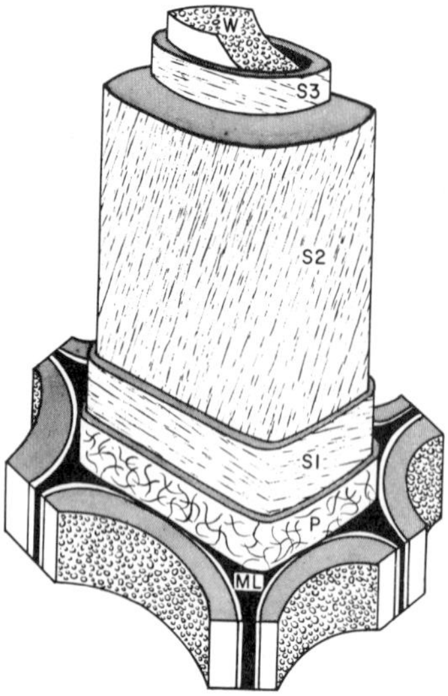

Figure 3
Diagram of generalized cell-wall organization for most wood cells. The primary wall (P) is the outer envelope and consists of randomly distributed microfibrils. S1, S2 and S3 are secondary wall layers with the indicated microfibrillar orientation. W is the warty layer which is found in some species. The middle lamella (ML) is the intercellular region (after Côté (1967). Reproduced by permission of University of Washington Press)

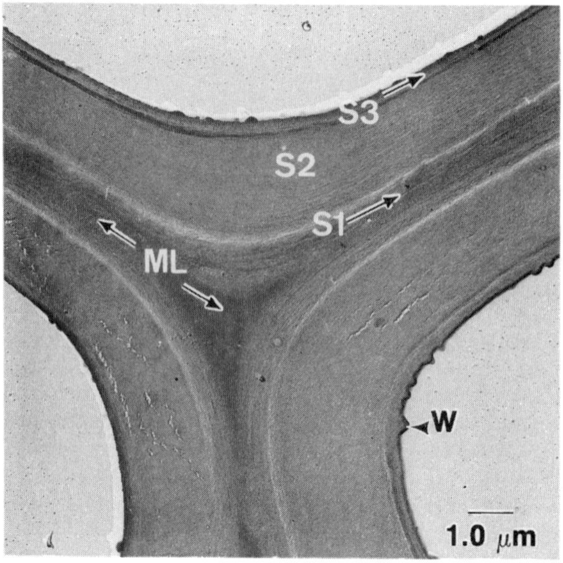

Figure 4
Transmission electron micrograph of a cross section of portions of three longitudinal tracheids in a southern yellow pine. The characteristic three-layered organization of the secondary wall with S1, S2 and S3 layers, the middle lamella (ML) and the warty layer found in hard pines can be observed

primary wall is shown with randomly distributed microfibrils. This is the case in all primary walls. They contain less cellulose and are rather loose-textured. The primary wall is very thin and may not be detectable in a sectional view.

The first layer of the secondary wall lines the primary wall and is called the outer or S1 layer. The microfibrils of this region are oriented nearly perpendicular to the long axis of the cell. A second layer, the middle or S2 layer, is oriented nearly parallel to the cell axis and usually is the thickest of the three secondary-wall layers. Finally, a third, the inner or S3 layer, is produced with an orientation similar to that of the S1. Much of the evidence for microfibrillar orientation was obtained with light microscopy and polarization optics; it has since been confirmed by both TEM and SEM. One additional thin layer can be found in some species and was first recognized through TEM. It is called the warty layer, and is generally designated as layer W. It is not a framework layer, as it contains no cellulose microfibrils and is quite amorphous.

Simultaneous with the synthesis of cellulose, hemicelluloses are produced and are thought to fill voids surrounding the slender microfibrils. In effect they serve as matrix substances.

Later, when the cell-wall organization is established, lignification takes place. Lignin is the encrusting sub-

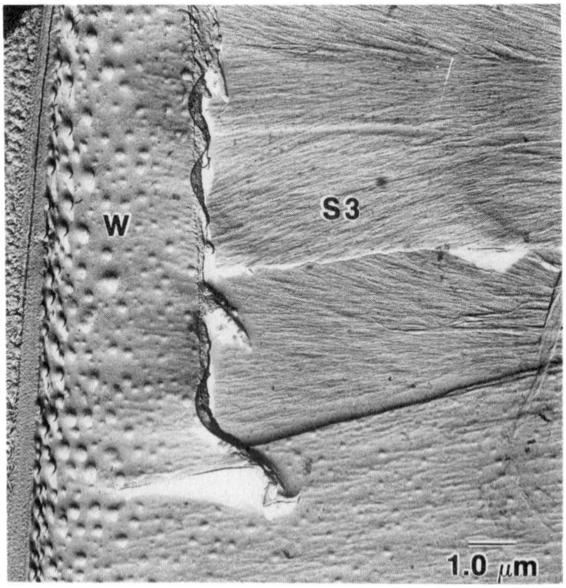

Figure 5
Transmission electron micrograph of a portion of the lumen lining of a southern pine longitudinal tracheid. The warty layer has been peeled away at the upper right revealing the microfibrillar nature of the S3 layer

stance of wood. Presumably its precursors are dispersed throughout the wall during its formation, because studies have found lignin to be rather evenly distributed in the secondary wall. However, at the junction between cells (the middle lamella), the concentration and packing density of lignin is much greater (Fig. 4).

The warty layer (Figs. 4 and 5) is extremely variable, when present. The warts range from very small, almost hairlike structures, to very large protuberances. They may be deposited on a microfibrillar background (the S3 cell-wall layer) or against a membranous substrate which coats the S3. Warts may be sparsely or very densely distributed and, when present in a wood, they often occur in the bordered pit chambers as well as in the cell lumens.

2. *Abnormal Wood Cell Walls*

Minor variations in cell-wall organization occur in normal wood, but they are not very marked changes and they affect only an occasional cell. When wood is abnormal, such as the reaction wood produced in leaning trees, the cell walls are very different.

In compression wood, the reaction wood of conifers, the tracheids are more rounded than polygonal in cross section, thus creating intercellular spaces. The tracheid walls lack an S3 layer, and the S2 is sculptured with deep fissures which reach nearly to the S1. These ultrastructural deviations, coupled with the modifications in distribution of the major chemical constituents, result in a wood with inferior mechanical-strength properties; the material is brash and lacks toughness.

The hardwood reaction wood—tension wood—contains large numbers of fibers which have an inner layer of virtually pure cellulose. This unlignified zone is produced last and lines the cell lumen with material that appears gel-like and swollen. It is called the gelatinous layer, and its cellulose microfibrils are oriented parallel with the cell axis. Gelatinous fibers, as these are called, can be normal components of trees which contain no reaction wood. However, in tension wood this same type of cell is termed tension fiber.

Tension fibers can have a normal three-layered secondary wall, and then contain the additional G layer. However, it is also possible for these special elements to lack an S3 and even an S2. The gelatinous layer is produced after the S1, S2 or S3 layer, evidently depending on the level of cambial activity. When a piece of tension wood is cut transversely using a sharp blade, examination of the tension-wood fibers reveals that the G layer is almost invariably pulled away from the remainder of the secondary wall in the direction of the cut. Lumber containing tension wood has a tendency to bow and twist upon drying. The surface of the rough-swan lumber will often be woolly, and when planed it may exhibit a silvery sheen.

3. *Surface Characteristics of Elements*

The ultrastructural features described to this point, with the exception of the warty layer, have been related to the lamellar nature of the cell wall—i.e., its internal structure. The surface characteristics of cell walls have also been examined with great interest, especially because of physiological implications.

As shown by several of the Figures, the walls lining the cell cavity are sculptured and perforated. Some of the recognizable features are common to both softwoods and hardwoods, while others are quite limited in occurrence. Although much of the sculpturing can be seen with the aid of a light microscope, the higher resolving power of SEM and TEM is required for full characterization or elucidation.

3.1 Helical Thickenings

Sculpturing of the walls of cells by helical thickenings occurs in both hardwoods and softwoods, but in only a relatively small number of species. Among the widely utilized coniferous woods, the tracheids of Douglas fir exhibit this striking feature, which is also known as spiral thickening (Fig. 6). The vessel elements of basswood are lined with a similar structure.

These ropelike thickenings (Fig. 7) were once considered to be tertiary in nature; that is, to be part of an additional wall formed after the secondary wall. Through TEM evidence, helical thickenings are now generally recognized as part of the S3 layer. They

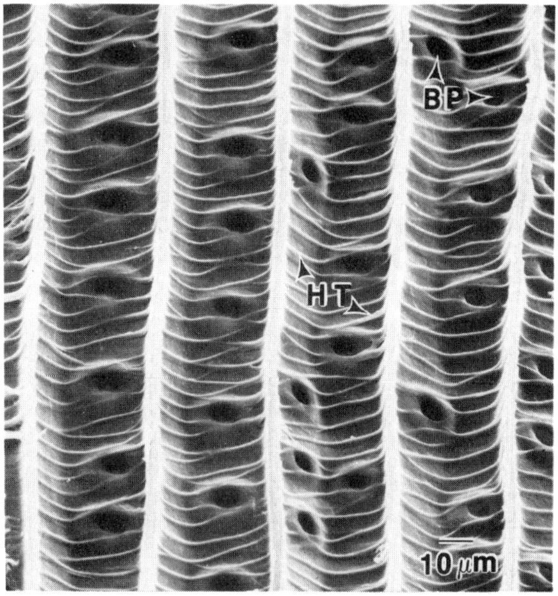

Figure 6
Scanning electron micrograph of the cell lumens of several longitudinal tracheids in Douglas fir. Helical thickenings (HT), which are part of the S3 layer, and bordered pits (BP), are prominent features

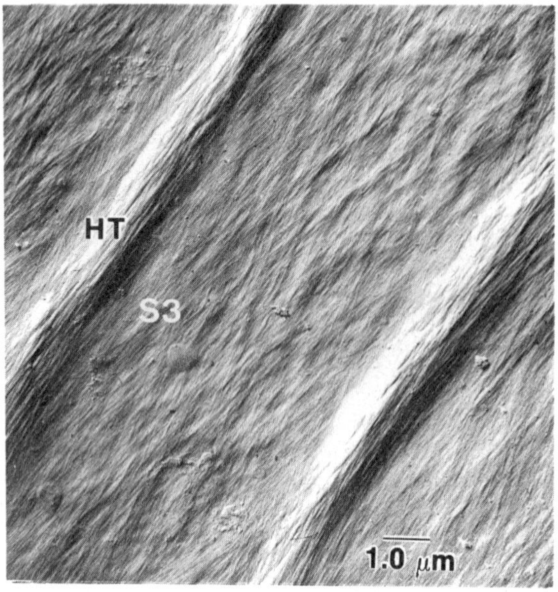

Figure 7
Transmission electron micrograph of the longitudinal tracheid lumen wall in Douglas fir. Note the microfibrillar nature of the S3 layer with its helical thickenings (HT)

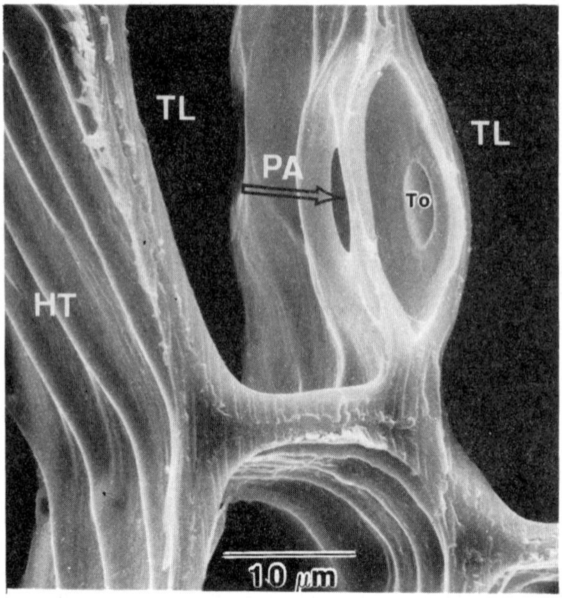

Figure 8
A cross section of Douglas fir wood observed with the SEM reveals the helical thickenings (HT) and a bordered-pit pair from a different perspective. The transverse cut through the pit chamber reveals the pit membrane and torus in the aspirated (sealed) state. A second bordered pit can be seen. TL is the tracheid lumen and To is the pit torus. The pit aperture is indicated by PA

consist of cellulose microfibrils and appear to be composed of the same materials as the remainder of the secondary wall.

3.2 Pitting

The longitudinal tracheid of coniferous wood is a very long and narrow element whose walls are punctuated by gaps or openings called pits (Fig. 6). Two types of pit can be noted in Fig. 2: the bordered pits which connect with a matching one on a neighboring tracheid to form a bordered-pit pair; and the half-bordered pit, which joins with a simple pit in ray parenchyma to form a half-bordered-pit pair. This system of pitting provides communication laterally and vertically in the tree, since coniferous tracheids have closed ends. Most of the pits on the tracheids are located on radial faces of these cells, which are square or polygonal in cross section. The half-bordered pits are also found on the radial walls of the tracheids, of course, since they lead to the rays which radiate from the pith to the cambium and bark.

The bordered-pit locations, as viewed from the cell lumen, are found in raised or domelike swellings which extend into the cell cavity (Figs. 6 and 8). The pit aperture is located at the center of this raised area, and leads into a pit canal whose length varies with cell-wall thickness.

The tracheid walls conform to the bands of ray tissue which are inserted between tracheids. The half-bordered pits are therefore found in these raised areas. Pit-aperture shapes vary according to genus and occasionally to species.

The cell-wall organization in the vicinity of pits appears designed to distribute stress around these cell-wall openings. The microfibrils form streamlined patterns round these secondary-wall gaps.

There are many similarities between hardwood bordered pits and those found in softwood tracheids. However, in vessel elements the cell-walls are frequently much more heavily pitted, with intervessel pits as well as openings to ray parenchyma, fibers and fiber tracheids (Fig. 1). The pits in libriform fibers are smaller and the pit canals are long, due to the thickness of the cell walls.

The designation of bordered-pit pair and half-bordered-pit pair applies to hardwoods as well as softwoods. Pits with an overhanging border, when matched with simple pits in the parenchyma, are termed half-bordered-pit pairs. The simple pit has no perceptible border in the secondary wall, but it is merely an opening or gap leading to the pit membrane in the middle-lamella region. All of the wall sculpturing or modification appears on the secondary cell wall.

3.3 Pit Membranes

The pit canals in pit pairs lead from the cell lumen to the pit membrane. The membrane is actually a three-layered structure composed of the primary walls of two adjacent cells with the middle lamella sandwiched between them. In living cells there are plasmodesmata, which penetrate from cell to cell through the pit membrane. However, in dead cells, which is the case in wood, the fine openings through the membrane are not readily seen even with the aid of an electron microscope. Diffusion can take place across the membrane but the movement of particulate materials cannot. In simple pit pairs, where two pits without overhanging secondary walls are joined, the membrane is of this type in both hardwoods and softwoods.

In half-bordered-pit pairs, the same kind of membrane exists for both categories of wood. Hardwood bordered-pit pairs are similarly equipped with a membrane with no visible openings (Fig. 9). In all of these cases it is assumed that communication between cells is through diffusion.

The bordered-pit pairs of softwood, however, have a unique type of membrane (Fig. 10). The primary wall type of organization, with randomly distributed microfibrils, is modified. The microfibrillar strands radiate from the center of the pit chamber to its periphery, where they then continue into the primary wall. In a number of species, including the pines, a central thickening called the torus is formed, usually with concentrically organized microfibrils. Real openings in the membrane surrounding the torus allow the flow of liquids from one tracheid to the adjacent one, provided that the torus and its supporting membrane remain in the median position. During the drying process, however, negative pressure may draw the torus against one of the apertures and thus effectively seal the pit pair and prevent fluid flow (Fig. 8). This condition is called pit aspiration, and aspirated pits adversely affect penetration of softwoods by pulping liquors and wood preservatives.

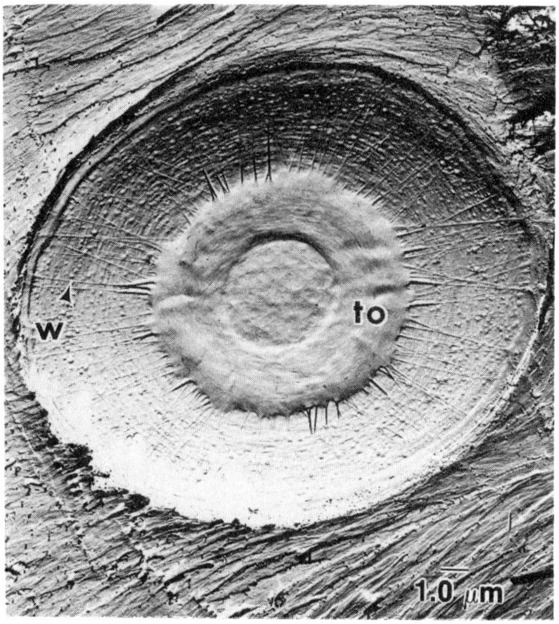

Figure 10
Transmission electron micrograph of a bordered pit with membrane and torus (to) in a tracheid of a southern yellow pine. The depressed center of the flexible torus seals the aperture underneath. Note the warts (w) in the pit chamber. The adjoining tracheid with the matching pit was removed to reveal this split surface

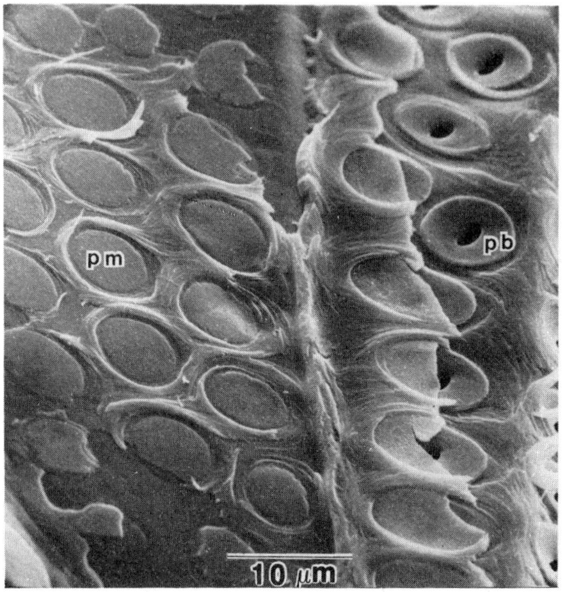

Figure 9
Intervessel pitting in red oak. The fractured cell walls allow a view of intact as well as ruptured pit membranes (pm). The pit borders (pb) are exposed at right. Matching bordered pits were torn away from the surface of the specimen

3.4 Perforation Plates

The individual vessel elements are joined end-to-end to form long, tubular, conductive structures. At the junctions of the vessel elements or segments, there is often a distinctive structure known as a perforation plate (Fig. 1). These plates can be ornate in design or very simple thickenings limited to the periphery of the vessel in the junction area. Like the helical thickenings, they are part of the secondary wall. However, fragments of the membranes which stretched across the junction may be retained. These are of primary-wall origin, and display the random microfibrillar structure.

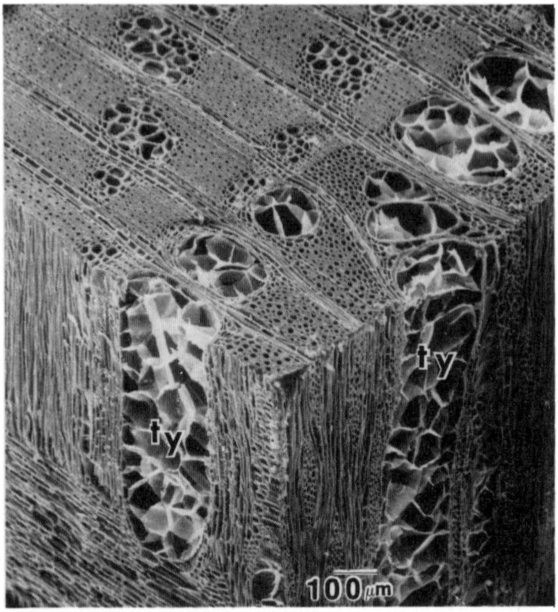

Figure 11
Scanning electron micrograph of a small cube of black locust wood, a hardwood whose vessels are filled with tyloses (ty). These membranous structures are of microfibrillar structure and proliferate from parenchyma cells through pits into the vessel cavities

3.5 Dentate Ray Tracheids

It was noted earlier that in the rays of some of the softwoods there are conductive elements found along with the ray parenchyma cells. These are ray tracheids whose walls have bordered pits similar to but smaller than those of longitudinal tracheids. The inner walls of ray tracheids of the hard pines such as the southern yellow pines have prominent projections which nearly bridge the cell cavity in some cases. The ultrastructure of these dentations is very complex, but the composition is similar to that of the secondary walls.

4. Cell Inclusions

Most woods have some cells which contain inclusions such as crystals, gums or resins. Such materials generally are not structural in nature, although their presence or absence may have diagnostic value.

One quite common form of structural inclusion is the tylosis, which is found only in hardwoods. Tyloses are membranous structures which project through pits into vessel cavities from adjacent parenchyma cells (Fig. 11). In fact, these balloonlike structures grow into the vessels and have a microfibrillar organization and wall structure that may be pitted.

Tyloses are important economically, because they occlude the vessels and prevent fluid flow longitudinally. White oak is utilized extensively for tight cooperage (e.g., in the wine and whiskey industry) because of the presence of tyloses. Red oak lacks tyloses, and liquids would seep from a container made of this species of wood.

See also: Macroscopic Anatomy

Bibliography

Côté W A (ed.) 1965 *Cellular Ultrastructure of Woody Plants.* Syracuse University Press, Syracuse, New York
Côté W A 1967 *Wood Ultrastructure—An Atlas of Electron Micrographs.* University of Washington Press, Seattle
Ledbetter M C, Porter K R 1970 *Introduction to the Fine Structure of Plant Cells.* Springer, Berlin
Meylan B A, Butterfield B G 1972 *Three-dimensional Structure of Wood: A Scanning Electron Microscope Study.* Syracuse University Press, Syracuse, New York

W. A. Côté
[State University of New York, Syracuse, New York, USA]

W

Weathering

The durability of wood in outdoor exposure is quite variable, depending on the environment. The surface of exposed wood erodes away at the rate of only 6–13 mm per century. Consequently, many examples exist of wood structures lasting for centuries. Yet in an unfavorable environment wood can completely lose its utility, and even vanish within a few years. This article reviews the nature of the weathering process of wood and wood products.

Freshly cut wood surfaces exposed to outdoor environments gradually change color, develop micro-checks which can grow into cracks, and become rough. Exposed pieces of wood may cup and warp, and surface fibers may loosen, usually from the low-density earlywood portions of the annual ring, leaving dense latewood as ridges. Completely weathered wood, when microorganisms are absent, has a soft silver-gray color on the surface. Just below the surface the wood is not affected by the weathering process. An accumulation of dirt and mildew may make the surface unsightly.

1. Weathering Factors

Natural outdoor weathering of wood is effected by a complex combination of chemical, mechanical and light energies. The weathering process differs according to local climatic conditions and the duration and severity of exposure to sun and rain. Different species of wood respond somewhat differently to outdoor exposure. Table 1 summarizes the relative effects of various energy forms that cause weathering of wood surfaces.

The action of water from rain or dew is a primary cause of weathering. The water produces moisture and as a result of restrained shrinking or swelling, stress gradients in the wood that lead to checks and warping. Water also leaches out extractive components from the wood, leading to color changes. Woods containing large amounts of extractives, such as redwood, may become bleached as the extractives are leached out. This bleached color is not permanent as chemical changes result in a yellowing or browning of the surface.

Absorption of ultraviolet light by and subsequent degradation of lignin is the primary cause of darkening. Eventually, solubilized lignin-degradation products are completely washed out by water. The silver-gray surface remaining consists primarily of cellulose components. The completely weathered gray layer is approximately 0.1 mm thick. Wood within about 2 mm of the surface is completely normal.

Another cause of graying of weathered wood is the presence of stain fungi which causes mildew on the surface. Again, water from rain or dew is necessary for the fungi to grow.

Table 1
Summary of relative effects of various energies that weather wood

	Indoor		Outdoor	
Energy form	Effect	Degree of effect	Effect	Degree of effect
Heat (moderate levels)	darkening of color	slight	darkening of color	slight
Light (visible and ultraviolet)	color change	slight	color change	severe
			chemical degradation	severe
Mechanical	wear	slight	wear	slight
			wind erosion	slight
			surface roughening	severe
			defiberization	severe
Chemical	staining	slight	surface roughening	severe
	discoloration	slight	defiberization	severe
	color change	slight	leaching	severe
			color change	severe
			strength loss	severe

2. Effect on Properties

The effect of weathering on the strength properties of wood depends on the extent of biological deterioration, if any. In the absence of decay, weathering is a surface effect, so that overall strength properties are hardly affected at all. Surface abrasion resistance is reduced, and toughness is lowered owing to some loss in flexibility. Other properties of solid wood are essentially unchanged.

Direct exposure of wood to weathering can produce conditions favorable to decay fungi that cause degradation or destruction of the cellulose and/or lignin that form the cell walls in wood. Decay fungi can completely destroy wood in time. Yet, in its early stages, decay is not readily detectable; toughness and bending properties are, however, severely reduced (see *Decay During Use*).

3. Weathering of Wood-Based Materials

The weathering process for wood products such as plywood, particleboard and fiberboard includes all the elements discussed for solid wood. In addition, the water resistance of the adhesive that binds these products is important. In plywood, surface checking occurs in the early stages of weathering. The checking is more pronounced in plywood made with lower-quality face veneers. Plywood strength properties depend on the type and durability of the adhesive used.

The aging of particleboard and fiberboard in outdoor exposure is primarily a function of the type and amount of adhesive used to bind the wood particles together. Deterioration takes place as a result of expansion of the wood particles from the compression set developed during manufacture, differential shrinkage due to moisture gradients, and breakdown of the adhesive. Wood particles can be loosened from the surface. If this happens, particles in the interior of the panel deteriorate, resulting in significant loss of strength. Phenolic adhesives appear to be best suited to wood-based products that will be exposed to weather during use (see *Adhesives and Adhesion*).

4. Protection from Weathering

Since weathering is primarily a surface effect (in the absence of decay), coatings to protect solid wood from weathering are used either for aesthetic reasons or to prevent mildew. Plywood in outdoor exposure should be protected with a coating or by overlaying with a weather-resistant sheet material. Particleboard and fiberboard should be protected by overlaying or coating in most climates.

Typical coatings used on wood and wood-based products include paints, varnishes, water repellents, pigmented penetrating stains, and preservative solutions.

See also: Protective Finishes and Coatings; Radiation Effects; Shrinking and Swelling; Surface Properties; Strength

Bibliography

Carll C G, Feist W C 1987 Weathering and decay of finished aspen waferboard. *For. Prod. J.* 37(4): 27–30

Feist W C, Hon D N S 1984 Chemistry of weathering and protection. In: Rowell R (ed.) 1984 *The Chemistry of Solid Wood*, Advances in Chemistry 207. American Chemical Society, Washington, DC, pp. 401–51

Hon D N S, Feist W C 1986 Weathering characteristics of hardwood surfaces. *Wood Sci. Technol.* 20: 169–83

Meyer R W, Kellogg R M (eds.) 1982 *Structural Use of Wood in Adverse Environments.* Van Nostrand Reinhold, New York

Miniutti V P 1967 *Microscopic Observations of Ultraviolet Irradiated and Weathered Softwood Surfaces and Clear Coatings*, USDA Forest Service Research Paper FPL 74. US Forest Products Laboratory, Madison, Wisconsin

Roux M L, Wozniak E, Miller E R, Boxall J, Böttcher P, Kropf F, Sell J 1988 Natural weathering of various surface coatings on five species at four European sites. *Holz Roh-Werkstoff* 46: 165–70

Sell J, Feist W C 1986 Role of density in the erosion of wood during weathering. *For. Prod. J.* 36(3): 57–60

Stamm A J 1964 *Wood and Cellulose Science.* Ronald Press, New York

Williams R S 1987 Acid effects on accelerated wood weathering. *For. Prod. J.* 37(2): 37–38

Williams R S, Feist W C 1985 Wood modified by inorganic salts: Mechanism and properties. *Wood Fiber Sci.* 17: 184–98

D. E. Lyon
[Mississippi State University, Mississippi State, Mississippi, USA]

Wood–Polymer Composites

Over the years many improvements in wood properties have been achieved by various treatments and modifications. A relatively recent development is the production of wood–polymer composites. These composites offer desirable aesthetic appearance and high compression strength, hardness and abrasion resistance; they also have improved dimensional stability.

1. Wood Modification

Traditional wood treatment include the use of tars, pitches, creosote, resins and salts to coat the surface or fill its porous structure. Table 1 illustrates the range of new treatments introduced during the period 1930 to 1960. Some of the monomers are of the condensation type and react with the hydroxyl groups in the wood, whereas other chemicals react with the hydroxyl groups to form cross-links. Another group of compounds simply bulk the wood by replacing the moisture in the cell walls.

Table 1
Wood modification processes (after Meyer and Loos 1969)

Modification	Details
Acetylation	Hydroxyl groups reacted with acetic anhydride and pyridine catalyst to form esters. Capillaries empty. Antishrink efficiency (ASE) about 70%
Ammonia treatment	Evacuated wood exposed to anhydrous ammonia vapor or liquid at 1 MPa. Bends in 1.25 cm stock up to 90°
Compreg process	Compressed wood–phenolic–formaldehyde composite. Dried treated wood compressed during curing to collapse cell structure. Relative density 1.3–1.4. ASE ~ 95%. Usually thin veneers for cutlery handles
Cross-linking	Catalyst 2% $ZnCl_2$ in wood then exposed to paraformaldehyde heated to 120°C for 20 min. ASE ~ 85%. Drastic loss of toughness and abrasion resistance
Cyanoethylation	Reaction with acrylonitrile (ACN) with NaOH catalyst at 80°C. Fungi resistant, impact strength loss
Ethylene-oxide treatment	High-pressure gas treatment, amine catalyst. ASE to 65%
Impreg process	Noncompressed wood–phenolic–formaldehyde composite. Thin veneers, soaked, dried and cured under mild pressure. Swells cell wall, capillaries filled. ASE ~ 75%. Used in modelling cars
Irradiation	Exposure to 10^6 rad of γ radiation gives slight increase in mechanical properties. Above this level cellulose is degraded and mechanical properties decrease rapidly. Low exposure used to temporarily inhibit growth of fungi
Ozone treatment	Gas-phase treatment degrades cellulose and lignin, pulping action
β-propiolactone treatment	β-propiolactone diluted with acetone, wood loaded and heated. Grafted polyester side chains on swollen cell wall cellulose. Carboxyl end groups reacted with copper or zinc to decrease fungi attack. Compression strength increased
Staybwood	Heat-stabilized wood. Wood heated to 150–300°C
Staypak	Heat-stabilized compressed wood. Wood heated to 320°C, then compressed, at 2.75–27.5 MPa, then cooled and pressure released. Used for handles and desk legs

During the 1960s treatment with vinyl-type monomers that could be polymerized into the solid polymer by means of free radicals was introduced (Siau et al. 1965). This vinyl polymerization is an improvement over the condensation reaction because the free-radical catalyst does not degrade the cellulose as do the acid and base catalysts used with other treatments. Vinyl polymers vary in their properties, ranging from soft rubber to hard brittle solids depending upon the groups attached to the carbon–carbon backbone. Examples of vinyl monomers used in wood–polymer composites include styrene, methyl methacrylate, vinyl acetate and acrylonitrile. In general, such vinyl polymers simply bulk the wood structures by filling the void spaces. The free radicals used for the polymerization reaction are usually produced via the use of temperature-sensitive catalysts or ^{60}Co γ radiation; in each case the vinyl polymerization is the same.

2. *Chemistry of the Vinyl Polymerization Process*

"Vazo" (Du Pont 1967) or 2,2′-azobisisobutyronitrile is preferred over peroxide catalysts because of its low decomposition temperature and its nonoxidizing nature. Vazo will not bleach dyes dissolved in the monomer during polymerization. Vazo decomposes as follows:

$$(CH_3)_3C{-}N{=}N{-}C(CH_3)_3 \rightarrow 2(CH_3)_3C\cdot + N_2 \quad (1)$$

The rapid decomposition of Vazo with increasing temperature means that the reaction can be initiated at the moderate temperature of 60°C. Since the decomposition rate of Vazo is negligible at 0°C, the catalyzed monomer can be stored safely for months at this temperature.

The use of radiation as a source of free radicals has many inherent complications, but it does have advantages. Since the monomer is not catalyzed, it can be stored at ambient temperature as long as the proper amount of inhibitor is maintained. The rate of free-radical generation is constant for a given amount of γ radiation and does not increase with temperature as with the heat-sensitive catalysts. When γ radiation passes through wood and vinyl monomer, it leaves behind a trail of ions and excited states, which in turn produce free radicals. These free radicals, like the catalyst free radicals, initiate the vinyl-monomer polymerization reaction (Meyer 1965).

3. *Impregnation Process*

Figure 1 represents the essential components of a system for the vinyl-monomer impregnation of wood. In this process, air is first evacuated from the wood vessels and cell lumens by means of a vacuum pump. The pump is then isolated from the system and the catalyzed monomer, containing cross-linking agents (Meyer 1968) and on occasion dyes, is introduced into the evacuated chamber through a reservoir at atmospheric pressure. The wood must be weighted so that it does not float in the monomer solution. In the

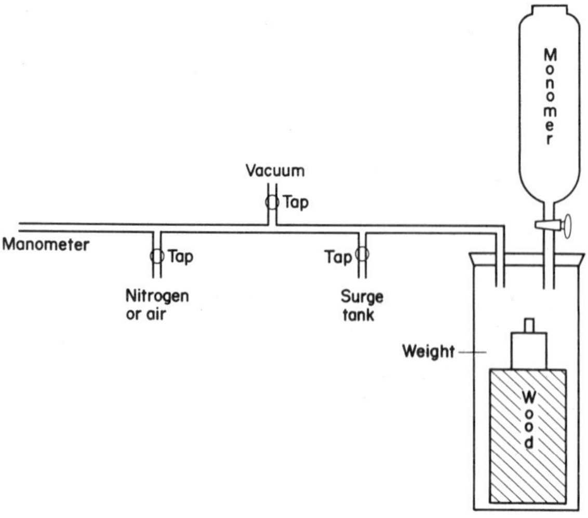

Figure 1
Essential components of a system for the vinyl-monomer impregnation of wood

radiation process the catalyst is omitted from the monomer. After the wood is covered with the monomer solution, air at atmospheric pressure is admitted, or dry nitrogen in the case of the radiation process. The monomer solution immediately flows into the evacuated wood structure to fill the void spaces. The time to fill the wood depends upon the structure of the wood and the viscosity of the monomer solution.

After the impregnation is complete, the wood–monomer is removed and placed in an explosion proof oven, or in the γ-radiation source, for curing. On a laboratory scale or in a small production unit the wood–monomer is wrapped in aluminum foil before placing in the curing oven at 60°C. In larger production units the wood–monomer is placed directly into the curing oven, usually in the basket which held the wood during impregnation. In the radiation-cure procedure the thin metal can, in which the wood was impregnated, is flushed with nitrogen and is lowered into a water pool next to the ^{60}Co source (Witt and Morrissey 1972). With high-vapor-pressure monomers, the wood surface is depleted to some extent by surface evaporation, but this depleted zone is usually removed by machining. Methyl methacrylate has a vapor pressure of 40 mm Hg at room temperature, whereas *t*-butyl styrene has a vapor pressure of only 0.8 mm Hg (Meyer 1981).

4. Monomers for Wood–Polymer Composites

Many different vinyl monomers have been used to make wood–polymers in recent years (Langwig et al. 1969), but methyl methacrylate appears to be the preferred monomer for both the catalyst–heat and radiation processes. In fact, methyl methacrylate is the only monomer that can be economically polymerized using γ radiation. On the other hand, all types of liquid vinyl monomers can be polymerized with Vazo or peroxide catalysts.

All vinyl monomers contain inhibitors to prevent premature polymerization during transport and storage. If these inhibitors are not removed before polymerization, the catalyst or radiation must generate enough free radicals to use up the inhibitor before polymerization can initiate. Wood also contains natural inhibitors, depending upon the species of wood. Monomers extract the soluble fractions from the wood structure and these extractives can inhibit the polymerization as well as causing excessive foaming under vacuum.

The polymerization of vinyl monomers is an exothermic reaction and a considerable amount of heat is released, above 75 kJ mol^{-1}. Since wood is an insulator due to its cellular structure, heat flow into and out of the wood–monomer is restricted. The temperature of the wood–monomer–polymer composite increases rapidly once the reaction is started, and can reach temperatures as high as 250°C in thick pieces (Duran and Meyer 1972). This increases the vapor pressure of the moisture in the cell walls and drives it out of the wood, causing shrinkage and distortion of the original shape. Wood–polymer composites cured by the catalyst–heat process must be machined to final shape after treatment. Soluble dyes can be added to the vinyl-monomer solution—these produce a three-dimensional depth of color not present in surface-finished wood (Meyer 1977, 1984).

5. Physical Properties and Commercial Applications

Improvements of the physical properties of wood–polymer composites are related to polymer loading. This, in turn, not only depends upon the permeability of the wood species, but also on the particular piece of wood being treated (Young and Meyer 1968). Sapwood is filled to a much greater extent than heartwood for most species. Table 2 gives a few examples of loading and physical-property improvement. Compressive strength, hardness and abrasion resistance are the three most improved properties. Bending, impact resistance, toughness, modulus of rupture, work to the proportional limit and other properties are improved to some extent, again depending upon the species involved (Langwig et al. 1968). Along with improved strength wood–polymer composites also have a desirable improved dimensional stability, depending on polymer loading.

Commercial production of wood–polymer composites for flooring began in the mid-1960s using the radiation process. One company, Perma-Grain Products, is still producing parquet flooring and other items using γ radiation. Commercial catalyst–heat

Table 2
Selected physical property data on wood–polymer composites (after Young and Meyer 1968)

Species		Polymer loading (%)	Density increase (%)	Compressive-strength increase (%)	Tangential-hardness increase (%)
Sugar maple	S	40	65	160	229
	H	38	58	125	200
Basswood	S	63	168	425	626
	H	62	160	288	505
Yellow birch	S	37	58	146	215
	H	31	43	56	120
Beech	S	36	53	201	261
	H	24	30	30	112
Red pine	S	51	100	636	523
	H	8	7	1	1

S = sapwood H = heartwood

wood–polymer production began in 1967 when the American Machine and Foundry Company produced novel billiard cues. Since 1967 the catalyst–heat system has found its way into many small specialty operations making wood–polymer-composite items, such as archery bows, golf clubs, bagpipes, flutes, guitar fret boards, drum sticks, jewelry, office-desk items and parquet flooring.

See also: Chemically Modified Wood

Bibliography

Du Pont 1967 *VAZO Vinyl Polymerization Catalyst*, Du Pont Product Bulletin. Du Pont, Wilmington, Delaware

Duran J A, Meyer J A 1972 Exothermic heat released during catalytic polymerization of basswood–methyl methacrylate composites. *Wood Sci. Technol.* 6: 59–66

Langwig J E, Meyer J A, Davidson R W 1968 Influence of polymer impregnation on mechanical properties of basswood. *For. Prod. J.* 18(7): 33–36

Langwig J E, Meyer J A, Davidson R W 1969 New monomers used in making wood–plastics. *For. Prod. J.* 19(11): 57–61

Meyer J A 1965 Treatment of wood–polymer systems using catalyst–heat techniques. *For. Prod. J.* 15: 362–64

Meyer J A 1968 The effect of crosslinking on the finishing properties of wood–plastics. *For. Prod. J.* 18(5): 89

Meyer J A 1977 Wood–polymer composites and their industrial applications. In: Goldstein I S (ed.) 1977 *Wood Technology: Chemical Aspects,* American Chemical Society Symposium Series No. 43. American Chemical Society, Washington, DC, pp. 301–25

Meyer J A 1981 Wood–polymer materials: State of the art. *Wood Sci.* 14(2): 49–54

Meyer J A 1982 Industrial use of wood–polymer materials: State of the art. *For. Prod. J.* 32(1): 24–29

Meyer J A 1984 Wood–polymer materials, In: Rowell R M (ed.) 1984 *The Chemistry of Solid Wood,* American Chemical Society Symposium Series No. 207. American Chemical Society, Washington, DC, Chap. 6, pp. 257–89

Meyer J A, Loos W E 1969 Processes of, and products from, treating southern pine wood for modification of properties. *For. Prod. J.* 19(12): 32–38

Siau J F, Meyer J A, Skaar C 1965 Dimensional stabilization of wood. *For. Prod. J.* 15: 162–66

Witt A E, Morrissey J A 1972 Economics of making irradiated wood–plastic products. *Mod. Plast.* 49: 78–82

Young R A, Meyer J A 1968 Heartwood and sapwood impregnation with vinyl monomers. *For. Prod. J.* 18(4): 66–68

J. A. Meyer
[State University of New York, Syracuse, New York, USA]

SYSTEMATIC OUTLINE OF THE ENCYCLOPEDIA

The Systematic Outline of the Encyclopedia, which is supplementary to the Subject Index, groups all articles into a number of broad fields, presenting a general overview of the contents of the Encyclopedia.

LIST OF CONTRIBUTORS

Contributors are listed in alphabetical order, together with their addresses. Titles of articles which they have authored follow in alphabetical order. Where articles are co-authored, this has been indicated by an asterisk preceding the article title.

Anthony, R. W.
Engineering Data Management
Oak Ridge Business Park
4700 McMurray Avenue, Building A
Fort Collins, CO 80528
USA
**Lumber: Behavior Under Load*

Arganbright, D. G.
Haldsworth Hall
Department of Forestry and Wildlife Management
University of Massachusetts
Amherst, MA 01003
USA
Drying Processes

Ashby, M. F.
Department of Engineering
University of Cambridge
Trumpington Street
Cambridge CB2 1PZ
UK
Cork

Bariska, M.
Department of Wood Science
University of Stellenbosch
Stellenbosch 7600
South Africa
Forming and Bending

Beall, F. C.
Forest Products Laboratory
University of California, Berkeley
1301 South 46th Street
Richmond, CA 94804
USA
Acoustic Emission and Acousto-Ultrasonic Characteristics

Bond, C. W.
64 Kelvin Road
Leamington Spa
Warwickshire CV32 7TQ
UK
Timbers of Africa
Timbers of Europe

Chow, S.
Canadian Forest Products Ltd
15th Floor
505 Burrard Street
Vancouver, BC V7X 1B5
Canada
Surface Properties

Côté, W. A.
Department of Wood Products Engineering
State University of New York
College of Environmental Science & Forestry
Syracuse, NY 13210
USA
Macroscopic Anatomy
Ultrastructure

Darr, D. R.
US Forest Service
Forest Inventory & Economics Research Staff
1621 North Kent Street
Arlington, VA 22209
USA
Future Availability

Dransfield, J.
Herbarium
Royal Botanic Gardens
Kew
Richmond
Surrey TW9 3AE
UK
Rattan and Cane

Eckelman, C. A.
Department of Forestry & Natural Resources
Purdue University
West Lafayette, IN 47907
USA
Glued Joints

Ellefson, P. V.
Department of Forest Resources
College of Chemistry
University of Minnesota
1530 North Cleveland Avenue
St Paul, MN 55108
USA
Industries Based on Wood in the USA: Economic

Structure in a Worldwide Context
Trade, Prices and Consumption of Timber in the USA

Feist, W. C.
US Forest Products Laboratory
One Gifford Pinchot Drive
Madison, WI 53705
USA
Protective Finishes and Coatings

Freas, A. D.
Professional Engineer
2618 Park Place
Madison, WI 53705
USA
Building with Wood

Glasser, W. G.
Department of Forest Products
Virginia Polytechnic Institute & State University
210 Cheatham Hall
Blacksburg, VA 24061
USA
Lignin-Based Polymers

Glomb, J. W.
Westvaco
290 Park Avenue
New York, NY 10017
USA
**Paper and Paperboard*

Goldsmith, D. F.
Toxic Substances Research and Teaching Programs
University of California
LEHR Facility
Davis, CA 95616
USA
Health Hazards in Wood Processing

Goldstein, I. S.
Department of Wood & Paper Science
North Carolina State University
PO Box 5488
Raleigh, NC 27650
USA
Cellulose: Nature and Applications
Chemically Modified Wood
Chemicals and Liquid Fuels from Wood

Goodell, B.
College of Forest Resources
University of Maine
Orono, ME 04469
USA
Biotechnology in Wood Processing

Goodman, J. R.
Civil Engineering Department
Colorado State University
Fort Collins, CO 80521
USA
**Lumber: Behavior Under Load*

Higuchi, T.
Wood Research Institute
Kyoto University
Uji
Kyoto 611
Japan
Bamboo

Hillis, W. E.
Division of Chemical and Wood Technology
Commonwealth Scientific and Industrial Research Organization
Private Bag No. 10
Clayton, Victoria 3168
Australia
**Timbers of Australia and New Zealand*

Hon, D. N.-S.
Wood Chemistry Laboratory
Department of Forestry
Clemson University
Lehotsky Hall
Clemson, SC 29634-1003
USA
Cellulose: Chemistry and Technology
Radiation Effects

Hoyle, R. J. Jr
College of Engineering
Washington State University
Pullman, WA 99163
USA
Lumber: Types and Grades

James, W. L.
US Forest Products Laboratory
PO Box 5130
Madison, WI 53705
USA
Electrical Properties

Johnson, J. A.
College of Forest Resources AR-10
University of Washington
303 Bloedell Hall
Seattle, WA 98105
USA
Structure, Stiffness and Strength

Keating, W. G.
Division of Chemical and Wood Technology
Commonwealth Scientific and Industrial Research Organization
Private Bag No. 10
Clayton, Victoria 3168
Australia
**Timbers of Australia and New Zealand*

Kellogg, R. M.
Forintek Canada Corporation
Western Forest Products Laboratory
6620 N W Marine Drive
Vancouver, BC V6T 1X2
Canada
Density and Porosity

Kubler, H.
126 Russell Laboratories
University of Wisconsin
Madison, WI 53706
USA
Thermal Properties

LeVan, S. L.
US Forest Products Laboratory
One Gifford Pinchot Drive
Madison, WI 53705
USA
Thermal Degradation

Lyon, D. E.
Forest Products Utilization Laboratory
Mississippi State University
PO Drawer FP
Mississippi State, MS 39762
USA
Weathering

Maeglin, R. R.
US Forest Products Laboratory
One Gifford Pinchot Drive
Madison, WI 53705
USA
Timbers of Canada and the USA

Maloney, T. M.
Department of Materials Science and Engineering
Washington State University
Pullman, WA 99163
USA
Particleboard and Dry-Process Fiberboard

Mark, R. E.
College of Environmental Science and Forestry
State University of New York
Syracuse, NY 13210
USA
Constituents of Wood: Physical Nature and Structural Function

Meyer, J. A.
Chemistry Department
State University of New York
College of Environmental Science & Forestry
Syracuse, NY 13210
USA
Wood–Polymer Composites

Miller, R. B.
US Forest Products Laboratory
One Gifford Pinchot Drive
Madison, WI 53705
USA
Timbers of Central and South America

Moore, H. B. Jr
Department of Entomology
North Carolina State University
Box 5215
Raleigh, NC 27650
USA
Deterioration by Insects and Other Animals During Use

Morrell, J. J.
Department of Forest Products
College of Forestry
Oregon State University
Corvallis, OR 97331
USA
Fumigation

Mosteiro, A. P.
Forest Products Research & Development Institute
College
Laguna 3720
Republic of the Philippines
**Coconut Wood*

Mulligan, D. D.
Westvaco
Covington Research Center
Covington, VA 24426
USA
**Paper and Paperboard*

Nicholas, D. D.
Forest Products Utilization Laboratory
Mississippi State University
P O Drawer FP
Mississippi State, MS 39762
USA
Preservative-Treated Wood

O'Halloran, M. R.
American Plywood Association
7011 South 19th Street
Tacoma, WA 98411
USA
Plywood
Structural-Use Panels

Ono, T.
Faculty of Engineering
Gifu University
Gifu City 501-11
Japan
Stringed Instruments: Wood Selection

Pellerin, R. F.
Wood Technology Section
Washington State University
Pullman, WA 99164
USA
**Nondestructive Evaluation of Wood and Wood Products*

Rocafort, J. E.
Forest Products Research & Development Institute
College
Laguna 3720
Republic of the Philippines
**Coconut Wood*

Rojo, J. P.
Forest Products Research & Development Institute
College
Laguna 3720
Republic of the Philippines
**Coconut Wood*

Ross, R. J.
Trus Joist Corporation
Boise
Idaho
USA
**Nondestructive Evaluation of Wood and Wood Products*

Sasaki, H.
Composite Wood Section
Wood Research Institute
Kyoto University
Uji
Kyoto 611
Japan
Laminated Veneer Lumber

Schniewind, A. P.
Forest Products Laboratory
University of California, Berkeley
1301 South 46th Street
Richmond, CA 94804
USA
Archaeological Wood
Deformation Under Load
Strength

Siau, J. F.
Department of Wood Products Engineering
SUNY College of Environmental Science & Forestry
Syracuse, NY 13210
USA
Fluid Transport

Simatupang, M. H.
Institut für Holzchemie und Chemische Technologie des Holzes
Leuschnerstrasse 9LB
205 Hamburg 80
FRG
Mineral-Bonded Wood Composites

Siopongco, J. O.
Forest Products Research & Development Institute
College
Laguna 3720
Republic of the Philippines
**Coconut Wood*

Skaar, C
Virginia Polytechnic Institute and State University
210 Cheatham Hall
Blacksburg, VA 24061
USA
Hygroscopicity and Water Sorption
Shrinking and Swelling

Smith, W. R.
College of Forest Resources
University of Washington
Seattle, WA 98195
USA
Acoustic Properties

Steiner, P. R.
Faculty of Forestry
University of British Columbia
2357 Main Hall
Vancouver, BC V6T 1W5
Canada
Adhesives and Adhesion

Suchsland, O.
Department of Forestry
Natural Resources Building, Room 210
Michigan State University
East Lansing, MI 48824
USA
Hardboard and Insulation Board

Suddarth, S. K.
Department of Forestry & Natural Resources
Purdue University
West Lafayette, IN 47909
USA
Design with Wood

Szymani, R.
Wood Machining Institute
PO Box 476
Berkeley, CA 94701
USA
Machining Processes

Takahashi, A.
Faculty of Agriculture
Shimane University
Matsue City 690
Japan
Timbers of Southeast Asia
Timbers of the Far East

Tillman, D. A.
24615 S.E. 45th Way
Issaquah, WA 98027
USA
Energy Generation from Wood

Timell, T. E.
Department of Chemistry
State University of New York
College of Environmental Science & Forestry
Syracuse, NY 13210
USA
Chemical Composition

Wibbens, R. P.
American Institute of Timber Construction
333 West Hampden Avenue
Englewood, CO 80110
USA
Glued Laminated Timber

Wilcox, W. W.
Forest Products Laboratory
University of California, Berkeley
1301 South 46th Street
Richmond, CA 94804
USA
Decay During Use

Wilkinson, T. L.
US Forest Products Laboratory
One Gifford Pinchot Drive
Madison, WI 53705
USA
Joints with Mechanical Fastenings

Youngs, R. L.
Department of Forest Products
Virginia Polytechnic Institute
& State University
210 Cheatham Hall
Blacksburg, VA 24061
USA
History of Timber Use

Zavarin, E.
Forest Products Laboratory
University of California, Berkeley
1301 South 46th Street
Richmond, CA 94804
USA
Surface Chemistry

Zicherman, J. B.
IFT Technical Services Inc
2550 Ninth Street
Suite 112
Berkeley, CA 94710
USA
Fire and Wood

Zivnuska, J. A.
Department of Forestry & Resource
Management
University of California
Berkeley, CA 94720
USA
Resources of Timber Worldwide

SUBJECT INDEX

The Subject Index has been compiled to assist the reader in locating all references to a particular topic in the Encyclopedia. Entries may have up to three levels of heading. Where there is a substantive discussion of the topic, the page numbers appear in ***italic bold*** type. As a further aid to the reader, cross-references have also been given to terms of related interest. These can be found at the bottom of the entry for the first-level term to which they apply. Every effort has been made to make the index as comprehensive as possible and to standardize the terms used.